Waste-to-Wealth

This book covers state-of-the-art resource recovery technologies from the different components of solid waste, such as plastics, e-waste, fly ash, sewage sludge, slag and their real applications. Furthermore, it explains various management strategies for agricultural waste, including the generation of bioenergy from agri-crop residue. It also highlights the recent technologies used in the management of industrial waste, their implementation at a large scale and the treatment of industrial effluent with the rationale synthetic approach, hybrid advanced oxidation process and bio-methanation.

Features:

- Provides a technical interpretation for creating wealth from waste by the experts in the research domain.
- Covers various aspects of waste management, current resource recovery and recycling trends.
- Presents a unique combination of municipal, agricultural and industrial waste management towards achieving a resilient smart city.
- Imparts knowledge of policies and regulations in different countries and their impacts on waste management.
- Illustrates various technologies for waste processing through case studies.

This book is aimed at researchers and policymakers in environmental engineering, waste management and clean energy.

Waste-to-Wealth

Resource Recovery and Value-Added Products for Sustainable Development

Edited by

Vinay Yadav and Shishir Shrotriya

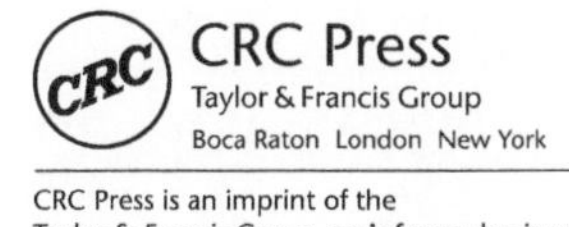

CRC Press
Taylor & Francis Group
Boca Raton London New York

CRC Press is an imprint of the
Taylor & Francis Group, an **informa** business

Designed cover image: © Shutterstock

First edition published 2025
by CRC Press
2385 NW Executive Center Drive, Suite 320, Boca Raton FL 33431

and by CRC Press
4 Park Square, Milton Park, Abingdon, Oxon, OX14 4RN

CRC Press is an imprint of Taylor & Francis Group, LLC

ISBN: 978-1-032-35609-9 (hbk)
ISBN: 978-1-032-35610-5 (pbk)
ISBN: 978-1-003-32764-6 (ebk)

DOI: 10.1201/9781003327646

Typeset in Times
by Newgen Publishing UK

The electronic version of this book was funded to publish Open Access through Taylor & Francis' Pledge to Open, a collaborative funding open access books initiative. The full list of pledging institutions can be found on the Taylor & Francis Pledge to Open webpage.

Contents

Foreword

Waste-to-Wealth is one of the most important Government of India missions, led by the Prime Minister's Science, Technology and Innovation Advisory council (PM STIAC), steered by the office of PSA.

The mission aims to identify, develop and deploy technologies for recycling and extracting resources of value from waste. In order to strengthen the activities under the mission, the Science & Technology wing of the Embassy of India, Moscow conducted an International Scientific webinar in March 2021, under the 'Waste-to-Wealth' theme.

I am pleased that the organizers further integrated the experts and compiled this focus book, titled 'Waste-to-Wealth: Resource Recovery and Value-Added Products for Sustainable Development'.

The contingencies of development can be secured only through sustainable practices and resource recovery. Useful technologies and practices are thus the steps necessary to contain the harm that has been already done to nature. This focus book presents contributions from researchers from around the world, including Russia, South Africa, China, Brazil and India towards sustainable resource recovery. The authors have communicated about the potential of waste management, bio-energy, upcycling, recycling, energy conversion, e-waste recovery and sustainable agricultural practices. These contributions are in alignment with the vision of greener and sustainable environment.

Leading experts from institutes from IITs, Russian Academy of Sciences (RAS), Council of Scientific and Industrial Research (CSIR) and NITs have contributed to this book with their work on resource recovery and technology implementation for conversion of waste to energy. It is interesting to note that most of the chapters have showcased technologies and use-cases with execution.

The lifecycle of resources, which are limited, does not have to end as waste, with the use of innovative technologies they can be converted to useful resources. The authors in this focus book have approached each topic with this mindset, grounding all concepts with concrete examples and evidences, and urging the reader to consider the technologies, techniques and approaches for a sustainable greener future.

This book is surely a work to refer, implement and treasure for the global scientific community and practitioners and will lead to the curious exploration of various alternatives and technology options that could be deployed to generate wealth from waste.

Prof. Ajay K Sood
Principal Scientific Adviser to the Government of India

Preface

The global generation of solid waste was 2.01 billion tonnes (BT) in 2016, which is estimated to increase to 2.59 BT in 2030 and 3.40 BT in 2050 by the World Bank. ~33% of this total solid waste generated in the globe is mismanaged and succumbs to open dumping or burning, which severely pollute the environment. Diversion of waste from open dumping or burning is crucial to recover value from it. Constant degradation of natural resources also enunciates the need for resource recovery through recycling, reuse and recovery of waste. The proposed book brings academicians and practitioners together to disseminate the knowledge of state-of-the-art resource recovery technologies and their real applications.

The proposed focus book will provide a resource for a large group of readers, including professionals, academicians and students, and technical personnel from waste management industries involved in studying contemporary techniques for resource recovery from waste. We recommend it to environmental engineers, scientists, mangers and policy makers. The chapters are evidence-based and are focused on practices that are executional and not just limited to ideas. Important facts and detailed figures are presented in most chapters to aid the readers' understanding. We believe that readers will benefit from this book and will explore various technology options that could be deployed to generate wealth from waste.

Dr. Vinay Yadav
Dr. Shishir Shrotriya

About the Editors

Vinay Yadav is currently working as an Assistant Professor at the Indian Institute of Technology Kharagpur (West Bengal). He earned his master's and doctoral degrees in Environmental Science and Engineering from IIT Bombay, Mumbai, and his bachelor's degree from Banaras Hindu University, Varanasi. He is also a recipient of the Award for Excellence in PhD Research for the year 2018–2020 for outstanding research contributions by IIT Bombay. Immediately after his PhD, he worked as the Marie Curie postdoctoral fellow in the Department of Technology, Management and Economics, Technical University of Denmark. His expertise include environmental management, optimization under uncertainty and interval analysis-based operations research techniques.

Shishir Shrotriya is currently Head, Centre for Maritime Economy & Connectivity (CMEC) at Research & Information System (RIS), New Delhi. During his international assignment he has been Counsellor (S&T) at the Embassy of India, Moscow. He received his PhD and MTech from IIT Delhi. He specializes in communication engineering, maritime technologies and integration. He has varied scientific and project management experience and has worked closely with international and Indian partners. He has worked in areas related to indigenous sea-vessel design & related technologies and has authored many papers on innovation concepts, creative problem-solving and systems integration. He is a member of various science and engineering professional bodies and has vast collaborative project experience with academia, industry, and research institutes. During his current assignment, he is actively involved in working towards the Maritime Amrit Kaal Vision (MAKV) 2047 and Maritime India Vision (MIV) 2030 with research work, stakeholder integration, bilateral multilateral projects and initiatives.

Contributors

Sabu Abdulhameed
Department of Biotechnology and
    Microbiology
Kannur University, Dr. Janaki Ammal Campus
Kannur, Kerala, India

M. I. Alymov
Merzhanov Institute of Structural
    Macrokinetics and Materials Science
Russian Academy of Science
Chernogolovka, Russia

Gangagni Rao Anupoju
Bioengineering and Environmental Sciences
    (BEES) Group
Department of Energy and Environmental
    Engineering (DEEE)
CSIR-Indian Institute of Chemical Technology
    (IICT)
Tarnaka, Hyderabad, India
and
Academy of Scientific and Innovative Research
    (AcSIR)
Ghaziabad, India

Vijayalakshmi Arelli
Bioengineering and Environmental Sciences
    (BEES) Group
Department of Energy and Environmental
    Engineering (DEEE)
CSIR-Indian Institute of Chemical Technology
    (IICT)
Tarnaka, Hyderabad, India
and
Academy of Scientific and Innovative Research
    (AcSIR)
Ghaziabad, India

T. V. Barinova
Academician Osipyan Str.
Moscow Region, Russia

Sameena Begum
Bioengineering and Environmental Sciences
    (BEES) Group
Department of Energy and Environmental
    Engineering (DEEE)

CSIR-Indian Institute of Chemical Technology
    (IICT)
Tarnaka, Hyderabad, India
and
Academy of Scientific and Innovative Research
    (AcSIR)
Ghaziabad, India

Andrea Bernardes
Department of Materials Engineering
Universidade Federal do Rio Grande do Sul
Brazil

Preeti Chaturvedi Bhargava
Aquatic Toxicology Laboratory
Environmental Toxicology Group
FEST Division
Council of Scientific and Industrial Research-
    Indian Institute of Toxicology Research
    (CSIRIITR)
Lucknow, Uttar Pradesh, India

Bhashkar Singh Bohra
Prof. Rajendra Singh Nanoscience and
    Nanotechnology Centre
Department of Chemistry
DSB Campus, Kumaun University
Nainital, India

Aleksandr Briukhanov
Federal Scientific Agroengineering Centre VIM
    (FSAC VIM)

Deepshi Chaurasia
Aquatic Toxicology Laboratory
Environmental Toxicology Group
FEST Division
Council of Scientific and Industrial
    Research-Indian Institute of Toxicology
    Research (CSIRIITR)
Lucknow, Uttar Pradesh, India

Mousumi Das M
Department of Biotechnology and
    Microbiology
Dr. Janaki Ammal Campus, Kannur University
Kannur, Kerala, India

Ridham Dhawan
Vikas Lifecare Limited
East Punjabi Bagh, New Delhi, India

S K Dhawan
Vikas Lifecare Limited
East Punjabi Bagh, New Delhi, India

Brajesh Kumar Dubey
Department of Civil Engineering
Indian Institute of Technology Kharagpur
Kharagpur, West Bengal, India
and
School of Environmental Science and
    Engineering
Indian Institute of Technology
Kharagpur, West Bengal, India

Ilia Eremeev
Topchiev Institute of Petrochemical Synthesis
Russian Academy of Sciences
Moscow, Russia

Anatoly A. Fomkin
M.M. Dubinin Laboratory of Sorption
    Processes
A.N. Frumkin Institute of Physical Chemistry
    and Electrochemistry RAS (IPCE RAS)
    Moscow, Russia

Vikas Garg
Vikas Ecotech Limited
East Punjabi Bagh, New Delhi, India

Sourja Ghosh
Water Technology Division, CSIR- Central
    Glass and Ceramic Research Institute
India

Georgy Golubev
Topchiev Institute of Petrochemical Synthesis
Russian Academy of Sciences
Moscow, Russia
and
Eurasian Patent Office
Moscow, Russia

Tao He
Shanghai Advanced Research Institute
Chinese Academy of Sciences
Shanghai, China

Hema Jha
P. K. Sinha Centre for Bioenergy and
    Renewables
IIT Kharagpur

Sudharshan Juntupally
Bioengineering and Environmental Sciences
    (BEES) Group
Department of Energy and Environmental
    Engineering (DEEE)
CSIR-Indian Institute of Chemical
    Technology (IICT)
Tarnaka, Hyderabad, India
and
Academy of Scientific and Innovative Research
    (AcSIR)
Ghaziabad, India

A. M. Kalinkin
Tananaev Institute of Chemistry - Subdivision
    of the Federal Research Centre "Kola
    Science Centre of RAS"
Apatity, Murmansk Region, Russia

E. V. Kalinkina
Tananaev Institute of Chemistry - Subdivision
    of the Federal Research Centre "Kola
    Science Centre of RAS"
Apatity, Murmansk Region, Russia

Denis Kalmykov
Topchiev Institute of Petrochemical
    Synthesis
Russian Academy of Sciences
Moscow, Russia

Neha Kamal
Aquatic Toxicology Laboratory
Environmental Toxicology Group
FEST Division
Council of Scientific and Industrial
    Research-Indian Institute of Toxicology
    Research (CSIRIITR)
Lucknow, Uttar Pradesh, India

Debapriya Kar
BITS Pilani
K.K. Birla Goa Campus
Goa, India

Elena V. Khozina
M.M. Dubinin Laboratory of Sorption
  Processes
A.N. Frumkin Institute of Physical Chemistry
  and Electrochemistry RAS (IPCE RAS),
  Moscow, Russia

Lueta-Ann de Kock
College of Science Engineering and
  Technology
Institute of Nanotechnology and Water
  Sustainability
University of South Africa
South Africa

E. A. Kruglyak
Tananaev Institute of Chemistry - Subdivision
  of the Federal Research Centre "Kola
  Science Centre of RAS"
Apatity, Murmansk Region, Russia

V. N. Lozhkin
St. Petersburg University of State Fire Service
  of EMERCOM of Russia
Moskovsky, St. Petersburg, Russia.

O. V. Lozhkina
St. Petersburg University of State Fire Service
  of EMERCOM of Russia
Moskovsky, St. Petersburg, Russia
and
Solomenko Institute of Transport Problems of
  the Russian Academy of Sciences
St. Petersburg, Russia

Haridas Madathilkovilakathu
Emeritus Professor
Department of Biotechnology and
  Microbiology
Kannur University, Dr. Janaki Ammal Campus
Kannur, Kerala, India

Swachchha Majumdar
Water Technology Division, CSIR- Central
  Glass and Ceramic Research Institute
India

Sergey Makaev
Topchiev Institute of Petrochemical Synthesis
Russian Academy of Sciences
Moscow, Russia

Pubali Mandal
Department of Civil Engineering
Birla Institute of Technology and Science
Pilani, India

Ilya E. Men'shchikov
M.M. Dubinin Laboratory of Sorption Processes
A.N. Frumkin Institute of Physical Chemistry
  and Electrochemistry RAS (IPCE RAS)
Moscow, Russia

Unnikrishna Menon
Department of Civil Engineering
Indian Institute of Technology Kharagpur
Kharagpur, West Bengal, India

Satchidananda Mishra
CV Raman Global University
Bhubaneswar, Odisha, India

Abhisek Mondal
Department of Infrastructure Engineering
The University of Melbourne
Department of Civil Engineering, IIT
  Kharagpur

S. V. Chinna Swami Naik
Department of Civil Engineering
Indian Institute of Technology
Roorkee, Uttarakhand, India

Sagarika Panigrahi
CV Raman Global University
Bhubaneswar, Odisha, India
and
National Institute of Technology Agartala
Agartala, Tripura, India

Mayank Pathak
Prof. Rajendra Singh Nanoscience and
  Nanotechnology Centre
Department of Chemistry
DSB Campus, Kumaun University
Nainital, India

Nanda Gopal Sahoo
Prof. Rajendra Singh Nanoscience and
  Nanotechnology Centre
Department of Chemistry,
DSB Campus, Kumaun University
Nainital, India

Upakrishnam S. V. Sanskrit
Department of Civil Engineering
IIT Kharagpur

Vijaykumar Sekar
National Institute of Technology Andhra
 Pradesh

Satya Brot Sen
Vinod Gupta School of Management
Indian Institute of Technology Kharagpur
Kharagpur, West Bengal, India

Ekaterina Shalavina
Topchiev Institute of Petrochemical Synthesis
Russian Academy of Sciences
Moscow, Russia

Andrey V. Shkolin
M.M. Dubinin Laboratory of Sorption
 Processes
A.N. Frumkin Institute of Physical Chemistry
 and Electrochemistry RAS (IPCE RAS)
Moscow, Russia

Anuradha Singh
Aquatic Toxicology Laboratory
Environmental Toxicology Group
FEST Division
Council of Scientific and Industrial Research-
 Indian Institute of Toxicology Research
 (CSIRIITR)
Lucknow, Uttar Pradesh, India

Shivani Singh
Aquatic Toxicology Laboratory
Environmental Toxicology Group
FEST Division
Council of Scientific and Industrial Research-
 Indian Institute of Toxicology Research
 (CSIRIITR)
Lucknow, Uttar Pradesh, India

Olga V. Solovtsova
M.M. Dubinin Laboratory of Sorption
 Processes
A.N. Frumkin Institute of Physical Chemistry
 and Electrochemistry RAS (IPCE RAS)
 Moscow, Russia

Jianfeng Song
Shanghai Advanced Research Institute
Chinese Academy of Sciences
Shanghai, China

Baranidharan Sundaram
National Institute of Technology Andhra
 Pradesh

Kumar Raja Vanapalli
Department of Civil Engineering
National Institute of Technology Mizoram
Aizawl, India

Eduard Vasilev
Topchiev Institute of Petrochemical Synthesis
Russian Academy of Sciences
Moscow, Russia
and
Eurasian Patent Office
Moscow, Russia

Nimisha Vijayan P
Department of Biotechnology and
 Microbiology
Dr. Janaki Ammal Campus, Kannur University
Kannur, Kerala, India

Alexey Volkov
Topchiev Institute of Petrochemical Synthesis
Russian Academy of Sciences
Moscow, Russia

Vinay Yadav
Vinod Gupta School of Management, Indian
 Institute of Technology Kharagpur
Kharagpur, West Bengal, India
and
School of Civil and Environmental Engineering
Nanyang Technological University
Singapore

Grigory V. Zyryanov
I.Ya. Postovsky Institute of Organic Synthesis,
 Ural Branch of the Russian Academy of
 Sciences

Abbreviations

ABA	abcisic acid
ALG	alginate
BES	bioelectrochemical systems
CAT	catalase
CCS	carbon capture and storage
CCU	carbon capture and utilization
CCUNET	CCU as negative emission technology
CCUS	carbon capture, storage, and utilization
e-waste	waste of electrical and electronic equipment
EOL	end of life
EoL	end-of-life
EOR	enhanced oil recovery
EPR	Extended producer responsibility
ET	ethylene
GHG	greenhouse gas
GIS	geographic information system
IPCC	Intergovernmental Panel on Climate Change
JA	jasmonic acid
LCA	life cycle assessment
LCC	life cycle costing
LED	light-emitting diodes
MFA	material flow analysis
nHAP	hydroxyapatite nanoparticle
NUE	nutrient utilizing efficiency
ROS	reactive oxygen species
RREs	rare earth elements
RWGS	reverse water gas shift reaction
SOD	superoxide dismutase
TCE	technology critical elements
TPP	tripolyphosphate
WEEE	waste of electrical and electronic equipment

1 Opportunities and Challenges in Resource Recovery from Waste

Unnikrishna Menon and Brajesh Kumar Dubey

1.1 INTRODUCTION

The global contribution towards development has brought major challenges in the field of waste management and there are massive hikes in waste generation as well as no signs of its reduction. Approximately 2.02 billion metric tonnes of solid wastes were generated in the year 2016, which is expected to rise to 2.59 billion metric tonnes by 2030 and 3.4 billion metric tonnes by 2050, which is about a 70% rise in three decades. The majority of wastes still go to landfill and only around 20% of waste is recycled per annum (Tiseo 2018; World Bank 2018). This huge pile of waste can create immense pressure on the municipal authority if it cannot be managed scientifically. A large amount of organic waste, paper waste, chemical wastes, industrial wastes, packaging wastes, e-wastes, etc., can lead to socio-economic and environmental problems, including air, land and water degradation (Bui et al. 2022; Sengupta et al. 2022). 100% recycling and recovery of solid wastes helps in ensuring the *Zero Waste (ZW)* concept. However, when there is immense dependence on natural resources and waste generation on one side, it becomes difficult to achieve the ZW objective. Consequently, most of the waste generated goes to landfill, especially in developing countries, though it is not preferred as the primary choice to manage waste. The motive of any waste producer should be to minimize the waste that is generated. But when the wastes are generated, reusing and recycling should be the next priority (Demirbas 2011). The material recovered should be used till the end of life, after which it shall be sent for disposal, say in landfill. Moreover, there are options to divert the waste from landfill so as to use it as a resource. Since the wastes generated are of different characteristics it would require different treatment techniques. This chapter analyzes the challenges and opportunities in converting waste into a resource considering factors like type of waste and the recommended approaches for managing different types of wastes.

1.2 CLASSIFICATION OF WASTES

Based on the chemical nature, wastes can be divided into organic and inorganic, which can be further divided into biodegradable and non-biodegradable and is shown in Figure 1.1.

Biodegradable wastes are those wastes that are decomposed naturally, such as food wastes, agricultural waste, paper industry wastes, vegetable wastes, etc., whereas non-biodegradable wastes are those that remain in the environment without undergoing any decomposition, such as glass, plastics, metals, nuclear wastes, etc. Both types of the mentioned wastes can be point source as well as non-point source. While point source refers to pollutants discharged from any single identifiable source viz. municipal and effluent wastewater treatment plants, discharge pipes, sewage leakages,

DOI: 10.1201/9781003327646-1

1

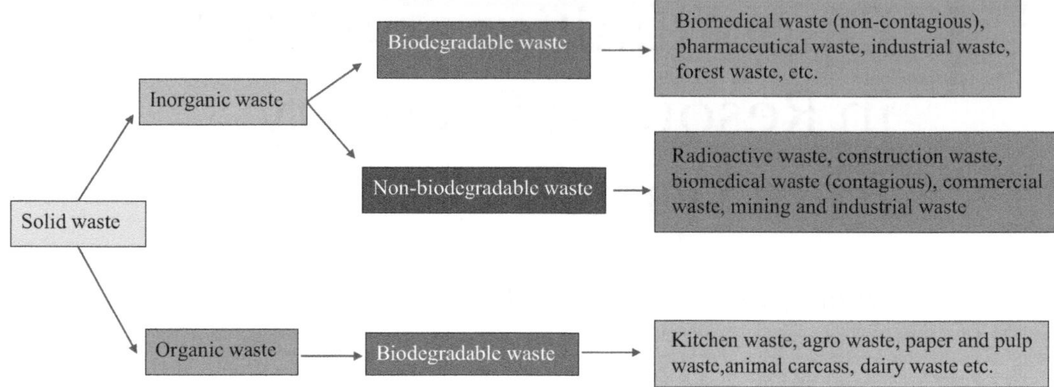

FIGURE 1.1   Classification of wastes.

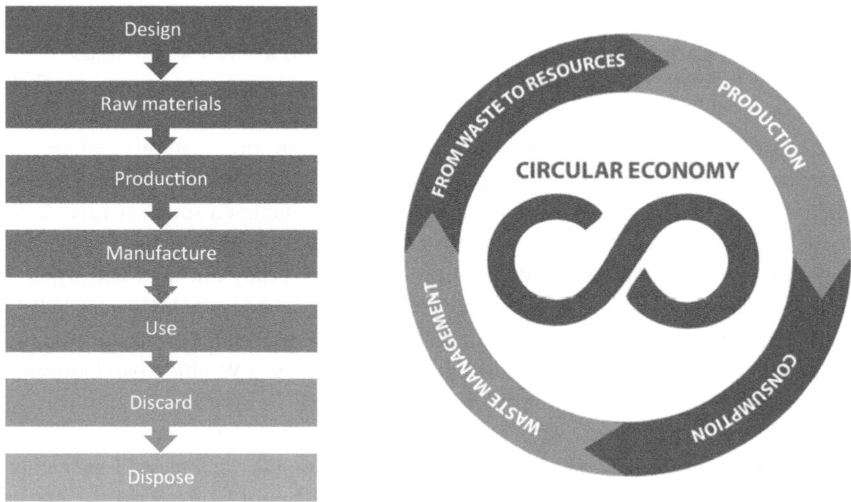

FIGURE 1.2   Representation of a linear and circular economy model.

etc., wastes from non-point sources are those which have many origins, such as agricultural fields, dumping grounds, etc. (Sengupta et al. 2022)

1.3   SIGNIFICANCE OF WASTE RECOVERY

Although most developed countries have a proper waste management system, many emerging countries face a lack of proper solid waste management (SWM) and a high rate of discarded waste as a result ofdumping and less waste collection, which consequently, reduces the reusing and recycling potential. Those economies will be forced to follow the linear model of material flow, which is a cradle-to-grave approach (Vignesh et al. 2021). In this model, the materials are discarded after use and the waste ends up in dump sites or illegal landfills and this can affect surroundings negatively.

In order to divert waste from landfill, it is recommended to practice the reduce, reuse, recycle (3R) model, which is also called the cradle-to-cradle approach, wherein a waste from one product serves as an input for another. Thus, the material flow takes place in a closed loop, ensuring the waste products are reused and recycled as seen in Figure 1.2. The materials recovered by following this model is called a circular economy (Winans et al. 2017). To further emphasize the circular economy, Kakwani and Kalbar (2020) recommended a reduce, reuse, recycle, reclaim, recover and restore (6Rs) strategy for water and wastewater resources. Therefore, any type of waste can be recovered sticking to a circular economy, be it biomass, plastics or even e-waste. The potential of a material to be reused and recycled depends on its ability to return to its original form. Reuse and recycling practices possess various advantages like preventing the requirement for extraction of additional raw materials to produce new materials, cutting down greenhouse gas emissions, reducing pollution and energy consumption.

Japan executed a circular economy as they utilized e-waste to manufacture the Tokyo Olympic medals in 2020. Approximately 79000 tonnes of e-wastes were collected including cell phone, digital cameras, laptops, etc., (Leader et al. 2017). Government intervention supported by community education and involvement made this project possible. Further, Kumar et al. (2020) considered different criteria for selecting a location for e-waste reuse and recycling operations, such as distance from residential areas, water bodies, financial support from local authorities, closeness to the waste collection point, etc. This not only maximized the recovery rate of valuable assets but also reduced the impact on the environment caused due to landfill disposal. However, SWM has not been evolved in many developing countries due to political, social and economic intervention.

1.4 REQUIREMENTS FOR A CIRCULAR ECONOMY AND RESOURCE RECOVERY

Success of any sustainable waste management system depends on the operational quality andrecovery efficiency from municipal solid wastes. Considering waste as a vital resource, the products generated via reuse and recycling proves to be the integral solution in most waste management systems. However, this needs to be complemented by socio-economic factors like communities, local and state authorities, financial support, stakeholders, etc. Policy aspects like closed loop supply chain activities, extended producer responsibility (EPR) can ensure valuable material recovery for reuse (Bui et al. 2022; Kumar et al. 2020). Fei et al. (2016) mentioned that integrating formal and informal sectors can facilitate proper recycling, making sure that the waste generated will be looping within the system.

1.4.1 SOCIETAL REQUIREMENTS

The society plays an imperative role in construction and operational strategies of recycling projects. They are the ones who uses various products, segregate the waste materials after use at the source itself. For example, a person who ordered food online and dispose the remains is recommended to separate the waste into two where the food waste goes into the bins that contains biodegradable waste, the food package in bins containing non-biodegradable waste. The former goes for composting and the latter goes to recycling. Public sentiments and satisfaction becomes decisive for empowering the future of a recycling or compost plant (Kheybari et al. 2019). Initiatives by ecologists, conservationists and volunteers for raising environmental awareness physically and through social media can induce pressure in the society, inspiring the public to shift gradually from a solely disposal to reuse and recycling practices. Activists, non-governmental organizations, residents' association require businesses and governments to act upon to develop the SWM across a particular region. Therefore, social awareness and evolution are the major drivers for transformation.

1.4.2 MUNICIPAL REQUIREMENTS

A municipality with sufficient resource in terms of infrastructure to manage waste, a skilled workforce, a well-defined logistic chain and recovery facilities that are accessible are some of the technical requirements for recycling. Municipal ecotechnological indicators like a road–rail network, transmission networks, water supply and disposal facilities need to be accommodated to back the sustainable concept of SWM. Moreover, authorities' understanding about population and budget helps to reduce the recycling gap, especially during the distribution of waste containers, waste collection routes and transportation networks. Further, the Internet of Things and smart device access keeps municipalities healthier and industrious. For instance, Thiruvananthapuram, the capital of Kerala state in India is one of the perfect examples in terms of the way waste is managed. There are approximately 414 aerobic compost bins at 54 different centers of the city, large-sized community-level biogas plants (28 numbers) have been set up in various markets, 78585 pipe composting units and various types of composting units, including portable models of biogas plants (1074 in total), are also being encouraged at a household level. Thiruvananthapuram Municipal Corporation (TMC) facilitates the treatment of organic wastes at a domestic level, thereby reducing the organic load at an industrial level. The compost formed can be used as a manure and it is given to the agricultural department. This boosts agricultural efficiency in the city. For non-biodegradable wastes, drop-off centers are located at different parts of the city from which these wastes will be sent to a material collection facility by an auto tipper. However, the municipal corporation announces specific dates for glass and e-waste collection. The wastes collected will be taken to a resource recovery facility (RRF) in a large covered truck at which segregation, sorting, etc., will take place before further processing or disposal. Moreover, the details of the RRFs in the city are given in the *Smart Trivandrum* application so that the public can also utilize these facilities. Waste plastic materials are sent to a shredding unit in Muttathara (one among ten plastic shredding units in Trivandrum). The shredded plastic is given to local bodies for road tarring by private agents. The facility has managed to sell plastic waste of about 29 tonnes for road tarring and 39 tonnes for other purposes out of an incoming 105.6 tonnes between June 2020 and August 2021. The materials that cannot be recycled or processed further are sent to an energy plant in Coimbatore (Tamil Nadu).

1.5   RECOVERY STATUS AND OPPORTUNITIES

There are many examples to cite when it comes to waste resource recovery. But what essentially involves recycling and recovery should be understood. The Waste Framework Directive (2008/98/EC) defines recycling and recovery as follows: **'recycling'** means any recovery operation by which waste materials are reprocessed into products, materials or substances whether for the original or other purpose. It includes the reprocessing of organic material but does not include energy recovery and the reprocessing into materials that are to be used as fuels or for backfilling operations. **'Recovery'** means any operation where the principal result is waste serving a useful purpose by replacing other materials which would otherwise have been used to fulfil a particular function, or waste being prepared to fulfil that function, in the plant or in the wider economy. However, under the REACH (Regulation for Registration, Evaluation, Authorization and Restriction of Chemicals), this distinction is not made. REACH only speaks of 'recovered substances', thus considering recycled and recovered substances in the same way as any other material (with a number of exceptions granted conditionally). The material recovery rates of some of the nations are given in Figure 1.3. The recovery here includes both recycling and composting.

As we can observe in Figure 1.3, resource recovery is most prominent in developed countries. However, an increase in resource recycling and recovery need not necessarily mean that the number of landfills or the load on landfills are low. It can also be due to the excess load of municipal solid waste in landfill due to which some part goes to recycling and recovery and if we consider the case of Italy, there were 180 landfills in 2013, which dropped to 123 in 2017. It was during the same time period the recovery rates in Italy rose from 40.8% to 52.4%., as shown in Figure 1.4. However, the number of landfills rose again to 131 in 2020, keeping up with the consistent contribution through

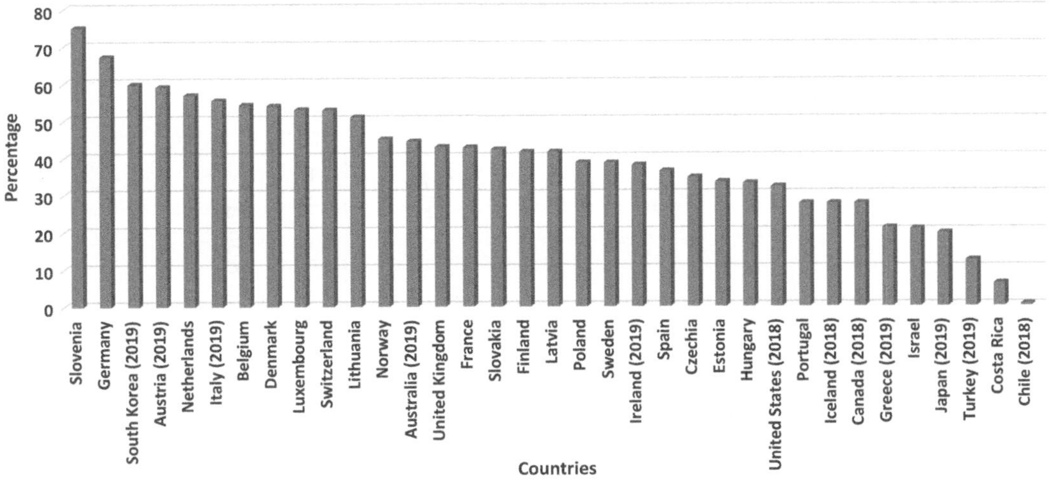

FIGURE 1.3 Municipal solid waste material recovery rates worldwide. (Tiseo 2022a.)

recycling practice as well. It accounts for approximately 5.8 million metric tonnes but it is still less than the 12 million metric tonnes in 2012. Italy belongs to one of the top recyclers in the European Union with a 51% recycling rate in 2019 and 55.4% total recovery. Approximately one-third of materials were recovered in 2020, even though landfilling is still represented as a relevant means in waste management as they accounted for almost 20% of all waste treatments in the same year. Slovenia, the country with the highest recycling rate (75%) also possesses landfill operations, which is higher than Germany, the second country with the highest recycling rate (67.1%), which is depicted in Figure 1.4. The rapid transition of Slovenia during the last decade is significant. There had been a massive jump in recycling rates from 2010 to 2020, where the rise is significant after 2013. Spain also gives priority to landfilling with rates of nearly 50% along with recycling with rates of around 36.4%, as shown in Figure 1.5.

In order for the nations to shine in the field of waste management, central and local governments should work together along with public consultation. It starts from each house, small towns, cities and spreads the practice nation-wide. Once the standards and protocols are established, developing and achieving something like new fuel recovery technology viz. waste-to-energy, fertilizers and chemicals becomes feasible. Its demonstration for various applications increases the trustworthiness among the public and it becomes widely accepted. This is especially important in developing countries. Similar is the case of the Kozhikode Municipal Corporation (Kerala-India). The existing solid waste management system in the Kozhikode Corporation faces challenges such as a lack of segregation, unsatisfactory waste treatment plant capacity and environmental issues due to landfill, etc. (Chithra and Yoonus 2021). However, the Kozhikode Municipal Corporation has already started to tackle these challenges through a waste-to-energy project. The first waste-to-energy plant of Kerala state in Njeliyanparamba is in the initial phase of construction. This project is expected to process 450 tonnes of solid waste and generate 6 MW of electricity per day. The process used here will be a controlled combustion method by which the wastes will be burned at high temperature. The steam thus produced will be used to work turbines and generate electricity, which will be sold to the state electricity board (*The Hindu* 2020). The state and local governments confirmed approval and the public accepted the first waste-to-energy project as it was already highly important to manage the waste in the district.

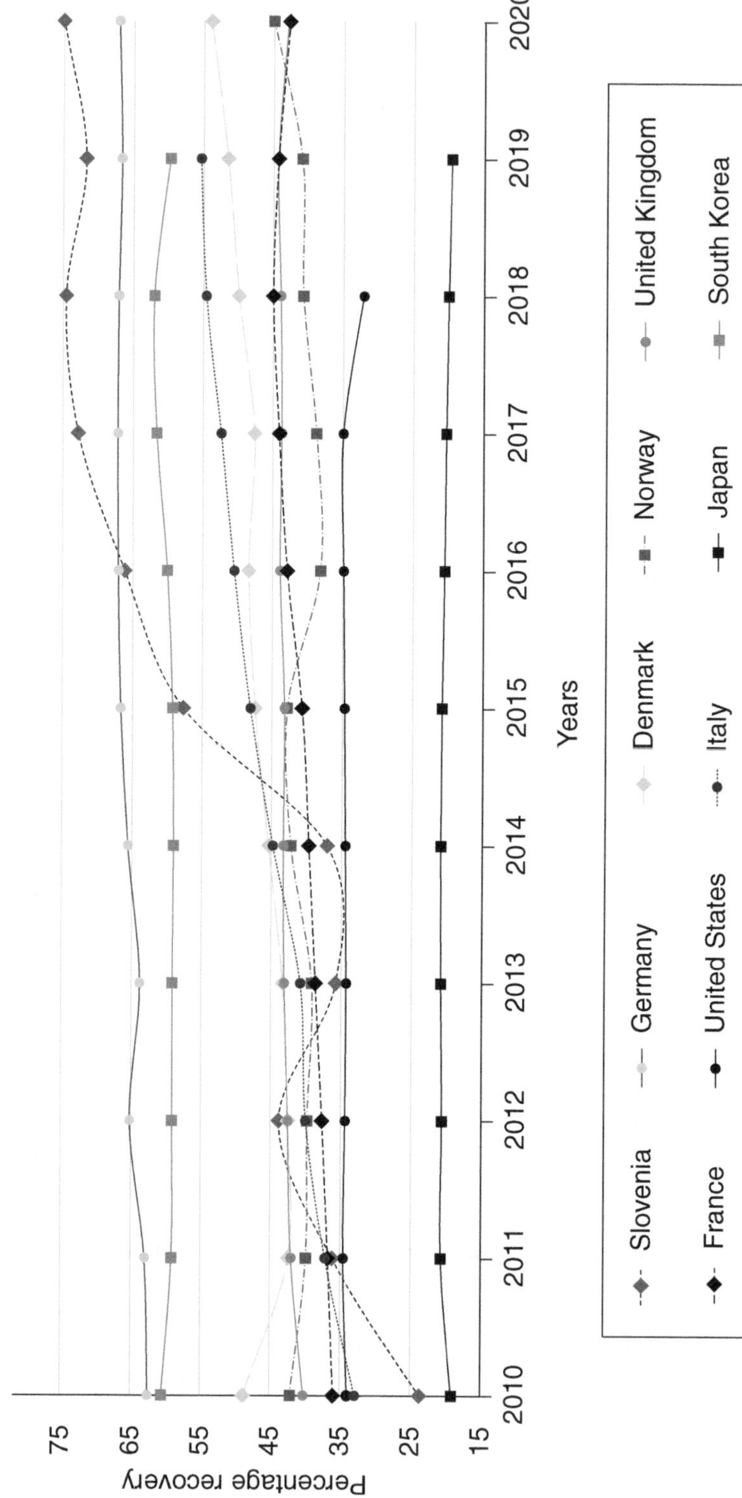

FIGURE 1.4 Yearly material recovery rates of different countries from 2010–2020*. (Tiseo 2022b.)

***Note: Data for South Korea is unavailable for 2020, data for the United States is unavailable for 2019 and 2020.**

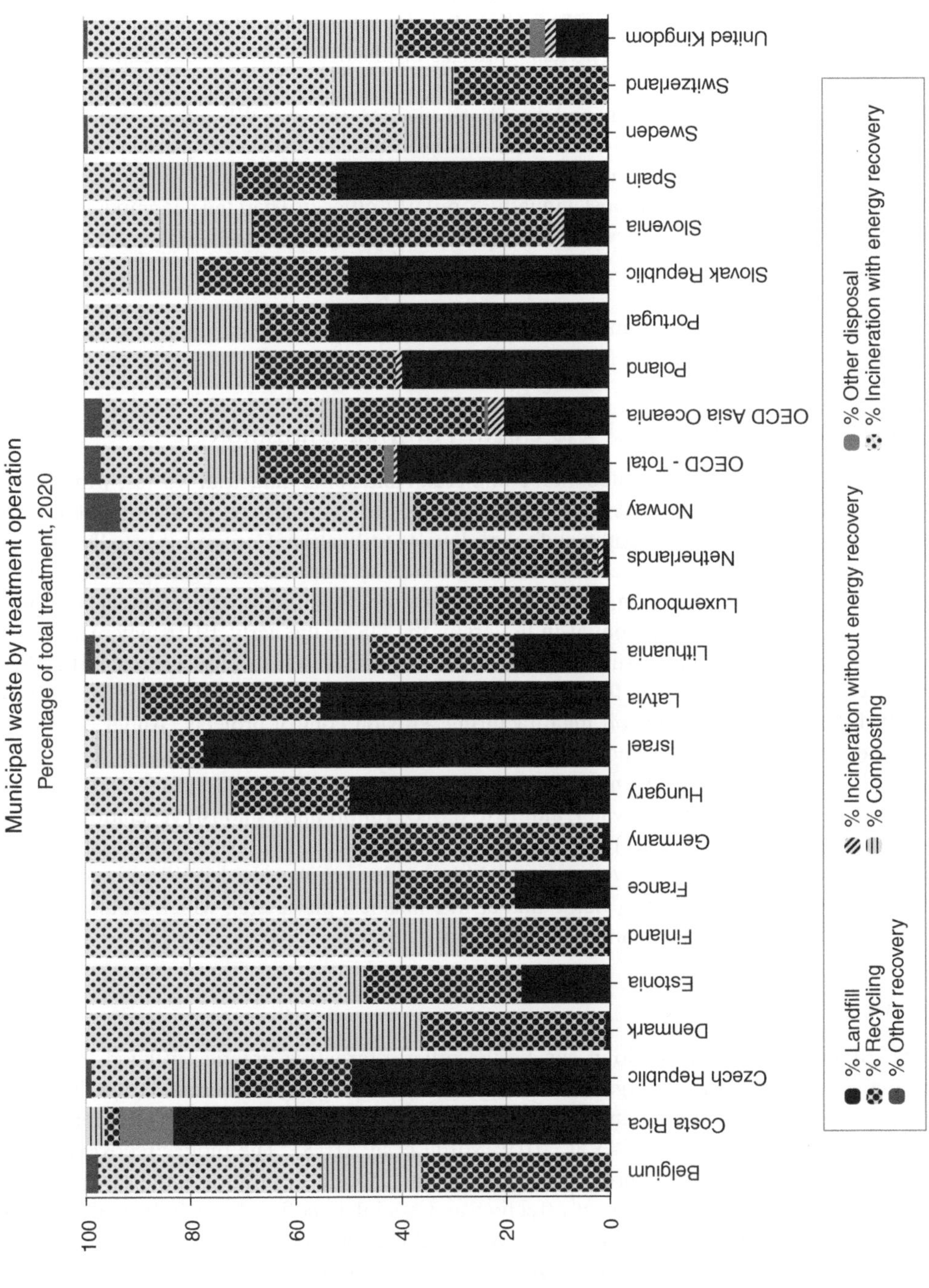

FIGURE 1.5 Municipal waste by treatment operation for different countries as of 2020. (OECD 2023.)

Segregation is the Key

The success of any waste management system depends on the waste segregation, which is the basic requirement to treat waste. Mixed waste will always affect the recovery efficiency and may even cause equipment failures. For example, biogas production from waste would be affected if there is the presence of metals, batteries or plastics in the waste, creating a negative impact on methane production (Aprilia et al. 2013). The wastes shall be segregated into wet (biodegradable organic wastes) and dry (glass, plastics, metals). When the organic wastes go for composting/aerobic digestion or anerobic digestion, the dry wastes possess the potential for recycling. In India, there are currently more than 70 composting facilities, generating around 4.3 million tonnes of compost annually. Similarly, cities in Thailand decompose around 1.26 million tonnes of food wastes to generate biogas and organic fertilizer (Pharino 2021).

Some of the techniques or opportunities to utilize the waste through recycling and recovery are discussed in the following sections.

1.5.1 Organic Matter Compost and Energy Potential Utilization

Organic wastes are comprised of residential, commercial and some industrial wastes, which can be used as an economic and ecological fuel. Though composting is practiced in several countries for numerous years, it is only a minor pathway for solid waste utilization. Compost is usually a marketable product for agricultural purposes. The net cost of refuse disposal composting is less than the costs for landfilling and incineration. Other than composting, this high organic content solid waste had been utilized for its energy potential in boilers in the US in the early 1970s where waste fuel replaced 13 to 20% of the coal burned in each boiler to generate electricity (Lesher 2014). Years went by as researches came up with the organic matter compost being used in highway construction projects, turf dressing, topsoil/subsoil manufacturing, erosion control projects (Antler et al. 2022; Waldron and Nichols 2009). Compost can also improve downstream water quality by retaining pollutants like heavy metals, nitrogen, phosphorous, oil and grease, herbicides, fuels and pesticides. It can also retain a large amount of water, aiding to reduce runoff and establish vegetation (US EPA 2017). According to Fan et al. (2021) heat can be recovered as reused via composting and is capable of meeting the energy demand. However, it should be made sure that the MSW for composting is segregated and is devoid of any recyclable matter like plastics, glass, etc., which would otherwise lead to the production of a large quantity of low-quality compost.

1.5.2 Anaerobic Digestion

Anaerobic digestion (AD) technology can be the other alternative practice utilized for organic waste management. AD converts organic waste by hydrolysis, acidogenic fermentation, hydrogen-producing acetogenesis, and methanogenesis to bioenergy (biogas) that can be transformed into electric energy and heat energy (Li et al. 2019; Mittal et al. 2018). The biogas recovery ensures reduction in greenhouse gases, improving sanitation, and indoor air pollution, thereby promoting sustainable practices bringing out cost-effective and social benefits. Like Italy and Germany, other countries have also developed an extensive technique for producing renewable energy from energy crops via biogas production through the AD process (Lindfors et al. 2020). China uses a solar energy system in biogas production during low-temperature conditions and has proven to be more efficient to accomplish the optimum temperature to carry out the AD process (Gaballah et al. 2020).

1.5.3 Plastics Recycling and Recovery

Plastics had been popular for use in road construction, the manufacture of tiles and other building materials (Awoyera and Adesina 2020; Kwabena et al. 2017; Vasudevan et al. 2012). Single-use plastics are now being used for manufacturing polybricks. Plastic wastes can be utilized for recovering resources like oil, benzene, hydrocarbons, styrene, hydrogen, carbon nanotubes, etc., by different techniques like wet torrefaction, hydrothermal carbonization, pyrolysis, gasification, hydrocracking, etc. (Qureshi et al. 2020; Salaudeen et al. 2019; Samal et al. 2021b; Shen 2020; Yao et al. 2018). These plastic wastes can be combined with biomass, such as saw dust, starch and other lignocellulosic substances as well to obtain the binding property to be used as fuel. There are studies on co-pyrolysis of waste low-density polyethylene and saw dust as well as polystyrene and eucalyptus biomass. Both the studies showed excellent thermal properties, energy yield and density, indicating the co-pyrolysis of plastics and biomass has the potential to be used as solid fuel (Samal et al. 2021a,b). Treatments like photodegradation, mechanochemical degradation, and thermo degradation practices have been used for plastic waste degradation (Yang et al. 2018; Zhang et al. 2021). However, the non-recyclable fraction of plastic waste recovered after mechanical treatment can be used as RDF (refuse-derived fuel) in cement, chemical or paper manufacturing plants (Onwosi et al. 2017).

1.5.4 Refuse-Derived Fuel

Refuse-derived fuel is an alternative fuel consisting mainly of combustible components of waste materials, such as textiles, non-recyclable plastics, labels, cardboard, paper and rubber (Rotter et al. 2011). It possesses a high calorific value, usually in the range of 11–25 MJ/kg original substance, and homogenous particle size (5–300 mm) (Sarc and Lorber 2013). RDF production consists of a multi-step process of separation technologies, such as sieving, sifting, grinding, and can be followed by briquetting or in the generation of RDF-fluff (Rajca et al. 2020). The type of fuel generated is generally cheaper, readily available, and relatively generates less CO_2 than conventional fuels, such as coal. It is estimated that many European countries produce 4–5 million tonnes of RDF from MSW every year (Gallardo et al. 2014). European countries have adopted mechanical treatment (MT), or mechanical biological treatment (MBT) plants' technologies for the production of a high calorific value fraction (HCVF) that can be utilized in RDF production. Pre-sorting solid waste prior to the incineration process would offer a new perspective in solid waste management that includes improved incinerator performance, and the direct revenues from recycled materials (Gallardo et al. 2014). Poland is one of the countries that is the primary benefactor of RDF and is known to use it in cement plants for over 15 years (Berardi et al. 2020).

1.5.5 Inert Waste Recycling

Inert waste is generated mainly from construction, excavation, demolition and glass processing activities. These wastes do not undergo physical, chemical or biological transformations (Menegaki and Damigos 2018; Sharma et al. 2020). Such types of wastes also contribute to environmental and health-related issues. Hence, obtaining value-added products via reuse and recycle is the best way to handle inert wastes. If we consider ash generated from various industrial and mining activities, it can be recycled to prepare geopolymers, which have great polymerization affinity. The recycled aggregates obtained from concrete and demolition waste can construct roads, cementitious materials, landscaping and asphalt concrete. Glass waste materials can also be recycled many times without significant alteration in their chemical characteristics, which also has the potential application to be used as an aggregate in construction where it can replace up to 15% of fine aggregates and 40% of coarse aggregates (Mohajerani et al. 2017; Mohammadinia et al. 2019).

1.5.6 CARBON-BASED MATERIALS

The fabrication of carbon material for environmental remediation, energy storage, composting additives, adsorbents, etc., has gained major research interest in recent years. The MSW, like sewage sludge, food waste, yard waste, agricultural residues and other lignocellulosic biomass, has been studied via different techniques, such as carbonization, gasification, pyrolysis, etc., for various purposes like soil amendments andenergy storage devices, such as lithium ion batteries, supercapacitors, hydrogen generation, etc. (Dulanja et al. 2020; Jaganathan et al. 2019; Senthil and Lee 2021; Sharma et al. 2019, 2021; Venna et al. 2022). However, based on parameters like reaction time, temperature, heating rate, etc., bio-oil can also be generated via fast pyrolysis. Excellent thermal properties, energy yield and density was exhibited by co-pyrolysis of plastics and biomass indicating its potential to be used as solid fuel (Samal et al. 2021a,b). Similarly, carbon-based adsorbents like graphene and its derivatives, activated carbon, carbon nanotubes and biochar are used for heavy metal removal from aqueous solutions. Activated carbons are widely used as adsorbents for water and wastewater treatment (Yang et al. 2019; Zhang et al. 2019).

1.5.7 INCINERATION

Incineration is a thermal waste treatment technique where there is a large-scale accumulation of heterogeneous (both organic and inorganic) waste needing a treatment. It is a controlled combustion process with the primary objective of volume reduction and energy recovery from the waste stream. Incineration is the most popular waste-to-energy technique, whereby heat produced from combustion can be recovered and converted to electric power. The organic content of waste is combusted and heat is produced, whereas the inorganic content contributes to the formation of ash (Rodríguez et al. 2014). Utilizing this technique can significantly reduce the load that needs to be sent to landfills. Temperature, time and turbulence are the three factors that determine the efficiency of an incinerator. End products of incineration include fly ash, heat and mixture of toxic gases. Countries like Japan have been using the ash byproduct to develop sewage bricks, slags, cement, etc. However, metals cannot be destroyed by this technique. Hence, separation/removal of metals are necessary from ash before sending it to landfill.

1.5.8 HEAVY METAL RECOVERY

There can be different sources for heavy metals from MSW, such as electronic goods, painting waste, electroplating waste, fertilizers and pesticides from yard waste, etc. (Esakku et al. 2014). The concentration of heavy metal can vary at all times. Waste components, like paper, will have low heavy metal concentration whereas discarded batteries will have high concentrations of heavy metals, even if the quantity of battery waste generated is less. Heavy metals should be separated from wastes before they reach landfills or any other disposal facility. One way to achieve this is through the incineration of MSW, where metals may be concentrated in the ash, and some amounts in emission control devices (Stewart and Lemieux 2003). Besides this, ash can also be melted at high temperatures to produce slag, which can be used for road construction, where some amount of metals remain in the newly formed secondary fly ash (Kuboňová et al. 2013). Recovery of heavy metals from ash includes thermal separation, chemical extraction, bioleaching and electrochemical processes. Bioleaching is one of the promising techniques for recycling heavy metals with minimum environmental impact but it is time consuming (Wang et al. 2021). Also, hydrometallurgy, electrometallurgy, pyrometallurgy and biometallurgy can be used for the recovery of heavy metals from the e-waste after dismantling plastics. A recent practical example of heavy metals' recovery was showcased by the Japanese government for the Tokyo Olympics 2020, where pyrometallurgy was used to recover metals and to produce medals from e-waste.

1.6 CHALLENGES

Developing nations are currently engaged in setting up the standards and capabilities for sustainable SWM (SSWM). Despite developing protocols, sometimes it gets difficult to execute or put into practice. Waste treatment technologies, informal sectors, policy restrictions and material flow analysis are some of the important indicators that require improvement. Even from Figure 1.3, one can comprehend that most countries prominent in recycling rates are developed countries, except South Korea, a developing country that is performing well as of 2020 with a recycling rate of nearly 60% in 2019. Developing countries like China (7–10%), Thailand (15%) and Costa Rica (6.1%) still need improvement in terms of recovery rates. The following are some of the hurdles for recovery, mostly in developing countries.

1.6.1 APPROACH OF PUBLIC AND MIXING UP OF WASTES

Generally solid wastes are heterogenous in nature and its composition varies from place to place and it is difficult to determine its physical and chemical characteristics, which is the reason why segregation is an inevitable step before disposal. This can help authorities to determine the types of wastes generated and decide the type of treatment required. The public plays a major part in segregation of wastes as they are also waste generators (Abdel-shafy and Mansour 2018). Presence of e-waste in the biodegradable wastes and occurrence of biodegradable wastes in the recyclable wastes produces the same poor-quality output after treatments, which is sufficient enough to affect the efficiency or even contribute to the closure of compost plants and recycling facilities. Since the product quality is compromised, it may not have its market value. Recruiting additional labour for waste segregation is another aspect the authorities need to look at when it comes to time consumption and economic feasibility. Scenarios still exist even if separate bins are kept for organic, inorganic, recyclable and used electronic goods, wastes get mixed up due to the lack of awareness or negligence of the public. There can even be situations of scattering of loose wastes on the roads when piled up on the side of the roads because animals dig into the waste piles and scatter it. It consequently ends in different varieties of wastes being mixed up due to animal interference, causing contamination. This mostly happens in developing and undeveloped countries. Although it is recommended to segregate the waste at the source itself, the wastes that are sent to MRFs can be both clean as well as dirty. 'Clean' MRF/'Pre-sorted' MRF deals with recyclable materials that have already been separated from other waste streams whereas 'dirty' MRF receives unsorted wastes and is contaminated, thus affecting the quality of the recycled product. Therefore, manual separation, mechanical separation or a combination of both are required to separate out the recyclables from contaminants, which otherwise would also affect the efficiency of the treatment systems.

Do You Know?

The Thiruvanathapuram Municipal Corporation (Kerala-India) transformed the dump yard in Erumakuzhy village into a park and garden, named Sanmathi. It took around 2567 man days to clear the dump yard.

The same Thiruvananthapuram city also consists of a landfill (dump yard) in Vilappil village, which had to shut down following protests from people living in the locality since the leachate from landfills started flowing into the nearby stream.

Lesson: Public acceptance is very important as landfill operation will potentially have negative impacts on human health as well as the value of land and property in the surrounding area. They include site topography, drainage, soils, geohydrology and adjacent land use.

1.6.2 Recovery Feasibility

It is well known that the recovery of materials, like paper plastic, glass, iron, lead, steel, copper, etc., will decrease the requirement of import funding for these materials and save energy (Marmolejo 2012). But due to a lack of trained professionals, proper technology, public awareness and high initial capital investments, proper recovery is not well applied. Even though a few large- and medium-scale solid waste treatment facilities imported from industrialized countries have been built and operated, rigorous mechanical and energy demands of these technologies forces the facilities to be closed. Moreover, the prices for the recovered materials also remains lower when compared to the segregation/reprocessing costs and this is more than the costs of virgin materials. This raises the necessity of subsidizing the recycling activities by the concerned authorities (Marmolejo et al. 2012). Now, in certain cases even if it is assumed that the treatment facilities are well equipped and the recovered products are profitable, there is still risk for the workers in this field. The ash or slag formed in the municipal solid waste incinerator need to be taken for chemical and metal recovery as loading into landfill without proper treatment may generate leachates that are toxic. Some metal recovery methods like hydrometallurgy involves the use of chemical reagents and a leaching process such as cyanide leaching, thiourea leaching, etc., and a leaching process using acids, such as nitric acid, hydrochloric acid, sulfuric acid, aquaregia, etc., possess a potential threat to the workers dealing with metal recycling and recovery. This is also applicable during e-waste recycling. Lead saturated fumes are released during the de-soldering of circuit boards. Inhalation of these chemical fumes results in adverse effects to the workers. Furans and dioxins may be released due to incomplete combustion when e-wastes are burned at lower temperatures. Further, open burning of e-waste can cause threat to informal workers as the toxic elements enter into their blood stream.

1.6.3 Climate-Solid Waste Relation

There is a frequent misconception that technological advancements are the remedy for unmanaged and increasing waste. But it is just one of many solutions while managing wastes. The technology should be chosen based on the geological and meteorological conditions also. Factors like climate becomes decisive while selecting a treatment technique to manage waste since it directly influences the waste management process and it should be regionally appropriate solution. For example, the presence of cold ambient conditions, especially during winter can hinder the composting and anerobic digestion process since it affects microbial activity, which consequently results in low compost quality (Xie et al. 2017). Several countries depend on landfills for waste disposal. However, if the landfills are located at high temperature and low-humidity zones, decomposition rates may become affected. Since the decomposition is a complex process, higher temperatures with lower moisture content will hamper the decomposition of wastes. It can also lead to higher evaporation losses with production of lesser but strong leachates. The situation aggravates if the wastes are dumped in dumpsites, as these strong leachates can contaminate soil as well as groundwater (Bebb and Kersey 2003; Nnaji and Utsev 2011). Such countries/provinces/states require alternativee solutions to be found that are economically and environmentally feasible.

1.6.4 Income Distribution of Population

In practice, recycling programmes emerge as an independent initiative for the Municipal SWM where the funding is usually allocated by local authorities. Also, allocating resources for efficient recycling with the available municipal budget largely depends on the income of residents. However, this practice is usually most benefitted for the people living in urban areas, whereas the population living in rural areas still tend to be far from attaining these resources (Araya-Córdova et al. 2021). Even if this imbalance between the rural and urban settlements is eliminated, there can still be fluctuations in the recycling practice if a nation is in a developing phase. According to a study by Medina (1997), increase in population income in a developing country tends to generate more waste whereas the increase in income tends to produce less waste in a developed country, making the recycling practice easier. Thus, generally income unevenness in a developing country can result in uncertainty in the type of recycling practice required due to varying wealth and demographic conditions.

1.6.5 Policies and Framework, Fund Allocation Implications

Although this chapter describes various practices for waste recovery, its establishment is complex and time consuming due to the government requirements for planning approval. Also, a lack of details and clarification of reuse and recycling policy restrictions and the absence of rules and regulations are some of the barriers for SWM development. Regulations for recycling and waste treatment technologies, residential and business payments for recycling and waste disposal are unclear. There is a gap in sustainable materials' policy focused on identifying specific waste resource. Clear regulatory architecture and policies where waste management firms and cities collaborate and advanced coalitions to encourage SSWM outcomes so as to keep up with the environmental standards are required, especially in developing countries. Even if we take the case of a developed country like the United States (US), there are not a lot of rules and regulations at the Federal level. The waste generated in the US is controlled at the state level (World Population Review 2021). It is also evident from Figure 1.4 that the recycling rate in the US is not that high when compared to other developed countries, though it is maintaining its recycling consistency.

Several developing countries pose the challenge of an inability to untap the technology potential, unavailability of data, outdated legal systems due to the gap between local and regional governments raise the questions on firms, leading to an inability to acquire funds to support reuse and recycling practices as well as SSWM planning and process. Therefore, proper communication between authorities at different levels is crucial to promote this CE practice.

1.7 POSSIBLE STRATEGIES FOR IMPROVED RESOURCE RECOVERY

1.7.1 Decentralization of Waste

Decentralized solid waste management aims at reducing the quantity of waste at source by installing a certain number of small waste management centers within a community. Different methods like composting, small-scale anaerobic digestion, recycling, etc., are usually adopted. The selection of various technologies depends on climate, techno-economic viability, land availability, capital investment, etc. Researches reveal that a decentralized system is more economic than a centralized system as the collection, production and use are all operated by the local community (Joshi and Seay 2019; Rathore et al. 2022). This not only reduces the burden of waste collection, transportation costs, reduction of load in landfills but also maximizes the processing efficiency on the waste treatment units, improves the aesthetics of the community and changes peoples' mindset. This system can also provide better employment and income opportunities for the informal sector of waste management. Most importantly, this system is environmentally benign, economical and makes the public aware about the social responsibility of managing waste (Joshi and Seay 2019).

However, there can be situations when difficulties arise during cold climates for composting and AD processes since lower temperature affects microbial activity. However, there are studies that may overcome such problems. Introduction of cold-tolerant microbial consortia with efficient hydrolytic activities have been reported to breakdown complex organic waste into nutrient-rich compost that can be utilized as fertilizers (Hou et al. 2017).

There are many examples from developing countries where decentralization has become a common practice. Thiruvananthapuram Municipal Corporation introduced a decentralized system for waste management, following local protests and shut down of landfill at Vilappilsala. Following the 2011 protests, the TMC introduced segregated collection of waste to ensure maximum efficiency. It formalized and institutionalized source-level composting and decentralized resource recovery as part of city waste management. For residential flats and gated communities, the TMC offered a 50% subsidy for setting up organic waste management *in situ*. For low-value non-recyclables such as laminates, households are encouraged to drop them off in TMC collection centers (Henam and Sambyal 2019a).

1.7.2 EXTENDED PRODUCER RESPONSIBILITY

Extended producer responsibility (EPR) is based on a 'polluter pays' principle, which has become an established environmental policy in several countries. Under EPR, producers will be responsible for the environmental impacts of their products such as plastics or electronic goods, starting from design till the post-consumer phase. This can eliminate the physical and financial burden on the public and municipalities to manage end-of-life products by increasing the recycling rate and reducing the load on disposal facilities (OECD 2016a). This policy can be voluntary or mandatory and is well established in Europe, Canada, Japan and South Korea. There are broadly four categories of EPR instruments to address specific aspects of waste management: product take-back requirements, economic and market-based instruments, regulations and performance standards and accompanying information-based instruments. The selection of instruments can vary across countries, region and industry, based on political and social, economic, legal and cultural context. If the government's priority is to improve waste collection, mandatory collection targets need to be introduced. If the collection and recycling target is the priority, a take-back policy is recommended, though the priorities can be voluntary or mandatory. For example, a take-back scheme was established in Columbia as per the legislation since the collection target is more important than recycling targets, whereas Austria, Germany and Sweden give priority to both collection and recycling. However, in South Korea, economic instruments like advanced disposal fees (ADF) are imposed on importers and producers of the products that are hazardous and difficult to recycle (OECD 2016b, 2019).

1.7.3 PUBLIC SENTIMENTS AND WASTE PREVENTION BEHAVIOUR

Despite the availability of recycling services, the success of a recycling program depends much more on the population that generates waste. Residents' behaviour in society towards recycling, their awareness on environmental issues and significance of waste segregation helps authorities to manage their waste in a better manner. To exemplify, the waste collection in Germany covers the cost of collection and management (Gellynck et al. 2011). Therefore, people will be forced to reduce the waste generation so as to keep the collection charges low. In the UK, there exists an indirect waste prevention policy by introducing high landfill tax, which started at £24/t on 2007 and reached £80/t on 2014. Besides this, the UK also introduced several other waste prevention schemes at the local scale viz. the Merseyside Waste Prevention Plan (2011–2041). This focused on changing the residents' behaviour and raising environmental awareness among the public and businesses (DEFRA 2013; *RESOURCES Merseyside 2011–2041* 2013) . The authorities should also be profound about the age, gender and economic status of people in the society. This aids authorities to understand the socio-economic behaviour. Coordination between local governments and various environmental groups in South Korea take credit for the successful implementation of the VBWF system across the nation by conducting feasibility studies and public hearings. This helped to change the waste generation patterns as the public became more aware about waste disposal and the behaviour of consumers and producers (Henam and Sambyal 2019b).

1.7.4 INCORPORATING TECHNOLOGY

The incorporation of GIS into the waste management system can also help in achieving realistic strategies, processing, control, and guide the regulators on how to deal with the solid waste stream. Artificial Intelligence (AI) combined with the 4R principle is a reliable method for waste reduction. Hua et al. (2021) designed one such model, gathering the information of garbage holders using sensor technologies and AI also claims to be a cost-effective way of managing waste. Another modelling study by Razzaq et al. (2021) brought up the interrelationships as a result of solid waste recycling towards economic growth, carbon emission and recovery efficiency. The results show that the increase in MSW recycling helps in reducing carbon emissions, improving recovery efficiency and stimulating economic growth in the long run. This tool will help the policymakers to counteract the carbon emissions and to create an economic value to the waste through recycling practices.

1.7.5 IMPROVED DATA RELIABILITY ON WASTE COMPOSITION

Challenges exist in terms of information and data on the leftover materials. Though waste materials reach the recovery or recycling facility, there may be some wastes that do not contribute towards these practices. For example, some plastics could be observed in animals' bodies, glasses and paper-based materials on beaches, etc. This can manipulate authorities' decisions on the number of composting and recycling plants required in a city. Underestimation or overestimation of any treatment facility can bring huge loss to the local authority, forcing the waste management firms to halt the operations and hinder further SSWM projects. This is where the necessity of the informal sector comes into the picture. They identify the materials that can be recovered from various locations that are inaccessible to the formal sectors, whether it is the beach, dumping ground or streets.

1.7.6 POLICY IMPLEMENTATION

Diverting the waste from landfill and its recycling, choosing the most efficient and economical policy measure are some of the major challenges of waste management policies as it involves a decision-making process and trade-offs between economic and environmental aspects (Di Maria et al. 2020). The waste management policy and guidelines consists of the following components: sector-specific norms, standards and procedures; compliance and enforcement (compulsory recycling); taxes and levies (differentiated VAT); environmental liability by self-responsiveness; green public procurement; fees and user charges (weight-based waste fee); subsidies/incentives (subsidies in some services); awareness campaigns (mandatory labeling on hazardous chemicals); information; and monitoring and feedback system (Tripathi et al. 2022). Governments and policymakers should choose policies based on relevance and proper assessment of the focus area where it requires the most. For example, factors like geography, climate, demography, urbanization, diversity of culture and consumption pattern, infrastructure, and technological capacity play an important role in dealing with the wider context of waste generation and the intervention required. The regional and provincial governments are supposed to take part in policy setting and action planning. Even in a unitary system, regional and sub-national governments set policy objectives and legislative frameworks, say, South Korea, one of the top recyclers in the world started managing waste three decades back. The South Korean government implemented a VBFS (volume-based fee system) in 1995, which includes two main objectives: to impose waste treatment costs on each polluter based on the amount of waste generated, and to *provide a free collection service for recyclable wastes*, thereby inducing a reduction in the generation of wastes at source and *encouraging the collection of recyclable wastes*. Residents pay different fees for the bags depending on the size and the regions in which they reside, while enjoying a free service for collection of recyclables. The law requires that residents properly sort out their waste and imposes fines of up to approximately \$1,000 for violations of the garbage disposal rules (Henam and Sambyal 2019b). It is also interesting to note that rewards will be given for those who report violators (up to \$235) (Belcher 2022).

1.7.7 Formal–Informal Sector Collaboration

In developing countries, recycling is practiced on a community basis or for monetary benefit. The wastes are collected by informal waste workers, which includes ragpickers and itinerant recyclable buyers. They divert the recyclable portion of MSW from reaching the landfill. These informal sectors also perform hauling wastes from households, scavenge wastes and discard the remaining portion at the dump sites whereas in an urban area, majorly the waste collection and transportation are done by formal sectors, usually by a private agent. Materials are recovered after sorting activities. Although they may not be able to achieve a maximum amount for recycling practices, a collaboration between formal and informal waste sectors can help in collecting maximum amount of waste (organic and inorganic) from a specific region. A separate facility should be set up for sorting so as to account for the type of waste received and to proceed for further processing. The recycling rates accounted by the informal sector is in the range of 20–50%, and are comparable to those achieved by modern waste management systems in developed countries (Wilson et al. 2009). This not only diverts waste from landfill but also advances the waste management sector of a nation. Since the workers in the informal sector work with less pay or without pay, they belong to poor, downgraded and vulnerable communities in the society. When supported and organized, this sector can attract investments, build business competitiveness, minimize material shortages, preserve natural resources, create jobs and save cities' money, etc. Also, formalizing their appearance helps them to boost their self-esteem, confidence, achieve healthy acceptance among the public as well as establishing informal recycling and advancement of sector integration in SSWM (Bui et al. 2022). Alappuzha (a city in India) Municipal Corporation has distinguished itself by its success in source-level segregation coupled with decentralized waste management. The engagement of the informal sector into the waste management system is one of many factors for their success. Similar is the case of Indore, where community engagement, municipal authorities trusting citizens and make them actively take part in cleaning up the city, employing women from marginalized communities, etc., tells the success story of Indore being the cleanest city in India (Atin Biswas 2021).

1.8 CONCLUSION

Numerous developing countries have been facing the issue of SWM, unable to practice the reuse/recycling practices due to a lack of potential strategies, data unavailability, unawareness about the type of waste a society generates and its fate, funds and treatment technologies. Sustainable waste management system becomes difficult due to highly multi-layered nature of material flows in both supply chain and SWM networks. Material flow and resource distribution are inevitable to create a closed-loop material movement and to balance industrial development. Although there are different options and opportunities to manage and recover resources from waste, various policies need to be set up by the national and provincial governments along with public participation in order to recover resources from waste and utilize it to its maximum. Besides this, it is also necessary to develop production and marketing strategies to accomplish the sustainability. Segregation and decentralized treatment are the key factors to look at for the success of any nation targeting to achieve zero waste and a circular economy. The public should be socially responsible, aware and properly educated to manage the waste. Also, the governments should also update the social, geological and economic status of the society so as to take monetary policy-based decisions as well as to take decisions on the treatment technologies that need to be adopted to a particular locality. The corporates, brands should also come into the limelight to deal the collection and recycling of the products that has reached end-of-life. When all the strategies are put in place, every society progresses and give a new face to the nation.

REFERENCES

Abdel-shafy, H. I., and M. S. M. Mansour. 2018. "Solid waste issue: Sources, composition, disposal, recycling, and valorization." *Egypt. J. Pet.*, 27 (4): 1275–1290. https://doi.org/10.1016/j.ejpe.2018.07.003.

Antler, S., L. Cooperband, C. S. Coker, M. Schwarz, and R. Rynk. 2022. "Chapter 2 – Enterprise planning." In *The composting handbook*, R. Rynk, ed., 27–49. Elsevier.

Aprilia, A., T. Tezuka, and G. Spaargaren. 2013. "Inorganic and hazardous solid waste management: Current status and challenges for Indonesia." *Procedia Environ. Sci.*, 17 (81): 640–647. https://doi.org/10.1016/j.proenv.2013.02.080.

Araya-Córdova, P. J., S. Dávila, N. Valenzuela-Levi, and Ó. C. Vásquez. 2021. "Income inequality and efficient resources allocation policy for the adoption of a recycling program by municipalities in developing countries: The case of Chile." *J. Clean. Prod.*, 309 (April): 127305. https://doi.org/10.1016/j.jclepro.2021.127305.

Atin Biswas, S. P. et al. 2021. Waste-wise cities: Best practices in municipal solid waste management. *Centre for Science and Environment and NITI Aayog*, New Delhi. www.niti.gov.in/sites/default/files/2021-12/Waste-Wise-Cities.pdf

Awoyera, P. O., and A. Adesina. 2020. "Case studies in construction materials plastic wastes to construction products: Status, limitations and future perspective." *Case Stud. Constr. Mater.*, 12: e00330. https://doi.org/10.1016/j.cscm.2020.e00330.

Bebb, J., and J. Kersey. 2003. *Potential impacts of climate change on waste management*. Environment Agency R&D Dissemination Centre, c/o WRc, Frankland Road, Swindon, Wilts.

Belcher, D. 2022. "In South Korea, an emphasis on recycling yields results." *New York Times*. www.nytimes.com/2022/05/21/business/south-korea-recycling.html.

Berardi, P., M. Fonseca, M. De Lurdes, and J. Maia. 2020. "Analysis of Portugal's refuse derived fuel strategy, with particular focus on the northern region." *J. Clean. Prod.*, 277: 123262. https://doi.org/10.1016/j.jclepro.2020.123262.

Bui, T., J. Tseng, M. Tseng, and M. K. Lim. 2022. "Resources, conservation & recycling opportunities and challenges for solid waste reuse and recycling in emerging economies: A hybrid analysis." *Resour. Conserv. Recycl.*, 177 (October 2021): 105968. https://doi.org/10.1016/j.resconrec.2021.105968.

Chithra, K., and S. Yoonus. 2021. "Decentralised solid waste management for Kozhikode corporation: A sustainable approach." *Environ. Waste Manag. Recycl.*, 4 (1): 1–4.

DEFRA. 2013. *Waste prevention programme: Summary of existing measures*. The National Archives, Kew, London.

Demirbas, A. 2011. "Waste management, waste resource facilities and waste conversion processes." *Energy Convers. Manag.*, 52 (2): 1280–1287. https://doi.org/10.1016/j.enconman.2010.09.025.

Di Maria, A., J. Eyckmans, and K. Van Acker. 2020. "Use of LCA and LCC to help decision-making between downcycling versus recycling of construction and demolition waste." *Advances in construction and demolition waste recycling: management, processing and environmental assessment*, F. Pacheco-Torgal, Y. Ding, F. Colangelo, R. Tuladhar, and A. Koutamanis, eds., 537–558. Woodhead Publishing.

Dulanja, P., S. You, A. Deshani, Y. Xia, A. Bhatnagar, S. Gupta, H. Wei, S. Kim, J. Kwon, D. C. W. Tsang, and Y. Sik. 2020. "Biochar-based adsorbents for carbon dioxide capture: A critical review." *Renew. Sustain. Energy Rev.*, 119 (October 2019): 109582. https://doi.org/10.1016/j.rser.2019.109582.

Esakku, S., K. Palanivelu, and K. Joseph. 2014. *Assessment of heavy metals in a municipal solid waste dumpsite*. (researchgate.net).

Fan, S., A. Li, C. J. N. Buisman, and W. Chen. 2021. "Heat potential, generation, recovery and utilization from composting: A review." *Resour. Conserv. Recycl.*, 175 (May): 105850. https://doi.org/10.1016/j.resconrec.2021.105850.

Fei, F., L. Qu, Z. Wen, Y. Xue, and H. Zhang. 2016. "How to integrate the informal recycling system into municipal solid waste management in developing countries: Based on a China ' s case in Suzhou urban area." *Resour. Conserv. Recycl.*, 110: 74–86. https://doi.org/10.1016/j.resconrec.2016.03.019.

Fernández Rodríguez, M. D., M. C. García Gómez, N. Alonso Blazquez, and J. V Tarazona. 2014. "Soil pollution remediation." *Encyclopedia of toxicology* (Third Ed.), P. Wexler, ed., 344–355. Academic Press. https://doi.org/10.1016/B978-0-12-386454-3.00579-0

Gaballah, E. S., T. Kh, S. Luo, and Q. Yuan. 2020. "Enhancement of biogas production by integrated solar heating system: A pilot study using tubular digester." *Energy*, 193: 116758. https://doi.org/10.1016/j.energy.2019.116758.

Gallardo, A., M. Carlos, M. D. Bovea, F. J. Colomer, and F. Albarr. 2014. "Analysis of refuse-derived fuel from the municipal solid waste reject fraction and its compliance with quality standards." *J. Clean. Prod.*, 83: 118–125. https://doi.org/10.1016/j.jclepro.2014.07.085.

Gellynck, X., R. Jacobsen, and P. Verhelst. 2011. "Identifying the key factors in increasing recycling and reducing residual household waste: A case study of the Flemish region of Belgium." *J. Environ. Manage.*, 92 (10): 2683–2690. https://doi.org/10.1016/j.jenvman.2011.06.006.

Henam, S., and S. Sambyal. 2019a. "Ten zero-waste cities: How Thiruvananthapuram cleaned up its act." *DownToEarth.* www.downtoearth.org.in/news/waste/ten-zero-waste-cities-how-thiruvananthapuram-cleaned-up-its-act-68539.

Henam, S., and S. Sambyal. 2019b. Ten zero-waste cities: How Thiruvananthapuram cleaned up its act." *DownToEarth.* www.downtoearth.org.in/waste/ten-zero-waste-cities-how-seoul-came-to-be-among-the-best-in-recycling-68585.

Hou, N., L. Wen, H. Cao, K. Liu, X. An, D. Li, H. Wang, X. Du, and C. Li. 2017. "Role of psychrotrophic bacteria in organic domestic waste composting in cold regions of China." *Bioresour. Technol.*, 236: 20–28. https://doi.org/10.1016/j.biortech.2017.03.166.

Hua, K., Y. Zhang, D. Li, and C. E. Montenegro-Marin. 2021. "Environmental planning based on reduce, reuse, recycle and recover using artificial intelligence." *Environ. Impact Assess. Rev.*, 86 (January 2021): 106492. https://doi.org/10.1016/j.eiar.2020.106492.

Jaganathan, V. M., O. Mohan, and S. Varunkumar. 2019. "Intrinsic hydrogen yield from gasification of biomass with oxy-steam mixtures." *Int. J. Hydrogen Energy*, 44 (33): 17781–17791. https://doi.org/10.1016/j.ijhydene.2019.05.095.

Joshi, C., and J. Seay. 2019. "Building momentum for sustainable behaviors in developing regions using Locally Managed Decentralized Circular Economy principles." *Chinese J. Chem. Eng.*, 27: 1566–1571. https://doi.org/10.1016/j.cjche.2019.01.032.

Kakwani, N. S., and P. P. Kalbar. 2020. "Review of Circular Economy in urban water sector: Challenges and opportunities in India." *J. Environ. Manage.*, 271 (July): 111010. https://doi.org/10.1016/j.jenvman.2020.111010.

Kheybari, S., M. Kazemi, and J. Rezaei. 2019. "Bioethanol facility location selection using best-worst method." *Appl. Energy*, 242 (March): 612–623. https://doi.org/10.1016/j.apenergy.2019.03.054.

Kocasoy, G. 2001. "Solid waste management in developing countries: Proposed amendments in the existing situation." *Eighth International Waste Management and Landfill Symposium*, 1–5 October, Cagliary, Italy.

Kuboňová, L., Š. Langová, B. Nowak, and F. Winter. 2013. "Thermal and hydrometallurgical recovery methods of heavy metals from municipal solid waste fly ash." *Waste Manag.*, 33 (11): 2322–2327. https://doi.org/10.1016/j.wasman.2013.05.022.

Kumar, A., P. Wasan, S. Luthra, and G. Dixit. 2020. "Development of a framework for selecting a sustainable location of waste electrical and electronic equipment recycling plant in emerging economies." *J. Clean. Prod.*, 277: 122645. https://doi.org/10.1016/j.jclepro.2020.122645.

Kwabena, J., V. N. Berko-Boateng, and T. Ama. 2017. "Case studies in construction materials use of waste plastic materials for road construction in Ghana." *Case Stud. Constr. Mater.*, 6: 1–7. https://doi.org/10.1016/j.cscm.2016.11.001.

Leader, A. M., X. Wang, and G. Gaustad. 2017. "Creating the 2020 Tokyo Olympic medals from electronic scrap: Sustainability analysis." *J. Miner. Met. Mater. Soc.*, 69 (9): 1539–1545. https://doi.org/10.1007/s11837-017-2441-4.

Lesher, R. 2014. "Resource Recovery from Solid Waste in the Future." *R. Swedish Acad. Sci.*, 3 (3): 156–160.

Li, Y., Y. Chen, and J. Wu. 2019. "Enhancement of methane production in anaerobic digestion process: A review." *Appl. Energy*, 240 (February): 120–137. https://doi.org/10.1016/j.apenergy.2019.01.243.

Lindfors, A., M. Gustafsson, S. Anderberg, M. Eklund, and M. Mirata. 2020. "Developing biogas systems in Norrköping, Sweden: An industrial symbiosis intervention." *J. Clean. Prod.*, 277: 122822. https://doi.org/10.1016/j.jclepro.2020.122822.

Marmolejo, Luis F., L. F. Diaz, P. T. Lozada, and M. Garcia. 2012. "Perspectives for sustainable resource recovery from municipal solid waste in developing countries: Applications and alternatives." In *Waste Management* Fernando, L., and Rebellon, M. (eds.). IntechOpen. 10.5772/52303. www.intechopen.com/chapters/40500.

Medina, M. 1997. "Effect of income on municipal solid waste generation rates for countries of varying levels of economic development: A model." *J. Solid Waste Technol. Manag.*, 24 (3): 149–155.

Menegaki, M., and D. Damigos. 2018. "A review on current situation and challenges of construction and demolition waste management." *Curr. Opin. Green Sustain. Chem.*, 13: 8–15. https://doi.org/10.1016/j.cogsc.2018.02.010.

Mittal, S., E. O. Ahlgren, and P. R. Shukla. 2018. "Barriers to biogas dissemination in India: A review." *Energy Policy*, 112 (October 2017): 361–370. https://doi.org/10.1016/j.enpol.2017.10.027.

Mohajerani, A., J. Vajna, T. Ho, H. Cheung, H. Kurmus, and A. Arulrajah. 2017. "Practical recycling applications of crushed waste glass in construction materials: A review." *Constr. Build. Mater.*, 156: 443–467. https://doi.org/10.1016/j.conbuildmat.2017.09.005.

Mohammadinia, A., Y. Choy, A. Arulrajah, and S. Horpibulsuk. 2019. "Strength evaluation of utilizing recycled plastic waste and recycled crushed glass in concrete footpaths." *Constr. Build. Mater.*, 197: 489–496. https://doi.org/10.1016/j.conbuildmat.2018.11.192.

Nnaji, C. C., and T. Utsev. 2011. "Climate Change and Waste Management: A Balanced Assessment." *J. Sustain. Dev. Africa*, 13 (May): 13–34.

OECD. (2022). "Waste: Waste - Municipal waste: generation and treatment", *OECD Environment Statistics* (database), https://doi.org/10.1787/data-00601-en (accessed on 18 November 2022).

OECD. 2016a. *OECD policy highlights – Extended producer responsibility.* OECD Publishing.

OECD. 2016b. *Extended producer responsibility.* OECD Publishing.

OECD. 2019. *Waste management and the circular economy in selected OECD countries.* OECD Publishing.

OECD. 2023. OECD Database. www.oecd-ilibrary.org/environment/data/oecd-environment-statistics/municipal-waste_data-00601-en

Onwosi, C. O., V. C. Igbokwe, J. N. Odimba, I. E. Eke, M. O. Nwankwoala, I. N. Iroh, and L. I. Ezeogu. 2017. "Composting technology in waste stabilization: On the methods, challenges and future prospects." *J. Environ. Manage.*, 190: 140–157. https://doi.org/10.1016/j.jenvman.2016.12.051.

Pharino, C. 2021. "Food waste generation and management: Household sector." *Valorization of agri-food wastes and by-products*, R. Bhat, ed., 607–618. Academic Press.

Qureshi, M. S., A. Oasmaa, H. Pihkola, I. Deviatkin, A. Tenhunen, J. Mannila, H. Minkkinen, M. Pohjakallio, and J. Laine-ylijoki. 2020. "Pyrolysis of plastic waste: Opportunities and challenges." *J. Anal. Appl. Pyrolysis*, 152 (October 2019): 104804. https://doi.org/10.1016/j.jaap.2020.104804.

Rajca, P., A. Poskart, M. Chrubasik, M. Sajdak, M. Zajemska, and A. K. Andrzej Skibinski. 2020. "Technological and economic aspect of refuse derived fuel pyrolysis." *Renew. Energy*, 161: 482–494. https://doi.org/10.1016/j.renene.2020.07.104.

Rathore, P., S. Chakraborty, M. Gupta, and S. P. Sarmah. 2022. "Towards a sustainable organic waste supply chain: A comparison of centralized and decentralized systems." *J. Environ. Manage.*, 315 (April): 115141. https://doi.org/10.1016/j.jenvman.2022.115141.

Razzaq, A., A. Sharif, A. Najmi, M. Tseng, and M. K. Lim. 2021. "Dynamic and causality interrelationships from municipal solid waste recycling to economic growth, carbon emissions and energy efficiency using a novel bootstrapping autoregressive distributed lag." *Resour. Conserv. Recycl.*, 166 (December 2020): 105372. https://doi.org/10.1016/j.resconrec.2020.105372.

RESOURCES Merseyside 2011–2041. 2013. Merseyside Recycling and Waste Authority, 6th Floor, North House, 17 North John Street, Liverpool. https://www.merseysidewda.gov.uk/wp-content/ uploads/2012/10/RESOURCES-MAIN-DOCUMENT.pdf.

Rotter, V. S., A. Lehmann, and T. Marzi. 2011. "New techniques for the characterization of refuse-derived fuels and solid recovered fuels." *Waste Manag. Res.*, 29 (2): 229–236. https://doi.org/10.1177/0734242X10364210.

Salaudeen, S. A., P. Arku, and A. Dutta. 2019. "Gasification of plastic solid waste and competitive technologies." *Plastics to energy*, S. M. Al-Salem, ed., 269–294. Elsevier.

Samal, B., K. Raja, and B. Kumar. 2021a. "Influence of process parameters on thermal characteristics of char from co-pyrolysis of eucalyptus biomass and polystyrene: Its prospects as a solid fuel." *Energy*, 232: 121050. https://doi.org/10.1016/j.energy.2021.121050.

Samal, B., K. R. Vanapalli, B. K. Dubey, J. Bhattacharya, S. Chandra, and I. Medha. 2021b. "Char from the co-pyrolysis of Eucalyptus wood and low-density polyethylene for use as high-quality fuel: Influence of process parameters." *Sci. Total Environ.*, 794: 148723. https://doi.org/10.1016/j.scitotenv.2021.148723.

Sarc, R., and K. E. Lorber. 2013. "Production, quality and quality assurance of refuse derived fuels (RDFs)." *Waste Manag.*, 33 (9): 1825–1834. https://doi.org/10.1016/j.wasman.2013.05.004.

Sengupta, D., B. K. Dubey, and S. Goel. 2022. "Wastes to Wealth for Bioenergy Generation." *Treatment and disposal of solid and hazardous wastes*, D. Sengupta, B. K. Dubey, and S. Goel, eds., 211–231. Springer.

Senthil, C., and C. W. Lee. 2021. "Biomass-derived biochar materials as sustainable energy sources for electro-chemical energy storage devices." *Renew. Sustain. Energy Rev.*, 137: 110464.

Sharma, H. B., S. Panigrahi, and B. K. Dubey. 2019. "Hydrothermal carbonization of yard waste for solid bio-fuel production: Study on combustion kinetic, energy properties, grindability and flowability of hydrochar." *Waste Manag.*, 91: 108–119. https://doi.org/10.1016/j.wasman.2019.04.056.

Sharma, H. B., S. Panigrahi, and B. K. Dubey. 2021. "Food waste hydrothermal carbonization: Study on the effects of reaction severities, pelletization and framework development using approaches of the circular economy." *Bioresour. Technol.*, 333 (February): 125187. https://doi.org/10.1016/j.biortech.2021.125187.

Sharma, V., A. Giri, S. Thakur, and D. Pant. 2020. "Resource recovery from inert municipal waste." *Current developments in biotechnology and bioengineering: Resource recovery from wastes*, S. Varjani, A. Pandey, E. Gnansounou, S. K. Khanal, and S. Raveendran, eds., 251–262. Elsevier.

Shen, Y. 2020. "A review on hydrothermal carbonization of biomass and plastic wastes to energy products." *Biomass Bioenergy*, 134 (January): 1–18. https://doi.org/10.1016/j.biombioe.2020.105479.

Stewart, E. S., and P. M. Lemieux. 2003. "Emissions from the Incineration of Electronics Industry Waste." *Int. Symp. Electron. Environ.*, 2003: 271–275.

The Hindu. 2020. "Work on Kerala's first waste-to-energy plant begins at Njeliyanparamba in Kozhikode." www.thehindu.com/news/cities/kozhikode/work-on-keralas-first-waste-to-energy-plant-begins-at-njeliyanparamba-in-kozhikode/article32955866.ece

Tiseo, I. 2018. "Projected generation of municipal solid waste worldwide from 2016 to 2050." *Statista.* www.statista.com/statistics/916625/global-generation-of-municipal-solid-waste-forecast/.

Tiseo, I. 2022a. "Municipal solid waste material recovery rates worldwide in 2020, by select country." *Statista.* www.statista.com/statistics/1052439/rate-of-msw-recycling-worldwide-by-key-country/.

Tiseo, I. 2022b. "Municipal solid waste material recovery rates worldwide from 2000 to 2020, by select country." *Statista.* www.statista.com/statistics/1317178/material-recovery-rates-worldwide-by-select-country/.

Tripathi, A., A. Prakash, and J. Prakash. 2022. "Economics and market of wastes." *Emerging trends to approaching zero waste: Environmental and social perspectives*, C. M. Hussain, S. Singh, and L. Goswami, eds., 319–338. Elsevier.

US EPA. n.d. "Reducing the impact of wasted food by feeding the soil and composting." www.epa.gov/sustainable-management-food/reducing-impact-wasted-food-feeding-soil-and-composting#success.

Vasudevan, R., A. R. Chandra, B. Sundarakannan, and R. Velkennedy. 2012. "A technique to dispose waste plastics in an ecofriendly way – Application in construction of flexible pavements." *Constr. Build. Mater.*, 28 (1): 311–320. https://doi.org/10.1016/j.conbuildmat.2011.08.031.

Venna, S., H. B. Sharma, D. Mandal, H. P. Reddy, S. Chowdhury, A. Chandra, and B. K. Dubey. 2022. "Carbon material produced by hydrothermal carbonisation of food waste as an electrode material for supercapacitor application: A circular economy approach." *Waste Manag. Res.*, 40 (10): 1514–1526. https://doi.org/10.1177/0734242X221081667.

Vignesh, K. S., S. Rajadesingu, and K. D. Arunachalam. 2021. "Chapter Four - Challenges, issues, and problems with zero-waste tools." *Concepts of advanced zero waste tools*, C. M. Hussain, ed., 69–90. Elsevier.

Waldron, K. W., and E. Nichols. 2009. "Composting of food-chain waste for agricultural and horticultural use." *Handbook of waste management and co-product recovery in food processing*, K. Waldron, ed., 583–627. Woodhead Publishing.

Wang, H., F. Zhu, X. Liu, M. Han, and R. Zhang. 2021. "A mini-review of heavy metal recycling technologies for municipal solid waste incineration fly ash." *Waste Manag. Res.*, 39 (9): 1135–1148. https://doi.org/10.1177/0734242X211003968.

Wilson, D. C., A. O. Araba, K. Chinwah, and C. R. Cheeseman. 2009. "Building recycling rates through the informal sector." *Waste Manag.*, 29: 629–635. https://doi.org/10.1016/j.wasman.2008.06.016.

Winans, K., A. Kendall, and H. Deng. 2017. "The history and current applications of the circular economy concept." *Renew. Sustain. Energy Rev.*, 68 (August 2016): 825–833. https://doi.org/10.1016/j.rser.2016.09.123.

World Bank. 2018. *Global municipal solid waste generation projection: 2016–2050.* World Bank.

World Population Review. n.d. *Recycling rates by country.* https://worldpopulationreview.com/country-rankings/recycling-rates-by-country.

Xie, X., Y. Zhao, Q. Sun, X. Wang, H. Cui, X. Zhang, Y. Li, and Z. Wei. 2017. "A novel method for contributing to composting start-up at low temperature by inoculating cold-adapted microbial consortium." *Bioresour. Technol.*, 238: 39–47. https://doi.org/10.1016/j.biortech.2017.04.036.

Yang, X., Y. Wan, Y. Zheng, F. He, Z. Yu, and J. Huang. 2019. "Surface functional groups of carbon-based adsorbents and their roles in the removal of heavy metals from aqueous solutions: A critical review." *Chem. Eng. J.*, 366 (February): 608–621. https://doi.org/10.1016/j.cej.2019.02.119.

Yang, Y., J. Wang, K. Chong, and A. V Bridgwater. 2018. "A techno-economic analysis of energy recovery from organic fraction of municipal solid waste (MSW) by an integrated intermediate pyrolysis and combined heat and power (CHP) plant." *Energy Convers. Manag.*, 174 (July): 406–416. https://doi.org/10.1016/j.enconman.2018.08.033.

Yao, D., Y. Zhang, P. T. Williams, H. Yang, and H. Chen. 2018. "Co-production of hydrogen and carbon nanotubes from real-world waste plastics: Influence of catalyst composition and operational parameters." *Appl. Catal. B Environ.*, 221 (September 2017): 584–597. https://doi.org/10.1016/j.apcatb.2017.09.035.

Zhang, F., Y. Zhao, D. Wang, M. Yan, J. Zhang, P. Zhang, T. Ding, L. Chen, and C. Chen. 2021. "Current technologies for plastic waste treatment: A review." *J. Clean. Prod.*, 282: 124523. https://doi.org/10.1016/j.jclepro.2020.124523.

Zhang, Z., Z. Zhu, B. Shen, and L. Liu. 2019. "Insights into biochar and hydrochar production and applications: A review." *Energy*, 171: 581–598. https://doi.org/10.1016/j.energy.2019.01.035.

2 Best Available Techniques for Organic Livestock Waste Management in Russia

Aleksandr Briukhanov, Ekaterina Shalavina and Eduard Vasilev

2.1 INTRODUCTION

The main task of the last 10 years has been to restore food security in Russia after the negative consequences of the 1990s. Thanks to special state programmes, we were able to restore food security for basic food products. By 2020, Russia was one of the top five pork-producing countries and one of the top ten milk, poultry meat and egg-producing countries. Russia has 100% self-sufficiency in pork, eggs and poultry meat. The average milk yield reached 6500 kg/cow/year and 84% self-sufficiency in milk. The president of Russia set a task of increasing agri-food exports to 45 billion US dollars by 2024. In 2020, it was about 25 billion US dollars. Academic Andrey Izmailov, speaking at the annual meeting of the Russian Academy of Sciences, said that the natural and biological potential of Russian production is 400 billion US dollars.

High indicators of agricultural production in Russia are achieved through the use of intensive technologies and the construction of large livestock complexes. This development method leads to big risks for the environment [1]. Today we already see key environmental challenges in agricultural production:

- a high level of secondary resources loss – ex. organic fertilizers – up to 70–80%;
- a high level of diffuse loading to the environment, ex., on water bodies – up to 150 kg N from ha and up to 10 kg P from ha [2, 3];
- a lack of special technological standards and criteria for environmental sustainability of agroecosystems;
- a degradation of soil quality of agricultural land – loss of humus, erosion, etc.;
- increased pollutant emissions, including greenhouse gases.

Researches into theotential risks of environmental pollution from agricultural production, which were conducted in North-West Russia, showed that utilization of organic waste leads to the main risk for the environment. Risk is above 85% (Figure 2.1).

Observing the main types of agricultural products, we can see that the total amount of manure is significantly more than all of the main agricultural products. Each previous year, there was about 580 million tons of manure, with the main portion being cattle manure contributing about 460 million tons (Figure 2.2).

The next largest is pig manure with about 78 million tons and poultry manure with about 40 million tons. But by observing statistics, we can see what is used as an organic fertilizer with less than 50% manure.

DOI: 10.1201/9781003327646-2

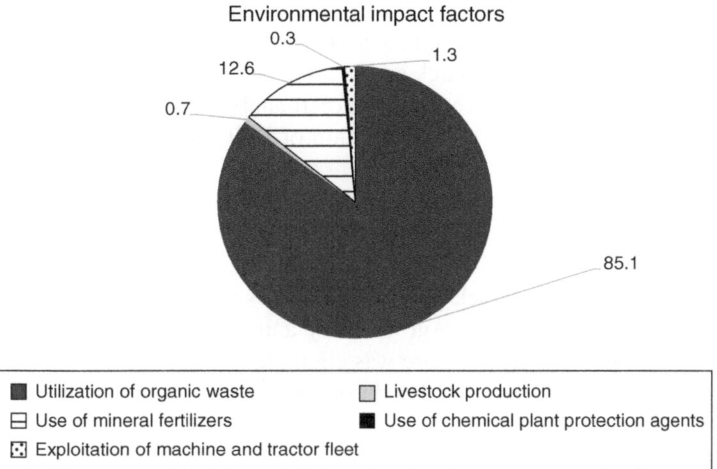

FIGURE 2.1   Potential risks of environmental pollution from agricultural production – an example of North-West Russia.

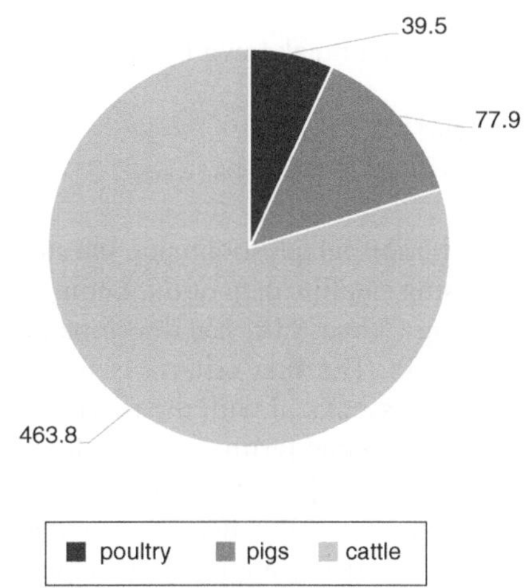

FIGURE 2.2   State of organic livestock waste utilization in Russia.

This situation raises questions and concerns about environmental safety. Inefficient use of organic waste can lead to nutrient loss: nitrogen – 2.2 million tons/year; phosphorus – 0.36 million tons/year.

To solve problems in intensive livestock production in 2017, best available technique (BAT) information and technical reference books were created on intensive pig and farm poultry rearing [4, 5]. BAT are determined based on special criteria:

- the lowest level of negative impact on the environment per unit of time or volume of products (goods) produced;
- economic efficiency of its implementation and operation;
- application of resource- and energy-saving methods;
- the period of its implementation;
- industrial implementation of this technology at two or more objects.

Best available techniques for organic livestock waste management are one of the key parts.

The goal of research is to improve the environmental safety and environmental quality of the rural territory by creating a system of design, technical, technological and managerial solutions and the infrastructure for utilization of livestock and poultry farm waste.

The tasks of the research are (1) scientific and expert analysis of the ecological status of livestock enterprises in the Leningrad Region; (2) justification of BATs for manure utilization; (3) development of BATs for manure utilization.

2.2   MATERIALS AND METHODS

The initial data on the status of livestock/poultry enterprises in the Leningrad Region were collected in cooperation with the Committee for Agroindustry and Fishery Complex within the Leningrad Region Government. A special questionnaire was circulated to agricultural enterprises.

The following information was considered for the environmental assessment of agricultural enterprises in the Leningrad Region:

- type of the enterprise (crop, livestock, poultry, mixed);
- profile of the enterprise (for pig farms – fattening, reproduction, combined breeding and feeding; for poultry factories – egg or broiler production; for cattle complexes – dairy or fattening);
- animal/poultry stock;
- area under agricultural crops (with a breakdown by cultivated crops), ha;
- use of mineral fertilizers, t/year;
- use of organic fertilizers, t/year;
- transfer of organic fertilizer to other enterprises, t/year.

To complete the task, the data from the special questionnaire, the survey outcomes of typical agricultural enterprises and the remote sensing satellite data of the Leningrad Region territories were used.

Federal Scientific Agroengineering Center VIM has developed a system for BAT formation for manure utilization. It has three criteria. The first criteria is environmental risks associated with the nutrient load. To identify the risks associated with the nutrient load from livestock and poultry complexes, the mass of nitrogen in the organic fertilizer was calculated for each district.

Using HELCOM Recommendations [6, 7] (170 kg/ha of nitrogen) and Management Directive for Agro-Industrial Complex "Recommended Practice for Engineering Designing of Systems for Animal and Poultry Manure Removal and Pre-application Treatment" the maximum application rate for each type of organic fertilizer, produced from cattle, pig or poultry manure, was calculated. With received application rates (t/ha), the known total area of agricultural land in each district and the cultivated crops, the potential for the complete use of animal/poultry manure in the districts of the Leningrad Region was determined.

The second criteria is environmental risks associated with intensive production. In the Baltic countries, the animal density of 1.5 LSU/ha of the agricultural area was adopted as a conditionally safe value [8, 9]. Exceeding the range of values of 1.5–1.7 LSU/ha can bring about the risk of accumulation of excessive amounts of nutrients in the soil and a significant increase in pollutant emissions into the atmosphere.

The third criteria is environmental risks associated with diffuse loads. To calculate the diffuse load on water bodies, the estimation method of the diffuse load from agricultural activities developed at IEEP was used [10]. An increase in the diffuse load can lead to the potential risk of leaching and migration of nutrients from the fields located in the catchment area of water bodies.

The status of available animal/poultry manure storages on the agricultural enterprises in the Leningrad Region was analyzed. When analyzing the agricultural enterprises in the Leningrad Region, special attention was paid to the machine and tractor fleet involved in the handling of animal/poultry manure and the organic fertilizer produced.

2.3   RESULTS

Preliminary consideration of the relevant data of the Leningrad Region showed that major environmental risks could arise during the organic waste utilization. This may be explained by the inefficient use of nutrients (nitrogen and phosphorus) in the intensive machine-based technologies and the lack of established interaction between the livestock and crop production sectors. The main task in developing proposals of how to remedy the environmental situation and to improve the ecological compliance of the region through the adequate technological and managerial decisions is to assess the agricultural enterprise as a whole and to identify the critically ineffective technological elements in place.

Our institute has developed a system for BAT formation for manure utilization.

Figures 2.3 and 2.4 show examples of the BAT formation pattern for cattle farms and poultry farms. There are the main technological processes that are applied depending on the specific conditions of the farms.

For cattle farms we can have more than 20 technological options. For poultry farms, there are more than 15 technological options. The basic processes are as follows: separation into fractions, composting (passive or active), long-term storing, transportation and application. Aerobic, anaerobic fermentation and drying are not very common in Russia now, but in the near future the number of farms that use these technologies will be higher.

As part of the work, the Russian catchment area of the Baltic Sea region was analyzed, and possible risks were calculated.

Calculations regarding the Leningrad Region show that the nutrient load on the environment in the districts is different. Most of the districts have the potential to apply additional nitrogen

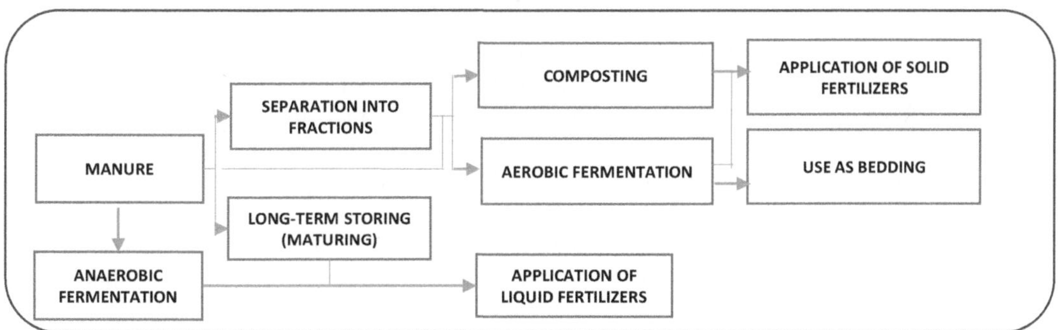

FIGURE 2.3   BAT formation pattern for cattle farms.

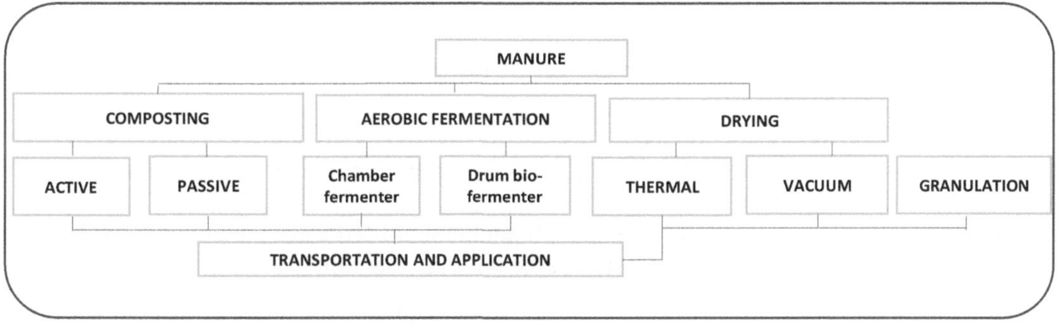

FIGURE 2.4   BAT formation pattern for poultry farms.

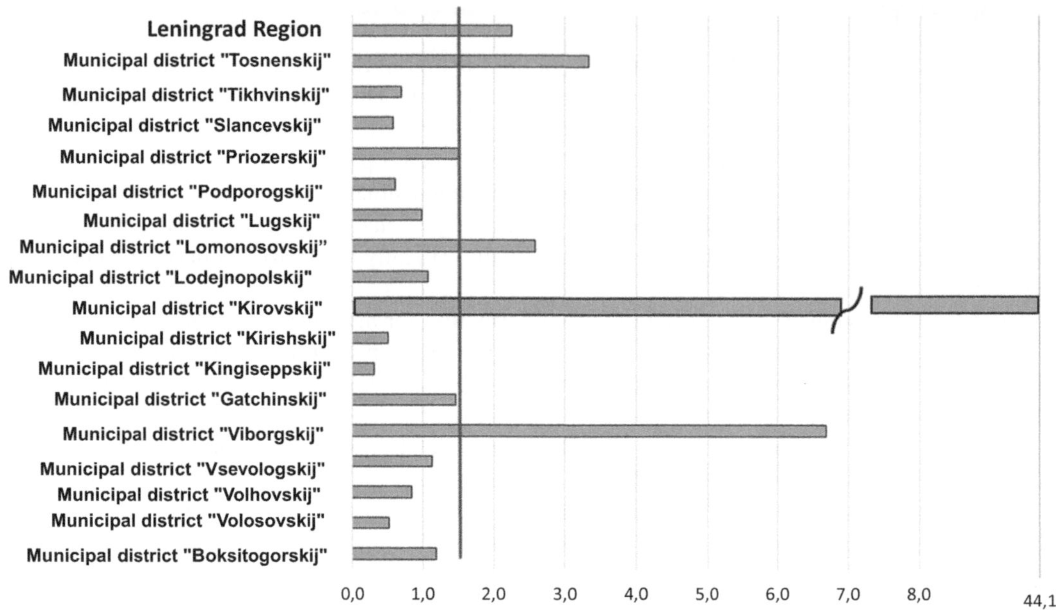

FIGURE 2.5   The average animal/poultry density in the Leningrad Region.

with organic fertilizers on the available agricultural land. Only two districts – Vyborgskij and Kirovskij Districts, have a shortage of land to apply the produced 2172.9 and 8433.2 t N/year, respectively, that leads to the possible environmental risks. However, the nutrient load in these two districts can be relieved by applying the produced organic fertilizers to the agricultural land in the neighbouring districts, as the total potential of nitrogen application in the Leningrad Region is positive: +8769.2 tons of nitrogen per year. The data obtained are also shown on the map of the Leningrad Region.

The case study (environmental risks associated with intensive production) showed that the average density of farm animals and poultry per unit of the cultivated land in the Leningrad Region is 2.2 LSU/ha. It means that the Leningrad Region exceeds the recommended value (Figure 2.5).

The highest environmental risks may occur in the Kirovskij District, with the animal/poultry density being 44.1 LSU/ha.

In general, three districts in the Leningrad Region feature high environmental risks, with the animal/poultry density being above 3.1 LSU/ha and 13 districts have low environmental risks in the case of further intensification of livestock and poultry farming.

The actual diffuse load of nitrogen and phosphorus on the catchment area from agricultural activities throughout the Leningrad Region was found to be 20.67 kg/ha for nitrogen and 1.24 kg/ha for phosphorus. At the same time, the recommended normative value, which does not generate environmental risks, for this region is N = 8.5–19.3 kg/ha, P = 0.07–2.03 kg/ha, depending on the soil type.

Expanding the existing enterprises and the construction of new complexes of intensive livestock and poultry farming may lead to the risk of excessive input of nitrogen and phosphorus into the water bodies from the catchment areas.

To reduce emissions into the environment, different types of fermenters are used (Figure 2.6).

We have been researching the aerobic fermentation for a long time.

- Aerobic solid-state fermentation has proved to be an efficient method of various organic waste processing, including a solid fraction of animal/poultry manure.

vertical fermenter

drum fermenter

drum fermenter

drum fermenter

chamber fermenter

FIGURE 2.6   Different types of fermenters.

- The closed-type fermenters minimize the impact of the external weather factors, such as precipitation, the outdoor air temperature and humidity (Figure 2.6).
- This has a positive effect on the nutrient loss, the end-product quality and the processing time, reducing it to several days.

The demonstration site at the pilot farm with modern and BAT complying technological solutions of organic waste management was created.

In Figure 2.7 the automated bio-fermenter for solid organic waste processing is shown, which was created at Institute for Engineering and Environmental Problems in Agricultural Production (IEEP) – branch of Federal Scientific Agroengineering Center VIM.

In Figure 2.8 machines for applying liquid organic fertilizers are shown. These machines are equipped with working elements to minimize emissions and control the application process. These machines were built by Joskin in Belgium according to our technical specifications.

This equipment is implemented on the demonstration site at the pilot farm.

As world experience shows, it is important to ensure monitoring of the generation of livestock waste [11–17]. Monitoring can help to organize logistics associated with organic fertilizer use [18]. For these tasks, we created an interactive programme. Now it is being tested in the Leningrad Region. The digital tool in the interactive programme form is designed for information support of the executive authorities and heads of agricultural enterprises in terms of improving the environmental compliance and production efficiency associated with preparation and field application of organic fertilizers.

All agricultural enterprises are shown on an interactive map and divided into five groups: cattle farms, pig farms, poultry farms (chickens), crop growing farms and mixed-type farms (cattle and pigs). Each group of agricultural enterprises has its own graphic designation. The interactive programme algorithm provides the export filter of individual groups of enterprises.

The functional objectives of the interactive programme are to receive the relevant source information on the region, agricultural organizations, applied animal/poultry manure handling technologies, and manure storage types; to visualize all agricultural organizations on a digital map: location, name, specialization, animal stock, available agricultural land; to calculate and display the current situation in agricultural enterprises: amount of organic fertilizer received, land sufficiency for all organic fertilizer application, and required volume of manure storages and composting pads; to calculate and display the forecast situation in agricultural organizations and to create the electronic passports of farms, districts and the whole Leningrad Region (Figure 2.9).

A very important function is the logistics of organic fertilizers' distribution from supplier farms to consumer farms considering the nutrient load standards and the data on the nutrient load distribution within the agricultural lands.

The limiting factor in the fertilizer application dose is total nitrogen (170 kilograms per hectare) and total phosphorus (25 kilograms per hectare). When one of the indicators reaches the limit value, the programme will give a signal. The indicator (total nitrogen or total phosphorus), the limit value of which is reached first, is considered the most significant in the calculation of the organic fertilizer application dose.

The results are displayed in the form of electronic passports of agricultural organizations, the districts and the entire region under study, taking into account the logistics of organic fertilizers' distribution and the data on the nutrient loads within the agricultural land

The programme can be used to justify plans for the modernization of livestock farms, also to generally assess the sustainability of agro-ecosystems based on nutrient balances.

Research best available techniques for organic livestock waste management in Russia is a very important direction for our institute. This year, we have a plan to open a Demonstration and Research Centre for the utilization of organic livestock waste in the Leningrad Region (Figure 2.10).

We will perform this as part of a project called EcoAgRas, whichis supported by the South-East Finland – Russia CBC Programme 2014–2020. The Demonstration and Research Centre will promote research on the impact of technology and machinery on the environment, also providing training for agricultural students and specialists.

Manure is not waste, it is a very valuable resource. By applying the best available techniques to return this resource to agricultural production, we save the environment, improve the soil and ensure the long-term sustainability of agroecosystems.

FIGURE 2.7   Automated bio-fermenters for solid organic waste processing.

FIGURE 2.8   Low-emission equipment for application of liquid organic fertilizers.

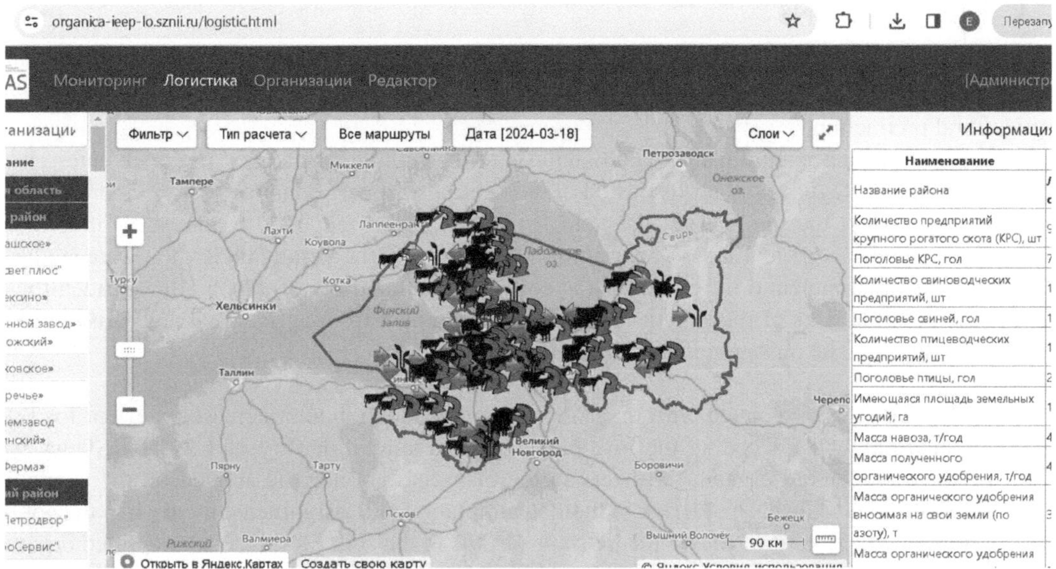

FIGURE 2.9 Interactive programme for monitoring livestock waste management and logistics associated with organic fertilizer use.

FIGURE 2.10 Demonstration and research centre for utilization of organic livestock waste (Leningrad Region).

2.4 CONCLUSIONS

The status of agricultural enterprises in the Leningrad Region was reviewed in terms of their environmental performance and proposals were prepared to improve the environmental compliance of agricultural production.

At the first stage, a scientific and expert analysis of the ecological status of livestock enterprises in the Leningrad Region was performed. For this purpose, questionnaires were elaborated and circulated to collect the initial data; the field trips were organized for the surveys of individual enterprises, and the satellite images of agricultural territories of the Leningrad Region were studied. This way, the environmental information was collected, the analysis of which allowed identification of the possible risks of the negative environmental impact from agricultural enterprises in the Leningrad Region. The most significant of them are as follows:

- ecological risks associated with intensification of production and increased nutrient load on the environment. Exceeding the safe range of farm animal density of 1.5–1.7 LSU/ha and production of more organic fertilizers in the districts than may be applied on the available agricultural land, leads to the risk of their accumulation and a significant increase in the application rate that, in turn, causes the nutrient surplus in the soil. Storing of non-applied organic fertilizers can lead to air pollution and higher diffuse loading on water sources due to the nutrients leaching with the storm runoffs;
- ecological risks associated with the diffuse loading. An increase in the diffuse load can lead to the risk of leaching and migration of nutrients from the fields located in the catchment area of the water bodies;
- insufficient capacity of manure storage facilities to accumulate all the animal/poultry manure produced and the lack of well-established logistics for the distribution of organic fertilizers between the livestock/poultry complexes and agricultural land of other farms increases the risk of organic fertilizers being applied on the fields in winter.

The following closely linked actions may contribute to reducing the risks of negative environmental impact from agricultural enterprises:

(1) Development and control over the fulfilment of recommendations on the introduction of the on-farm environmental monitoring and control system over the production processes. This system implies the approval and observance of local regulatory acts, such as Technological Regulations. Technological Regulations are the main tool in establishing the on-farm system of animal/poultry manure handling, which should be taken into consideration under transition to a BAT system.

(2) Development of recommendations on the introduction of the regional system for monitoring and coordination of organic and mineral fertilizers' production and application. These recommendations can be used by the authorities to assess the environmental situation, to establish relationships between agricultural enterprises regarding the effective use of organic fertilizers when forecasting the development and operation of agricultural production.

(3) Creation of a demonstration site at the pilot farm with modern and BAT-complying technological solutions of organic waste management. The demonstration site will launch a working facility, where the specialists of all agricultural enterprises can be trained to improve their environmental literacy and to further introduce state-of-the-art technological solutions.

ACKNOWLEDGEMENTS

The research was performed under the support of the projects "Introduction of the eco-logical system of agriculture is the basis for sustainable development of border rural area – EcoAgRas" of the South-East Finland – Russia CBC 2014–2020 Programme (Grant contract No 17086-LIP1601-KS1441) and "Water-driven rural development in the Baltic Sea Region – WaterDrive" of the Interreg Baltic Sea Region Programme.

REFERENCES

1. Briukhanov, A.Yu., Vasilev, E.V., Kozlova, N.P., Shalavina, E.V., Subbotin, I.A., and Lukin, S.M., "Environmental Assessment of Live-563 Stock Farms in the Context of BAT System Introduction in Russia", *Journal of Environmental Management*, Vol. 246, pp. 283–288, 2019. DOI: 10.1016/j.jenvman.2019.05.105

2. Subbotin, I.A., and Vasilev, E.V., "Forecasting Model the Complex Negative Impact of Agricultural Production Technologies on Water Bodies", *Engineering Technologies and Systems*, Vol. 31, No. 2, pp. 227–240, June 2021. DOI: 10.15507/2658-4123.031.202102.227-240

3. Danilov-Danilyan, V.I., Venitsianov, E.V., and Belyaev, S.D., "Some Problems of Reducing the Pollution of Water Bodies from Diffuse Sources", *Water Resoures,* Vol. 47, pp. 682–690, 2020. DOI: 10.1134/S0097807820050048

4. Information and technical BAT reference book "Intensive rearing of pigs", 2017. (In Russian) Available at: www.burondt.ru/NDT/NDTDocsDetail.php?UrlId=1138&etkstructure_id=1872 (accessed on 10 February 2022)

5. Information and technical BAT reference book "Intensive rearing of farm poultry", 2017. (In Russian). Available at: www.burondt.ru/NDT/NDTDocsDetail.php?UrlId=1140&etkstructure_id=1872 (accessed on 10 February 2022)

6. Convention on the protection of the marine environment of the Baltic sea area, 1992 (Helsinki Convention). Available at: www.helcom.fi/documents/about%20us/convention%20and%20commitments/helsinki%20convention/1992_convention_1108.pdf (accessed on 10 February 2022)

7. Measures aimed at the reduction of emissions and discharges from agriculture. HELCOM recommendation 24/3. Adopted 25 June 2003 having regard to Article 20, Paragraph 1 b) of the Helsinki Convention. Available at: www.helcom.fi/Recommendations/Rec%2024-3.pdf (accessed on 10 February 2022)

8. Stocking rates of land capability for agriculture classes. Available at: www.hutton.ac.uk/sites/default/files/files/ladss/Stocking-Rates-LCA.pdf (accessed on 10 February 2022)

9. Climate change and agricultural development: Adapting polish agriculture to reduce future nutrient loads in a coastal watershed. Available at: https://link.springer.com/article/10.1007/s13280-013-0461-z (accessed on 10 February 2022)

10. Briukhanov, A., Alexeev, L., Oblomkova, N., and Subbotin, I., "Results of field studies on nitrogen and phosphorus input from agricultural production to selected water bodies in western Dvina", *Proceedings of 17th International Scientific Conference Engineering for Rural Development*, pp. 285–291, 2018.

11. Sun, Z., "An Introduction to Intelligent Analytics Ecosystems", *PNG UoT BAIS*, Vol. 6, No. 3, pp. 1–11, 2021.

12. Karabutov, N., "Frameworks in Problems of Structural Identification Systems", *International Journal of Intelligent Systems and Applications (IJISA)*, Vol. 9, No. 1, pp. 1–19, 2017. DOI: 10.5815/ijisa.2017.01.01

13. Sharda, R., and Kalgotra, P., "The Blossoming Analytics Talent Pool: An Overview of the Analytics Ecosystem", In: *INFORMS Analytics Body of Knowledge* (J. J. Cochran, ed.). Chapter 9, pp. 311–326, 2018.

14. Omary, S., and Sam, A., "Web-Based Two-Way Electricity Monitoring System for Remote Solar Mini-Grids", *International Journal of Engineering and Manufacturing (IJ EM)*, Vol. 9, No. 6, pp. 24–41, 2019. DOI: 10.5815/ijem.2019.06.03

15. Bhavikatti, S. Sadanand, P., Mukta, P., Pradeep, V., and Shailaja S.M., "Automated Roof Top Plant Growth Monitoring System in Urban Areas", *International Journal of Engineering and Manufacturing (IJEM)*, Vol. 9, No. 6, pp. 14–23, 2019. DOI: 10.5815/ijem.2019.06.02

16. Akwu, S. et al., "Automatic plant Irrigation Control System Using Arduino and GSM Module", *International Journal of Engineering and Manufacturing (IJEM)*, Vol. 10, No. 3, pp. 12–26, 2020. DOI: 10.5815/ijem.2020.03.02

17. Mwemezi, K., and Sam, A., "Development of Innovative Secured Remote Sensor Water Quality Monitoring & Management System: Case of Pangani Water Basin", *International Journal of Engineering and Manufacturing (IJEM)*, Vol. 9, No. 1, pp. 47–63, 2019. DOI: 10.5815/ijem.2019.01.05

18. Briukhanov, A., Dorokhov, A., Shalavina, E., Trifanov, A., Vorobyeva, E., and Vasilev, E., "Digital Methods for Agro-Monitoring and Nutrient Load Management in the Russian Part of the Baltic Sea Catchment Area", *IOP Conference Series: Earth and Environmental Science*, Vol. 578, p. 012011, 2020. DOI: 10.1088/1755-1315/578/1/012011. https://www.elibrary.ru/item.asp?id=44389031

3 Bioenergy from Organic Waste

Translation of Technology from Laboratory to Land

Sameena Begum, Sudharshan Juntupally,
Vijayalakshmi Arelli and Gangagni Rao Anupoju

3.1 INTRODUCTION

Ever increasing demand for energy in various sectors due to uncontrolled and rapid growth in population and industrialization has led to the search for alternative forms of renewable energy sources. Considering the adverse impacts on the environment due to excess exploitation of natural energy reserves like coal, oil and gas to fulfil the current energy demand, and the escalating increase in solid waste generation globally has led the valorization of biodegradable organic wastes using waste-to-energy technologies to gain substantial interest [1]. This paradigm shift to explore waste-to-energy technologies not only meant to fulfil the excessive energy demand but also to resolve the issue of waste mismanagement. As per the estimates of the world bank in 2016, about 2.02 billion metric tonnes (BMT) of municipal solid waste (MSW) is generated globally, which is projected to increase by 28.2%, i.e., 2.59 BMT by 2030 and a further increase by 31.3% by the end of 2050, which makes a cumulative increase in MSW generation by 68.3% by 2050 against the current generation rate. About 40–45% of the MSW is biodegradable and has the potential to generate different forms of renewable energy [2, 3]. The MSW in India is mismanaged either by abandoning it on site or open landfilling it, due to the lack of suitable waste segregation and treatment technologies, despite having huge energy generation potential. In this context, it is worthwhile to mention that an anaerobic digestion process has been considered to be a promising technology to convert biodegradable organic waste to bioenergy.

The Government of India (GoI) through the Ministry of New and Renewable Energy (MNRE) has taken initiatives to indigenize the biogas technology and developed small-scale biogas plants for the treatment of animal manures, such as cattle dung, livestock litter, etc. [4]. Small-scale biogas plants suitable for the treatment of cattle manure known as "gobar gas plants" are very popular in India and it is estimated that the total number of such biogas plants installed in India were approximately 4.54 million in 2012, which showed an increasing figure compared to the 1990s (1.23 million) but a declining number despite having an estimated potential of installing 12.34 million digesters [5, 6]. These biogas plants are based on digester models that were developed in the early 1960s in India were centred on two types of digesters, i.e., fixed masonry dome type and floating drum type. A cattle dung-based floating drum type biogas unit named Khadi and Village Industries Commission (KVIC) was the first digester evolved to treat cattle dung in rural India. This technology was approved by the MNRE, GoI and promoted under the National Programme on Biogas Development (NPBD). Although it was a popular technology, the drawbacks were scum formation, low methane yield, poor organic matter biodegradation, large reactor volumes, longer

DOI: 10.1201/9781003327646-3

35

TABLE 3.1
Comparative Assessment of Slow-Rate and High-Rate digesters [8, 10]

Conventional Slow-Rate Digesters	Advanced High-Rate Digesters
Longer HRT	Short HRT
Large reactor volume	Compact digester
Low organic loading rates between 1 and 3 kg VS/m³/day	High organic loading rates between 1 and 10 kg VS/m³/day
Low volatile solids' reduction	High volatile solids' reduction
Suitable for animal manure treatment	Suitable for any biodegradable organic waste
Small treatment capacities between 50 and 200 kg/day	High treatment capacities up to 100 tons/day
Manual operation	Partial to fully mechanized operation
Scum formation and choking issues	No scum formation and choking problems
No provision for mixing and temperature maintenance	Efficient mixing and temperature management

hydraulic residence time (HRT) and the requirement of paints for coating to prevent corrosion. All these limitations, technical as well as economic, resulted in the implementation and continuous operation of these digesters across India. In addition to these limitations, these conventional digesters were designed to treat only small quantities of cattle manure in the range of 50 to 250 kg per day [7, 8]. In contrast, the majority of the European countries, such as the United Kingdom, Netherlands, Germany, Denmark etc., who pioneered the high-rate biomethanation or the advanced biogas technology based on anaerobic digestion principle has installed several biogas plants to convert biodegradable organic waste to biogas [9]. The Indian biogas digester models were regarded as slow-rate digesters or conventional digesters due to their inferior performance and engineering design compared to the high-rate digesters developed by the European countries that are way superior in performance and ease of handling [10]. A comparative assessment of the novel features of slow-rate and high-rate digesters is summarized in Table 3.1. The idea of implementing the advanced high-rate digester-based biogas plants has gained substantial importance in the Indian market as well, but its implementation in the real field has slowed down due to barriers, such as availability of indigenous technologies, poor return on capital investments and public awareness. Therefore, the necessity for the development of remunerative and cost-effective advanced biodigesters incorporating the novel features of high-rate biomethanation technology has increased tremendously. Accordingly, different research organizations in India have taken up initiatives aimed at the development of high-rate anaerobic digesters that could meet the standards set by the advanced digesters in European countries [11].

3.1.1 BIOMETHANATION PROCESS

Biomethanation or anaerobic digestion is a complex biological process wherein the biodegradable organic waste is converted to biogas and nutrient-rich organic fertilizer. The AD is a combination of biological processes that occur in a sequence, namely, hydrolysis, acidogenesis, acetogenesis and methanogenesis. The first biochemical reaction that occurs in the biomethanation process is the hydrolysis wherein the complex polymeric molecules, such as carbohydrates, proteins, lipids, cellulose, hemicellulose and lignin, are converted to simple monomers, such as glucose, amino acids, glycerol, fructose and xylose, respectively, by the action of hydrolytic bacteria (Eq. 3.1). Subsequently, the monomers are converted to organic acids (short-chain carboxylic acids) by the acidogens in the second biochemical reaction (Eq. 3.2). The short-chain carboxylic acids preferably acetic acid is converted to acetate, hydrogen and carbon dioxide by the acetogens in the third biochemical pathway (Eq. 3.3) whereas in the final step that is the methanogenesis, the acetate, hydrogen and carbon dioxide are converted to methane and carbon dioxide by the hydrogenotrophic methanogens and acetoclastic methanogens, respectively (Eq. 3.4).

$$(C_6H_{10}O_5)n + nH_2O \rightarrow n\ C_6H_{12}O_6 + n\ H_2 \tag{3.1}$$

$$C_6H_{12}O_6 \rightarrow 3CH_3COOH \tag{3.2}$$

$$CH_3CH_2OH + 2H_2O \leftrightarrow CH_3COO^- + 3H_2 + H^+ \tag{3.3}$$

$$CH_3COOH \rightarrow CH_4 + CO_2 \tag{3.4}$$

The process of biomethanation is mainly dependant on the activity of microbes, which is again dependant on the critical operational and environmental parameters such as pH, temperature, residence time, organic loading rate, solids' consistency and reactor configuration. The output from the biomethanation process is the biogas, which can be utilized for combined heat and power applications whereas the digested slurry can be dried and used as a solid fertilizer or the liquid digestate can be directly used as a liquid fertilizer as the digested slurry upon digestion is rich in nutrients (N, P and K) that are required for plant growth. The typical composition of biogas generated from organic wastes is 55–70% methane, 30–45% carbon dioxide, 0.5–1% moisture, less tham 0.5% of hydrogen sulphide, and some traces of nitrogen. The composition of biogas depends on numerous factors, such as the type of waste, operational strategy of the anaerobic digester, source and ecology of microbial culture, residence, temperature, pH, additives, OLR, etc.

3.2 DEVELOPMENT OF ADVANCED HIGH-RATE ANAEROBIC DIGESTERS: SIGNIFICANCE AND INDIGENIZATION

It is very well known that the anaerobic digesters currently employed for solid waste treatment are broadly classified into two distinct categories, such as conventional (slow-rate) and high-rate digesters [8]. It could be observed that the high-rate digesters are much superior to conventional digesters in terms of all the important operating and design parameters as shown in Table 3.1. However, experiences in European countries showed that these designs have been applied for very high capacities of waste treatment, i.e., 50 ton/day and above [12]. Efforts to implement high-rate biomethanation technologies for the generation of biogas and bio manure from organic solid waste in India were initiated by MNRE, GoI and United Nations Development Programme (UNDP) jointly through the Global Environmental Facility (GEF) fund. In this project, a good number of biogas plants were installed in India based on high-rate biomethanation technologies with the technical assistance of European technology partners for the treatment of variety of organic with an intent to replicate the same technologies in India [13]. Subsequent to this programme, it was observed that similar European technologies could not be replicated in India due to various reasons and therefore, the success rate of this programme was much limited. Hence, CSIR-IICT comprehended these developments and extensively worked on high-rate anaerobic digestion of organic wastes for the development of technology indigenously. Application of indigenous high-rate biomethanation technology for decentralized remunerative organic waste treatment is also one of the endeavours of CSIR-IICT so that the technology could be made available to many users across India.

One of the major drawbacks of the conventional digesters is the lack of proper reactor contents' mixing mechanism, which is highly essential for superior performance of anaerobic systems and this has been observed to be the main reason for failure of most of the conventional digesters due to the formation of scum at the top and accumulation of inorganic matter at the bottom of the digester [14]. In order to overcome this, digesters were designed with mechanical mixers popularly known as continuous stirred tank digesters (CSTD). However, CSTDs also failed in many cases [15]. In addition, poor biogas and methane yield, low organic matter removal rates, long HRT and large reactor volume, low organic or VS loading rates (OLR or VSLR) are the other main drawbacks of existing conventional slow-rate digesters. Therefore, CSIR-IICT conceived the idea to develop a self-mixed anaerobic digester (SMAD) taking the above issues into consideration. SMAD was developed, tested in laboratory and then patented (Patent Grant Number: 289243) in the year 2007. The core idea of the SMAD is that mixing of the slurry inside the digester takes place without the use of mechanical mixers and the differential pressure across the two chambers in the digester creates pressure difference across the reactor, which assists in self mixing of the slurry inside the digester [16]. This specific design feature of SMAD was useful in mixing the digester contents without the need of external power and then de-alienated the problem of scum formation. The other salient features of SMAD includes short HRT, enhanced VSLR, high organic matter removal efficiency and higher methane yield.

3.2.1 Self-Mixed Anaerobic Digester (SMAD) for the Treatment Poultry Litter: Development and Demonstration

Initially, the motive to develop SMAD was to provide a feasible solution to the poultry industry for the safe disposal of poultry litter (PL), which is generated in large quantities during intensive poultry farming. The poultry litter containing biodegradable organic and inorganic matter with 75–80% moisture is an amenable substrate for biomethanation [16]. PL contains 1.22–1.63% of nitrogen, 0.89–1.04% of phosphorus, 1.34–1.7% of potassium and many micronutrients, such as zinc, copper, iron and selenium. In order to evaluate the performance of SMAD in laboratory, a multi-stage high-rate biomethanation process based on novel SMAD and up-flow sludge blanket (UASB) reactor was used to treat PL with a total solids (TS) consistency of 10%. The configuration of the multi-stage unit included SMAD I followed by SMAD II and UASB reactor. Experiments were carried out in a different sequence of configurations to assess the impact of sequential unit operation. The multi-stage high-rate biomethanation unit resulted in the VS reduction of 58%, methane yield of 0.16 m^3/kg$_{(VS\ reduced)}$ at a VS loading rate of 3.5 kg VS/m^3/day at an HRT of 13 days from PL as reported by [16]. These results indicated that the multi-stage configuration with SMAD improved the anaerobic digestion process efficiency of PL in comparison to the conventional digesters [16]. However, it was learnt through the aforementioned laboratory studies that multi-stage configurations are suitable for treating large quantities of PL and hence its applicability for treating smaller quantities of organic waste in the order of 500 kg are limited [16–20]. Therefore, a single stage reactor with a suitable mixing arrangement for the treatment of smaller quantity of PL ranging between 100–500 kg/day is required to provide a solution to the poultry industry. A pilot-scale SMAD system with 8 m^3 of reactor volume and 6 m^3 of biogas holder for the treatment of 100 kg of poultry litter every day was designed and installed at Sri Venkateshwara Veterinary University (SVVU), Rajendranagar (presently known as P. V. Narasimha Rao Telangana Veterinary University) in 2007 with the financial support of the Department of Biotechnology (DBT), GoI and SVVU. The pilot plant was installed, commissioned and operated successfully for 15 months for the treatment of PL. The biogas generated from the plant was used to light up the brooders in the poultry farming area. The volatile solids' (VS) reduction of 60% and biogas generation of 0.2 m^3/kg VS was obtained at an HRT of 24 days and VSLR of 3 kg VS/m^3/day at the stable performance of the pilot plant and it was handed over to SVVU.

3.2.2 Anaerobic Gas Lift Reactor (AGR): Concept to Commissioning and Commercialization

It was observed during the operation of the biogas plant at SVVU based on SMAD for the generation of biogas and bio manure from PL that necessary improvements in the design of the digester are required along with certain additional features in terms of the following aspects. Therefore, extensive research again on the development of another high-rate biomethanation technology were initiated with an aim to develop a robust technology that is suitable to not only a single waste like PL but also to any other biodegradable organic waste. It is very well known that the C/N ratio of the substrate plays very critical role for the optimal generation of biogas and bio manure. The C/N ratio of PL is low whereas it is high for food waste in comparison to the optimal value required for efficient biomethanation to occur, i.e., in the range of 25 to 30. From the practical observations during the plant operation at SVVU University, it was learnt that excess formation of ammoniacal nitrogen could deteriorate the performance of the digester as PL is an ammonia liberating substrate [16, 21]. Therefore, development of an improved version of a high-rate digester is required, which should be suitable for any organic waste, for instance, food waste is required.

Food waste is considered as one of the ideal substrates for biogas generation as it is easily biodegradable by the microorganisms. Unlike the PL, food waste contains high C/N ratio, which results in rapid acidification leading to the drop in pH [22]. In addition, if the solid waste contains a high amount of sulphate that could result in the formation of sulphide inside the digester, which is inhibitory beyond the limit of 100 to 150 mg/L [23]. Therefore, in either of the cases, the digester needs an inhibition control mechanism for excess ammoniacal nitrogen and sulphide. It is also essential that hydraulic residence time (HRT) and solids' residence time (SRT) should be delinked such that SRT is longer and HRT is shorter. The number of mixing cycles in SMAD are dependent on the generation of biogas in the reactor, which are not sufficient for the thorough mixing of the digesters. Hence, it is planned to change the design features of the digester to enhance the number of mixing cycles in the new digester. It is also planned to control the temperature of the digester either at mesophilic or thermophilic as it is one of the critical parameters to enhance the performance of the digester. After intensive research efforts, an improved version of SMAD with high-rate anaerobic digestion features was developed and named the anaerobic gas lift reactor. The anaerobic gas lift reactor was given as the mixing of the reactor contents inside the digester is based on the biogas recirculation wherein the slurry is being lifted up by the biogas under pressure enabling the proper mixing. Pre- and post-processing mechanisms were also integrated in the main digester to provide a complete solution for waste treatment and valorization. The AGR technology was developed, tested and patented (Patent Grant Number: 307102) considering the aforesaid issue [24]. Therefore, extensive efforts were taken to develop a high-rate biomethanation reactor for the treatment of a variety of organic wastes because in the high-rate anaerobic digestion apart from maintenance of all the important parameters (pH, food to microorganism ratio, C/N ratio, temperature, inlet slurry concentration, etc.), gentle mixing of the slurry is highly essential for efficient organic matter degradation and biogas yield. Therefore, the core design of AGR is based on mixing of influent slurry in the reactor without the need of an external mechanical mixer and centred on liquid and gaseous hydrodynamics through pressure management across the digester. The reactor design encompasses a liquid and gas distribution system at the bottom, two stage three settlers in the middle and bottom of the reactor, baffler settler in the middle of the reactor, liquid outlet gutter at the top, a liquid down comer and gas riser inside the reactor. The schematic of the AGR is described in detail, which was reported by Begum et al. [25]. The advantages of the AGR are a possibility for scaling up to any capacity, short HRT, elimination of scum formation, de-linking of HRT and SRT, improved biogas production and methane content in biogas.

3.2.2.1 Configuration and Working of AGR Technology

The anaerobic gas lift reactor is a vertical cylindrical tank with an aspect ratio of 2:1 with two compartments, which are hydraulically connected. Upon commissioning of the reactor on site, the reactor is inoculated with 30–40% of mixed microbial inoculum followed by organic waste loading. The influent from the feed preparation tank is pumped into the bottom compartment of the reactor using an efficient distribution system, and mixed with anaerobic biomass. Most of the organic components are converted to methane and carbon dioxide in the lower part of the reactor. Sufficient autogenerated biogas pressure is created inside the reactor in the bottom compartment, which lifts the slurry upwards through the riser into the upper compartment. To separate the biogas, biomass and the slurry, a three-phase separator comprising of baffles, an inverted cone and a downer pipe on top of the reactor is provided. Biogas is liberated from the three-phase separator while the slurry returns through the downer to the bottom of the system. Provision of the three-phase separator mechanism alleviates the biomass washout as the biogas falls back to the reactor. The digested slurry is collected from the recycling port wherein the digested slurry is partially recirculated back to the reactor to ensure proper mixing and pH conditions. This action takes place continuously at regular intervals based on the set pressure. Similarly, the biogas produced from the top of the reactor is then partially recycled and injected at the bottom of the reactor to boost the pressure inside the lower compartment. The recirculation of biogas under pressurized conditions results in the increase in the number of mixing cycles periodically [25]. Due to this phenomenon, the total slurry inside the digester gets mixed thoroughly. This mixing enhances the volatile solids' degradation rate (VSDR), improves heat and mass transfer between particles thereby increases the biogas production and reduces the formation of scum in the digester. The number of mixing cycles per day depends upon the biogas production potential of organic waste, the set pressure value, temperature, VSLR and VSDR. The number of mixing cycles could be enhanced by recycling the biogas back to the digester. The three-phase separator mechanism delinks the solids' retention time (SRT) and HRT so that active biomass could be retained as much as possible to enhance the microbial activity, whereas the liquid could be pushed out to handle higher throughput. AGR could be operated either at ambient, mesophilic or thermophilic temperature conditions and a provision was made for measuring slurry temperature of the reactor outlet. The biogas from the reactor is collected in the biogas balloons at ambient temperature and pressure conditions. Depending on the type of biogas utilization, biogas is either utilized directly for LPG replacement or it is converted to electricity.

3.2.2.2　Demonstration and Performance Assessment of High-Rate Biomethanation Plant Based on AGR

The first pilot biogas plant based on AGR technology with a capacity of 1000 kg/day is installed and operated in Pedda Shivanoor, Chegunta Mandal, Medak Dist in Telangana state. The digester is tested extensively in the laboratory and after achieving promising results, the same was licensed to engineering companies (M/s Ahuja engineering services private limited (AESPL)) for commercialization. The AGR technology was then scaled up to 1000 kg/day treatment capacity and a semi-commercial biogas plant was installed near Hyderabad to assess the performance of the plant at bigger scale before commercialization in the real field. The motive to install the AGR-based biogas plant was to treat PL, which has low C/N ratio, and provide a solution to the poultry farms for safe disposal of poultry litter as well as to create a source of energy generation. This means, the plant is installed near the poultry farms where huge volumes of PL is generated on a daily basis. The PL generated in the poultry farms was treated in the pilot biogas plant and the biogas generated from the plant was converted to electricity to power the brooders and the digested slurry was used as organic fertilizer by the farmers. The plant was initially operated for the treatment of PL to generate biogas-based electrical power. About 1000 kg of PL was fed to the digester every day, which resulted in the generation of 65 to 80 m^3/day of biogas and 100 to 130 kg/day of bio manure (digested slurry). The biogas was used for combined heat and power (CHP) applications. Part of the biogas generated was used to operate the brooders in the poultry farm and the remaining biogas is used to generate electricity. The AGR-based biogas plant installed near the poultry farms provided multiple benefits to the poultry farmers apart from waste management. The digested slurry was solar dried and used as bio manure (organic fertilizer). The plant was operated with PL and other mixed organic wastes for 48 weeks with promising results [25]. Organic solid wastes having optimum C/N ratio in the range of 20 to 30 or C/N ratio as low as 6 are suitable for the treatment using AGR technology at an HRT ranging between 15–30 days depending on the organic waste with 65% VS reduction at a VS loading rate up to 4 kg VS/ (m³.day). Electrical power of 104 kWh/day was generated from the biogas produced (65–80 m³/day) 1000 kg of PL and other mixed organic wastes every day and 100–130 kg of bio manure per day was generated. The plant was operated under stable conditions and is in continuous operation since 2012. The biogas-based power was used to operate water pumps in the agriculture fields (around 6 acres of land was watered using biogas-based electrical power) and the bio manure was applied to the fields as a fertilizer for growing crops. This full-scale experience [25, 26] comprehended that the decentralized off-grid power plant based on AGR for the treatment of multiple feedstocks could be a remunerative option to the farmers and other stake holders. Francesca Girotto [26] has reviewed the success stories of AGR technology based on the published articles and published a review article "A Glance at the World" in *Waste Management* journal.

3.2.3　Commercialization and Success Stories of AGR Technology

3.2.3.1　Model I: Waste to Energy from Kitchen to Kitchen, Biogas Plant at CSIR-IICT, Hyderabad

In order to demonstrate the working of a biogas plant based on food waste to the interested stake holders, a biogas plant based on AGR was installed inside CSIR-IICT campus for the treatment and concomitant generation of biogas and bio manure from kitchen waste. About 250–300 kg of kitchen waste (cooked food waste, uncooked food waste, vegetable and fruit peels) was generated in the kitchen, which was subjected to biomethanation every day. About 30–32 m³ of biogas was generated every day, which was used as a cooking fuel directly in the same kitchen. The kitchen saved about 14 kg of LPG every day due to biogas-based cooking. Approximately 30–35 kg of dry bio manure is generated every day, which is used for in campus landscaping.

3.2.3.1.1　*Biogas Plants Based on Food Waste at the Akshaya Patra Foundation (TAPF), an NGO*

Kitchens with large-scale operations are discovering that converting their food and vegetable waste into bio energy not only replaces the non-renewable fuel but also resolves the issue of waste disposal and its mismanagement. Further, biogas plant installation at the source of waste generation can improve operation and profits. A biogas plant at source can deliver the heat needed to replace LPG/PNG for cooking, partly or fully while supplying a nutrient-rich liquid organic biofertilizer as a by-product. In this endeavour, CSIR-IICT along with its project executing company has installed a biogas plant at one of the kitchens of The Akshaya Patra Foundation (TAPF) at Bellary in Karnataka state, which is an NGO that serves food to school children under mid-day meal program (Figure 3.1). The plant is in operation since 2015, which aims to treat about 1000 kg of food waste generated in the kitchen every day. About 120–140 m³ of biogas and 150–200 kg of bio manure is generated every day. The biogas generated from food waste is used to replace about 58–60 kg of LPG in the kitchen, which is used for cooking applications. Therefore, the food waste is converted to biogas, which is used in the same kitchen and the bio manure is used as an organic fertilizer to grow vegetables and fruits, which are again used in the same kitchen. Therefore, waste to energy from kitchen to kitchen is a suitable phrase for this kind of onsite waste treatment. The detailed description and the performance output of the plant is published by Kuruti et al. [22]. They have reported the operational issues encountered and the strategies that were employed to retrofit and ensure the health of the digester. Further an insight on the techno-economic aspects of the biogas plant with a simple return on investment (ROI) is also discussed. It was reported that the ROI for a 1000 kg/day biogas plant based on AGR is 3.89 years, which is lucrative and compelled the TAPF to install such a biogas plant at many of their other kitchens across India. To date, 16 biogas plants have been installed and operated for the treatment of food waste alone with capacities ranging from 150 kg/day – 1 ton/day at different

FIGURE 3.1   Biogas plant at Bellary in Karnataka State for the generation of biogas and bio manure from food waste.

locations in India including TAPF. The intangible benefits associated with this technology includes avoiding the transportation of waste, reduction in the cost of waste disposal, reduction is usage of conventional energy, employment generation, reduction of GHGs etc. This technology is a replicable model, which can be adopted by the bulk food waste generators, such as restaurants, industrial canteens, large kitchens, etc.

3.2.3.1.2   Operational Problems and Solutions Achieved for a Food Waste-Based Biogas Plant

It was understood during the operation of the biogas plant for the biomethanation of food waste through AGR technology that the rapid acidification phenomena of food waste is the major bottleneck that may result in the failure of the biodigester. It was interesting to note that the food waste slurry tends to acidify within 12 h resulting in the drop in pH due to its inherent nature of putrefaction. Due to the rapid acidification of the feed slurry inside the digester, the biogas production decreased and within a few days, the reactor soured, resulting in the digester failure. The sudden drop in the pH occurred due to the accumulation of volatile fatty acids inside the digester. To retrieve the health of the digester, it was kept under recycle mode for a week and later on the reactor was fed with mixed feedstocks (cooked and uncooked food waste, vegetable and fruit waste, rice gruel water) along with the addition of required quantity of pH boosters. Due to the addition of pH booster, proper management of recirculation liquid and feed to the digester with mixed feedstocks, the digester was stabilized and the production of biogas was gradually improved. Within two weeks, the performance of the digester was normalized and there was notable biogas production and bio

FIGURE 3.2 A 5 TPD biogas plant for the generation of biogas based off-grid power and bio manure from organic fraction of MSW and leachate.

manure on a day-to-day basis. This problem was used advantageously in reducing the HRT of the reactor in the subsequent plants by designing a pre-acidification tank for the feed slurry before pumping it to AGR [22].

3.2.3.2 Model II: Biogas Plants Based on Organic Fraction of MSW to Power

It is evident from the reports that promising results have been achieved from the operation of the biogas plants of different capacities (150 kg/day to 1 tons/day) for the treatment of livestock waste (animal manures), kitchen waste and food waste [22, 25–28]. This prompted CSIR-IICT to install another biogas plant of 5 ton/day capacity for the treatment of organic fraction of MSW and leachate in 2018 at Hyderabad Integrated MSW Limited (HIMSWL) Jawahar Nagar, Hyderabad in Telangana State for the generation of biogas based off-grid electricity and bio manure (Figure 3.2). The project was funded by the Indo-US Science and Technology Forum (IUSSTF), New Delhi, partially supported (provision of land, civil works, etc.) by beneficiary, M/s HIMSWL and it was installed by M/s AESPL. The biogas plant produces about 300 units of electricity per day along with the replacement of 15–20 kg of LPG and 5–6 KL of liquid fertilizer every day. The power generated from the plant is being used by M/s HIMSWL though off-grid mode and bio manure is being sold to farmers. One of the interesting aspects of this model plant is the landfill leachate is co-digested along with organic fraction of MSW. The detailed description on working of the plant including the performance outputs is published by Jayanth et al. [28].

3.2.3.3 Model III: Biogas Plants Based on Market and Vegetable Waste to Power

The agriculture marketing committees operated by the state governments in India are the major hub for the farmers to sell their produce (fruits and vegetables) to wholesale and retail traders. In this process, due to the lack of cold storage facilities and refrigerated transport systems in India, about 30% of this produce gets wasted annually. The vegetable and fruit waste generated in market yards gets transported to the local landfills, which not only involve labour and transportation costs, but it also results in greenhouse gas emissions and pollutes the environment.

FIGURE 3.3   The 10 TPD biogas plant for the generation of biogas and bio manure from market and vegetable waste at Bowenpally in Telangana State.

Keeping this as a major concern and the motive to demonstrate a sustainable, remunerative waste-to-energy model plant, CSIR-IICT as technology provider, Department of Biotechnology (DBT) as project funder and Department of agriculture marketing, Government of Telangana as project beneficiary a joint project was taken up and installed by M/s AESPL at Dr. B. R Ambedkar Vegetable Market yard in Bowenpally, Secunderabad in Telangana State for the generation of biogas-based electricity and bio manure from 10 tons per day of market and vegetable waste (Figure 3.3). The plant is currently treating 10 tons of mixed market and vegetable waste per day and generating 450 to 550 m³/day of biogas. The biogas is being used for combined heat and power (CHP) applications by providing electrical energy to the same market yard through off-grid mode. Part of the biogas is also supplied to the canteen in the same market yard, which is replacing 25 to 30 kg of LPG per day. The biogas generated from the 10 tons of market vegetable waste meets the daily electricity needs of the market yard along with providing them LPG savings in their canteen. The market yard at Bowenpally can thus save an expenditure of Rs. 3 lakhs towards power bills and Rs. 2 lakhs for the waste disposal costs every month. The plant currently has also generated employment for eight people who operate the plant every day. The success story of this plant resulted in the installation of about eight more plants at different locations in Telangana State.

3.2.4 NATIONAL RECOGNITION OF AGR TECHNOLOGY

The biogas plant installed in Bowenpally vegetable market yard is currently 2 years old wherein about 10 tons of market waste is converted to electricity and cooking fuel every day. The biogas plant installed in the market yard is considered a model plant for all the other bulk waste generating markets across India. Swachh Bharath Mission or Abhiyan, a programme initiated by the PM of India in 2016 wherein the prime motto is to treat and valorize the organic waste at the source of generation. That means, implementation of decentralized waste management approaches. As the biogas plant at Bowenpally vegetable market is installed and operated at the source of waste generation to valorize the waste, the biogas plant has received appraisal from the prominent leaders of the nation. One of the remarkable achievements of this technology is the appraisal of the technology by the Prime Minister of India in his Mann ki Baat in 2021 as this technology aligns with Swachh Bharath Mission.

3.2.5 TECHNO-COMMERCIAL ASPECTS OF THE BIOGAS PLANTS AGAINST THEIR CAPACITIES

The techno-commercial assessment of the biogas plants with different capacities ranging from 50 kg/day to 10,000 kg/day is tabulated and presented in Table 3.2. It is interesting to note that the simple payback period or the return on investment for the biogas plants is decreasing from 10.7 years to 2.34 years as the treatment capacity is increasing from 50 kg/day to 10,000 kg/day (Table 3.2). A common yet interesting observation from Table 3.2 is that the ROI for the plants with capacities ranging from 50 kg/day to 500 kg/day is relatively higher because of less revenue from the sale of LPG and bio manure, operating cost and the net revenue while the plants with capacities above 500 kg/day is promising indicating that the biomethanation systems for the treatment of large quantities of waste is highly remunerative. It is also appealing that the footprint area required for the installation of biogas plant, power consumption to operate the plant is less, which makes the AGR technology quite interesting.

3.2.6 FUTURE SCOPE IN THE BIOGAS INDUSTRY

The high-rate biomethanation technology has given fruitful results, which is evident from the number of installations. To date, the biogas plants have been installed for the treatment of food waste, market and vegetable waste, kitchen waste, organic fraction of MSW and landfill leachate. Foreseeing the necessity and significance of treatment of other wastes such as mulberry waste, agriculture residues (rice husk, rice straw, corn stalk, etc.), press mud, bagasse, palm oil industry waste, etc., in addition to the existing wastes, extensive research at laboratory scale initially is required. Upon successful validation of the technology, it can be scaled up and implemented on field to achieve remunerative economic and environmental benefits. Therefore, the state and central government/federations' interference in increasing the funding towards R&D on biomass conversion to bioenergy is essential. Further, the Ministry of Petroleum and Natura Gas (MoPNG), Government of India is planning to install a good number of compressed biogas (CBG) plants from various organic waste through the Sustainable Alternative Towards Affordable Transportation (SATAT) [29] programme. *Ex situ* biogas upgradation methods and the *in situ* biogas generation with methane > 75 % is required to utilize the biogas as CBG. Therefore, extensive research of different strategies to improve methane in biogas should be carried out to develop a technology that not only converts organic waste to biogas but improves the methane content in biogas by 10–20% from 60%, which is the most common composition that can be obtained from any anaerobic digester. Academic and research institutions in association with industrial partners must take up this job for their mutual benefit and timely development of novel anaerobic digester systems that can produce methane-rich biogas. Further, in lieu of biogas, technologies for the conversion of organic waste to value-added

TABLE 3.2

Techno-Commercial Aspects of the Biogas Plants with Different Capacities Ranging from 50 kg/day to 10,000 kg/day

Quantity of Organic waste (kg/day)	Foot print Area Required for Plant Installation (m²)	Power Consumption (kWh/day)	Average Biogas and Bio Manure Generation per Day		Equivalent LPG Replacement (Kg/day)	Revenue due to LPG/ annum @ INR 1.2S/kg (330 days)	Revenue due to Sale of bio manure/ annum @ 0.133$/kg (330 days)	Gross Revenue/ annum ($)	Approximate Operating Cost/annum ($)	Net Revenue/ annum ($)	Approximate Capital Cost ($)	Payback Period or Return on Investment (Years)
			Biogas (m³/day)	Bio manure (Kg/day)								
50	10	2	6	8	2	950	330	1,280	533	747	8,000	10.7
100	12	4	12	15	5	1,901	660	2,561	533	2,027	14,666	7.23
150	12	8	18	23	7	2,851	1,012	3,863	533	3,330	21,333	6.4
200	15	10	24	30	10	3,801	1,320	5,121	533	4,588	26,666	5.8
250	20	10	30	45	12	4,752	1,980	6,732	666	6,065	33,333	5.49
500	26	12	60	75	24	9,504	3,300	12,804	800	12,004	53,333	4.44
750	30	15	90	112	36	14,256	4,928	19,184	2,666	16,517	66,666	4.03
1,000	50	20	120	150	48	19,008	6,600	25,608	3,333	22,274	86,666	3.89
5,000	600	60	600	750	240	95,040	33,000	1,28,040	26,666	1,01,373	3,40,000	3.35
10,000	1,200	120	1,200	1,500	480	1,90,080	66,000	2,56,080	40,000	2,16,080	5,06,666	2.34

Note: 1 USD = 75 INR

products, such as short-chain carboxylic acids, ethanol, methanol, etc., must be taken into consideration for future developments as the value of these value-added chemicals is high compared to biogas.

3.3 CONCLUSIONS

The chapter described the need for developing the high-rate biomethanation technologies to overcome the bottlenecks associated with the conventional slow-rate digesters. This chapter also presents the journey of rate biomethanation technology that was developed and commercialized by CSIR-IICT for the generation of biogas and bio manure from a variety of organic wastes. The high-rate biomethanation technology based on AGR performed on a par with the European advanced high-rate digesters, which is evident from the success stories of AGR technology. In summary, it can be concluded that among the existing technologies for waste treatment, anaerobic digestion (AD) plays a key role in reducing the waste along with the generation of renewable bio energy as it mitigates direct and indirect GHG emissions, recycles nutrients in the form of organic fertilizers, prevents nitrogen leakage into groundwater and avoids the spread of harmful diseases through landfilling. The success stories of AGR technology have been reviewed based on the published articles and an article 'A Glance at the World', assessing the technology has been published by the editor of the *Waste Management* journal. Valorization of organic solid waste by implementing decentralized biogas solutions in rural and urban India resolves the need for energy along with a potential solution for the disposal of wastes. Biogas solutions are generally versatile, flexible and cost efficient from a societal perspective and when adapted to local conditions may contribute to sustainable development as well.

ACKNOWLEDGEMENTS

The authors are grateful to the funding agencies Department of Biotechnology (DBT), Government of India, Department of Science and Technology (DST), Government of India, Indo-US Science and Technology Forum (IUSSTF), Department of Agriculture Marketing (DAM); Government of Telangana, Council of Scientific and Industrial Research (CSIR), Government of India towards their financial support for the installation of biogas plants based on AGR technology. The authors express their sincere thanks to the project executing company and the licensees of the Technology M/s Ahuja Engineering Services Private Limited (AESPL), M/s Nyrmalya Bio engineering Solutions Private Limited (NYBES) and M/s Khar Energy Optimizers (KEO). The authors are thankful to the Director, CSIR-Indian Institute of Chemical Technology (IICT) for his continuous encouragement in carrying out this work.

REFERENCES

[1] Kumar, A., & Samadder, S. R. (2017). A review on technological options of waste to energy for effective management of municipal solid waste. *Waste Management, 69*, 407–422.

[2] Sharholy, M., Ahmad, K., Vaishya, R. C., & Gupta, R. D. (2007). Municipal solid waste characteristics and management in Allahabad, India. *Waste Management, 27*(4), 490–496.

[3] Begum, S., Shruti, A., Gangagni Rao, A., Kiran, G., Bharath, G., Kranti, K., Swamy, Y. V., & Devendar, K. A. (2016). Significance of decentralized biomethanation systems in the framework of municipal solid waste treatment in India, *Current Biochemical Engineering, 3*, 1.

[4] MNRE - Ministry of New and Renewable Energy - Biomass Power/Cogen, (2017). Mnre.gov.in, 2017. [Online]. Available at: http://mnre.gov.in/schemes/grid-connected/biomass-powercogen.

[5] Thomas, P., Soren, N., Rumjit, N., George, J. J., & Saravanakumar, M. (2017). Biomass resources and potential of anaerobic digestion in Indian scenario. *Renewable and Sustainable Energy Reviews, 77*, 718–730.

[6] Lohan, S. K., Dixit, J., Kumar, R., Pandey, Y., Khan, J., Ishaq, M., Modasir, S., & Kumar, D. (2015). Biogas: A boon for sustainable energy development in India's cold climate. *Renewable and Sustainable Energy Reviews, 43*, 95–101.

[7] Rao, A. G., Gandu, B., Sandhya, K., Kranti, K., Ahuja, S., & Swamy, Y. V. (2013). Decentralized application of anaerobic digesters in small poultry farms: Performance analysis of high rate self mixed anaerobic digester and conventional fixed dome anaerobic digester. *Bioresource Technology, 144*, 121–127.

[8] Begum, S., Anupoju, G. R., Sundergopal, S., Bhargava, S. K., Jegatheesan, V., & Eshtiaghi, N. (2018). Significance of implementing decentralized biogas solutions in India: A viable pathway for biobased economy. *Detritus, 1*(1), 75.

[9] IEA, (2021). An introduction to biogas and biomethane – Outlook for biogas and biomethane: Prospects for organic growth – Analysis. Available at: www.iea.org/reports/outlook-for-biogas-and-biomethane-prospects-for-organic-growth/an-introduction-to-biogas-and-biomethane [Accessed 6 April 2021].

[10] Nsair, A., Onen Cinar, S., Alassali, A., Abu Qdais, H., & Kuchta, K. (2020). Operational parameters of biogas plants: A review and evaluation study. *Energies, 13*(15), 3761.

[11] Estonian Biogas Association (EBA), (2016). Production and use. Available at: http://eestibiogaas.ee/tootmine-ja-kasutamine/ [Accessed 14 March 2022].

[12] Levaggi, L., Levaggi, R., Marchiori, C., & Trecroci, C. (2020). Waste-to-Energy in the EU: The effects of plant ownership, waste mobility, and decentralization on environmental outcomes and welfare. *Sustainability, 12*(14), 5743.

[13] Deodhar, V., & Van den Akker, J. (2005). Final review of the undp/gef project ind/92/g32 development of high-rate biomethanation processes as means of reducing greenhouse gas emission.

[14] Afridi, Z., & Qammar, N. (2020). Technical challenges and optimization of biogas plants. *Chembioeng Reviews, 7*(4), 119–129. doi: 10.1002/cben.202000005

[15] Al-Dahhan, M., Varma, R., Karim, K., Vesvikar, M., Hoffman, R., Klasson, T., Wintenberg, A. L., Alexander, C. W., Clonts, L. and Olson, N., (2007). Improved biomass utilization through remote flow sensing. (No. DE/FC36/01GO11054). Washington University, St. Louis.

[16] Rao, A. G., Prakash, S. S., Joseph, J., Reddy, A. R., & Sarma, P. N. (2011). Multi stage high rate biomethanation of poultry litter with self-mixed anaerobic digester. *Bioresource Technology, 102*(2), 729–735.

[17] Nuri, A., Pepi, U., & Richard, E.S. (2001). Effect of process configuration and substrate complexity on the performance of anaerobic processes. *Water Research, 35*, 817–829.

[18] Kelleher, B. P., Leahy, J. J., Henihan, A. M., O'Dwyer, T. F., Sutton, D., & Leahy, M. J. (2002). Advances in poultry litter disposal technology – A review. *Bioresource Technology, 83*, 27–36.

[19] Salminen, E., & Rintala, J. (2002). Anaerobic digestion of organic solid poultry slaughter waste – A review. *Bioresource Technology, 83*, 13–26.

[20] Sakar, S., Yetilmezsoy, K., & Kocak, E. (2009). Anaerobic digestion technology in poultry and livestock waste treatment—A literature review. *Waste Management & Research, 27*(1), 3–18.

[21] Rajagopal, R., Massé, D. I., & Singh, G. (2013). A critical review on inhibition of anaerobic digestion process by excess ammonia. *Bioresource Technology, 143*, 632–641.

[22] Kuruti, K., Begum, S., Ahuja, S., Anupoju, G. R., Juntupally, S., Gandu, B., & Ahuja, D. K. (2017). Exploitation of rapid acidification phenomena of food waste in reducing the hydraulic retention time (HRT) of high-rate anaerobic digester without conceding on biogas yield. *Bioresource Technology, 226*, 65–72.

[23] Pol, L. W. H., Lens, P. N., Stams, A. J., & Lettinga, G. (1998). Anaerobic treatment of sulphate-rich wastewaters. *Biodegradation, 9*(3), 213–224.

[24] Gangagni Rao, A., Swamy, Y.V. (2012). Anaerobic gas lift reactor for the treatment of organic solid wastes having high nitrogen (low C/N ratio), 2012, (PT-609/0207NF2012).

[25] Begum, S., Ahuja, S., Anupoju, G. R., Kuruti, K., Juntupally, S., Gandu, B., & Ahuja, D. K. (2017). Process intensification with inline pre and post processing mechanism for valorization of poultry litter through high rate biomethanation technology: A full scale experience. *Renewable Energy, 114*, 428–436.

[26] Begum, S., Ahuja, S., Anupoju, G. R., Ahuja, D. K., Arelli, V., & Juntupally, S. (2020). Operational strategy of high rate anaerobic digester with mixed organic wastes: Effect of co-digestion on biogas yield at full scale. *Environmental Technology, 41*(9), 1151–1159.

[27] Girotto, F. (2017). A glance at the world. *Waste Management, 70*, I–III.

[28] Jayanth, T. A. S., Mamindlapelli, N. K., Begum, S., Arelli, V., Juntupally, S., Ahuja, S., ... & Anupoju, G. R. (2020). Anaerobic mono and co-digestion of organic fraction of municipal solid waste and landfill leachate at industrial scale: Impact of volatile organic loading rate on reaction kinetics, biogas yield and microbial diversity. *Science of The Total Environment, 748*, 142462.

[29] MoPNG, (2021). Compressed Bio Gas. Ministry of Petroleum and Natural Gas, Government of India. Available at: https://mopng.gov.in/en/refining/compressed-bio-gas

4 Cyanobacterial Degradation of Pesticides

Nimisha Vijayan P, Mousumi Das M,
Haridas Madathilkovilakathu and Sabu Abdulhameed

4.1 INTRODUCTION

Agriculture is an important economic sector in all parts of the world. The growing population and increased food demand have built a huge pressure on the agriculture system in order to produce more crops both in quality and quantity. Hence, there is a need for sustainable crop protection. In this context, pesticides have played a major role in ensuring food security (Kole et al. 2019). Over the years, these synthetic chemicals have become vital for modern agriculture systems (Tudi et al. 2021). It has been estimated that approximately 45% of annual crop production gets affected because of the attack of various pests (Sharma et al. 2019). With the use of pesticides, crop loss can be controlled to a great extent and a significant increase in crop yield can be achieved. Farmers are compelled to use different kinds of agrochemicals for the improved production of agriculture crops.

Use of pesticides have resulted in increased crop yield and have reduced the spread of vector-borne diseases. One third of the total agricultural production depends upon the use of pesticides (Tudi et al. 2021). But their indiscriminate use has caused several problems in the environment apart from the development of resistance in pests (Vijayan and Abdulhameed 2020). Pesticides have emerged as a principal pollutant of air, water and soil thereby affecting many non-target organisms including humans. Pesticides may pose deleterious effects on animals, beneficial microorganisms, non-target plants, birds, fishes bringing an overall imbalance in the ecosystem (Assey et al. 2021). Pesticides poses a great threat to the biodiversity.

Although there are umpteen benefits for pesticides, their deleterious effects have raised a huge concern over the safety of using these chemicals. Considering the various adverse effects of pesticides, it has become mandatory to eliminate these toxic pesticides from the environment by a safe and ecofriendly manner. It is not possible to completely abolish these chemicals from the environment, but the problems associated with the pesticides can be managed to a great extent. Although there are various physico-chemical methods exists for the removal of pesticides, the generation of toxic intermediates and costly nature limits the use of these methods. Hence, researchers are more focused on the use of biological methods for the degradation of pesticides.

Different microbial strains are found to take part in the biodegradation of pesticides like bacteria, fungi, actinomycetes and cyanobacteria. Out of these microorganisms, cyanobacteria have emerged as a potential candidate in the field of pesticide degradation. Their widespread appearance in the polluted area and various other advantages like autotrophic nature and nitrogen fixation makes them a good choice for pesticide degradation. Researchers are now more focused to investigate the full potential of cyanobacteria for an efficient removal of pesticides from the environment. Although cyanobacteria are exploited in various other fields, their use as bioremediating agents needs much more attention. Degyadation of pesticides using cyanobacteria will be an economically viable and environmentally friendly solution.

DOI: 10.1201/9781003327646-4

4.2 PESTICIDES – A GROWING CONCERN

Pesticides are chemical compounds that are mainly used to protect the crops from various diseases, pests and weeds. Pesticides act by controlling, repelling or destroying the attacking pests. Agriculture is the main source of chemical pesticides. But these pesticides are also used for various other purposes like household applications and in forestry. In order to ensure improved global food production, more and more powerful pesticides are being developed (Rajmohan et al. 2020). Increased demand for agro products and climate change have caused a drastic increase in the application level of pesticides (Odukkathil and Vasudevan 2013). Pesticides can be used as liquid sprays, as seed treatment, application as granules or it can be incorporated into the soil (Mahmood et al. 2016). After reaching their target area, they can make their entry into the environment either through dispersion, microbial breakdown, absorption, adsorption, volatilization or through leaching into ground water and surface water (Mahmood et al. 2016; Assey et al. 2021). Pesticide residues can be seen in food products like fruits and vegetables and drinking water (Assey et al. 2021). The absorption and desorption of pesticides from soil particles determines the bioavailability and mobility of the pesticides (Rajmohan et al. 2020).

Pesticides are classified into different groups. Based on the target species, they are classified into insecticides, fungicides, herbicides, nematicides, etc. Among these different classes, insecticides are more toxic followed by fungicides and herbicides (Mahmood et al. 2016). The persistent nature and degradation rate of pesticides are determined by its chemical composition (Bose et al. 2021) Based on their chemical composition pesticides are grouped into organochlorines, organophosphates, carbamates and pyrethroids (Arya et al. 2017).

When the harmful effects of pesticides outclass its beneficial effects then it becomes a problem of utmost importance. Both terrestrial and aquatic ecosystems are affected by the use of pesticides. Pesticides cause contamination of the aquatic system by drift, agriculture runoff and leaching through the soil thereby making their entry into the nearby waterbodies. Pesticides cause serious effects to aquatic organisms like fishes and aquatic plants (Mahmood et al. 2016). The amount of dissolved oxygen and biological oxygen demand decreases in water bodies as a result of pesticide pollution (Rani and Dhania 2014). Pesticides may also cause a decrease in the population of beneficial insects like bees and beetles (Mahmood et al. 2016). Contamination of soil with pesticides causes adverse effects on the food quality and on the agriculture system (Tudi et al. 2021).

4.3 IMPACT OF PESTICIDES ON HUMAN HEALTH

Pesticides can make their entry into the human body through different ways like inhalation, ingestion and penetration through the skin. But most people are affected by consuming pesticide contaminated food. Effects of pesticides on humans may range from acute effects like vomiting, headache, dizziness, irritation to the eyes and skin to chronic effects like cancer, neurologic effects, diabetes, respiratory diseases and genetic disorders (Mahmood et al. 2016; Assey et al. 2021). Figure 4.1 illustrates some of the chronic diseases associated with pesticide exposure (Mostafalou and Abdollahi 2013)

4.4 BIODEGRADATION OF PESTICIDES

There is a huge demand for the development of effective methods that ensure the removal of pesticides in a safe manner. An ideal method must guarantee the complete destruction of the compound without the generation of any toxic intermediates (Parte et al. 2017). The conventional techniques including adsorption and advanced oxidation methods like photodegradation, fenton reaction, photocatalysis, ultrasound assisted remediation, etc., are highly expensive and have several disadvantages (Vijayan and Abdulhameed 2020; Parte et al. 2017). Biodegradation is a microorganism-mediated process by which complete breakdown of a toxic chemical into non-toxic products occurs (Vijayan et al.

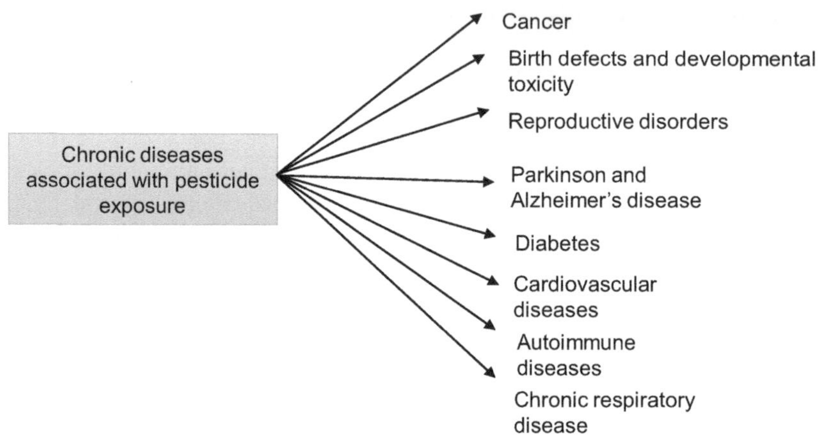

FIGURE 4.1 Various chronic diseases associated with exposure to pesticides.

2020b). Through biodegradation, microorganisms derive energy and other nutrient requirements by converting complex organic compounds into simpler inorganic products (Bose et al. 2021). Microorganisms play a significant role in the degradation of pesticides (Arya et al. 2017) and complete mineralization of the pesticides occurs without the generation of any toxic intermediates (Vijayan et al. 2020b). Microorganisms present in the pesticide contaminated sites are adapted or develop natural mechanisms of degradation due to the continuous exposure of the organisms to the pollutant (Bose et al. 2021).

Microorganisms can interact with the pesticide by both physical and chemical contacts. By inducing some structural changes to the pesticide, microorganisms can either completely degrade the pesticide or they can make the pesticide non-toxic (Bose et al. 2021). Microbial degradation of pesticides in soil depends on various factors like temperature, pH and availability of microbial flora, availability of pesticides, physiological status of the organisms and survival and multiplication of the pesticide degrading microorganism at the polluted site (Arya et al. 2017; Rani and Dhania 2014). Researchers have isolated different pesticide degrading microbial strains like bacteria, fungi, actinomycetes and algae (Huang et al. 2018). Microbial degradation involves reactions like oxidation, reduction, dehalogenation, dehydrogenation and hydrolysis, etc. (Huang et al. 2018). Microbial degradation is mainly mediated by enzymes. A wide variety of enzymes are involved in microbial degradation like monooxygenases, oxidoreductases, dioxygenases, haloalkane dehalogenases, hydrolases, carboxylesterases, etc. (Sharma et al. 2016).

4.5 CYANOBACTERIA – A PRECIOUS BIORESOURCE

Cyanobacteria are one of the most abundant groups of organisms on the Earth. These autotrophic organisms are found in a wide variety of habitats including marine and fresh water and terrestrial habitats (Chittora et al. 2020). They are the oldest photosynthetic organisms existing on Earth about 2.6–3.5 billion years ago (Hedges et al. 2001). Some of the species have the capacity to carry out atmospheric nitrogen fixation. A special structure called heterocyst help them to carry out nitrogen fixation. But some unicellular and filamentous cyanobacteria are able to fix atmospheric nitrogen without the involvement of this specialized structure (Singh et al. 2019). They can be seen in highly polluted environments (Mona et al. 2020). They exist in unicellular, filamentous, benthic or colonial forms (Singh et al. 2019). They are an excellent source of various bioactive compounds like antibacterial, antifungal antiprotozoal, antiviral and anticancer compounds, which are of pharmaceutical significance (Lau et al. 2015; Kumar et al. 2019). In addition to pollutant removal from water and

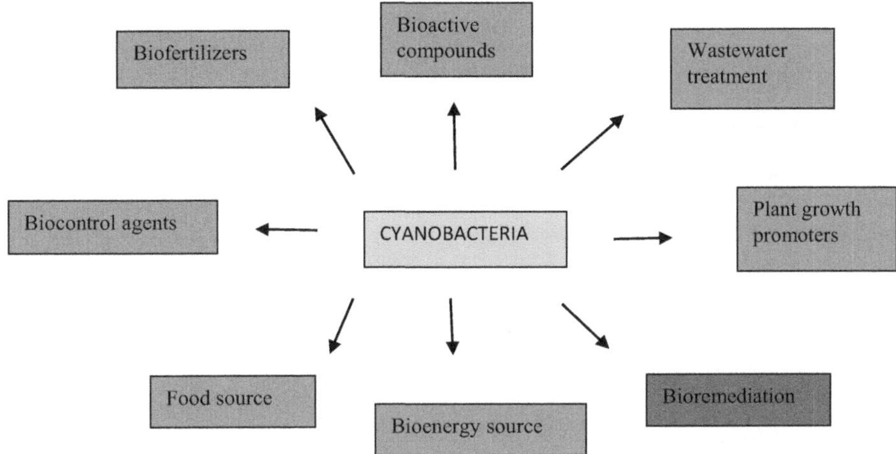

FIGURE 4.2 Various applications of cyanobacteria.

soil, cyanobacteria help to increase the fertility of soil and enhance the quality of the soil like mineral and nutrient composition and water holding capacity (Kondi et al. 2022).

Unlike other heterotrophic bacteria, cyanobacteria need minimal nutritional requirements eliminating the need for a complex media for their growth (Lau et al. 2015). Cyanobacteria finds wide applications in agriculture because it increases soil porosity, soil biomass, excretes growth promoting substance and decreases soil salinity (Malyan et al. 2020). The extracellular substances released by cyanobacteria play a significant role in modulating the pH, redox activity, temperature, methane generation and ammonia volatilization (Prasanna et al. 2008). Cyanobacteria are a very good source of various value-added products like pigments, vitamins, enzymes, biopolymers, isoprene, etc. The pigments phycobiliproteins and carotenoids are widely used in the bioindustry. Several strains of cyanobacteria have been found to produce the biodegradable polymer polyhydroxyalkanoate (Lau et al. 2015). Some metabolites from cyanobacteria finds application as biopesticides in agricultural fields (Singh et al. 2017). Figure 4.2 depicts applications of cyanobacteria in various fields.

4.6 CYANOBACTERIA AS BIOREMEDIATING AGENTS

Cyanobacteria can be used for bioremediation either in its natural form, mutant or genetically engineered forms. Their widespread distribution in the contaminated sites made them develop a resistant and selective nature against environmental contaminants. A decline in the level of pollutants will not have a prominent effect on the growth and metabolic activity of these organisms (El Bestawy et al. 2007). They are successfully used in waste water treatment processes for the removal of nitrogen and phosphorus, heavy metals and textile dyes (Kumar and Singh 2017). Compared to other microorganisms, the special characteristics of cyanobacteria like nitrogen fixation, adaptation to different environmental conditions, production of extracellular polysaccharides and high multiple rates makes them unique in the field of bioremediation (Malyan et al. 2020). The cyanobacterial biomass generated after bioremediation can be further used for the extraction of high value products, which is an added advantage in comparison with the other heterotrophs (Singh et al. 2019).

Cyanobacteria are used to remove pollutants from industrial effluents like textiles, paper mills, pharmaceuticals, sugar mills, food industries and distilleries (Malyan et al. 2020). The hydrophilic and hydrophobic moieties present in the cell wall of cyanobacteria provides adsorption sites for various organic contaminants and pollutants (Tiwari et al. 2017b). They can accumulate and degrade different kinds of pollutants like crude oil, phenol and catechol, pesticides and other xenobiotics (El Bestawy et al. 2007). Immobilized cyanobacteria can also be used for the bioremediation purposes.

TABLE 4.1
Pesticides and the Cyanobacteria Involved in their Degradation

Pesticide	Degrading Cyanobacteria	Reference
Anilofos	*Synechocystis* sp *strain PUPCCC64*	Singh et al. 2013
Fenamiphos	*Nostoc muscorum, Anabaena* sp.	Cáceres et al. 2008
Malathion	*Anabaena oryzae, N. muscorum, S. platensis*	Kuritz and Wolk 1995
Glyphosate	*Anabaena* sp., *Lyngbya* sp., *Microsystis* sp., *Nostoc* sp.	Forlani et al. 2008
Methyl parathion	*Micocystis novacekii*	Fioravante et al. 2010
	Fischerella sp.,	Tiwari et al. 2017a
	Anabaena sp.	Barton et al. 2004
Lindane	*Oscillatoria* sp., *Syncechococcus* sp., *Nodularia* sp., *Nostoc* sp.,	El- Bestawy et al. 2007
	Cyanothece Sp., *Anabaena cylindrical*	Zhang et al. 2012
	Anabaena azotica	
Monocrotophos	*Aulosira fertilissima*	Subramanian et al. 1994
Dichlorovos	*Nostoc muscorum*	
Phosphomidon		
Quinalphos		
Chlorpyrifos	*Synechocystis* sp. Strain PUPCC64	Singh et al. 2011
	Phormidium valderianum	Palanisami et al. 2009
	Spirulina platensis	Thengodkar and Sivakami 2010
Pyridiphenthion	*N. muscorum*	Hamed et al. 2021
Endosulfan	*Anabaena sp.* PCC 7120	Lee et al. 2003

Several polymers are available for the immobilization of cyanobacterial strains. Alginate and carrageenan are the widely used polymers for the immobilization of cyanobacterial strains (Singh et al. 2019). Immobilized cyanobacteria were found to be efficient in the removal of heavy metals from industrial effluents (El-Bestawy 2019). Table 4.1 lists different pesticides and the cyanobacterial strains involved in their degradation. The two widely used classes of pesticides are organophosphorus and organochlorine pesticides. Hence the biodegradation of these two classes of pesticides by cyanobacteria will be discussed in detail.

4.6.1 CYANOBACTERIAL DEGRADATION OF ORGANOPHOSPHORUS PESTICIDES

Organophosphorus pesticides are widely used in paddy fields (Kumar et al. 2021). These groups of pesticides possess high mammalian toxicity and it mainly acts by suppressing the action of acetylcholine esterase (Jiang et al. 2019). Organophosphorus pesticides are esters of phosphoric acid that includes aliphatic, phenyl and heterocyclic derivatives. These chemicals find application in controlling spider mites, sucking, chewing and boring insects, aphides, etc. (Parte et al. 2017). Longterm exposure to organophosphorus pesticides poses a serious risk to human health. Some of the organophosphorus pesticides are highly persistent and they can remain in the environment for a very long time, thereby accumulating in the organisms through the food chain (Jiang et al. 2019). Their residues can be found in the environment and research is being carried out on the biodegradation of organophosphorus pesticides. Cyanobacteria have proved to be an excellent organism in the degradation of organophosphorus pesticides.

Tiwari et al. (2017b) studied a new isolate of *Syctonema* capable of degrading methyl parathion. The organism was found to utilize methyl parathion as a source of phosphorus. Presence of para nitrophenol confirmed the degradation of methyl parathion since it is produced during the initial breakdown of methyl parathion by phosphotriesterase enzyme. This yellow colour product is widely used as an indicator of methyl parathion degradation. Chemosorption was the first step in the removal of methyl parathion by the cyanobacterial strain followed by its complete degradation. A paddy field cyanobacterium *Fischerella* sp. was found to be involved in the biodegradation of methyl parathion. The pesticide was found to interact with the –OH group on the cell surface. Para nitrophenol was present in the medium which confirmed the pesticide degradation. Complete removal of methyl parathion occurred through biosorption, bioaccumulation followed by biodegradation (Tiwari et al. 2017a).

Fenamiphos is a widely used organophosphorus insecticide causing widespread contamination of soil and aquatic bodies. Cáceres et al. 2008 reported the degradation of fenamiphos by cyanobacteria. Five different species of tested cyanobacteria (*Nostoc* sp. MM1, *Nostoc* sp. MM2, *Nostoc* sp. MM3, *Nostoc muscorum*, *Anabaena* sp.) were able to transform fenamiphos into fenamiphos sulfoxide (FSO), which was then hydrolyzed into fenamiphos sulfoxide phenol (FSOP). Among the tested cyanobacterial species, *Anabaena* was found to be more effective in degrading fenamiphos.

A chlorpyrifos degrading cyanobacterium *Synechocystis* sp. PUPCC 64 was isolated by Singh et al. (2011). The unicellular cyanobacterium tolerated chlorpyrifos upto 15 mg/L. 3, 5, 6 trichloro 2 pyridinol was found to be the major degradation product. The degradation product was detected in the cell extract, biomass wash and in the medium which proved the fact that the pesticide degradation occurred both intracellularly as well as extracellularly. In another study, the same strain was found to tolerate and mineralize the herbicide anilofos. The strain utilized anilofos as a source of phosphorus (Sigh et al. 2013). An alkaline phosphatase enzyme capable of chlorpyrifos degradation was isolated from the cyanobacterium *Spirulina platensis*. The organism was grown in the presence of chlorpyrifos and the enzyme activity was detected in the cell-free extracts. The purified enzyme degraded chlorpyrifos transforming it into 3, 5, 6 trichloro 2 pyridinol (Thengodkar and Sivakami 2010).

Phormidium valderianum, a marine cyanobacterium was able to grow in the presence of chlorpyrifos. Exposure to chlorpyrifos increased the activity of different enzymes like catalase, polyphenol oxidase, superoxide dismutase, esterase and glutathione S transferase. Esterase A was used by the organism for metabolizing chlorpyrifos (Palanisami et al. 2009).

Vijayan et al. (2020a) isolated a chlorpyrifos degrading cyanobacterial strain *Coleofasciculus chthonoplastes*. Degradation of chlorpyrifos was studied by different analytical techniques like thin-layer chromatography (TLC), Fourier transform infrared spectroscopy, gas chromatography mass spectrometry (GC-MS) and liquid chromatography mass spectrometry (LC-MS/MS). Extracted fractions on the TLC plate showed additional bands when compared with the parent pesticide. A difference in the IR spectrum of the test samples and the parent compound revealed structural changes occurred to chlorpyrifos during the course of its degradation. Further evidence was given by GC-MS and LC-MS/MS analyses. A new peak with the mass value corresponding to 3, 5, 6 trichloro 2 pyridinol was observed in the LC-MS spectrum, which was further confirmed by studying the fragmentation pattern by MS/MS analyses. GC-MS analyses showed no trace of chlorpyrifos or its degraded product thereby confirming the complete degradation of chlorpyrifos by the organism. The isolated strain can thus be successfully used to remediate the chlorpyrifos contaminated sites.

Barton et al. (2004) reported the reductive transformation of methyl parathion by *Anabaena* sp. The study illustrated the transformation of methyl parathion to *o,o,*- dimethyl *o-p-* aminophenyl thiophosphate by the reduction of nitro group. The transformation of methyl parathion took place under aerobic photosynthetic conditions. Methyl parathion degradation by the soil isolates of cyanobacteria was reported by Megharaj et al. (1994). The strains were found to use the pesticide as a source of phosphorus and nitrogen. Degradation of methyl parathion was indicated by the presence of *p*-nitrophenol in the medium, which was further degraded into nitrite.

Malathion degrading cyanobacterial strain was isolated by Ibrahim et al. (2014) Among the three tested filamentous cyanobacterial strains, *Nostoc muscorum* was found to tolerate different concentrations of malathion and it was found to be the efficient strain for biodegradation of malathion. An increase in the rate of cyanobacterial growth was observed when the phosphorus deficient medium was supplemented with malathion, thereby proving the efficiency of the strain to utilize the pesticide as a source of phosphorus.

Hamed et al. (2021) reported a cyanobacterial strain *Nostoc muscorum* capable of degrading pyridaphenthion, a widely used organic thiophosphate insecticide. The strain accumulated high amounts of the pesticide and the degradation product was found to be 6-hydroxy-2-phenylpyridazin-3(2 H)-one. Compared to the other tested strain Anabaena *laxa*, *N. muscorum* experienced less toxicity due to improved antioxidant defence system. Hence, *N. muscorum* would be an efficient strain to remediate pyridaphenthion contaminated soils.

4.6.2 Cyanobacterial Degradation of Organochlorine Pesticides

Organochlorine (OC) pesticides are synthetic agrochemicals, which belongs to a group of chlorinated hydrocarbon derivatives. They are extensively used in the agricultural fields to control the weeds and insect pests. These compounds are characterized by high lipid solubility, low aqueous solubility and low polarity (Jayaraj et al. 2016). Most of the organochlorine compounds like DDT and dieldrin, are highly toxic due to their slow rate of degradation, higher stability, and long half-life. They are very persistent in the environment and can migrate and get accumulated in the upper trophic levels of the food chain (Sharma et al. 2019). They mainly cause toxicity by stimulating the central nervous system (Jayaraj et al. 2016).

γ-hexachlorocyclohexane (lindane or γ-HCH) is a commonly used organochlorine pesticide in the agriculture. Residues of lindane found in the soil and water are highly toxic to humans (Zhang et al. 2020). Due to its toxicity, it has been prohibited in most countries; however, it can still be detected in soil and rice because of its high persistence (Badea et al. 2009; Yao et al. 2007). Singh (1973) reported that isolates of *Plectonema boryanum*, *Aulosira fertilissima*, and *Cylindrospermum* sp. from paddy fields were tolerant to commercial preparations of lindane in concentrations up to 80 pg/ml. Similarly, El-Bestawy et al. (2007) demonstrated the potential use of various cyanobacteria, including *Oscillatoria*, *Synechococcus*, *Nodularia*, *Nostoc*, and *Cyanothece*, in the bioremediation of lindane-contaminated effluents. All of the tested species were found to be highly efficient in the degradation of lindane and showed good resistance to its toxicity. The filamentous nitrogen-fixing *Anabaena* sp. strain, PCC 7120 and *Nostoc muscorum* were studied for their lindane degrading ability. The two strains of cyanobacteria were able to degrade lindane by dechlorinating it into pentachlorocyclohexane, which was then converted into a mixture of trichlorobenzenes. The presence of nitrate in the medium influenced the dichlorination process. Lindane dichlorination was affected by ammonium and darkness which proved the role of nitrate reduction system in the dichlorination process (Kuritz and Wolk 1995; Kuritz 1998).

The nitrogen-fixing cyanobacterium *Anabaena azotica* was demonstrated to be capable of biodegrading lindane (Zhang et al. 2012). Their study showed that temperature, irradiance and nitrogen source are important for lindane dehalogenation by *A. azotica*. The effective degradation of lindane by *A. azotica* under different conditions suggested the wide application of this strain in the bioremediation of paddy soils. This process may contribute to improving ecosystem and crop health by using this microbe as a bioremediation tool. Further study related to its key enzymes and photosynthetic functions would be more useful in the field of bioremediation.

Lee et al. (2003) reported the biotransformation of endosulfan by *Anabaena* sp. The study was carried out using *Anabaena* sp. PCC 7120 and *Anabaena flos-aquae*. Endosulfan diol and small amounts of endosulfan lactone, endosulfan hydroxyether and endosulfan sulfate were produced by *Anabaena* sp. PCC 7120. *Anabaena flos-aquae* biotransformed endosulfan by producing endosulfan diol and endosulfan sulfate. The results of the study suggested that the two cyanobacterial species could participate in the detoxification of endosulfan.

4.7 FUTURE PERSPECTIVES

All the aforementioned reports on the cyanobacterial degradation of organophosphate and organochlorine pesticides gives evidence for the potential role of cyanobacteria in the remediation of pesticide polluted sites. Application of genetic engineering technology will be helpful to enhance the biodegradation potential of cyanobacteria. The role of pesticide degrading enzymes of cyanobacteria warrants further study, which will make an important contribution to the enzymatic pesticide degradation. Studies regarding biodegradation of pesticides using immobilized cyanobacteria will find more applications in this field. Although there are few studies reporting the use of immobilized cyanobacteria for the waste water treatment, no studies have been undertaken in this regard for pesticide degradation. The use of cyanobacterial consortia consisting of two or more pesticide degrading strains will offer more applications in the field of bioremediation

Mass cultivation of cyanobacteria is required for its large-scale application in the contaminated sites. There are different methods for the mass production of cyanobacteria like cultivation using sunlight in open and closed systems and cultivation using artificial light in a closed system. All these methods have both pros and cons. The first two methods are comparatively cost effective because of the natural light source energy. The main drawback of open systems is the chance for contamination with other non-target organisms. Several photobioreactors have been designed which resemble a conventional fermenter in principle. But the high cost of these photobioreactors compared to the open and closed systems limited its use for commercial applications. Hence the future prospect will be the development of economically feasible photobioreactors and the production of efficient genetically engineered strains for the better use of these multifaceted bioagents in various fields including the biodegradation of pesticides.

4.8 CONCLUSION

A high level of pesticide application in agriculture and other fields created problems, such as pollution in the environment. Pesticides have benefited the humans because they increased crop productivity all over the world. But over the passage of time, the harmful effects of these chemicals surpassed their beneficial effects. So an urgent need to mitigate the problems associated with pesticides occurred. Various conventional methods are in use for the removal of pesticides from the environment. But the expensive nature of these methods and the generation of toxic intermediates have limited the use of these methods. In this regard, degradation using microorganisms will be a clean approach compared to the other existing methods. The ability of various microorganisms to degrade pesticides has been widely studied. To circumvent the problem of pesticide pollution, degradation using cyanobacteria will be a promising method. Their adaptability, nitrogen fixing capacity and metabolic flexibility offer enormous advantages in the field of biodegradation. They can improve soil health and fertility thereby representing an important group of organisms for a sustainable agriculture. Different cyanobacterial strains having great potential to degrade various organophosphorus and organochlorine pesticides have been isolated and studied by researchers. Studies relating to the molecular aspects of these organisms must be carried out to reveal the full potential of cyanobacteria in the field of pesticide degradation.

REFERENCES

Arya R, Kumar R, Mishra NK, Sharma AK (2017). Microbial flora and biodegradation of pesticides: Trends, scope, and relevance. In: Kumar R, Sharma A, Ahluwalia S (eds), *Advances in Environmental Biotechnology*, Springer, Singapore, pp. 243–263.

Assey GE, Mgohamwende R, Malasi WS (2021) A review of the impact of pesticides pollution on environment including effects, benefits and control. *J Pollut Eff Cont* 9:282.

Badea SL, Vogt C, Weber S, Danet AF, Richnow HH (2009) Stable isotope fractionation of γ-hexachlorocyclohexane (Lindane) during reductive dechlorination by two strains of sulfate-reducing bacteria. *Environ Sci Technol* 43:3155–3161.

Barton JW, Kuritz T, O'Connor LE, Ma CY, Maskarinec MP, Davison BH (2004) Reductive transformation of methyl parathion by the cyanobacterium *Anabaena* sp. Strain PCC 7120. *Appl Microbiol Biotechnol* 65:330-5.

Bose S, Kumar PS, Vo DV, Rajamohan N, Saravanan R (2021) Microbial degradation of recalcitrant pesticides: a review. *Environ Chem Lett* 19:3209–28.

Cáceres TP, Megharaj M, Naidu R (2008) Biodegradation of the pesticide fenamiphos by ten different species of green algae and cyanobacteria. *Curr Microbiol* 57:643–6.

Chittora D, Meena M, Barupal T, Swapnil P, Sharma K (2020) Cyanobacteria as a source of biofertilizers for sustainable agriculture. *Biochem Biophys Rep* 22:100737.

El-Bestawy E (2019) Efficiency of immobilized cyanobacteria in heavy metals removal from industrial effluents. *Desalination Water Treat* 159:66–78.

El-Bestawy EA, Abd El-Salam AZ, Mansy AE (2007) Potential use of environmental cyanobacterial species in bioremediation of lindane-contaminated effluents. *Int Biodeterior Biodegradation* 59:180–92.

Fioravante IA, Barbosa FA, Augusti R, Magalhães SM (2010) Removal of methyl parathion by cyanobacteria *Microcystis novacekii* under culture conditions. *J Environ Monit* 12:1302–6.

Forlani G, Pavan M, Gramek M, Kafarski P, Lipok J (2008) Biochemical basis for a wide spread tolerance of cyanobacteria to the phosphonate herbicide glyphosate. *Plant Cell Physiol* 49:443–456.

Hamed SM, Hozzein WN, Selim S, Mohamed HS, AbdElgawad H (2021) Dissipation of pyridaphenthion by cyanobacteria: insights into cellular degradation, detoxification and metabolic regulation. *J Hazard Mater* 402:123787.

Hedges SB, Chen H, Kumar S, Wang DY, Thompson AS, Watanabe H (2001). A genomic timescale for the origin of eukaryotes. *BMC Evol Biol* 1:4

Huang Y, Xiao L, Li F, Xiao M, Lin D, Long X, Wu Z (2018) Microbial degradation of pesticide residues and an emphasis on the degradation of cypermethrin and 3-phenoxy benzoic acid: a review. *Molecules* 23:2313.

Ibrahim WM, Karam MA, El-Shahat RM, Adway AA (2014) Biodegradation and utilization of organophosphorus pesticide malathion by cyanobacteria. *Biomed Res Int* 2014:392682.

Jayaraj R, Megha P, Sreedev P (2016) Organochlorine pesticides, their toxic effects on living organisms and their fate in the environment. *Interdiscip Toxicol* 9:90–100.

Jiang B, Zhang N, Xing Y, Lian L, Chen Y, Zhang D, Li G, Sun G, Song Y (2019) Microbial degradation of organophosphorus pesticides: novel degraders, kinetics, functional genes, and genotoxicity assessment. *Environ Sci Pollut Res Int* 26:21668–81.

Kole RK, Roy K, Panja BN, Sankarganesh E, Mandal T, Worede RE (2019) Use of pesticides in agriculture and emergence of resistant pests. *Indian J Animal Health* 58:53–70.

Kondi V, Sabbani V, Alluri R, Karumuri TS, Chawla P, Dasarapu S, Tiwari ON (2022). Cyanobacteria as potential bio resources for multifaceted sustainable utilization. In: Singh HB, Vaishnav A (eds), *New and Future Developments in Microbial Biotechnology and Bioengineering*, Elsevier, Amsterdam, Netherlands, pp73–87.

Kumar A, Singh JS (2017) Cyanoremediation: a green-clean tool for decontamination of synthetic pesticides from agro-and aquatic ecosystems. In: Singh J., Seneviratne G. (eds), *Agro-Environmental Sustainability*, Springer, Cham, pp. 59–83.

Kumar J, Singh D, Tyagi MB, Kumar A (2019) Cyanobacteria: applications in biotechnology. In: Mishra AK, Tiwari DN, Rai AN (eds), *Cyanobacteria*, Academic Press, Cambridge, USA, pp. 327–346.

Kumar M, Yadav AN, Saxena R, Paul D, Tomar RS (2021) Biodiversity of pesticides degrading microbial communities and their environmental impact. *Biocatal Agric Biotechnol* 31:101883.

Kuritz T (1998) Cyanobacteria as agents for the control of pollution by pesticides and chlorinated organic compounds. J *Appl Microbiol* 8:186S-192S.

Kuritz T, Wolk CP (1995) Use of filamentous cyanobacteria for biodegradation of organic pollutants. *Appl Environ Microbiol* 61:234–8.

Lau NS, Matsui M, Abdullah AA (2015) Cyanobacteria: photoautotrophic microbial factories for the sustainable synthesis of industrial products. *Biomed Res Int* 2015:754934.

Lee SE, Kim JS, Kennedy IR, Park JW, Kwon GS, Koh SC, Kim JE (2003) Biotransformation of an organo-chlorine insecticide, endosulfan, by *Anabaena* species. *J Agric Food Chem* 51:1336–40.

Mahmood I, Imadi SR, Shazadi K, Gul A, Hakeem KR (2016) Effects of pesticides on environment. In: Hakeem K, Akhtar M, Adullah S (eds), *Plant, Soil and Microbes*, Springer, Cham, pp. 253–269.

Malyan SK, Singh S, Bachheti A, Chahar M, Sah MK, Kumar A, Yadav AN, Kumar SS (2020) Cyanobacteria: a perspective paradigm for agriculture and environment. In: Rastegari AA, Yadav AN, Yadav N (eds), *New and Future Developments in Microbial Biotechnology and Bioengineering*, Elsevier, Amsterdam, pp. 215–224.

Megharaj M, Madhavi DR, Sreenivasulu C, Umamaheswari A, Venkateswarlu K (1994) Biodegradation of methyl parathion by soil isolates of microalgae and cyanobacteria. *Bull Environ Contam Toxicol* 53:292–297.

Mona S, Kumar V, Deepak B, Kaushik A (2020) Cyanobacteria: the eco-friendly tool for the treatment of industrial wastewaters. In: Bharagava RN, Saxena G (eds), *Bioremediation of Industrial Waste for Environmental Safety*, Springer, Singapore, pp. 389–413.

Mostafalou S, Abdollahi M (2013) Pesticides and human chronic diseases: evidences, mechanisms, and perspectives. *Toxicol Appl Pharmacol* 268:157–77.

Odukkathil G, Vasudevan N (2013) Toxicity and bioremediation of pesticides in agricultural soil. *Rev Environ Sci Biotechnol* 12:421–44.

Palanisami S, Prabaharan D, Uma L (2009) Fate of few pesticide-metabolizing enzymes in the marine cyano-bacterium *Phormidium valderianum* BDU 20041 in perspective with chlorpyrifos exposure. *Pestic Biochem Physiol* 94:68–72.

Parte SG, Mohekar AD, Kharat AS (2017) Microbial degradation of pesticide: a review. *Afr J Microbiol Res* 11 :992–1012.

Prasanna R, Jaiswal P, Kaushik BD (2008). Cyanobacteria as potential options for environmental sustainability—promises and challenges. *Indian J Microbiol* 48:89–94.

Rajmohan KS, Chandrasekaran R, Varjani SA (2020) Review on occurrence of pesticides in environment and current technologies for their remediation and management. *Indian J Microbiol* 60:125–38.

Rani K, Dhania G (2014). Bioremediation and biodegradation of pesticide from contaminated soil and water—a noval approach. *Int J Curr Microbiol App Sci* 3:23–33.

Sharma A, Kumar V, Shahzad B, Tanveer M, Sidhu GPS, Handa N, Kohli SK,Yadav P, Bali AS, Parihar RD, Dar OI, Singh K, Jasrotia S, Bakshi P, Ramakrishnan M, Kumar S, Bhardwaj R, Thukral AK (2019) Worldwide pesticide usage and its impacts on ecosystem. *SN Appl Sci* 1:1446.

Sharma AN, Pankaj PK, Gangola S, Kumar G (2016) Microbial degradation of pesticides for environmental cleanup. In: Bharagava RN, Saxenaeditor G (eds) *Bioremediation of Industrial Pollutants*, Write & Print Publications, New Delhi, pp. 178–205.

Singh DP, Khattar JI, Kaur M, Kaur G, Gupta M, Singh Y (2013) Anilofos tolerance and its mineralization by the cyanobacterium *Synechocystis* sp. strain PUPCCC 64. *PLoS One* 8:e53445.

Singh DP, Khattar JI, Nadda J, Singh Y, Garg A, Kaur N, Gulati A (2011) Chlorpyrifos degradation by the cyanobacterium *Synechocystis* sp. strain PUPCCC 64. *Environ Sci Pollut Res Int* 18:1351–9.

Singh JS, Kumar A, Singh M (2019) Cyanobacteria: a sustainable and commercial bio-resource in production of bio-fertilizer and bio-fuel from waste waters. *Environ Sustain Indic* 3–4:100008.

Singh PK (1973). Effect of pesticides on blue-green algae. *Archiv Mikrobiol* 89: 317–320.

Singh R, Parihar P, Singh M, Bajguz A, Kumar J, Singh S, Singh VP, Prasad SM (2017) Uncovering potential applications of cyanobacteria and algal metabolites in biology, agriculture and medicine: Current status and future prospects. *Front Microbiol* 8:515.

Subramanian G, Sekar S, Sampoornam S (1994) Biodegradation and utilization of organophosphorus pesticides by cyanobacteria. *Int Biodeterior Biodegrad* 33:129–143.

Thengodkar RR, Sivakami S (2010) Degradation of chlorpyrifos by an alkaline phosphatase from the cyano-bacterium *Spirulina platensis*. *Biodegradation* 21:637–44.

Tiwari B, Chakraborty S, Srivastava AK, Mishra AK (2017a) Biodegradation and rapid removal of methyl para-thion by the paddy field cyanobacterium *Fischerella* sp. *Algal Res* 25:285–96.

Tiwari B, Singh S, Chakraborty S, Verma E, Mishra AK (2017b) Sequential role of biosorption and biodegrad-ation in rapid removal degradation and utilization of methyl parathion as a phosphate source by a new cyanobacterial isolate *Scytonema* sp. BHUS-5. *Int J Phytoremediation* 19:884–93.

Tudi M, Daniel Ruan H, Wang L, Lyu J, Sadler R, Connell D, Chu C, Phung DT (2021) Agriculture development, pesticide application and its impact on the environment. *Int J Environ Res Public Health.* 18:1112.

Vijayan NP, Ali SH, Madathilkovilakathu H, Abdulhameed S (2020a) Chlorpyrifos-degrading cyanobacterium–Coleofasciculus chthonoplastes isolated from paddy field. *Int J Environ Stud.* 77:307–17.

Vijayan NP, Madathilkovilakathu H, Abdulhameed S (2020b) Biodegradation of Pesticides. In: Sáenz-Galindo A, Castañeda-Facio AO, Rodríguez-Herrera R (eds), *Green Chemistry and Applications*, CRC Press, Boca Raton, pp. 215–231.

Vijayan PN, Abdulhameed S (2020) Cyanobacterial Degradation of Organophosphorus Pesticides. In: Zakaria Z, Boopathy R, Dib J (eds), *Valorisation of Agro-industrial Residues- Volume I: Biological Approaches*, Springer, Cham, pp. 239–255.

Yao FX, Yu GF, Bian YR, Yang XL, Wang F, Jiang X (2007) Bioavailability to grains of rice of aged and fresh DDD and DDE in soils. *Chemosphere* 68:78–84.

Zhang H, Hu C, Jia X, Xu Y, Wu C, Chen L, Wang F (2012) Characteristics of γ-hexachlorocyclohexane biodegradation by a nitrogen-fixing cyanobacterium, *Anabaena azotica. J Appl Phycol* 24:221–5.

Zhang W, Lin Z, Pang S, Bhatt P, Chen S (2020) Insights into the biodegradation of lindane (γ-hexachlorocyclohexane) using a microbial system. *Front Microbiol* 11:522.

5 Upcycling of Plastic Waste

S.V. Chinna Swami Naik, Kumar Raja Vanapalli and Brajesh Kumar Dubey

5.1 INTRODUCTION

Plastics are ubiquitous in today's world. Plastic is a broad term for a variety of polymer materials that are suitable for moulding and may be easily processed into various shapes. The production of plastic on a global scale has increased to 368 million metric tonnes in 2019. The widespread usage of plastics in recent years has resulted in a massive rise in the production of PW leading to an escalating environmental problem. With the current trajectory, over 12000 Mt of PW is predicted to be dumped into the environment by 2050 (Roy et al., 2021). Further, the emissions from plastic waste were 1.7 giga tonnes (Gt) of CO_2-equivalents in 2015 and are anticipated to be more than quadrupled by 2050. With the current PW generation estimates, greenhouse gas (GHG) emissions from PW might account for up to 15% of the global carbon budget by 2050 (Chen & Jin, 2019).

The major types of plastics are polyethylene terephthalate (PET), linear-low-density polyethylene (LLDPE), low-density polyethylene (LDPE) and high-density polyethylene (HDPE), polypropylene (PP), polystyrene (PS) and polyvinyl chloride (PVC). While PET, polyamides, polycarbonates, polyurethanes and polyesters are made by condensation polymerization, polyolefins (PE, PS, PVC and PP) are made through addition polymerization. Figure 5.1 depicts the major types of plastics produced globally with roughly a 75% market share. Low-cost commodity thermoplastics, which are mostly utilized for packaging purposes comprise the majority of plastics manufactured globally accounting for 46% of the global PW output. For instance, PE produces the largest percentage of PW (69%) as well as 63% of all packaging garbage. This significant waste production may be explained by the packing plastic materials because of their shorter life spans as compared to those of products used in other industries, such as construction, transportation and other consumer goods.

Mismanaged PW poses a global health hazard to people, animals and marine life due to their resilient nature, which allows them to persist in the environment for hundreds of years and produce a massive accumulation of toxins. With the ubiquity of single-use plastics, particularly food packaging and biomedical waste, humanity is now confronting an unprecedented PW issue. Such polymers are discarded after a short period of usage, taking up significant space in landfills. Inadequate solid waste management also results in enormous amounts of plastics leaking into the environment. In the marine environment, such waste further disintegrates into micro and nano plastics, which have a negative influence on the terrestrial ecosystem (Wong et al., 2020). Scientists currently consider marine plastic pollution to be a planetary boundary danger because of the abundance of micro/nano plastics and the irreversible effects these materials have on the global environment.

Given the mis-effects of plastic pollution over the long term, it is now necessary to develop sustainable waste management strategies. The waste management hierarchy serves as a framework for modern waste management systems. While prevention is the most preferred management choice, waste minimization, reuse and recycling are depicted as the next choices in the order of preference.

DOI: 10.1201/9781003327646-5

61

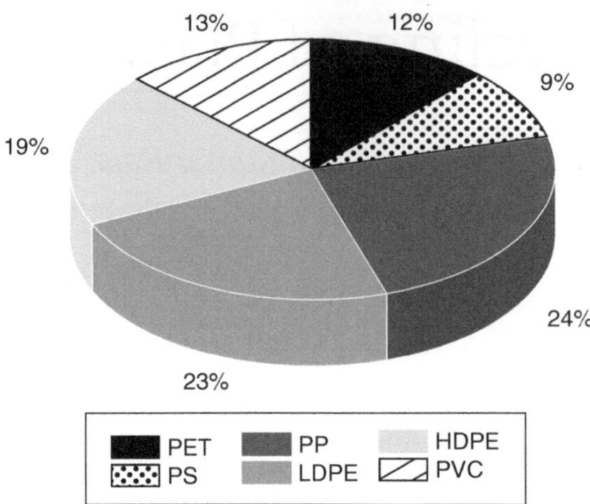

FIGURE 5.1 Worldwide production of commodity plastics modified from Abu-Thabit et al. (2021).

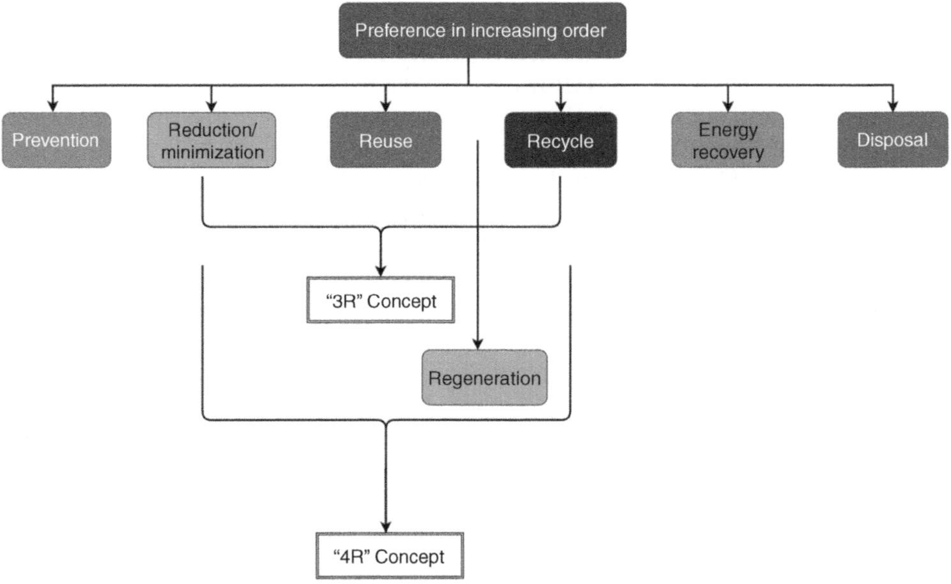

FIGURE 5.2 Waste management hierarchy and the "3R" and "4R" concepts. (Modified from Roy et al., 2021.)

In addition to economic and resource loss, waste disposal into the environment has associated problems and environmental risks. So, it is only suggested in scenarios of no alternative options. The three principles of sustainable waste management practices are reduce, reuse and recycle are often referred to as the "3R" principle (Roy et al., 2021). The idea of "4R", incorporates "repurpose" or "regeneration" as a part of recycling as illustrated in Figure 5.2. Reusing and recycling are often associated with material degradation, which leads to reduced economic value after each cycle. So, to facilitate improved economic viability, transformation of materials by regaining their useful capabilities, adding value to waste materials, and investigating high-value application areas needs to be considered. Upcycling of waste is a related concept which refers to the valorization of waste material into value-added goods with potentially higher value than its raw feed. Unlike traditional material flow where degraded materials cycle back into the environment, non-biodegradability of plastics with a linear material flow, makes them susceptible to pose harmful threats to human health and environment. So, design of circular economy-based waste management technologies with recovery of economically valuable by-products are the need of the hour (Stadler & de Vries, 2021).

While governments are looking for sustainable technologies for upgrading the existing waste management systems in developing countries, researchers are also eager to investigate novel ways to utilize waste plastics with greater economic value to encourage the commercial adoption of upcycling methods. This book chapter has evaluated these methods of PW upcycling along with their associated major challenges and opportunities.

5.2 CONVENTIONAL TECHNIQUES OF PLASTIC WASTE MANAGEMENT

Depending on their technoeconomic feasibilities and resources available, different countries adopt different approaches for PW management. Mechanical recycling, incineration and landfilling are three of the most prominently adopted techniques used for PW management.

5.2.1 Mechanical Recycling

Mechanical recycling entails processing of PW into a product without any change in its inherent chemical structure and properties of the original product. All thermoplastics may be mechanically recycled through re-extrusion via a closed loop process. However, this approach is not techno economically feasible to all types of polymers. For ease of separation and identification of plastics, each kind of plastic is assigned a unique resin identification number that depicts the types of plastics depending on their resin type, making recycling easier and thereby boosting the quality of the recovered output. This is necessary to increase the quality and value of the recycled product. When different types of polymers are melted combined, they tend to separate phase wise and set in layers, such as oil and water. All these phase barriers invariably cause less strength in the output material, lowering its quality, and so these blends have limited uses.

According to Figure 5.3, the most recycled polymers are PET and HDPE, whereas PS and PVC are the lowest recycled. Only a very tiny percentage of recycled PVC waste is collected from packaging waste streams, the majority is gathered from construction waste streams (such as pipes, floors, fittings, etc.). PVC and PS goods contain hazardous phthalate plasticizers that can leach out when they come into touch with food, posing a major risk to both human health and the environment. Also, plastic recycling through unconventional ways, as is common in many undeveloped and developing nations, causes a variety of respiratory and dermatological issues due to hazardous gases, particularly hydrocarbon inhalation and exposure to residues generated during the process.

5.2.2 Waste to Energy

PW is a perfect feedstock for energy recovery, owing to its high calorific value and dry nature of the waste. Although the framework of PW to energy can improve the management of PW in rising and developing nations, but their implementation is complex and requires considering several unique circumstances. Energy recovery from plastic waste can either be through direct heat recovery from incineration/co-incineration with other solid fuels combined by air pollution control or by converting this heat energy to electricity. Incineration aids in the volume reduction of waste along with its ability to handle diverse mixed plastics irrespective of their contamination levels. However, the technology should be matured to help in complete thermal oxidation of waste so as to prevent the release of harmful toxins including heavy metals, persistent organic pollutants (POP), etc., into the air and persist in ash waste residues. Moreover, when working with mixed plastic waste the air pollution control equipment should be able to handle gases such as styrene (from PS) and vinyl chloride, and HCl (from PVC), dioxins and furans, etc., which pose severe health risks. Although popular as a quick fix technology, incineration has not been of commercial scale in most

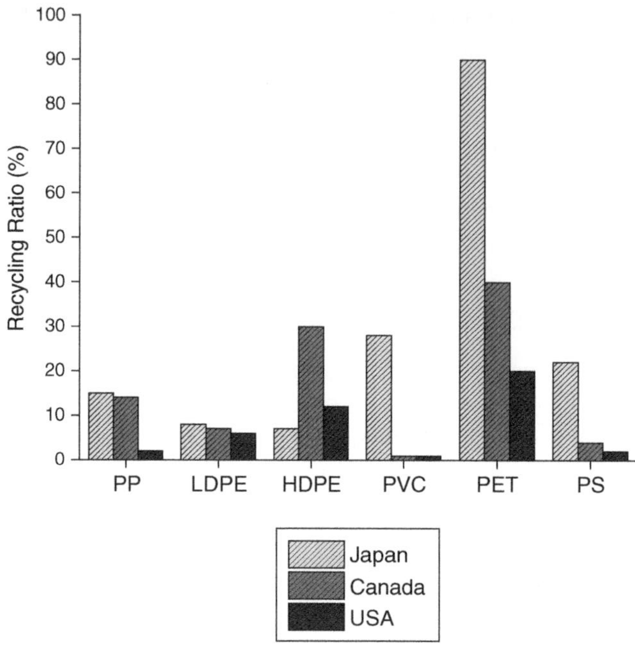

FIGURE 5.3 Plastic recycling percentages in the USA, Japan and Canada. (Modified from Abu-Thabit et al. 2021.)

of the developing countries, because of the problems of constant supply of waste with good thermal characteristics, and technoeconomic constraints associated with it.

5.2.3 LANDFILLING

Landfilling is one of the oldest methods of municipal solid waste disposal. The non-biodegradability and highly voluminous nature of plastic waste makes it an unsustainable choice for disposal in a landfill. Moreover, for such a highly demanded material, and its thermal characteristics, disposal of PW would be loss of resources and energy. Moreover, unsanitary disposal in waste dumps is constrained by the well-known negative impacts of PW on the surrounding atmosphere. Pollution of the land, air and water is brought on by non-biodegradable PW and toxic chemicals, which is extremely concerning from an ecological standpoint. Given the increased use of single-use plastics and their littering, another significant factor is the availability of land for this sort of garbage disposal. Moreover, unsanitary disposal of PW is considered as an aesthetic and visual nuisance, while it may leach hazardous chemicals into the ground water and can pollute surrounding surface water bodies. PW also interferes with the land reclamation through gardening causing in an irrecoverable loss of land as a resource during its working and even post closure. In both short and long terms, sanitary/unsanitary landfilling cannot be considered as a sustainable waste management solution because it has several disadvantages as shown in Table 5.1, but always be treated as a humble last management option. So, many countries are contemplating alternative waste management strategies in the current environment, such as reuse, recycling or energy recovery.

5.3 PLASTIC WASTE UPCYCLING TECHNIQUES

Plastic waste upcycling has received a lot of attention recently as a potentially effective way to achieve economic legitimacy by turning PW into goods of additional value. More than 150 new

TABLE 5.1
Different Types of Plastic Waste Management Techniques their Advantages and Disadvantages

Plastic Waste Management Technique	Advantages	Disadvantages
Mechanical Recycling	• Low cost • Less plastic is sent to landfill • Resources used to produce virgin polymers are reduced • Infrastructure and process with low technological requirements • Possibility of decentralized processing • High industrial maturity • More efficient in terms of greenhouse gas emissions • Creates jobs	• Processing cost • Recycled products are often of lesser quality • Down cycling • Improper separation by manpower according to the resin codes • Washing and melting plastic requires waste and energy consumption • Emission of VOCs • Heterogeneity of waste
Waste to Energy	• Energy generation • High volume of waste reduction • Superior disease control • Metal recycling • Destroys all contaminants • Plant can be located near to residential areas	• Toxic emissions • Promotes waste production • Ash management • Capital cost • Requires a constant supply of waste • Flue gas purification
Landfilling	• Corban sequestration • Reduced transportation costs • Easy to construct new landfill • Low cost comparatively • Can be used for temporary storage space	• Ground water pollution • Soil pollution • Loss of resources • Land acquisition • Energy loss • Space consumption • Material loss

chemical recycling businesses are already operating in Europe, helping to advance the goal of the European Union to recycle 50% of plastic packaging by 2025 and 55% by 2030 (Choi et al., 2022). However, the main obstacle in achieving commercial success for valorization of PW into new by-products of value is achieving economic viability. Some of the most prominent upcycling techniques including thermal and chemical routes are discussed below in detail.

5.3.1 Thermal Upcycling Technique

Although mechanical recycling has a lower carbon footprint, the market for recycled plastics in new applications is still small. As a result, historically, thermal approaches for plastics upcycling have gained prominence, which included depolymerization or chain scission. This includes PW conversion into carbon or doped carbon materials, value-added small molecules, functionalization at the polymer chain's periphery for new purposes, and the insertion of vulnerable linkers to facilitate degradation.

5.3.1.1 Carbonization

Carbonization is the recovery of carbonaceous materials through thermal conversion of PW. This involves a series of activities including pre-treatment, carbonization and post-treatment.

Pre-treatment is a feedstock stabilization process which specifically establishes the cross-linked structures in polymer materials and is frequently carried out to maximize the carbonization of polymer waste. Cross-linking thermal stabilization is typically carried out in an atmosphere of air. For instance, Qiao et al. (2006) used thermal stabilization before the carbonization stage for a PVC-based pitch. Also, Choi et al. (2017) identified the above-mentioned stability criteria while stabilizing waste PE through establishing crosslinking between linear low-density PE (LLDPE) by heat treatment to produce good strength.

The subsequent step of carbonization involves the recovery of carbonaceous solid residues from organic polymers, majorly through pyrolysis or hydrothermal carbonization, which involves increasing concentration of the element carbon. It is described as a complicated process in which there are several simultaneous processes, such as condensation, dehydrogenation, hydrogen transfer and isomerization. It has been demonstrated that the carbonization temperature has an impact on how well the recommended application performs or exhibits the required qualities. For instance, Danmaliki & Saleh (2017) have depicted the influence of pyrolysis temperature pyrolysis duration on the yields and formation of gas, liquid and solid materials from used tire rubber with an emphasis on the effect of the process on the carbon recovery. The yield, and physicochemical characteristics of carbon materials such as porosity, surface properties, are hugely dependent on various pyrolysis/process parameters (Shen, 2020). Hydrothermal carbonization, on the other hand, is a process in which low moisture and oxygen content PW have typically been processed thermochemically at higher temperatures to recover fuels with greater heating values, which may then be hydrothermally transformed into solid fuels (Sharma et al., 2020). Pyrolysis of chlorinated polymers causes chlorinated chemicals affecting the quality of the oil, which in turn pollutes the environment. However, hydrothermal carbonization also has the added advantage of dehalogenation from the polymers. Figure 5.4 illustrates the many forms of carbon compounds that may be produced from the plastics under various reaction conditions and their applications.

The most typical form of post-treatment for upcycling PW is activation. Using chemical and thermal activation techniques, enhanced development of pores in carbon compounds is engineered. Applications of porous carbon materials include carbon dioxide (CO_2) capture, pollution adsorbents and battery electrodes. Because CO_2 is produced during the heat interaction between carbon and

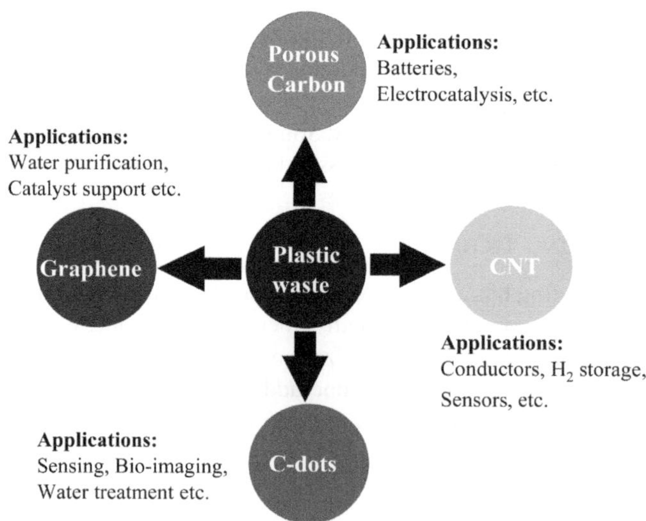

FIGURE 5.4 Applications of carbon materials derived from carbonization of PW modified from Chen et al. (2021).

activation agents, pores are made by removing a piece of the carbon cluster (He et al., 2013). Although activation takes place in an inert environment, the presence of activation chemicals also causes the oxidation to happen.

Physical activators like steam or CO_2 and chemical activators like KOH, NaOH and H_3PO_4 fall into two groups, respectively. Jehanno et al. (2019) studied the effect of temperature and length of the CO_2 activation process on the structure of the activated carbon that was created and showed how various chemical activation agents may control pore size and porosity. Esfandiari et al. (2012) investigated the impact of temperature, flow rate and holding duration during the activation stage on the adsorption capacity of activated carbon, while converting waste PET into porous carbon for CF_4 adsorption.

5.3.1.2 Pyrolysis

Pyrolysis is the thermal degradation of PW polymers into syngas, liquid oil and solid chars at relatively high temperatures in an inert atmosphere. Due to a wide range of feedstock tolerance and adaptability in process parameters, pyrolysis has gained significant prominence in recent years. It can be either performed with/without the use of a catalyst (Chen & Yan, 2020). Unlike conventional pyrolysis, catalytic pyrolysis involves the use of an active catalyst that helps in the reduction of the reaction temperature and enhances the selectivity towards the required outputs. Pyrolysis is majorly focused on the recovery of liquid and gaseous fuels because of their high proportion in the output (Vanapalli et al., 2021). Due to their high energy density, these by-products can be harnessed and further processed to be excellent fuel substituents. This makes valorizing each product fraction separately without intricate purifications more cost-effective. The solid carbons with further post-processing may also be used in a variety of applications such as energy storage, environmental absorbents and others.

5.3.1.2.1 Production of Gas and Liquid Fuels by Pyrolysis

Pyrolysis helps in the recovery of several hydrocarbons potentially used as a chemical feedstock or as energy. This minimizes the dependency on non-renewable fossil fuels which helps in solving the landfilling problem. To create a fuel that satisfies the requirements outlined in standards or specifications, transport fuels like diesel or gasoline are mixed from different refinery fractions. Technologies for pyrolysis can be set up to create intermediate transport fuel precursors from polyolefins or final fuel that is suitable for engines. The infrastructure of a current oil refinery can then update intermediaries. Pyrolysis reactors include several batch and continuous setups, including fixed bed, fluidized bed, conical sprouted bed reactor, microwave assisted technology, etc. While batch reactors can be used for dealing with small amounts of waste, continuous reactors demand constant supply of waste feed. The major influential process parameters of pyrolysis include residence time, temperature, pressure, heating rate and feed characteristics, which influence the by-product yield and characteristics. According to Quesada et al. (2020) liquid oils were the main results of the pyrolysis of several plastic mixes in a fixed-bed reactor at roughly 550°C under N_2 (60–75 wt.%). The pyrolysis of PE film produced liquid oils with properties that were most comparable to those of commercial diesel fuel, but with a greater viscosity and waxier look.

The physical characteristics of the gas products produced by the thermal pyrolysis from different feedstocks, like PE, PS, PP, PET and their combinations at various temperatures in a vertical dual mode chamber reactor, have been thoroughly studied by Honus et al. (2018).

High temperature and long residence were reported to maximize gas production in pyrolysis process (Sharuddin et al., 2016). In general, the pyrolysis process favors only 5–20 % gas production from polyolefins and PS plastics. The gas composition depends on the composition of feedstock, while its yields are majorly dependent on the temperature and other process parameters. Wyss et al. (2021) studied individual pyrolysis of HDPE, LDPE, PP, PS, PET and PVC and found the major gas components of pyrolysis were hydrogen, butane, butene, methane, propane, propene, ethane and ethene. While PVC created hydrogen chloride, PET also produced other gas components such as carbon dioxide and carbon monoxide. Additionally, the gas created during the pyrolysis process has a substantial calorific value. According to Armenise et al. (2021), the gas generated by the pyrolysis of only PE and PP had a high calorific value between 42 and 50 MJ/kg. Therefore, there was a great potential for the pyrolysis gas as a heating source in pyrolysis industrial plants. Additionally, if isolated from other gas constituents, ethene and propene can be used as chemical feedstock for the manufacture of polyolefins.

Recovery of solid carbons from pyrolysis and their applicability in various fields has been discussed in the previous section. After post-processing, activated carbons can be recycled into the system as catalysts. For instance, seven different forms of activated carbon have been looked at by Kizza et al. (2022) as a catalyst for the pyrolysis of PE. This study depicted that alkanes and aromatic hydrocarbons in the jet-fuel range as well as H_2-rich gas products were produced in large quantities. Strong acidity encouraged the creation of aromatics whereas mild acidity encouraged the development of alkanes. Apart from carbons, the utility and influence of other catalysts on the production of char has also been studied by previous literature. Kim et al. (2019) investigated the effects of the HZSM-11 catalyst and the flow atmospheres a larger yield of solid chars with acceptable heating values of around 26 MJ kg^{-1} were produced by catalytic pyrolysis under CO_2 flow. PET waste were converted into porous carbons using a sequential thermal pyrolysis-base activation technique carried out by Yuan et al. (2020). For CF_4 greenhouse gas absorption at ambient settings, the porous carbons were used as selective and efficient materials. Similarly, Zhang et al. (2020) heated the pyrolytic carbon with KOH at 700°C for one hour after initially pyrolyzing the PET waste at 650°C.

5.3.1.3 Gasification

Gasification is a thermochemical process that produces syngas because of interactions between the fuel and the gasification agent. The syngas consists mostly of H_2 CO, CO_2, N_2 and minor compounds like C_2H_4, CH_4, C_2H_6, etc. H_2S, NH_3 and tars may also be present in trace concentrations. The methods used for plastic gasification are essentially the same ones that have been created for other gasifier feedstocks, such as biomass and coal. The process of gasification is broken down into the following steps: drying, pyrolysis, cracking and reforming reactions in the gas phase, and heterogeneous char gasification (He et al., 2013). Depending on the properties of the feedstock and the gasification circumstances, these phases' importance for the process success and their kinetics will vary. Waste plastics often have substantially lower moisture contents than other routinely gasified feedstocks like biomass and coal. However, waste plastic's unique properties, particularly their low heat conductivity, sticky behaviour, high volatile content and spectacular tar production, prevent their treatment by conventional gasification methods and present a significant obstacle for the process implementation.

Environmental benefits of gasification over alternative technologies, including incineration of PW are considerable. During the incineration process, hazardous compounds such as heavy metals (Hg or Cd), dioxins, furans, HCl, SO_2, NOx and HF can be discharged into the environment. This may cause the dangerous emission of furans, dioxins, and other polychlorinated biphenyls (Shen, 2020). On the other hand, gasification facilities, which operate at lower temperatures than combustion and use a sub-stoichiometric amount of oxidant, create less of the above-mentioned pollutants, specifically NOx.

By converting organic material thermochemically at high temperatures, gasification enables the production of chemicals, including adhesives, methane, fatty acids, surfactants, ethylene, detergents and plasticizers in addition to syngas for energy production. There are various types of biomass gasification processes, including oxygen gasification, air gasification, steam gasification, supercritical water gasification and carbon dioxide gasification (Jia et al., 2021). The process of supercritical water gasification is usually carried out between 600 and 650°C and at a pressure of around 30 MPa. Water acts as a potent oxidant over 600°C. Oxidation of carbon atoms result in the formation of CO_2. The removal of microplastics from the marine environment is a fascinating application

(Röttger et al., 2017). Because of the high-water content of the plastics retrieved from the ocean, other procedures like pyrolysis and gasification are not an appropriate option. Doing so would need a drying stage, which would greatly raise the cost of recycling.

In co-gasification, two or three carbonaceous fuels are simultaneously transformed into a gas with usable heating value, one of which must be a fossil fuel like coal, fuel oil, or PW. This technique appears to be highly promising since it can lower GHG emissions by substituting part of the carbon from fossil fuels with carbon from bio-based sources. In co-pyrolysis processes involving plastic and biomass, it has been found that biomass devolatilizes at temperatures between 200 and 400°C whereas polymers like PS, PE, PP, PET and PVC devolatilize at higher temperatures between 300 to 500°C. Interactions between the volatiles in plastic and the char in biomass are made feasible by the differing devolatilization of the materials.

5.3.2 CHEMICAL UPCYCLING TECHNIQUES

Various chemical processes for upcycling plastic into value-added compounds include hydrogenolysis, solvolysis and photocatalysis. Plastic solvolysis has been widely researched upon along with thermal pyrolysis in recent years in the areas of plastic recycling. Similarly, a rapid development has been observed in recent years, particularly in the field of catalytic chemical reactions. Plastic hydrogenolysis will be introduced first due to recent significant advancements. Given the structural resilience of plastics, effective catalyst development is critical for hydrogenolysis to selectively create valuable compounds, and innovative heterogeneous metal catalysts will be explored.

5.3.2.1 Solvolysis

The solvent molecules present in excessive amounts during the solvolysis of PW act through nucleophilic substitutions to depolymerize the plastic chains (Rorrer et al., 2019). Various solvent systems, including polyols, methanol, water, ammonia and amines, have been used for plastic solvolysis. Methanol was the best solvent for solvolysis since other solvents could be used, but the efficiency decreased with bigger alcohol molecules. Typically, these systems are heated to a high temperature and include a catalyst to speed up the depolymerization. The ionic liquids' lactate anion, which reacted with the molecules of methanol to produce the active oxyanion intermediates, originally caused the polycarbonate polymers to disintegrate (Saito et al., 2020). The polycarbonate was subsequently broken down into shorter, more soluble oligomers by the active intermediates by nucleophilic attack, which was followed by the activated methanol's depolymerization of the oligomers into monomeric bisphenol A and dimethylcarbonate.

5.3.2.2 Hydrogenolysis

A catalytic chemical process called hydrogenolysis uses H_2 to dissolve the C-C or C-O bond in feedstocks. The hydrogenolysis method can selectively produce useful oligomeric or monomeric chemicals from used PW with the careful design of an effective catalyst (Bunescu et al., 2017). For plastic hydrogenolysis, a variety of homogeneous and heterogeneous catalytic systems have been investigated thus far in real-world applications, heterogeneous catalysts are simpler to recover and simpler to separate products from PW. Commercial polyolefins, such as PE and PP plastics, underwent catalytic hydrogenolysis as early as 1998 (Kumar et al., 2020). To transform PE and PP into a variety of short-chain alkanes, a precise, highly electrophilic silica-supported Zr-H catalyst was created. Low-M_w PE was cleaved at 150°C under about 1 bar H_2.

For the hydrocracking of waste HDPE into gasoline-range products at a reasonably high temperature of 375°C under roughly 70 bar H_2 and a reaction duration of one hour, Ni and NiMo sulfides/ HZSM-5 and Si-Al support were used. The catalysts could be utilized again after being recalcined and resulfurized and were not readily poisoned by contaminants like N, S and others found in the PW (Jehanno et al., 2019). PE hydrocracking using Pt catalysts on various supports. The utilized catalysts' acidity and pore size had a combinational impact on the process. The distributions of the products were also significantly influenced by the reaction temperature. The most popular catalysts for the hydrogenolysis of PW are supported noble metal nanoparticles.

5.3.2.3 Photocatalysis

While employing photocatalysis to upcycle plastic is still in its infancy as an emerging subject, photocatalysis has long been used in many parts of chemical processes (Guo et al., 2019). Utilizing solar energy to process PW offers a simple and environmentally friendly method of valorizing the feedstocks into chemicals and fuels with additional value. In addition to aiding in plastic decomposition in the environment, photocatalysis may also start redox processes that produce beneficial chemical compounds. PW undergo solid phase photodegradation under solar irradiation because they are typically immediately exposed to ambient air and sunlight. In comparison to liquid-phase systems with the photocatalyst and plastics suspended in solution, the solid phase has a considerably higher possibility of reactive oxygen species (Parrino et al., 2018). It is worth mentioning that the deterioration of composite plastics begins at the interface between the plastics and the photocatalyst and subsequently spreads to the plastics' matrix via the diffusion of active oxygen created on the photocatalyst's surface. A semiconductor material is used as a photocatalyst in a typical photochemical process, producing *in situ* photoexcited electron-hole pairs as the active sites to promote the redox reactions.

5.3.2.3.1 PW to H_2 by Photo-Reforming

An early study on photocatalytic PW conversion may be traced back to 1981, when a Pt/TiO_2 photocatalyst was used to transform PVC into H_2 gas in water at room temperature. However, the performance is very poor, and recent research with novel heterogeneous catalysts created has enhanced the product yield. Uekert et al. (2019) created CdS/CdOx quantum dots in 2018 to convert diverse plastic feedstocks such as poly (lactic acid) (PLA), polyurethane (PU) and PET in 10 M NaOH aqueous solution at room temperature under atmospheric N_2 and solar light.

5.3.2.3.2 Plastics into C_2 Chemicals

In addition to H_2, photocatalytic conversion of PW can result in the synthesis of C_2 compounds as a critical product. Jiao et al. (2020) proposed a new photocatalytic method for converting waste plastics into acetic acid (CH_3COOH) via a CO_2 intermediate via a sequential C-C bond scission and cross-linking pathway without the usage of sacrificial chemicals. The polymers, which included PP, PVC and PE were initially totally degraded into CO_2 as the major product utilising single-unit cell thick Nb_2O_5 films under good natural environment situation after 90 hours, respectively. The produced CO_2 was then upgraded further to yield CH_3COOH over a longer reaction period.

5.3.3 CHEMO-BIOTECHNOLOGICAL TECHNIQUE

5.3.3.1 Upcycling Plastic Waste to PHAs

The most intriguing value-added products that may be created using waste plastics that have been depolymerized as a cost-effective source of carbon feedstock are biodegradable polyhydroxyalkanoates (PHAs). The creation of PHAs is anticipated to support a sustainable and circular economy by offering a bioplastic that is renewable and sustainable as an alternative to plastics made from petroleum, reproducing monomers from biologically recycled PHAs with the help of microorganisms, and reducing environmental pollution due to PHAs' complete biodegradability. Thus, the manufacture of high-value PHAs would increase the cyclic usage of plastics as well as aid in the disposal of PW.

Guzik et al. (2014) revealed in 2014 that a two-step chemo-biotechnological approach may valorize PE waste to produce biodegradable mcl-PHA. For the microbial fermentation of PHA, the solid waxy fraction of the pyrolyzed waste PE was employed as a carbon source. To create the value-added PHA bioplastic, *C. necator* was used to bio-upcycle the pyrolyzed nonoxygenated PE wax (N-PEW). Since it is created by thermal pyrolysis without the presence of air, the N-PEW often comprises a combination of hydrophobic carbon chains. According to Radecka et al. (2016) by pyrolyzing PE waste in the presence of air, oxygenated polyethylene wax (O-PEW) is created by oxidative degradation. As a result, O-PEW has a variety of oxygen-containing hydrophilic characteristics that make it easier to employ them as emulsifiers and additives in a variety of applications.

The first instance of using oxidized PP waste as an additional carbonaceous source for the bacterial fermentation of *C. necator* H16 and formation of PHA was described by Johnston et al. (2017). Using a glass reactor and a two-phase gas-solid system, pyrolysis of the PP waste film was conducted. Ozone and oxygen were injected into the reactor as the oxidizing agent. When PP is depolymerized in supercritical water, a variety of branching cycloalkanes, alkenes, alkanes and aromatic compounds are created.

Shake flask tests were the first to use pyrolitic styrene oil as the only source of carbon and energy for *Pseudomonas putida* growth and PHA production (Ward et al., 2006). In shaking flask studies, sodium ammonium phosphate (SAP), a limiting nitrogen source, was added to the growth environment to encourage PHA accumulation. According to Johnston et al. (2019) prodegraded PS (PS0), which is made up of a mixture of high-impact polystyrene and general-purpose polystyrene underwent oxidative pyrolysis. The pyrolysis product was then used as a carbon source for the fermentation of the *C. necator* strain and the subsequent biosynthesis of PHA.

The PET waste may be converted into bioplastics was published by Kenny et al. (2008). As carbon feedstock, they employed the PET hydrolysis solid fraction. For this, PET was heated at 450°C in a fluidized reactor without air, resulting in three fractions: gas, liquid and solid. The sodium hydroxide is used to treat solid content, which also boosts the Terephthalic acid (TA) fraction by further hydrolyzing the produced oligomers. Several co-feeding techniques were put to the test in a fed-batch system by Kenny et al. (2012) to improve TA utilization, biomass productivity and PHA output. For growth and/or the manufacture of PHA, waste glycerol, a by-product of biodiesel, was employed.

5.4   UPCYCLED PRODUCT APPLICATIONS

Table 5.2 depicts different applications of output value-added products from conversion of PW to products like graphene sheets, porous carbon compounds, carbon nano particles, carbon dots, activated carbon, biofuels, biodegradable polymers, three-dimensional printing filaments, acids, detergents, plasticizers, and different types of batteries. By using different techniques like pyrolysis, hydrothermal carbonization, carbonization, solvolysis, photo catalysis, gasification and hydrogenolysis. This table explains the properties, their uses and applications of value-added products generated from different techniques as per previous research.

5.5   CONCLUSION AND FUTURE PROSPECTS

The development of an effective management plan is essential to reducing the harmful effects that PW have on the environment because their production is quickly growing. Numerous innovative methods for polymer upcycling have been developed in recent years. These are usually either functionalizing polymers or converting them into more costly speciality polymers. However, when we consider the likelihood that these strategies will really be used, we must draw the conclusion that the situation is less promising. The issue is that a more expensive product often also has a much smaller market, therefore these methods will have little effect on cutting down on the excess of plastic waste. PP and PE presently hold a combined around 50% of the market, although the bulk of upcycling strategies focus on polymers that are already recycled, such as PET.

TABLE 5.2

Different Types of Plastics their Applications and Upcycling Products

Polymer Name	Properties & Applications	Products of Upcycling & Process	Reference
Polyethylene terephthalate (PET) 1 PETE	Because of its properties like moisture resistance, transparent, lightweight, robust, water barrier, and oxygen permeability PET bottles are extensively used in the manufacture of food and beverage containers. PET is also utilised in X-ray, other photographic films, magnetic tapes, printing sheets, electrical insulation, and photographic films, as well as magnetic tapes.	For producing graphite from PET by carbonization, catalytic graphitization, and liquid phase exfoliation for producing potential graphene sheets. PET waste upcycling to potassium diformate, terephthalic acid and H_2 generation by electrocatalysis utilising a bifunctional electrocatalyst in KOH electrolyte. Waste PET was employed as the foundation for porous carbon compounds by carbonization technique. PET monomer can be used to convert into fatty acid derivatives such as hydroxyalkanoyloxy alkanoates Due to their surface-active characteristics, these substances can then be utilised in the catalytic conversion to produce drop-in biofuels as well as in the polymerization to produce bio-based poly (amide urethane). The upcycling of PET waste into items often made from oil is a tactic to address the PET waste by PET hydrolysis solid fraction. Converting PET into carbon spherules using methods devoid of catalyst and solvent. A carbon nanomaterial made from PET is a porous carbon nanosheet. By using molten salts and waste PET. PET being converted into PHA and a bio-based poly(amideurethane) by carbonization technique. PET waste was hydrolytically pyrolyzed to produce terephthalic acid (TPA) solid residue. Utilizing carbon rich TPA as a substrate for a biotechnological procedure. From PET waste activated carbon was produced by carbonising PET waste at 800°C and activating it with CO_2 at 975°C, which is a physical activation process. Porous N-doped carbon (NPC) was created using a process with a strong crosslinking structure, PET was combined with predetermined quantities of melamine and $ZnCl_2/NaCl$ in the first phase. Following the crosslinking structures dehydration and decarbonylation, $ZnCl_2/NaCl$ breaks down the weak bonds to produce a more thermally stable carbon framework.	(Ko et al., 2020). (Zhou et al., 2021) (Gong, Michalkiewicz, et al., 2014) (Welsing et al., 2021), (Adibfar et al., 2014). (Kenny et al., 2008) (Li et al., 2016). (Pol et al., 2009) (Joseph Berkmans et al., 2014) (Kenny et al., 2008) (Sarda et al., 2022) (Esfandiari et al., 2012) (Song et al., 2022) (Ghosh et al., 2013) (Gardea et al., 2014) (El-Sayed & Yuan, 2020) (Slater et al., 2019)

		Density functional theory was utilized to create disodium terephthalate from PET waste for Li/Na storage. In a microwave oven, PET waste was saponified to create disodium terephthalate.	
		PET may be converted to poly(aryl ether sulfone-amide), a high-performance thermoplastic material, using aminolysis with 4-aminophenol followed by nucleop hilic aromatic substitution polymerization.	
		PET and PLA may both be employed as ink sources to create value-added metal-organic frameworks using one-pot or two-step procedures, employing the respective monomers as linkers.	
High-density polyethylene (HDPE)	HDPE contains higher levels of crystallinity and stronger intermolecular tensions, which contribute to its higher tensile strength and harder texture. Because of this, in rigid packaging such as milk bottles, shampoo bottles, yoghurt containers, detergent bottles, and agricultural pipes, HDPE is frequently utilized.	Waste PE plastics may also be used to make carbon nanoparticles, straight, and helical carbon nanotubes for anodes in lithium electrochemical cells and porous carbon for gas collection, electrode materials, and hazardous dye removal by using catalytic carbonization.	(Kong & Zhang, 2007). (Pol et al., 2009). (Celik et al., 2019) (Tennakoon et al., 2020)
		By combining HDPE from various sources with Pt nanoparticles supported on $SrTiO_3$ at 300 °C under 12 bars of hydrogen, they were able to create low molecular weight oligomers with low polydispersities.	(X. Chen et al., 2021) (Jie et al., 2020)
		By using microwave-initiated catalytic deconstruction, it has been possible to convert HDPE and LDPE into high-purity H_2 and high-value carbon materials.	(Tennakoon et al., 2020)
		A catalyst such as $Ru/Pt/C/SiO_2$ has been used to upcycle HDPE toward lubricant-range alkanes, Jet-fuel, lubricant-range hydrocarbons, and diesel with good yield (about 75%). Adapted a method that mimics the depolymerization of biomacromolecules into atom-precise fragments by enzymes precedes selective hydrogenolysis.	(Jiao et al., 2020) (Röttger et al., 2017) (Mejia et al., 2020)
		HDPE cross-linking functional polymers with pendant dioxaborolane units using a bis-dioxaborolane, HDPE has been transformed into high-performance vitrimers.	
		HDPE waste combined with low-molecular-weight PP may be transformed into 3D printing filaments with qualities equivalent to polylactic acid (PLA).	

(continued)

TABLE 5.2 (Continued)
Different Types of Plastics their Applications and Upcycling Products

Polymer Name	Properties & Applications	Products of Upcycling & Process	Reference
Polyvinyl Chloride (PVC) 3 PVC	PVC is less rigid, stronger against IMPACTS, easier to extrude or mould, less temperature- and chemical-resistant, and often has lower ultimate tensile strength it is used in electronics, construction, health care etc.	PVC has been upcycled into porous carbon, activated carbon fibres, carbon aerogels, and Amorphous carbon containing SO_3H as a catalyst for heterogeneous acids. The complete mineralization of PVC into CO_2 was accomplished by photocatalytic degradation employing single-unit-cell-thick Nb_2O_5 films as photocatalyst by oxidative C—C breakage induced by O_2 and OH radicals, and the Nb_2O_5 catalyst enables a one-pot conversion of the polymers to acetic acid (CH_3COOH).	(Qiao et al., 2007) (Suganuma et al., 2012) (Jiao et al., 2020)
Low-density polyethylene (LDPE) 4 LDPE	Soft, flexible, and light weight plastic is known as LDPE. LDPE is renowned for its flexibility, durability, and corrosion resistance at low temperatures. It is not appropriate for applications requiring structural strength, high temperature resistance, or stiffness. Used in plastic bags, six-pack rings, a range of containers, dispensing containers, milk bottles and a range of moulded laboratory apparatus etc.	Conversion of LDPE to long chain di-alkyl aromatics, di-alkyl naphthalene and the products may be readily sulfonated to produce detergents. These reactions take place at 280°C and were catalyzed by platinum. Waste PE may be utilized to create PLA plasticizers. Both HDPE and LDPE are preferentially oxidised by HNO_3 to create succinic, glutaric, and adipic acid when exposed to microwave radiation. LDPE may be effectively converted to liquid fuels and waxes by tandem catalytic cross-alkane metathesis (CAM) at 175°C. A two-step procedure is used to convert LDPE to CNTs. Pyrolysis is used to first turn polyolefin polymers into gas products, which are subsequently changed into CNTs using catalysts made of Ni or Fe. LDPE followed by 900°C pyrolysis can obtain a product as an interlayer in high-performance lithium-sulfur batteries.	(Bäckström et al., 2017) (Beydoun & Klankermayer, 2020) (Bäckström et al., 2017) (Beydoun & Klankermayer, 2020) (X. Jia et al., 2016) (Bazargan & McKay, 2012) (Yao et al., 2018) (J. Jia et al., 2021) (P. J. Kim et al., 2018)

Polymer	Description	Upcycling	References
Polypropylene (PP) 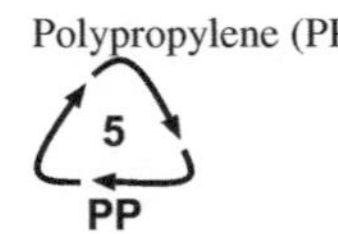	PP is a semi-crystalline material with low density, high degrees of hardness and stiffness, flexibility, and exceptional chemical resistance. Its usage in goods including packaging, toys, medical devices, lab equipment automobile parts, and for cleaning, bleaches, furniture agents, and first-aid supplies is made possible by these characteristics.	Carbon nanotubes are mostly created from waste PP. The creation of graphene flakes from PP was also explored 0D carbon generated from PP dots. The fuel oil formed by the thermal pyrolysis of PP in the presence of air primarily included branched fatty alcohols. Using PP and polyacrylonitrile as the carbon source, a flash Joule heating procedure enables gram-scale, bottom-up synthesis of high-quality flash graphene. By grafting maleic anhydride (MA) utilising di-cumyl peroxide as a free-radical initiator and epoxy-anhydride curing with a di-functional epoxy, PP to high-performance vitrimers have been obtained. PP waste, which is constituted of acrylonitrile butadiene styrene, was effectively transformed into filaments equivalent to virgin materials. Using block copolymers as intermediates, recycled PP may be remoulded into water pipes with better flexural qualities than virgin PP without losing its impact capabilities.	(Sharon et al., 2015) (Gong, Liu, et al., 2014) (Algozeeb et al., 2020) (Luong et al., 2020) (Kar et al., 2020) (Nur-A-Tomal et al., 2020) (Matias et al., 2020) (Lessard et al., 2020)
Polystyrene (PS)	Due to its rigidity, light weight, transparency, and ease of moulding PS is used in a wide variety of products, including home appliances, hard packaging, electronics, toys, and building materials like insulation foam and plumbing. It is also used in the medical industry for petri dishes and tissue culture containers.	Perfluoroalkyl radicals are produced from trifluoroacetic anhydride (TFAA) by pyridine-N-oxide oxidation driven by blue light using $Ru(bpy)_3 Cl_2$ as a photocatalyst Styrene monomers are produced when PS is pyrolyzed, which may be utilised to create virgin PS or other products with added value like fuel oil, which was then employed as a carbon source for microbial fermentation and PHA formation. Used PS cups create carbon materials for Na-ion batteries. By employing heat in a stainless reactor with air to turn old PS cups into carbon compounds such as carbon nanoparticles, carbon nanosheets, and hollow carbon nanotubes are more examples of recycled waste PS. Reactive processing used to functionalize PS with azidoformate grafting agents results in vitrimers with strong creep resistance at room temperature and high mechanical characteristics. The one-pot conversion of PS to multi-walled carbon nanotubes (MWCNTs) made possible by the $FeAlO_x$ catalyst allowed for the quick, easy and effective manufacture of CNTs without the need of gas intermediates.	(Lewis et al., 2019) (Ward et al., 2006) (Johnston et al., 2017) (Kong & Zhang, 2007) (Röttger et al., 2017) (Breuillac et al., 2020) (Luong et al., 2020)

(continued)

TABLE 5.2 (Continued)
Different Types of Plastics their Applications and Upcycling Products

Polymer Name		Properties & Applications	Products of Upcycling & Process	Reference
Others 7 **Others**	polyurethane (PU)	Polyurethanes are employed in high flex fatigue applications because of its strong tensile, tear resistance, and superior electrical insulating qualities.	Production of carbon dots and carbon aerogels from PU by the photocatalysis technique. The hydrogenation of PU to diol, diamine, and methanol can be catalysed by the Ru catalytic system. A dehydrogenative coupling technique can be utilised to manufacture original polyamides (PA) from diol and diamine. PU plastic has been photoreformed without the need of metal using a CdS/CdOx quantum dot catalyst. This catalytic system achieves the highest rates of H_2 generation from PU PU converted into filaments or films that exhibit stiff and elastic thermoset mechanical characteristics when combined with a conventional thermoset material, allowing for high-temperature processing via a dynamic carbamate-exchange reaction.	(dela Cruz et al., 2019) (Potts et al., 2011) (Kumar et al., 2020) (Kawai & Sakata, 1981) (Uekert et al., 2018) (Breuillac et al., 2020)
	Polyesters	Thermoplastic polyester polybutylene terephthalate (PBT) is yet another significant material. It is used as an engineering plastic that is frequently found in home appliances such coffeemakers, hair dryers, irons, and plug outlets.	Conversion of a PBT to poly(butylene-1,4 cyclohexanedicarboxylate) (PBC) by hydrogenation method. PBC is a biodegradable polyester that may be turned into polyurethanes to create shape-memory materials and utilized as a scaffold in tissue engineering. Polyesters to polyether polyols by Hydrogenation, combined use of an aluminium triflate as a Lewis acid with a ruthenium triphos complex to catalyse the etherification of the intermediate diols and hydrogenation of the ester groups. Polyesters/cotton blend textiles may be further transformed to fiber-reinforced composites by co-depositing polyethylenimine and catechol to improve the fibre matrixs interfacial bonding.	(Yu et al., 2019) (Liu et al., 2019) (Chen et al., 2021) (Sharma et al., 2020)

Polyoxymethylene (POM)	Cogwheels are one example of an application that calls for stiffness and where swelling brought on by environmental moisture may be harmful.	POM depolymerizes quickly and quantitatively to formaldehyde above its critical temperature (T_c) of 125°C in the presence of strong acids, which function as catalysts to cleave acetal bonds and create active cationic end groups. Formaldehyde produced *in situ* is condensed with diols to create cyclic acetals.	(Schirmeister & Mülhaupt, 2022)
Tires	NR/SBR mixes waste tire compounds to improve hardness, rebound resilience, tensile, and tear strength. used as fuel because of their high heating value.	The carbon black particles from scrap tires by the oxygen reduction process for an alkaline fuel cell was catalysed by tire-derived char. Na-ion batteries that used carbon materials produced from used tires as anode electrode materials. The tire was stabilised in the investigation by sulfonation at 120 °C. To produce the anode material for Li-ion batteries, performed pyrolysis on used tires in an inert environment.	(Naskar et al., 2014) (Li et al., 2017) (Passaponti et al., 2019) (Jiao et al., 2020) (Chen et al., 2020)

Another viable strategy appears to be the conversion of polymers into small-scale compounds other than monomers for polymers. But it could be essential to develop new uses for these substances. The only remaining methods to prevent PW from ending up in landfills are carbonization, gasification and pyrolysis into liquid fuel mixes and burning it to produce power or fuel to industries. The chemical upcycling of PW highlights the "plastic-based refinery" idea, which uses the waste plastic as a platform feedstock to create high-value monomeric or oligomeric chemicals, reintroducing the waste plastic into a circular economy. To valorize various plastic feedstocks into value-added chemicals, materials or fuels, several chemical ways to upcycle PW are introduced in each section. These processes include photocatalysis, hydrogenolysis, and others.

Due to the wide range of practical uses of carbonaceous goods made from PW, including carbon fibres, carbondots, carbon tubes, absorbents for water purification and electrodes for energy storage, several research have sought to upcycle PW into carbonaceous materials. Future options for successful upcycling of PW into carbonaceous materials are recommended considering the findings. Utilizing readily available carbon sources from non-recyclable PW enables the production of value-added PHA bioplastics and lowers the cost of feedstock, which significantly aids in the circular economy approach to solving the plastic pollution problem. The current attempts to valorize plastic and lignocellulosic wastes to create PHAs utilizing a biotechnological approach and a two-step process are highlighted in this study. In fact, it is still very difficult to combine sustainability, recyclability, usability and affordability optimally. As a result, once they are manufactured on a big scale, sustainable polymers also need the accompanying recycling, upcycling and degrading techniques.

REFERENCES

Abu-Thabit, N. Y., Pérez-Rivero, C., Uwaezuoke, O. J., & Ngwuluka, N. C. (2021). From waste to wealth: upcycling of plastic and lignocellulosic wastes to PHAs. *Journal of Chemical Technology and Biotechnology*, *97*(12), 3217–3240. https://doi.org/10.1002/jctb.6966

Adibfar, M., Kaghazchi, T., Asasian, N., & Soleimani, M. (2014). Conversion of Poly(Ethylene Terephthalate) Waste into Activated Carbon: Chemical Activation and Characterization. *Chemical Engineering & Technology*, *37*(6), 979–986. https://doi.org/10.1002/ceat.201200719

Algozeeb, W. A., Savas, P. E., Luong, D. X., Chen, W., Kittrell, C., Bhat, M., Shahsavari, R., & Tour, J. M. (2020). Flash Graphene from Plastic Waste. *ACS Nano*, *14*(11), 15595–15604. https://doi.org/10.1021/acsnano.0c06328

Anuar Sharuddin, S. D., Abnisa, F., Wan Daud, W. M. A., & Aroua, M. K. (2016). A review on pyrolysis of plastic wastes. *Energy Conversion and Management*, *115*, 308–326. https://doi.org10.1016j.enconman.2016.02.037

Armenise, S., SyieLuing, W., Ramírez-Velásquez, J. M., Launay, F., Wuebben, D., Ngadi, N., Rams, J., & Muñoz, M. (2021). Plastic waste recycling via pyrolysis: A bibliometric survey and literature review. *Journal of Analytical and Applied Pyrolysis*, *158*, 105265. https://doi.org/10.1016/j.jaap.2021.105265

Bäckström, E., Odelius, K., & Hakkarainen, M. (2017). Trash to Treasure: Microwave-Assisted Conversion of Polyethylene to Functional Chemicals. *Industrial & Engineering Chemistry Research*, *56*(50), 14814–14821. https://doi.org/10.1021/acs.iecr.7b04091

Bazargan, A., & McKay, G. (2012). A Review – Synthesis of Carbon Nanotubes from Plastic Wastes. *Chemical Engineering Journal*, *195–196*, 377–391. https://doi.org/10.1016/j.cej.2012.03.077

Beydoun, K., & Klankermayer, J. (2020). Efficient Plastic Waste Recycling to Value-Added Products by Integrated Biomass Processing. *ChemSusChem*, *13*(3), 488–492. https://doi.org/10.1002/cssc.201902880

Bunescu, A., Lee, S., Li, Q., & Hartwig, J. F. (2017). Catalytic Hydroxylation of Polyethylenes. *ACS Central Science*, *3*(8), 895–903. https://doi.org/10.1021/acscentsci.7b00255

Celik, G., Kennedy, R. M., Hackler, R. A., Ferrandon, M., Tennakoon, A., Patnaik, S., LaPointe, A. M., Ammal, S. C., Heyden, A., Perras, F. A., Pruski, M., Scott, S. L., Poeppelmeier, K. R., Sadow, A. D., & Delferro, M. (2019). Upcycling Single-Use Polyethylene into High-Quality Liquid Products. *ACS Central Science*, *5*(11), 1795–1803. https://doi.org/10.1021/acscentsci.9b00722

Chen, X., & Jin, F. (2019). Photocatalytic Reduction of Carbon Dioxide by Titanium Oxide-Based Semiconductors to Produce Fuels. *Frontiers in Energy*, *13*(2), 207–220. https://doi.org/10.1007/s11 708-019-0628-9

Chen, X., Liu, Y., & Wang, J. (2020). Lignocellulosic Biomass Upgrading into Valuable Nitrogen-Containing Compounds by Heterogeneous Catalysts. *Industrial & Engineering Chemistry Research*, *59*(39), 17008–17025. https://doi.org/10.1021/acs.iecr.0c01815

Chen, X., Wang, Y., & Zhang, L. (2021). Recent Progress in the Chemical Upcycling of Plastic Wastes. *ChemSusChem*, *14*(9), 4137–4151. https://doi.org/10.1002/cssc.202100868

Chen, X., & Yan, N. (2020). A Brief Overview of Renewable Plastics. *Materials Today Sustainability*, *7–8*, 100031. https://doi.org/10.1016/j.mtsust.2019.100031

Chen, Y., Wu, X., Wei, J., Wu, H., & Fang, J. (2021). Reuse Polyester/Cotton Blend Fabrics to Prepare Fiber Reinforced Composite: Fabrication, Characterization, and Interfacial Properties Evaluation. *Polymer Composites*, *42*(1), 141–152. https://doi.org/10.1002/pc.25814

Choi, D., Jang, D., Joh, H.-I., Reichmanis, E., & Lee, S. (2017). High Performance Graphitic Carbon from Waste Polyethylene: Thermal Oxidation as a Stabilization Pathway Revisited. *Chemistry of Materials*, *29*(21), 9518–9527. https://doi.org/10.1021/acs.chemmater.7b03737

Choi, J., Yang, I., Kim, S. S., Cho, S. Y., & Lee, S. (2022). Upcycling Plastic Waste into High Value-Added Carbonaceous Materials. *Macromolecular Rapid Communications*, *43*(1), e2100467. https://doi.org/ 10.1002/marc.202100467

Danmaliki, G. I., & Saleh, T. A. (2017). Effects of Bimetallic Ce/Fe Nanoparticles on the Desulfurization of Thiophenes Using Activated Carbon. *Chemical Engineering Journal*, *307*, 914–927. https://doi.org/ 10.1016/j.cej.2016.08.143

dela Cruz, Ma. I. S., Thongsai, N., de Luna, M. D. G., In, I., & Paoprasert, P. (2019). Preparation of Highly Photoluminescent Carbon Dots from Polyurethane: Optimization Using Response Surface Methodology and Selective Detection of Silver (I) Ion. *Colloids and Surfaces A: Physicochemical and Engineering Aspects*, *568*, 184–194. https://doi.org/10.1016/j.colsurfa.2019.02.022

El-Sayed, E.-S. M., & Yuan, D. (2020). Waste to MOFs: Sustainable Linker, Metal, and Solvent Sources for Value-Added MOF Synthesis and Applications. *Green Chemistry*, *22*(13), 4082–4104. https://doi.org/ 10.1039/D0GC00353K

Esfandiari, A., Kaghazchi, T., & Soleimani, M. (2012). Preparation and Evaluation of Activated Carbons Obtained by Physical Activation of Polyethyleneterephthalate (PET) Wastes. *Journal of the Taiwan Institute of Chemical Engineers*, *43*(4), 631–637. https://doi.org/10.1016/j.jtice.2012.02.002

Gardea, F., Garcia, J. M., Boday, D. J., Bajjuri, K. M., Naraghi, M., & Hedrick, J. L. (2014). Hybrid Poly(Aryl Ether Sulfone Amide)s for Advanced Thermoplastic Composites. *Macromolecular Chemistry and Physics*, *215*(22), 2260–2267. https://doi.org/10.1002/macp.201400267

Ghosh, D., Giri, S., & Das, C. K. (2013). Preparation of CTAB-Assisted Hexagonal Platelet Co(OH)$_2$ / Graphene Hybrid Composite as Efficient Supercapacitor Electrode Material. *ACS Sustainable Chemistry & Engineering*, *1*(9), 1135–1142. https://doi.org/10.1021/sc400055z

Gong, J., Liu, J., Wen, X., Jiang, Z., Chen, X., Mijowska, E., & Tang, T. (2014). Upcycling Waste Polypropylene into Graphene Flakes on Organically Modified Montmorillonite. *Industrial & Engineering Chemistry Research*, *53*(11), 4173–4181. https://doi.org/10.1021/ie4043246

Gong, J., Michalkiewicz, B., Chen, X., Mijowska, E., Liu, J., Jiang, Z., Wen, X., & Tang, T. (2014). Sustainable Conversion of Mixed Plastics into Porous Carbon Nanosheets with High Performances in Uptake of Carbon Dioxide and Storage of Hydrogen. *ACS Sustainable Chemistry & Engineering*, *2*(12), 2837–2844. https://doi.org/10.1021/sc500603h

Guo, Q., Ma, Z., Zhou, C., Ren, Z., & Yang, X. (2019). Single Molecule Photocatalysis on TiO$_2$ Surfaces. *Chemical Reviews*, *119*(20), 11020–11041. https://doi.org/10.1021/acs.chemrev.9b00226

Guzik, M. W., Kenny, S. T., Duane, G. F., Casey, E., Woods, T., Babu, R. P., Nikodinovic-Runic, J., Murray, M., & O'Connor, K. E. (2014). Conversion of Post Consumer Polyethylene to the Biodegradable Polymer Polyhydroxyalkanoate. *Applied Microbiology and Biotechnology*, *98*(9), 4223–4232. https://doi.org/ 10.1007/s00253-013-5489-2

He, C., Giannis, A., & Wang, J.-Y. (2013). Conversion of Sewage Sludge to Clean Solid Fuel Using Hydrothermal Carbonization: Hydrochar Fuel Characteristics and Combustion Hehavior. *Applied Energy*, *111*, 257–266. https://doi.org/10.1016/j.apenergy.2013.04.084

Honus, S., Kumagai, S., Fedorko, G., Molnár, V., & Yoshioka, T. (2018). Pyrolysis Gases Produced from Individual and Mixed PE, PP, PS, PVC, and PET—Part I: Production and Physical Properties. *Fuel, 221*, 346–360. https://doi.org/10.1016/j.fuel.2018.02.074

Jehanno, C., Pérez-Madrigal, M. M., Demarteau, J., Sardon, H., & Dove, A. P. (2019). Organocatalysis for Depolymerisation. *Polymer Chemistry, 10*(2), 172–186. https://doi.org/10.1039/C8PY01284A

Jia, C., Xie, S., Zhang, W., Intan, N. N., Sampath, J., Pfaendtner, J., & Lin, H. (2021). Deconstruction of High-Density Polyethylene into Liquid Hydrocarbon Fuels and Lubricants by Hydrogenolysis Over Ru Catalyst. *Chem Catalysis, 1*(2), 437–455. https://doi.org/10.1016/j.checat.2021.04.002

Jia, J., Veksha, A., Lim, T.-T., & Lisak, G. (2021). Weakening the Strong Fe-La Interaction in A-Site-Deficient Perovskite via Ni Substitution to Promote the Thermocatalytic Synthesis of Carbon Nanotubes from Plastics. *Journal of Hazardous Materials, 403*, 123642. https://doi.org/10.1016/j.jhazmat.2020.123642

Jia, X., Qin, C., Friedberger, T., Guan, Z., & Huang, Z. (2016). Efficient and Selective Degradation of Polyethylenes into Liquid Fuels and Waxes Under Mild Conditions. *Science Advances, 2*(6), e1501591. https://doi.org/10.1126/sciadv.1501591

Jiao, X., Zheng, K., Chen, Q., Li, X., Li, Y., Shao, W., Xu, J., Zhu, J., Pan, Y., Sun, Y., & Xie, Y. (2020). Photocatalytic Conversion of Waste Plastics into C 2 Fuels under Simulated Natural Environment Conditions. *Angewandte Chemie International Edition, 59*(36), 15497–15501. https://doi.org/10.1002/anie.201915766

Jie, X., Li, W., Slocombe, D., Gao, Y., Banerjee, I., Gonzalez-Cortes, S., Yao, B., AlMegren, H., Alshihri, S., Dilworth, J., Thomas, J., Xiao, T., & Edwards, P. (2020). Microwave-Initiated Catalytic Deconstruction of Plastic Waste into Hydrogen and High-Value Carbons. *Nature Catalysis, 3*(11), 902–912. https://doi.org/10.1038/s41929-020-00518-5

Johnston, B., Jiang, G., Hill, D., Adamus, G., Kwiecień, I., Zięba, M., Sikorska, W., Green, M., Kowalczuk, M., & Radecka, I. (2017). The Molecular Level Characterization of Biodegradable Polymers Originated from Polyethylene Using Non-Oxygenated Polyethylene Wax as a Carbon Source for Polyhydroxyalkanoate Production. *Bioengineering, 4*(4), 73. https://doi.org/10.3390/bioengineering4030073

Johnston, B., Radecka, I., Chiellini, E., Barsi, D., Ilieva, V. I., Sikorska, W., Musioł, M., Zięba, M., Chaber, P., Marek, A. A., Mendrek, B., Ekere, A. I., Adamus, G., & Kowalczuk, M. (2019). Mass Spectrometry Reveals Molecular Structure of Polyhydroxyalkanoates Attained by Bioconversion of Oxidized Polypropylene Waste Fragments. *Polymers, 11*(10), 1580. https://doi.org/10.3390/polym11101580

Joseph Berkmans, A., Jagannatham, M., Priyanka, S., & Haridoss, P. (2014). Synthesis of Branched, Nano Channeled, Ultrafine and Nano Carbon Tubes from PET Wastes Using the Arc Discharge Method. *Waste Management, 34*(11), 2139–2145. https://doi.org/10.1016/j.wasman.2014.07.004

Kar, G. P., Saed, M. O., & Terentjev, E. M. (2020). Scalable Upcycling of Thermoplastic Polyolefins into Vitrimers Through Transesterification. *Journal of Materials Chemistry A, 8*(45), 24137–24147. https://doi.org/10.1039/D0TA07339C

Kawai, T., & Sakata, T. (1981). Photocatalytic Hydrogen Production From Water by the Decomposition of Poly-Vinylchloride, Protein, Algae, Dead Insects, and Excrement. *Chemistry Letters, 10*(1), 81–84. https://doi.org/10.1246/cl.1981.81

Kenny, S. T., Runic, J. N., Kaminsky, W., Woods, T., Babu, R. P., Keely, C. M., Blau, W., & O'Connor, K. E. (2008). Up-Cycling of PET (Polyethylene Terephthalate) to the Biodegradable Plastic PHA (Polyhydroxyalkanoate). *Environmental Science & Technology, 42*(20), 7696–7701. https://doi.org/10.1021/es801010e

Kenny, S. T., Runic, J. N., Kaminsky, W., Woods, T., Babu, R. P., & O'Connor, K. E. (2012). Development of a Bioprocess to Convert PET Derived Terephthalic Acid and Biodiesel Derived Glycerol to Medium Chain Length Polyhydroxyalkanoate. *Applied Microbiology and Biotechnology, 95*(3), 623–633. https://doi.org/10.1007/s00253-012-4058-4

Kim, H. T., Kim, J. K., Cha, H. G., Kang, M. J., Lee, H. S., Khang, T. U., Yun, E. J., Lee, D.-H., Song, B. K., Park, S. J., Joo, J. C., & Kim, K. H. (2019). Biological Valorization of Poly(Ethylene Terephthalate) Monomers for Upcycling Waste PET. *ACS Sustainable Chemistry & Engineering, 7*(24), 19396–19406. https://doi.org/10.1021/acssuschemeng.9b03908

Kim, P. J., Fontecha, H. D., Kim, K., & Pol, V. G. (2018). Toward High-Performance Lithium–Sulfur Batteries: Upcycling of LDPE Plastic into Sulfonated Carbon Scaffold via Microwave-Promoted Sulfonation. *ACS Applied Materials & Interfaces, 10*(17), 14827–14834. https://doi.org/10.1021/acsami.8b03959

Kizza, R., Banadda, N., & Seay, J. (2022). Qualitative and Energy Recovery Potential Analysis: Plastic-Derived Fuel Oil versus Conventional Diesel Oil. *Clean Technologies and Environmental Policy*, 24(3), 789–800. https://doi.org/10.1007/s10098-021-02028-9

Ko, S., Kwon, Y. J., Lee, J. U., & Jeon, Y.-P. (2020). Preparation of Synthetic Graphite from Waste PET Plastic. *Journal of Industrial and Engineering Chemistry*, 83, 449–458. https://doi.org/10.1016/j.jiec.2019.12.018

Kong, Q., & Zhang, J. (2007). Synthesis of Straight and Helical Carbon Nanotubes from Catalytic Pyrolysis of Polyethylene. *Polymer Degradation and Stability*, 92(11), 2005–2010. https://doi.org/10.1016/j.polymdegradstab.2007.08.002

Kumar, A., von Wolff, N., Rauch, M., Zou, Y.-Q., Shmul, G., Ben-David, Y., Leitus, G., Avram, L., & Milstein, D. (2020). Hydrogenative Depolymerization of Nylons. *Journal of the American Chemical Society*, 142(33), 14267–14275. https://doi.org/10.1021/jacs.0c05675

Lessard, J. J., Scheutz, G. M., Sung, S. H., Lantz, K. A., Epps, T. H., & Sumerlin, B. S. (2020). Block Copolymer Vitrimers. *Journal of the American Chemical Society*, 142(1), 283–289. https://doi.org/10.1021/jacs.9b10360

Lewis, S. E., Wilhelmy, B. E., & Leibfarth, F. A. (2019). Upcycling Aromatic Polymers Through C–H Fluoroalkylation. *Chemical Science*, 10(25), 6270–6277. https://doi.org/10.1039/C9SC01425J

Li, Y., Adams, R. A., Arora, A., Pol, V. G., Levine, A. M., Lee, R. J., Akato, K., Naskar, A. K., & Paranthaman, M. P. (2017). Sustainable Potassium-Ion Battery Anodes Derived from Waste-Tire Rubber. *Journal of The Electrochemical Society*, 164(6), A1234–A1238. https://doi.org/10.1149/2.1391706jes

Li, Y., Paranthaman, M. P., Akato, K., Naskar, A. K., Levine, A. M., Lee, R. J., Kim, S.-O., Zhang, J., Dai, S., & Manthiram, A. (2016). Tire-Derived Carbon Composite Anodes for Sodium-Ion Batteries. *Journal of Power Sources*, 316, 232–238. https://doi.org/10.1016/j.jpowsour.2016.03.071

Liu, X., de Vries, J. G., & Werner, T. (2019). Transfer Hydrogenation of Cyclic Carbonates and Polycarbonate to Methanol and Diols by Iron Pincer Catalysts. *Green Chemistry*, 21(19), 5248–5255. https://doi.org/10.1039/C9GC02052G

Luong, D. X., Bets, K. v., Algozeeb, W. A., Stanford, M. G., Kittrell, C., Chen, W., Salvatierra, R. v., Ren, M., McHugh, E. A., Advincula, P. A., Wang, Z., Bhatt, M., Guo, H., Mancevski, V., Shahsavari, R., Yakobson, B. I., & Tour, J. M. (2020). Gram-Scale Bottom-Up Flash Graphene Synthesis. *Nature*, 577(7792), 647–651. https://doi.org/10.1038/s41586-020-1938-0

Matias, Á. A., Lima, M. S., Pereira, J., Pereira, P., Barros, R., Coelho, J. F. J., & Serra, A. C. (2020). Use of Recycled Polypropylene/Poly(Ethylene Terephthalate) Blends to Manufacture Water Pipes: An Industrial Scale Study. *Waste Management*, 101, 250–258. https://doi.org/10.1016/j.wasman.2019.10.001

Mejia, E. B., Al-Maqdi, S., Alkaabi, M., Alhammadi, A., Alkaabi, M., Cherupurakal, N., & Mourad, A.-H. I. (2020). Upcycling of HDPE Waste using Additive Manufacturing: Feasibility and Challenges. *2020 Advances in Science and Engineering Technology International Conferences (ASET)*, 1–6. https://doi.org/10.1109/ASET48392.2020.9118269

Naskar, A. K., Bi, Z., Li, Y., Akato, S. K., Saha, D., Chi, M., Bridges, C. A., & Paranthaman, M. P. (2014). Tailored Recovery of Carbons from Waste Tires for Enhanced Performance as Anodes in Lithium-Ion Batteries. *RSC Advances*, 4(72), 38213. https://doi.org/10.1039/C4RA03888F

Nur-A-Tomal, M. S., Pahlevani, F., & Sahajwalla, V. (2020). Direct Transformation of Waste Children's Toys to High Quality Products using 3D Printing: A Waste-to-Wealth and Sustainable Approach. *Journal of Cleaner Production*, 267, 122188. https://doi.org/10.1016/j.jclepro.2020.122188

Parrino, F., Bellardita, M., García-López, E. I., Marcì, G., Loddo, V., & Palmisano, L. (2018). Heterogeneous Photocatalysis for Selective Formation of High-Value-Added Molecules: Some Chemical and Engineering Aspects. *ACS Catalysis*, 8(12), 11191–11225. https://doi.org/10.1021/acscatal.8b03093

Passaponti, M., Rosi, L., Savastano, M., Giurlani, W., Miller, H. A., Lavacchi, A., Filippi, J., Zangari, G., Vizza, F., & Innocenti, M. (2019). Recycling of Waste Automobile Tires: Transforming Char in Oxygen Reduction Reaction Catalysts for Alkaline Fuel Cells. *Journal of Power Sources*, 427, 85–90. https://doi.org/10.1016/j.jpowsour.2019.04.067

Pol, S. v., Pol, V. G., Sherman, D., & Gedanken, A. (2009). A Solvent Free Process for the Generation of Strong, Conducting Carbon Spheres by the Thermal Degradation of Waste Polyethylene Terephthalate. *Green Chemistry*, 11(4), 448. https://doi.org/10.1039/b819494g

Potts, J. R., Dreyer, D. R., Bielawski, C. W., & Ruoff, R. S. (2011). Graphene-Based Polymer Nanocomposites. *Polymer*, 52(1), 5–25. https://doi.org/10.1016/j.polymer.2010.11.042

Qiao, W. M., Song, Y., Yoon, S.-H., Korai, Y., Mochida, I., Yoshiga, S., Fukuda, H., & Yamazaki, A. (2006). Carbonization of Waste PVC to Develop Porous Carbon Material Without Further Activation. *Waste Management*, 26(6), 592–598. https://doi.org/10.1016/j.wasman.2005.06.010

Qiao, W. M., Yoon, S. H., Mochida, I., & Yang, J. H. (2007). Waste Polyvinylchloride Derived Pitch as a Precursor to Develop Carbon Fibers and Activated Carbon Fibers. *Waste Management*, 27(12), 1884–1890. https://doi.org/10.1016/j.wasman.2006.10.002

Quesada, L., Calero, M., Martín-Lara, M. Á., Pérez, A., & Blázquez, G. (2020). Production of an Alternative Fuel by Pyrolysis of Plastic Wastes Mixtures. *Energy & Fuels*, 34(2), 1781–1790. https://doi.org/10.1021/acs.energyfuels.9b03350

Radecka, I., Irorere, V., Jiang, G., Hill, D., Williams, C., Adamus, G., Kwiecień, M., Marek, A., Zawadiak, J., Johnston, B., & Kowalczuk, M. (2016). Oxidized Polyethylene Wax as a Potential Carbon Source for PHA Production. *Materials*, 9(5), 367. https://doi.org/10.3390/ma9050367

Rorrer, N. A., Nicholson, S., Carpenter, A., Biddy, M. J., Grundl, N. J., & Beckham, G. T. (2019). Combining Reclaimed PET with Bio-based Monomers Enables Plastics Upcycling. *Joule*, 3(4), 1006–1027. https://doi.org/10.1016/j.joule.2019.01.018

Röttger, M., Domenech, T., van der Weegen, R., Breuillac, A., Nicolaÿ, R., & Leibler, L. (2017). High-performance Vitrimers from Commodity Thermoplastics through Dioxaborolane Metathesis. *Science*, 356(6333), 62–65. https://doi.org/10.1126/science.aah5281

Roy, P. S., Garnier, G., Allais, F., & Saito, K. (2021). Strategic Approach Towards Plastic Waste Valorization: Challenges and Promising Chemical Upcycling Possibilities. *ChemSusChem*, 14(9), 4007–4027. https://doi.org/10.1002/cssc.202100904

Saito, K., Jehanno, C., Meabe, L., Olmedo-Martínez, J. L., Mecerreyes, D., Fukushima, K., & Sardon, H. (2020). From Plastic Waste to Polymer Electrolytes for Batteries through Chemical Upcycling of Polycarbonate. *Journal of Materials Chemistry A*, 8(28), 13921–13926. https://doi.org/10.1039/D0TA03374J

Sarda, P., Hanan, J. C., Lawrence, J. G., & Allahkarami, M. (2022). Sustainability Performance of Polyethylene Terephthalate, Clarifying Challenges and Opportunities. *Journal of Polymer Science*, 60(1), 7–31. https://doi.org/10.1002/pol.20210495

Schirmeister, C. G., & Mülhaupt, R. (2022). Closing the Carbon Loop in the Circular Plastics Economy. *Macromolecular Rapid Communications*, 43(13), 2200247. https://doi.org/10.1002/marc.202200247

Sharma, H. B., Sarmah, A. K., & Dubey, B. (2020). "Hydrothermal Carbonization of Renewable Waste Biomass for Solid Biofuel Production: A Discussion on Process Mechanism, the Influence of Process Parameters, Environmental Performance and Fuel Properties of Hydrochar." *Renewable and Sustainable Energy Reviews, 123*, 109761. https://doi.org/10.1016/j.rser.2020.109761

Sharma, K., Khilari, V., Chaudhary, B. U., Jogi, A. B., Pandit, A. B., & Kale, R. D. (2020). Cotton Based Composite Fabric Reinforced with Waste Polyester Fibers for Improved Mechanical Properties. *Waste Management*, 107, 227–234. https://doi.org/10.1016/j.wasman.2020.04.011

Sharon, M., Mishra, N., Patil, B., Mewada, A., Gurung, R., & Sharon, M. (2015). Conversion of Polypropylene to Two-Dimensional Graphene, One-Dimensional Carbon Nano Tubes and Zero-Dimensional C-Dots, All Exhibiting Typical sp^2-Hexagonal Carbon Rings. *IET Circuits, Devices & Systems*, 9(1), 59–66. https://doi.org/10.1049/iet-cds.2014.0117

Shen, Y. (2020). A Review on Hydrothermal Carbonization of Biomass and Plastic Wastes to Energy Products. *Biomass and Bioenergy*, 134, 105479. https://doi.org/10.1016/j.biombioe.2020.105479

Slater, B., Wong, S.-O., Duckworth, A., White, A. J. P., Hill, M. R., & Ladewig, B. P. (2019). Upcycling a Plastic Cup: One-Pot Synthesis of Lactate Containing Metal Organic Frameworks from Polylactic Acid. *Chemical Communications*, 55(51), 7319–7322. https://doi.org/10.1039/C9CC02861G

Song, C., Zhang, B., Hao, L., Min, J., Liu, N., Niu, R., Gong, J., & Tang, T. (2022). Converting Poly(Ethylene Terephthalate) Waste into N-Doped Porous Carbon as CO2 Adsorbent and Solar Steam Generator. *Green Energy & Environment*, 7(3), 411–422. https://doi.org/10.1016/j.gee.2020.10.002

Stadler, B. M., & de Vries, J. G. (2021). Chemical Upcycling of Polymers. *Philosophical Transactions of the Royal Society A: Mathematical, Physical and Engineering Sciences*, 379(2209), 20200341. https://doi.org/10.1098/rsta.2020.0341

Suganuma, S., Nakajima, K., Kitano, M., Hayashi, S., & Hara, M. (2012). sp3-Linked Amorphous Carbon with Sulfonic Acid Groups as a Heterogeneous Acid Catalyst. *ChemSusChem*, 5(9), 1841–1846. https://doi.org/10.1002/cssc.201200010

Tennakoon, A., Wu, X., Paterson, A. L., Patnaik, S., Pei, Y., LaPointe, A. M., Ammal, S. C., Hackler, R. A., Heyden, A., Slowing, I. I., Coates, G. W., Delferro, M., Peters, B., Huang, W., Sadow, A. D., & Perras, F. A. (2020). Catalytic Upcycling of High-Density Polyethylene via a Processive Mechanism. *Nature Catalysis*, *3*(11), 893–901. https://doi.org/10.1038/s41929-020-00519-4

Uekert, T., Kasap, H., & Reisner, E. (2019). Photoreforming of Nonrecyclable Plastic Waste over a Carbon Nitride/Nickel Phosphide Catalyst. *Journal of the American Chemical Society*, *141*(38), 15201–15210. https://doi.org/10.1021/jacs.9b06872

Uekert, T., Kuehnel, M. F., Wakerley, D. W., & Reisner, E. (2018). Plastic Waste as a Feedstock for Solar-Driven H_2 Generation. *Energy & Environmental Science*, *11*(10), 2853–2857. https://doi.org/10.1039/C8EE01408F

Vanapalli KR, Bhattacharya J, Samal B, Chandra S, Medha I, Dubey BK. (2021). Optimized Production of Single-Use Plastic-Eucalyptus Wood Char Composite for Application in Soil. *Journal of Cleaner Production*, *278*, 123968. https://doi.org/10.1016/j.jclepro.2020.123968

Ward, P. G., Goff, M., Donner, M., Kaminsky, W., & O'Connor, K. E. (2006). A Two Step Chemo-biotechnological Conversion of Polystyrene to a Biodegradable Thermoplastic. *Environmental Science & Technology*, *40*(7), 2433–2437. https://doi.org/10.1021/es0517668

Welsing, G., Wolter, B., Hintzen, H. M. T., Tiso, T., & Blank, L. M. (2021). Upcycling of hydrolyzed PET by microbial conversion to a fatty acid derivative. *Methods in Enzymology*, *648*(2021), 391–421. https://doi.org/10.1016/bs.mie.2020.12.025

Wong, S. L., Nyakuma, B. B., Wong, K. Y., Lee, C. T., Lee, T. H., & Lee, C. H. (2020). Microplastics and Nanoplastics in Global Food Webs: A Bibliometric Analysis (2009–2019). *Marine Pollution Bulletin*, *158*, 111432. https://doi.org/10.1016/j.marpolbul.2020.111432

Wyss, K. M., Beckham, J. L., Chen, W., Luong, D. X., Hundi, P., Raghuraman, S., Shahsavari, R., & Tour, J. M. (2021). Converting plastic waste pyrolysis ash into flash graphene. *Carbon*, *174*, 430–438. https://doi.org/10.1016/j.carbon.2020.12.063

Yao, D., Zhang, Y., Williams, P. T., Yang, H., & Chen, H. (2018). Co-Production of Hydrogen and Carbon Nanotubes from Real-World Waste Plastics: Influence of Catalyst Composition and Operational Parameters. *Applied Catalysis B: Environmental*, *221*, 584–597. https://doi.org/10.1016/j.apcatb.2017.09.035

Yu, L., Chen, Y., Tan, S., Zhang, W., & Song, R. (2019). Potassium Chloride-Assisted Synthesis of Hollow Carbon Spheres from Pure or Mixed Polyolefins Containing Silica Templates. *Nanoscience and Nanotechnology Letters*, *11*(8), 1084–1092. https://doi.org/10.1166/nnl.2019.2985

Yuan, X., Cho, M.-K., Lee, J. G., Choi, S. W., & Lee, K. B. (2020). Upcycling of Waste Polyethylene Terephthalate Plastic Bottles into Porous Carbon for CF4 Adsorption. *Environmental Pollution*, *265*, 114868. https://doi.org/10.1016/j.envpol.2020.114868

Zhang, F., Zeng, M., Yappert, R. D., Sun, J., Lee, Y.-H., LaPointe, A. M., Peters, B., Abu-Omar, M. M., & Scott, S. L. (2020). Polyethylene Upcycling to Long-Chain Alkylaromatics by Tandem Hydrogenolysis/Aromatization. *Science*, *370*(6515), 437–441. https://doi.org/10.1126/science.abc5441

Zhou, H., Ren, Y., Li, Z., Xu, M., Wang, Y., Ge, R., Kong, X., Zheng, L., & Duan, H. (2021). Electrocatalytic Upcycling of Polyethylene Terephthalate to Commodity Chemicals and H2 Fuel. *Nature Communications*, *12*(1), 4679. https://doi.org/10.1038/s41467-021-25048-x

6 High-Energy Methane Storage Systems Based on Nanoporous Carbon Adsorbent from Biomass Wastes

Ilya E. Men'shchikov, Elena V. Khozina, Olga V. Solovtsova, Andrey V. Shkolin and Anatoly A. Fomkin

6.1 INTRODUCTION

Depletion of world oil and coal reserves as well as the environmental issues related to their large-scale combustion are gradually reducing the share of these fuels in the global energy sector by increasing natural gas (NG) consumption. NG contains more than 70% methane and is the cleanest fossil fuel. Russia has the largest NG reserves in the world: 37.4 trillion cubic meters or ~20% of global reserves [1]. There is considerable scope for expanding NG utilization: portable energy systems, onboard vehicular applications, electricity generating plants, etc. The main drawbacks when using NG is the low energy density under normal conditions: its value is only 0.12% of that of gasoline [2]. In this regard, there is a need for special technologies for NG storing and transportation. At the present time, two options are available: compressed NG (CNG) obtained by compressing at 20–25 MPa to less than 1% of its standard atmospheric volume, and liquified NG (LNG) produced by cooling down to below the boiling temperature of 111.7 K. The CNG production requires expensive multistage high-pressure compressors, and CNG is stored in the heavy thick cylindric or spherical steel tanks. The CNG tank contains 240 m^3(STP)/m^3 at 20 MPa and 293 K. The energy density of CNG at 25 MPa is 9.2 MJ/L as compared with 34.2 MJ/L for gasoline [3]. LNG is stored and transported in cryogenic heat-insulated tanks to keep the fuel cold (~113 K). The energy density of LNG (22.2 MJ/L) exceeds almost twice that of CNG at 25 MPa, but it is just around 65% that of gasoline [3]. The benefits of the high energy density of LNG are largely eliminated by the complexity and high cost of cryogenic equipment and safety risks associated with destruction of storage vessels, spills, and explosions of gas clouds [4].

The problems described above for LNG and CNG could be solved by storing NG in an adsorbed state (ANG) at moderate pressures and room temperatures [5, 6]. The ANG technology is based on physical adsorption phenomenon when methane molecules are bound to an adsorbent through a weak van der Waals interaction at relatively low pressures. Adsorption of gas molecules in pores with a size commensurate with their dimensions (2–3 molecular diameter) proceeds via the volume filling of the pores due to an overlap of the adsorption-force fields of the opposite pore walls [7]. This process provides an increased packing density of adsorbed molecules comparable with that in a liquid state. Thus, when employing the ANG technology, one can attain the CNG density at 20 MPa at ~1/5 of the gas pressure of CNG [5, 8, 9]. In addition to improved energy efficiency, ANG ensures

DOI: 10.1201/9781003327646-6

higher fire and explosion safety due to the bound state of gas molecules and can be employed in a wide range of gas systems: from NG-powered vehicles to mobile pipelines and stationary NG terminals [10, 11]. The progress in ANG technology implies the application of efficient adsorbents that meet the high requirements on specific adsorption capacity, cyclic stability, energy efficiency, and safety [8–13]. Fabrication of efficient porous material for ANG storage should be based on a target for storage and deliverable capacity that makes ANG equivalent to CNG. At present, the most mentioned target was established by the US Department of Energy's (DOE) program called the methane opportunities for vehicular energy (MOVE) program [14]. In 2012, the DOE set a methane storage target to achieve a volumetric capacity of 263 $m^3(STP)/m^3$ (volume of methane at the STP conditions of $T = 273$ K and $P = 0.1$ MPa per unit volume of adsorbent) at a pressure of 3.5 (6.5) MPa and 298 K. This DOE target is considered as a "milestone" for researchers [9, 12, 13, 15, 16].

As follows from a rich set of experimental data [9, 12, 13, 15–17], activated carbons (AC) are recognized as one of the most promising adsorbents for utilizing in ANG systems. The skeleton of AC is composed of distorted graphene-like layers of carbon atoms in sp2 configuration with defects (free edge sites, heteroatoms, sp3 carbon atoms) stacked in a turbostratic crystallites and an amorphous aliphatic-aromatic carbon phase [18–20]. The crystallites are formed by a few (2–5) parallel graphite layers and are interconnected in a random way. The ratio of the amorphous and crystalline phases is determined by precursor and carbonization conditions. During the activation process, the disordered matter and graphite layers burn out, resulting in a well-developed 3D hierarchical porosity [16, 18–21]. The pores of AC are formed by voids between the randomly arranged stacks of crystallites and cracks inside and parallel to the graphene-like layers. Thus, the porous structure of AC includes three types of pores according to the general classification of pores according to their width [22–24]:

- Micropores with sizes < 2 nm;
- Mesopores with sizes in a range from 2 to 50 nm;
- Macropores with sizes > 50 nm.

Micropores in ACs generally contribute to a major part of the porosity and determine its wide application in the adsorption-based technologies since it causes the excellent adsorption performance. Mesopores allow for mass transfer of gas molecules into and out of the micropores, and, hence, they are crucial for adsorption kinetics. Thus, the 3D hierarchical porosity of AC meets the requirements of high adsorption capacity, fast adsorption kinetics, and complete release of adsorbates at increasing temperatures and decreasing pressures, i.e., easy regeneration. One should mention the excellent physiochemical properties of ACs as hydrophobicity, mechanical strength, thermal stability and high resistance to chemical attack [25–27].

When searching an appropriate carbon adsorbent for ANG storage, an important criterion is the choice of raw materials that is based on considerations such as the following:

- relatively high yield of a high-quality carbon adsorbent;
- low content of inorganic matter;
- availability and abundance of raw material.

In practice, reasonably priced materials as coal, peat, polymers and various biomass residue waste are suitable precursors for preparing AC. It should be noted that for ACs the numerous studies revealed that the final textural properties, including the porous structure of the adsorbent, to a certain extent depends on the intrinsic physicochemical properties (phase and chemical compositions) of the raw material and methods of activations [17, 28, 29].

The use of biomass residues as a precursor for AC is one of the ways to mitigate tons of waste and thereby avoid the related risks to the environment. From the other side, the resultant AC-based adsorbents are also intended for technological applications that solve many ecological problems including the abovementioned ANG technology.

In addition to the physicochemical properties, the adsorption performance of an adsorbent in the ANG system is determined by its packing density. Indeed, to design a high-performing ANG storage system, the void space, and, consequently, the contribution from a gas phase in the ANG tank loaded with adsorbent must be minimized. For this purpose, powdered and granulated adsorbents, including AC, are compacted with and without binders [29, 30].

The efficiency of the ANG system greatly depends on thermal effects arising during the exothermic/endothermic adsorption/desorption processes [31–34]. Indeed, the heat release/absorption effects during the fueling/delivery processes change the temperature of the adsorbent, thereby lowering the gas storage performance below the isothermal performance [32, 33] and increasing the duration of the charge/discharge processes [34]. On the other hand, the heat of adsorption depends on the strength of the adsorbate-adsorbent interactions and is used to get deep insight into the adsorption process [34].

In the present work, we focused on investigating the prospects of a carbon adsorbent prepared from coconut shell by steam activation for ANG application. For this purpose, we studied its porous structure using the low-temperature adsorption of nitrogen. Then, the granules of the AC sample were compacted under pressure of 30 Pa with the polymer binder to prepare a monolith carbon adsorbent with improved packing density. Methane adsorption onto these carbon adsorbents was measured over a wide temperature range and at pressures up to 20 MPa to compare their adsorption capacity for ANG applications. We also explore the factors affecting the performance of a full-scale ANG system loaded with the monolith carbon adsorbent, namely: cyclability of the adsorbent, kinetics of methane adsorption and thermal effects, which are essential for the heat management of the system.

6.2 EXPERIMENTAL DETAILS

6.2.1 MATERIALS

6.2.1.1 Adsorbent

It was shown that due to low ash content, ACs can be prepared from coconut shell by steam activation [35]. In the present work, we studied C-1 carbon adsorbent prepared in three stages: (1) carbonization of the crushed precursor material at 870 K, (2) steam activation of the char at 1123–1273 K, and (3) size fractionation of the granules (three stages). The duration of steam activation was 1.5 h. C-1 carbon consisted of the granules with a size ranged from 0.1 to 1.7 mm.

To maximize the space occupied by adsorbent material in the ANG storage tank, we prepared the monolith blocks by shaping the granules of C-1 with a polymer binder: 70/30 butadiene–styrene. The polymer binder was taken on the assumption that its macromolecules would not penetrate into micropores and, thereby, the degradation of the microporosity would be minimal. The granulated C-1 adsorbent was crushed until the average particle size was no more than 500 microns. The size of the C-1 particles was monitored by a set of sieves with apertures ranging from 100 to 1000 microns. Then, the crushed carbon was stirred with 15% binder aqueous solution to a homogeneous elastic paste. Then, the paste was placed into a matrix with a punch and subjected to compaction under a pressure for 5 min. Based on the analysis of the experimental data on the changes in the porous structure of the carbon adsorbent after compaction under a pressure within a range from 10 to 100 MPa, we revealed that the optimal pressure was 30 MPa. Then, the blocks were dried in an oven for 20 h at the temperature of 413 K. The final monolith blocks contained 7.75 ± 0.25 wt.% of the binder. We fabricated monolith blocks of carbon adsorbent M-C1 in the form of a cylindrical pellet and a hexagonal prism (see Figure 6.1). The density, ρ [kg/cm^3], was calculated from the geometric volume and weight. The standard Brinell test [36] was used for evaluating the hardness of the M-C1 blocks before and after cyclic testing when the samples were subjected to the adsorption/desorption cycles.

FIGURE 6.1 Monolith blocks of C-1 carbon adsorbent prepared by compacting with the polymer binder under the pressure of 30 MPa.

6.2.1.2 Adsorbate

We used high-purity (99.999%) methane purchased from the Moscow Gas Processing Plant (Russia) as the adsorptive. Physicochemical properties of methane are as follows: molar mass $\mu = 16.0426$ g/mol; critical temperature $T_{cr} = 190.56$ K; critical pressure $p_{cr} = 4.599$ MPa [37]. The compressibility of gaseous methane was calculated from the state equations with the virial coefficients provided in the handbook [38].

6.2.2 METHODS

6.2.2.1 Characterization of the Porous Structure of Carbon Adsorbents

The parameters of the porous structure of the adsorbents were calculated from the data on *adsorption of nitrogen vapors* at 77 K measured on a Quantachrome Autosorb iQ multifunctional surface area analyzer (Quantachrome Instruments, US). We used the low-temperature nitrogen adsorption data for calculating the structural and energy characteristics of the carbon adsorbent: micropore volume, W_0 [cm³/g], characteristic energy of adsorption, E_0 [kJ/mol], and half-width of slit-like micropores, x_0 [nm]. The isotherms of nitrogen adsorption were plotted in coordinates of the well-known Dubinin–Radushkevich (D–R) equation [7] in a linear form:

$$\ln a = \ln a_0 - \left(\frac{A}{E_0}\right)^2, \tag{6.1}$$

where a [mmol/g] is the adsorption value at specified temperature and pressure; A [kJ/mol] is the differential molar work of adsorption; a_0 [mmol/g] is the limiting adsorption value. The differential molar work of adsorption is determined as follows:

$$A = RT \ln\left(\frac{f_s}{f}\right). \tag{6.2}$$

Here f and f_s are the fugacities of an equilibrium phase and saturated vapour of an adsorptive, respectively. At low pressures, fugacity f is equal to the vapour pressure, and f_s coincides with the saturated vapour pressure.

The limiting adsorption was evaluated from the Dubinin–Nikolaev equation [18]:

$$a_0 = a_0^0 \exp\left[-\alpha(T - T_0)\right], \qquad (6.3)$$

where a_0^0 [mmol/g] is the limiting adsorption at the boiling temperature T_0 [K]; α [1/K] is the thermal coefficient of the limiting adsorption [41]. The value of a_0^0 is calculated as follows:

$$a_0^0 = W_0 \times \rho_a \qquad (6.4)$$

Here, ρ_a is the density of the liquid phase of adsorptive at the boiling temperature.

The structural and energy characteristics, W_0 and E_0, were calculated from the isotherm of nitrogen adsorption at 77 K plotted in the D–R equation coordinates on assumption that the density of adsorbate in micropores is equal to that of liquid. The half-width of micropores was evaluated from the relationship [18]:

$$x_0 = 12 / E_0. \qquad (6.5)$$

The Brunauer–Emmet–Teller (BET) [39] equations were used to calculate the specific BET surface area, S_{BET} [m²/g].

6.2.2.2   Methane Adsorption Measurements

The equilibrium and kinetic methane adsorption measurements were carried out using the combined method on a custom-made volumetric-gravimetric setup designed in IPCE RAS [40]. Its scheme and detailed description were provided in recent works [41, 42]. The equilibrium methane adsorption was measured up to 30 MPa and at the temperatures of 193, 213, 233, 253, 273, 303, 313, 323 and 333 K. Prior to the methane adsorption measurements, the adsorbent was regenerated by thermal evaporation at 673 K to the residual pressure of no more than 1 Pa.

The value of absolute methane adsorption onto the adsorbent, a, was evaluated as follows:

$$a = \frac{\left(N - \left(V - V_{ads}\right) \times \rho_g\right)}{\left(m_0 \mu\right)}. \qquad (6.6)$$

Here, N [g] is the amount of methane injected into the measuring unit; V [cm³] is the entire volume of the system; V_{ads} [cm³] is the volume of the adsorbent with micropores calculated as a sum of the adsorbent skeleton volume determined from helium picnometry measurements, V_{He} [43], and the product $m_0 \times W_0$, where the value of W_0 was found from the standard low-temperature nitrogen adsorption data by the D–R equation [7] and m_0 [g] is the mass of the regenerated adsorbent; μ [g/mmol] is the molar mass of methane; ρ_g [g/cm³] is the density of a gaseous phase at specified pressure P and temperature T.

The accuracy of methane adsorption measurements was $\pm$ 1.5 % at a confidence level of 0.95 for most experimental values. The uncertainty of adsorption measurements was calculated following the standard requirements [44].

The kinetics characteristics of methane adsorption/desorption onto initial C-1 adsorbent were studied to find out its feasibility for utilizing in the ANG system that assumes a high rate of gas fuelling/delivery process. For this purpose, we examined the changes in the mass of the adsorbent over time, $m(t)$, after a rapid inlet of methane up to the pressure of 7.0 or 20 MPa. The methane desorption kinetics curves were obtained by a rapid pressure release from 7.0 or 20 MPa. The kinetics experiments were carried out at 303 K. The gas inlet and pressure release were performed in a time ~ 1 s, we assumed that the equilibrium pressure in the system maintained immediately.

6.2.2.3 Adsorption Performance of the ANG System

The deliverable volumetric methane capacity of the M-C-1 monolith adsorbent, i.e., the amount of methane delivered from an ANG tank when the pressure decreased from 7 or 20 to 0.1 MPa, was measured on a custom-made ANG bench designed in IPCE RAS. The scheme of the bench and the measurement procedure were described in detail recently [42]. It should be noted that the pressure-release process was carried out at a rate of about 10 L/min until the excess pressure of 20 and 7 MPa decreased to 0.1 MPa. The experiments were carried out at 233, 293 and 323 K. The FQI gas counter readings were recorded only after equilibrium was reached. The relative error in measuring the amount of methane by FQI was ±1%.

The ANG bench was calibrated to evaluate a dead volume of the system. For this purpose, we mounted a non-sorbing mockup (MO) with a volume equal to that of the stack of the M-C-1 blocks inside the ANG tank and carried out the adsorption experiment at the same thermodynamic conditions. The correction to the deliverable storage capacity was calculated as the difference between the FQI readings: $V_{\text{FQI-ANG}}$ and $V_{\text{FQI-MO}}$. Prior the measurements, three cycles of methane displacement adsorption up to 25 MPa were carried out to prepare the ANG bench.

The deliverable volumetric capacity, V_A [m³(STP)/m³], was evaluated, as follows:

$$V_A = \frac{V_{ds} - V_g}{V_{AC}} \tag{6.7}$$

Here, V_{ds} is the volume of desorbed methane calculated from the FQI readings; V_g is the volume of the residual gas in the system, its value is calculated from the residual pressure and temperature, taking into account the volume occupied by the adsorbent in the ANG vessel, V_{AC}, including the space between the M-C-1 blocks.

Cyclic methane adsorption/desorption experiments for simulating the charging/discharging processes of the ANG system were carried out at 303 K using a custom-made testing bench, the scheme of which is shown in Figure 6.2.

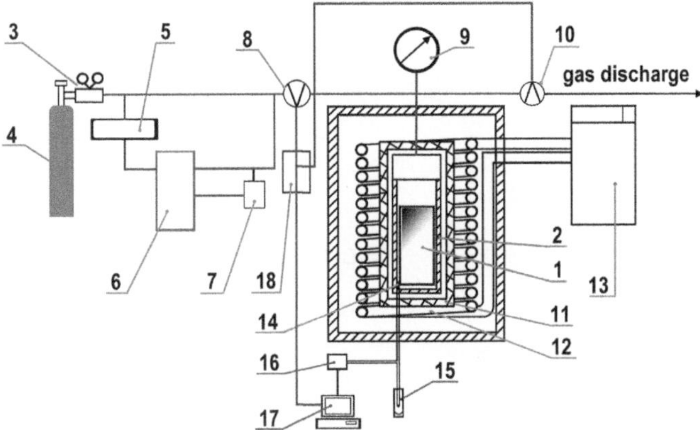

FIGURE 6.2 Scheme of the experimental ANG testing bench for cyclic adsorption/desorption (charging/discharging) processes. *1* – monolith adsorbent; *2* – high-pressure ANG tank (adsorber); *3* – double-reduction gear unit; *4* – gas cylinder (methane supplier); *5* – low pressure receiver for supplying methane to compressor *6*; *7* – high pressure receiver; *8* – electromagnetic valve; *9* – manometer; *10* – electromagnetic valve; *11* – adiabatic shell of the calorimeter; *12* – heat-exchange unit; *13* – cryo-thermostat; *14* – thermocouple unit; *15* – Dewar vessel; *16* – high-precision multimeter; *17* – PC; *18* – electromagnetic valve control unit.

Adsorbent (*1*) was placed into a high-pressure separable adsorber (*2*) made of stainless steel. Methane was supplied from the cylinder (*4*) through the two-stage reducer (*3*) to the compressor (*6*) maintaining a pressure of 7–20 MPa in the receiver (*7*). The required pressure is supplied through the electromagnetic valve (*8*) to the adsorber. The manometer (*9*) monitored the pressure in the adsorber. The pressure is released by electromagnetic valve (*10*). The number of the adsorption/desorption cycles was 5000.

The *heat effects* arising in the ANG system during the fuelling/refuelling processes were estimated using the same adsorption bench that ensured the achievement of the required pressure in the adsorber. Maximum changes in the temperature of the adsorber and the period of time to reach the equilibrium temperature were estimated.

According to the scheme in Figure 6.2, the adsorber was surrounded by the adiabatic shell (*11*) of the calorimeter, so that the heat losses were minimized. In turn, the adiabatic shell was inserted into the heat-exchange unit (*12*) that in combination with the cryo-thermostat (*13*) created a temperature within a range from 233 to 323 K. The differential thermal couple (*14*) was used to measure the temperature in the adsorber relative to the triple point for water, which was maintained by the Dewar vessel with ice (*15*). The thermocouple voltage was registered by means of the high-precision multimeter (*16*) connected with PC (*17*). The period of time to reach the equilibrium temperature was fixed by PC in real time.

6.3 RESULTS AND DISCUSSION

6.3.1 EFFECT OF COMPACTION ON THE POROUS STRUCTURE OF C-1 ACTIVATED CARBON

Figure 6.1 shows the M-C-1 monolith blocks prepared by compacting the mixture of C-1 granules and the polymer binder under a pressure of 30 MPa. This pressure was selected as optimal based on the changes in the density of the initial C-1 before and after compacting under different pressures within a range of 10–100 MPa (see Figure 6.3) and the assessments of degradation of the porous structure (Table 6.1).

The most intense increase in the packing density (by ~ 1.5 times) of the adsorbent was observed during compacting under pressures of up to ~ 30–40 MPa. At the same time, the porous structure parameters of the resulting adsorbent changed almost insignificantly compared to the powdered

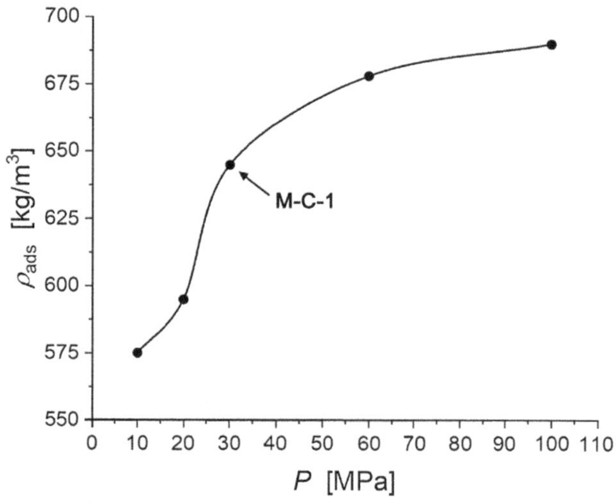

FIGURE 6.3 The density of carbon adsorbent C-1 compacted with the polymer binder versus the compaction pressure.

TABLE 6.1
The Parameters of the Porous Structure Determined by the D–R [7] and BET Equations [35], and the Packing Density, ρ, of Initial Powdered Carbon C-1, and Monolith Block M-C-1 Prepared by Compacting with the Polymer Binder under 30 MPa

Adsorbent	D–R Equation			BET	
	W_0, cm³/g	E_0, kJ/mol	x_0, nm	S_{BET}, cm²/g	ρ, kg/m³
C-1	0.62	19.7	0.61	1110	433
M-C-1	0.54	18.6	0.65	1080	628

material (see Table 6.1). These data indicate that the amount of the polymer binder was sufficient to protect the microporosity from degradation during the compaction under 30 MPa.

6.3.2 The Kinetics Characteristics of Methane Adsorption in C-1

The rates of the fuelling/delivery processes in the C-1/methane adsorption system were estimated from the kinetic curves of methane adsorption/desorption. Figure 6.4 shows the changes in the mass of C-1 over time after a rapid inlet of methane up to the pressure of 7 or 20 MPa (adsorption) and a rapid pressure release from 7 or 20 MPa (desorption) at 303 K.

The kinetic curves were used to quantify a period of time, $\tau_{0.5}$ [s], from the beginning of the experiment until the moment when the amount of adsorbed methane onto C-1 reaches 50% of the equilibrium adsorption capacity. The value of $\tau_{0.5}$ was used to calculate the coefficient of diffusion, D [cm²/s] [45]:

$$D = \left(K \cdot R^2\right) / \left(\pi^2 \cdot \tau_{0.5}\right). \tag{6.7}$$

Here, R is the radius of an adsorbent granule, K is the coefficient determined by the shape of the adsorbent granules (for spherical granules $K = 0.308$).

As follows from the data in Table 6.2, the adsorption process runs faster than desorption. The increase in pressure led to an acceleration of the adsorption/desorption processes, and almost had no effect on the residual mass of methane in pores of C-1 upon desorption. The kinetic experiments also demonstrated that C-1 is applicable for the use in the ANG system. Indeed, as follows from the kinetic curves in Figure 6.4a and b, the times required to reach 95% of methane adsorption capacity (20 and 12 s at 7.0 and 20 MPa, respectively) are much shorter than the time spent to charge or discharge the ANG system that depends on the compressor facilities and gas fitting.

6.3.3 Adsorption Performance of the ANG System Loaded with Granulated C-1 and Compacted M-C-1 Carbon Adsorbents

6.3.3.1 Adsorption Capacity of the ANG System

Figure 6.5 shows the experimental isotherms of methane adsorption on C-1 adsorbent measured over the wide temperature range of 193–333 K and at pressures up to 30 MPa. It was found that methane adsorption was reversible and increased with pressure.

The isotherms of methane adsorption were approximated by the equation derived by Bakaev using statistical thermodynamics for a model of adsorption in cavities considered as quasi-independent subsystems of the grand canonical ensemble [46].

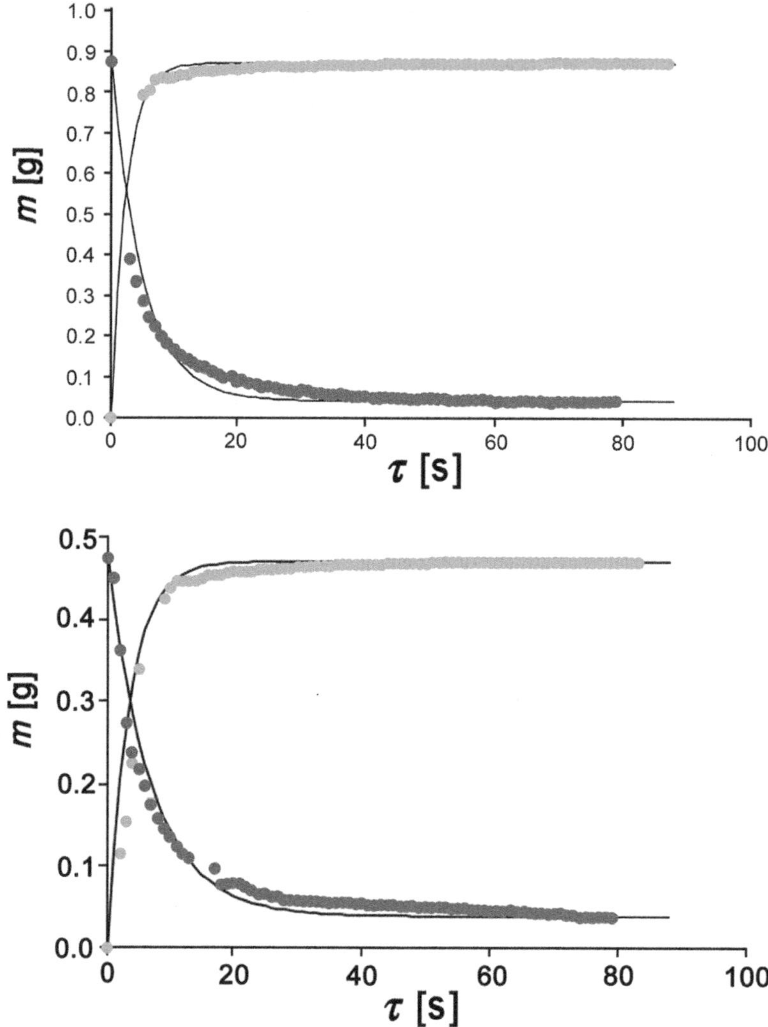

FIGURE 6.4 Kinetic curves of methane adsorption (light grey symbols)/desorption (dark grey symbols) onto C-1 at rapid inlet/release of methane up to a pressure of 7.0 MPa (a) and 20 MPa (b). The temperature of measurements was 303 K. Solid lines are the results of spline approximation.

$$a(P) = \frac{k_0\left(k_1 P + 2k_2 P^2 + 3k_3 P^3\right)}{1 + k_1 P + k_2 P^2 + k_3 P^3} \tag{6.8}$$

where k_0 characterizes an adsorption system, k_1, k_2, k_3 are the temperature-dependent and numerically adjusted coefficients, and P is the equilibrium pressure expressed in Pa.

This formula was successfully used for describing adsorption of different gases on zeolites [46], polymer adsorbents [47], AC [48, 49] and MOFs [42].

Experimental data on the equilibrium methane adsorption onto C-1 measured over the wide temperature and pressure ranges (see Figure 6.5) were used to evaluate two parameters, which are essential for designing the ANG system with AC as an adsorbent: specific volumetric capacity of the ANG system and the isosteric differential molar heat of adsorption. The latter is discussed in the next section.

TABLE 6.2
Kinetic Characteristics of Methane Adsorption/Desorption onto C-1 at the Pressures of 7.0 and 20 MPa and at 303 K: the Period of Time During Which the Amount of Adsorbed/ Desorbed Methane onto C-1 Reached 50% of Equilibrium Adsorption Capacity, $\tau 0.5$ [s], Effective Coefficient of Methane Diffusion, D [cm²/s], Calculated by Eq. (6.7), and the Maximum and Residual Masses of Adsorbed Methane onto C-1: Mads., [g] and Mds, [g], respectively

Process	Mass of adsorbent, [g]	Pressure, [MPa]	$\tau_{0.5}$, [c]	$D \times 10^5$ [cm²/s]	$M_{ads.}$, [g]	M_{ds}, [g]
Adsorption	2.673	7.0	2.426	13.51	0.470	–
		20	1.594	20.57	0.871	–
Desorption	2.673	7.0	4.852	6.76	–	0.037
		20	3.466	9.46	–	0.040

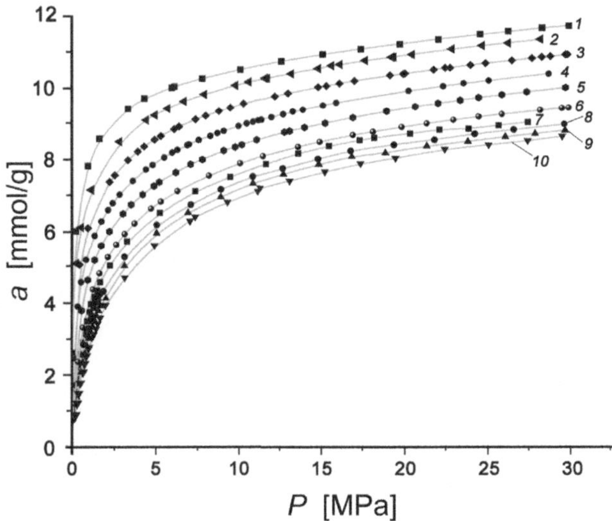

FIGURE 6.5 Dependence of absolute methane adsorption onto initial carbon adsorbent C-1 (symbols) on pressure at the temperatures, K: 193 (*1*), 213 (*2*), 233 (*3*), 253 (*4*), 273 (*5*), 293 (*6*), 303 (*7*), 313 (*8*), 323 (*9*) and 333 (*10*). The solid lines are the results of fitting the experimental data by Eq. (6.8).

Figure 6.6a and b demonstrates the specific volumetric capacity of C-1, V_F [m³(STP)/m³], for methane as a function of pressure for the temperatures ranging from 193 to 323 K (a) compared with that for a CNG tank without adsorbent at 193, 273, and 323 K (b). The comparison allowed one to evaluate the P,T-conditions corresponding to the superiority in the methane storage capacity of the ANG tank with C-1 over the CNG storage system. The region of the ANG storage superiority over CNG expands with increasing temperature towards the higher pressures.

The ANG system with C-1 achieved the DOE target 2012 of 266 m³(STP)/m³ [14] only at high pressure $P = 20$ MPa and low temperatures ≤ 233 K. At the same time, the C-1 sorption properties compare well with those of the other ACs reported in the literature [6, 12, 50–53] and summarized in Table 6.3.

It is important to highlight that the volumetric storage capacity is relevant for a subsequent ANG application of porous materials due to the necessity to adsorb a maximum amount of the target methane molecule in a minimum volume. As follows from Table 6.3, an increase in the porosity (specific BET surface and micropore volume) of an adsorbent is accompanied by a decrease in its packing density, and, as a consequence, does not guarantee the excellent volumetric capacity.

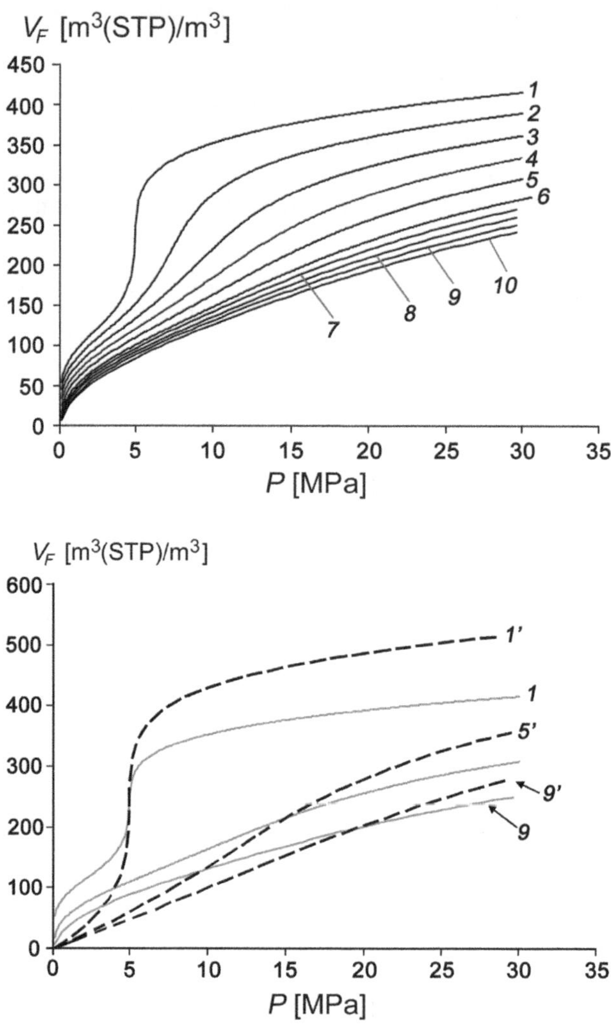

FIGURE 6.6 The dependence of total specific volume of methane, V_F, stored in the ANG system with C-1 as an adsorbent on pressure at the temperature, K: 193 (*1*), 213 (*2*), 233 (*3*), 253 (*4*), 273 (*5*), 293 (*6*), 303 (*7*), 313 (*8*), 323 (*9*) and 333 (*10*), shown by solid black (a) and grey (b) lines, respectively. Dashed black lines show the data for a CNG system at the same P,T-conditions (b): 193 (*1'*), 273 (*5'*) and 323 K (*9'*).

We used the data on Figure 6.6, to calculate the specific deliverable volumetric capacity of the ANG system loaded with C-1 granulated adsorbent, V_A(ANG/C-1) [m³(STP)/m³], at the isothermal conditions as a difference between the values of specific volume of adsorbed methane at 7 (20) and 0.1 MPa. Table 6.4 shows the results of the calculations, which were compared with the deliverable capacity, V_A(ANG/M-C-1), which was experimentally measured for the ANG system loaded with the high-density monolith blocks with an almost similar porosity. In the table, the performance of the ANG systems loaded with the carbon adsorbents is compared with that of the CNG cylinder.

Analysis of Table 6.4 revealed that the performance of both the ANG systems loaded with C-1 and M-C-1 exceeds that for CNG at 7 MPa within the temperature range from 233 to 333 K. The data also indicated a noticeable contribution of the increased packing density of the adsorbent to the deliverable capacity of the ANG system.

TABLE 6.3

Textural Parameters (Packing Density, Specific BETSurface, Micropore Volume) and Total Volumetric Methane Capacity of ACs

Adsorbent [Ref]	ρ, g/cm^3	S_{BET}, m^2/g	W_0, cm^3/g	V_f, m^3(STP)/m^3 (T, P-conditions)
MAXSORB III [50]	0.156	3140	0.179	60 (298 K, 3.5 MPa)
NUCHAR-SA [51]	0.34	1600	1.15	52 (299 K, 3.5 MPa)
PX-21 [6]	0.28	2671	0.93	24 (298 K, 3.5 MPa)
Kansai Maxsorb [6]	0.27	2671	0.93	127(298 K, 3.5 MPa)
California GMS-70 [6]	0.41	1502	0.57	112 (298 K, 3.5 MPa)
Darco©AC [52]	0.36	651	0.131	145 (298 K, 3.5 MPa)
LMA-738 [12]	0.53	3290	1.1	165 (298 K, 20 MPa)
PPAC1:3_10G1 [53]	0.43	2163	0.91	101 (298 K, 4 MPa)
C-1 this work	0.43	1110	0.63	129 (303 K, 7 MPa)

TABLE 6.4

The Specific Deliverable Volumetric Capacities of the ANG System, V_A, Loaded with Compared with that of CNG at Isothermal Conditions

T, K	V_A (ANG -C-1), m^3(STP)/m^3		V_A (ANG-M-C-1), m^3(STP)/m^3		V_A (CNG), m^3(STP)/m^3	
	7 MPa	20 MPa	7 MPa	20 MPa	7 MPa	20 MPa
193	267.5	333.7	n.a.	n.a.	387.7	484.7
213	155.9	309.9			184.5	433.2
233	124.6	282.9	166.2	258.3	123.8	378.1
253	114.5	256.7			100.4	324.2
273	108.8	233.8	n.a.	n.a.	86.5	277.8
293	103.9	214.0			76.9	241.3
303	102.5	206.4			73.1	226.7
313	98.9	198.3	146.2	226.1	69.6	213.7
323	96.8	191.9.	133.8	210.5	66.6	202.3
333	93.4	185.3.	n.a.	n.a.	63.9	192.2

6.3.3.2 Effect of Cyclic Operation on the Compacted Monolith Carbon Adsorbent

The practical implementation of an adsorbent for ANG systems requires the data on its cyclability. Table 6.5 enabled us to assess the resistance of M-C-1 monolith carbon adsorbent to a set of methane adsorption/desorption cycles via the changes in the parameters of its porous structure, density, and hardness (HB [kg/mm^2]) after 5000 cycles.

In our view, the cyclability of an adsorbent depends, on the one hand, on the mechanical properties of the material determined by its phase and chemical composition and, on the other hand, on the magnitude of the adsorption-induced deformation [54–56]. It is obvious that the effect of 5000 cycles on the porous structure of M-C-1 monolith adsorbent is negligible. Only a slight decrease in the hardness of M-C-1 was observed. We attributed these results to a protective impact of the polymer binder. Coudert et al. revealed that in composite adsorbent materials when the adsorbing core is surrounded by an elastic binder, the presence of the binder reduces the magnitude of the adsorption-induced stresses and strains [56].

TABLE 6.5

Textural Properties of M-C-1 Before and After 5000 Methane Adsorption/Desorption Cycles at $T = 293$ K

State	W_0, cm³/g	E_0, kJ/mol	x_0, nm	ρ, g/cm³	S_{BET}, cm²/g	HB, kg/mm²
Before test	0.54	18.6	0.65	433	1080	0.51
After test	0.54	18.5	0.66	433	1070	0.42

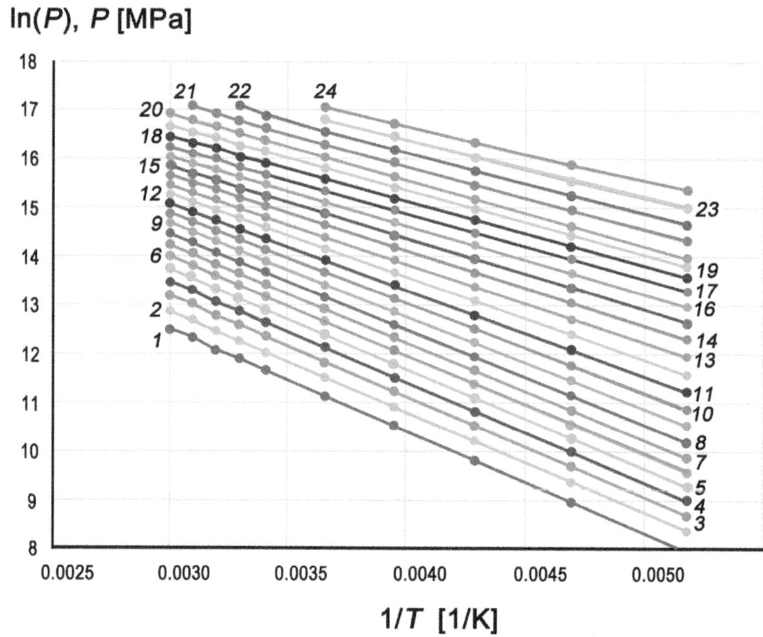

FIGURE 6.7 The isosters of methane adsorption onto C-1 at the values of methane adsorption, mmol/g: *1* – 1.23; *2* – 1.60; *3* – 1.97; *4* – 2.34; *5* – 2.71; *6* – 3.08; *7* – 3.45; *8* – 3.82; *9* – 4.19; *10* – 4.56; *11* – 4.93; *12* – 5.30; *13* – 5.67; *14* – 6.04; *15* – 6.41; *16* – 6.78; *17* – 7.15; *18* – 7.52; *19* – 7.89; *20* – 8.26; *21* – 9.02; *22* – 9.40; *23* – 9.78; *24* –10.16; *25* – 10.92. Symbols mark the experimental data, and the solid straight lines are the linear approximation.

6.3.3.3 Heat Effects in the ANG System

We used a set of experimental adsorption isotherms measured over the wide ranges of temperature and pressures to plot the isosteres in $\ln(P) = f(1/T)$ coordinates for different values of methane adsorption on C-1 adsorbent (see Figure 6.7) and explore the temperature dependence of methane adsorption under sub- and supercritical conditions.

As follows from Figure 6.7, the isosteres of methane adsorption on C-1 are well approximated by linear functions over the entire range of temperatures, including supercritical conditions ($T > T_{cr} = 190.11$ K). The linearity of adsorption isosteres in the region, where gases show the non-ideality behaviours, is indicative of a specific state of a highly dispersed adsorbate in micropores. This state ensures the accumulation of methane molecules in micropores without undergoing any phase transition over wide intervals of sub- and supercritical temperatures and pressures [57–60].

The experimental isosteres were used to evaluate the differential isosteric molar heat of methane adsorption onto C-1, q_{st} [kJ/mol], via the well-known formula derived by Bakaev [61]:

$$q_{st} = -R \cdot Z \cdot \left[\frac{\partial(\ln P)}{\partial(1/T)} \right]_a \cdot \left[1 - \left(\frac{\partial v_{ads}}{\partial a} \right)_T / v_g \right] - \left(\frac{\partial P}{\partial a} \right)_T \cdot \left[v_{ads} - T \cdot \left(\frac{\partial v_{ads}}{\partial T} \right)_a \right]$$ (6.9)

where $Z = P \times v_g/(RT)$ is the coefficient of compressibility of the equilibrium gas phase at pressure P [Pa] and temperature T [K]; v_g is the specific gas phase volume [m^3/kg]; R is the universal gas constant [J/(mol·K]; $v_{ads} = V_{ads}(P,T)/m_0$ is the reduced volume of the adsorbent [cm^3/g]; and V_{ads} and m_0 are the volume and mass of the regenerated adsorbent, respectively. Thus, the Bakaev Eq. (6.9) most fully takes into account the factors which affect the value of differential molar isosteric heat of adsorption: isothermal adsorption-induced deformation $(\partial v_{ads}/\partial a)_T$, temperature isosteric deformation $(\partial v_{ads}/\partial T)_a$, the slopes of the isotherm of adsorption $(\partial P/\partial a)_T$ and isosteres $[\partial \ln P/\partial(1/T)]_a$, and the non-ideality of a gas phase Z.

It was shown in [60] that in the conditions under consideration, the corrections for adsorption-stimulated deformation of carbon adsorbents upon methane adsorption do not exceed ~ 2–3 % and can be ignored in calculating q_{st}. We used the data by Novikova [62] to calculate the maximal value of isosteric temperature deformation $(\partial v_{ads}/\partial T)_a$ and found that the term $T \times (\partial v_{ads}/\partial T)_a$ was much lower than v_{ads} in the studied temperature and pressure intervals. Therefore, we used Eq. (6.9) without corrections for the adsorption-stimulated and thermal deformations of adsorbent.

It should be noted that the differential isosteric heat of adsorption is an important thermodynamic parameter, whose initial value is determined by the adsorbate-adsorbent interactions. In general, variations of the differential isosteric molar heat of gas adsorption during the adsorption process, $q_{st}(a)|_{T=const}$, reflect the changes in the relationship between the adsorbate-adsorbent and adsorbate-adsorbate interactions [57–60]. Figure 6.8 shows how the differential isosteric molar heat of methane adsorption onto C-1 during the adsorption process up to 20 MPa at temperatures ranged from 193 to 273 K. It is seen that at the initial stages of adsorption, up to a ~ 1 mmol/g, the values of $q_{st}(a)$ gradually decreased from 18.4 to ~ 17.0 kJ/mol, indicating the filling of various high-energy centers of adsorption (micropores and heteroatoms) with methane molecules. Subsequent behaviours of the $q_{st}(a)$ functions are determined by the contributions from the methane–methane interactions, leading to the formation methane associates [57–60].

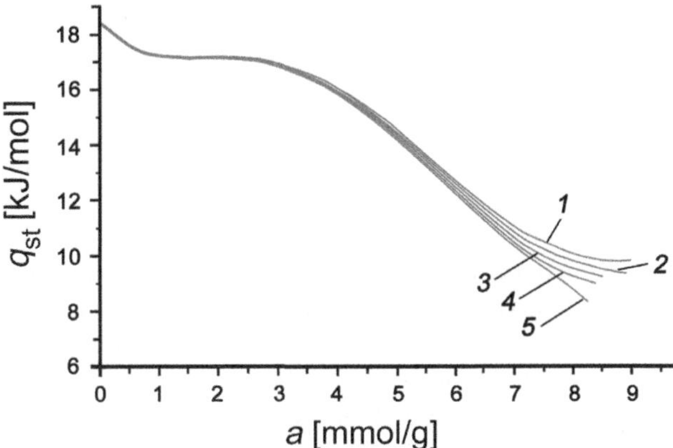

FIGURE 6.8 The variations in the isosteric differential molar heat of adsorption onto the C-1 adsorbent with increasing methane uptake up to the pressure of 20 MPa at a temperature, K: 193 (*1*), 213 (*2*), 233 (*3*), 253 (*4*) and 273 (*5*).

For practical applications, the differential isosteric heat of adsorption is used to calculate the integral heat of adsorption released/absorbed during the adsorption/desorption process:

$$Q_{int}(T) = \int_0^a q_{st}(a)_{T=const}\, da \tag{6.10}$$

The released heat affects the performance of the ANG systems through an increase in the temperature in the adsorber that, on the plus side, accelerates the mass transfer but, on the minus side, it reduces the methane uptake. Therefore, to optimize the fuelling/delivery process, it is necessary to estimate the value of the heat released in the ANG system during adsorption or the heat of adsorption – Q [kJ/kg]. The value of released heat Q can be quantified directly via the measurements of a change of the temperature, ΔT [K], in the ANG tank during its fueling to specified pressure performed by the thermocouple of the adiabatic calorimeter:

$$Q(P) = \Delta T(P,T) \cdot \left(C_0(P) + C_{tank} + C_g(P) \right) \tag{6.11}$$

Here, $C_0(P)$ [kJ/(kg$_{ads}$ · K)] is the heat capacity of the adsorbent/methane system at a specified pressure and temperature, $C_{tank}(T)$ [kJ/(kg · K)] is the heat capacity of the empty ANG tank in the calorimeter, including corrections for various heat losses; $C_g(P,T)$ is the heat capacity of methane in a gas phase at the specified pressure and temperature.

The sum of these parameters can be considered as an effective constant heat capacity, $C_{eff}(P,T)$, for the ANG tank with the adsorbent in the adiabatic calorimeter under the specified pressure and temperature:

$$Q(P) = \Delta T(P,T) \cdot C_{eff} \tag{6.11a}$$

Knowing the effective heat capacity $C_{eff}(P,T)$ for the ANG tank with a porous carbon material in the calorimeter at the specified P,T-conditions, one can evaluate the heat effects for the ANG systems with other porous ACs. This statement is true under the assumption that the heat capacity of carbon adsorbents does not depend on their porosity.

In the present study, we found the $C_{eff}(P,T)$ values using Eqs. (6.10) and (6.11a). First, we measured the maximum values of ΔT_{ads} reached in the ANG tank loaded with M-C-1 monolith blocks (a so-called calibration sample with a mass of 55 g) in the adiabatic calorimeter after a period of time, τ_{max} [s], at three magnitudes of methane pressures (7, 10 and 20 MPa) and three temperatures (233, 293 and 323 K). The measurements were carried out using the adsorption bench (see its scheme in Figure 6.2). Next, we calculated the values of Q according to Eq. (6.10) using the data on $q_{st}(a)$ for the C-1-methane adsorption system at the same P,T-conditions. It should be recalled that the structural and energy parameters of C-1 and M-C-1 almost coincide (see Table 6.1). Thus, the effective constant $C_{eff}(P,T)$ can be easy calculated by inserting the known values ΔT_{max} and Q_{int} into Eq. (6.11a). The results of these calculations as well as the values of ΔT_{max}, τ_{max} and Q_{int} are listed in Table 6.6.

The data in Table 6.6 testified that the heating-up of the ANG system is maximal under adiabatic conditions and low temperatures. The increase in temperature of the adsorption process leads to the acceleration of the heat transfer. At all pressures, the increase in temperature also has an impact on the effective constant heat capacity of the system, reducing its value. At the same time, the increase in pressure has less effect on C_{eff}, especially at high pressures.

6.4   SUMMARY: ANG STORAGE SYSTEM PROTOTYPES

The obtained data are essential for selecting the optimal carbon adsorbent, calculating the fuelling/delivery processes, and designing the heat management of the ANG tank intended to improve its

TABLE 6.6
The Maximum Values of ΔT_{max} Measured in the ANG Tank Loaded with M-C-1 Monolith Blocks (Calibration Adsorbent) During Methane Adsorption and the Period of Time to Reach These Values (τ_{max}), Integral Heat of Methane Adsorption (Q_{int}), and Effective Constant (C_{eff}) of the ANG Tank with M-C-1 in the Adiabatic Calorimeter at Various P,T-Conditions

T, K	ΔT_{max}, K	τ_{max}, s	Q_{int}, kJ/mol	C_{eff}, kJ/(kg$_{ad\,s}$ × K)
$P = 7$ MPa				
233	6.81	300	159.75	23.5
293	6.38	240	130.30	20.4
323	6.00	180	117.07	19.5
$P = 10$ MPa				
233	7.92	220	160.14	20.2
293	7.69	200	135.39	17.6
323	7.23	150	123.88	17.1
$P = 20$ MPa				
233	8.31	200	160.84	19.4
293	8.00	170	138.35	17.3
323	8.95	120	129.54	14.5

FIGURE 6.9 Functionalized monolith carbon adsorbent M-C-1 that fills the prototype of the ANG system for the gas-powered vehicle for a car.

performance. The results of the present study were used for developing the prototypes of ANG storage systems for gas-powered vehicles for a car (see Figure 6.9) with the monolith carbon adsorbent prepared from plant waste – coconut shells.

The monolith carbon adsorbent with increased packing density does not need extraordinary P,T-conditions for high volumetric adsorption capacity. It exhibited excellent cyclability and mechanical strength. The porous structure ensures the fast rates of mass and heat transfers. Thus, the ANG system loaded with the monolith carbon adsorbent operating at moderate pressures makes it possible to move from bulky CNG cylinders to a flat adsorber (shown in the centre of Figure 6.9), which seems to be more comfortable and ergonomic.

FUNDING

The research was carried out within the State Assignment of the Russian Federation (Project No. 122011300053-8).

REFERENCES

1. Natural Gas by Country 2022. Available online: https://worldpopulationreview.com/country-rankings/natural-gas-by-country (accessed on July 11, 2022)

2. Fossil and Alternative Fuels - Energy Content. Net (low) and gross (high) energy content in fossil and alternative fuels. Available online: www.engineeringtoolbox.com/fossil-fuels-energy-content-d_1298.html (accessed on June 12, 2022)

3. Khan, M.I.; Yasmin, T.; Shakoor, A. Technical overview of compressed natural gas (CNG) as a transportation fuel. *Renew. Sust. Energ. Rev.* 2015, 51, 785–797.

4. Mokhatab, S.; Mak, J.Y.; Valappil, J.V.; Wood, D.A. (eds.) LNG Safety and Security Aspects. In: *Chapter 9 in Handbook of Liquefied Natural Gas;*. Mokhatab, S.; Mak, J.Y.; Valappil, J.V.; Wood, D.A. (eds.) Gulf Professional Publishing, 2014, pp. 359–435, ISBN 9780124045859 https://doi.org/10.1016/B978-0-12-404585-9.00009-X

5. Wegrzyn, J.; Gurevich, M. Adsorbent storage of natural gas. *Appl. Energy.* 1996, 55, 71–83

6. Menon, V.C.; Komarneni, S. Porous adsorbents for vehicular natural gas storage: A review. *J. Porous Mater.* 1998, 5, 43–58.

7. Dubinin, M.M. Physical adsorption of gases and vapors in micropores. In: *Progress in Surface and Membrane Science*; Cadenhead, D.D., Ed.; Academic Press: London, 1975, pp. 1–69.

8. Broom, D.P.; Thomas, K.M. Gas adsorption by nanoporous materials: Future applications and experimental challenges. *MRS Bull.* 2014, 38, 412–421.

9. Kumar K.V.; Preuss K.; Titirici M.M.; Rodríguez-Reinoso F. Nanoporous materials for the onboard storage of natural gas. *Chem. Rev.* 2017, 117, 1796–1825.

10. Cook, T.L.; Komodromos, C.; Quinn, D.F.; et al. Adsorbent storage for natural gas vehicles. In: *Carbon Materials for Advanced Technologies*, 1st ed.; Burchell, T.D., Ed.; Elsevier Science Ltd: Oxford, UK, 1999, pp. 269–302.

11. Solar, C.; Blanco, A.G.; Vallone, A.; et al. Adsorption of methane in porous materials as the basis for the storage of natural gas. In: *Natural Gas*; Potocnik, P., Ed.; Sciyo/InTech: Rijeka, Croatia, 2010. pp. 205–249.

12. Casco M.E.; Martínez-Escandell, M.; Gadea-Ramos, E.; et al. High-pressure methane storage in porous materials: are carbon materials in the pole position? *Chem. Mater.* 2015, 27, 959–964.

13. Tsivadze, A.Yu.; Aksyutin, O.E.; Ishkov, A.G.; et al. Porous carbon-based adsorption systems for natural gas (methane) storage. *Russ. Chem. Rev.* 2018, 87, 950–983.

14. ARPA-EMOVEProgram Overview. Available online: https://arpa-e.energy.gov/sites/default/files/documents/files/MOVE_ProgramOverview.pdf.

15. Mahmoud, E. Evolution of the design of CH4 adsorbents. *Surfaces*, 2020, 3, 433–466.

16. Conchi, O.; Ania, C.O.; Raymundo-Piñero, E. Nanoporous carbons with tuned porosity. In *Nanoporous materials for Gas Storage*, 1st ed.; Kaneko, K.; Rodríguez-Reinoso, F., Eds.; Springer Nature Pte Ltd: Singapore, 2019, p. 712.

17. Men'shchikov, I.E.; Shiryaev, A.A.; Shkolin, A.V. et al. Carbon adsorbents for methane storage: genesis, synthesis, porosity, adsorption, *Korean J. Chem. Eng.* 2021, 38, 276–291.

18. Dubinin, M.M. Mikroporistie sistemy yglerodnikh adsorbentov (Microporous systems of carbon adsorbents). In: *Uglerodnie adsorbenty I ikh promishlennoe primenenie (Carbon Adsorbents and Their Industrial Applications)*; Plachenov, T.G., Ed.; Nauka: Moscow, 1983, pp. 186–192. (in Russian).

19. Fenelonov, V.B. *Poristyi uglerod (Porous Carbon)*. Izd. IK RAN: Novosibirsk, Russia, 1995, p. 518 (in Russian).

20. Marsh, H.; Rodriguez-Reinoso, F. *Activated carbon*. Elsevier: Oxford, 2006.

21. Menéndez-Díaza, J.A.; Martín-Gullón, I. Types of carbon adsorbents and their production. In: *Activated Carbon Surfaces in Environmental Remediation (Interface Science and Technology Series, 7)*; Bandosz, T., Ed.; Academic Press: New York, USA, 2006, p. 148.

22. Dubinin, M.M. The potential theory of adsorption of gases and vapors for adsorbents with energetically nonuniform surfaces, *Chem. Rev.* 1960, 60, 235–241.

23. Sing, K.S.W.; Everett, D.H.; Haul, R.W.; et al. Reporting physisorption data for gas/solid systems with special reference to the determination of surface area and porosity. *Pure Appl. Chem.* 1985, 57, 603–619.

24. Thommes, M.; Kaneko, K.; Neimark, A.V.; et al. Physisorption of gases, with special reference to the evaluation of surface area and pore size distribution. *Pure Appl. Chem.* 2015, 87, 1051–1069.

25. Rodríguez-Reinoso, F. Activated carbon and adsorption. In: *Encyclopedia of Materials: Science and Technology*; Buschow, K.H.J., Cahn, R.W., Flemings, M.C., Ilschner, B., Kramer, E.J., Mahajan, S., Veyssière, P., Eds.; Elsevier: Amsterdam, The Netherlands, 2001, pp. 22–34.

26. Marsh, H.; Heintz, E.; Rodriguez-Reinoso, F. *Introduction to Carbon Technologies.* University of Alicante: Alicante, 1997.

27. Mukhin, V.M.; Tarasov, A.V.; Klushin, V.N. *Active carbon of Russia*, Metallurgiya: Moscow, 2000 (in Russian).

28. Achaw, O.-W.; Afrane, G. The evolution of the pore structure of coconut shells during the preparation of coconut shell-based activated carbons, *Micropor. Mesopor. Mater.* 2008, 112, 284–290.

29. Giraldo, L.; Moreno-Pirajon; J.C.; Novel activated carbon monoliths for methane adsorption obtained from coffee husks. *Mater. Sci. Appl.* 2011, 2, 331–339.

30. Solovtsova, O.V.; Chugaev, S.S.; Men'shchikov, I.E.; et al. High-density carbon adsorbents for natural gas storage. *Colloid J.* 2020, 82, 719–726.

31. Chang, K.J.; Talu, O. Behavior and performance of adsorptive natural gas storage cylinders during discharge. *Appl. Therm. Eng.* 1996, 16, 359–374.

32. Talu, O. Physical chemistry and engineering for adsorptive gas storage in nanoporous solids, Chapter 4. In *Nanoporous Materials for Gas Storage*; Kaneko, K.; Rodríguez-Reinoso, F., Eds.; Springer Nature Pte Ltd: Singapore, 2019, p. 6590.

33. Shkolin, A.V.; Fomkin, A.A.; Men'shchikov, I.E. et al. Monolithic microporous carbon adsorbent for low-temperature natural gas storage. *Adsorption*, 2019, 25, 1559–1573.

34. Strizhenov, E.M.; Chugaev, S.S.; Men'shchikov, I.E.; Shkolin, A.V.; Zherdev, A.A. Heat and mass transfer in an adsorbed natural gas storage system filled with monolithic carbon adsorbent during circulating gas charging. *Nanomaterials*, 2021, 11, 3274.

35. Rangari, P.J.; Chavan, P. A review on preparation of activated carbon from coconut shell. Int. *J. Innov. Res. Sci. Eng. Technol.* 2017, 6(4), 5829–5849.

36. Tabor, D. *The hardness of metals.* Oxford University Press Inc.: New York, 1951.

37. Lemmon E.W.; Huber, M.L.; McLinden, M.O. NIST. Standard Reference Database 23: Reference Fluid Thermodynamic and Transport Properties-REFPROP. Version 9.1., 2013.

38. Sychev, V.V.; Vasserman, A.A. Zagoruchenko, V.A. et al. *Thermodynamic properties of methane.* Izd. Standartov: Moscow, 1979 (in Russian).

39. Brunauer, S.; Emmett, P.H.; Teller, E. Adsorption of gases in multimolecular layers. *J. Am. Chem. Soc.* 1938, 60, 309–319.

40. Pribylov, A.A.; Serpinskii, V.V.; Kalashnikov, S.M. Adsorption of gases by microporous adsorbents under pressures up to hundreds of megapascals. *Zeolites*, 1991, 11, 846–849.

41. Fomkin, A.A.; Shkolin, A.V.; Men'shchikov, I.E.; et al. Measurement of adsorption of methane at high pressures for alternative energy systems, *J. Meas. Techn.* 2016, 58, 1387–1391.

42. Solovtsova O.V., Men'shchikov I.E., Shkolin A.V., et al. ZrBDC-Based Functional Adsorbents for Small-Scale Methane Storage Systems. *Adsorp. Sci. Techn.* 2022, 2022, Article ID 4855466.

43. Fomkin, A.A.; Seliverstova, I.I.; Serpinskii, V.V. Determination of the parameters of the microprobe structure of solid adsorbents. Communication 1. Method of determination of the specific volume of totally microporous adsorbents. *Russ. Chem. Bull.* 1986, 35, 256259.

44. GOST 34100.3-2017/ISO/IEC Guide 98-3:2008. Part 3. Uncertainty of measurement. Part 3. Guide to the expression of uncertainty in measurement. Available online: https://files.stroyinf.ru/Data/651/65118.pdf (accessed on 18 May, 2021).

45. Kel'tsev, N.V. *Osnovy adsorbtsionnoi praktiki (Fundamentals of Adsorption Practice).* Khimiya: Moscow, USSR, 1976, p. 90 (in Russian).

46. Bakaev, V.A. The statistical thermodynamics of adsorption equilibriums in the case of zeolites. Dokl. *Acad. Nauk SSSR [Dokl.Chem. (Engl. Transl.)]*, 1966, 167, 369372.

47. Pribylov, A.A.; Murdmaa, K.A. Adsorption of gases onto Polymer Adsorbent MN-270 in the region of supercritical temperatures and pressures. *Prot. Met. Phys. Chem. Surf.* 2020, 56, 115–121.

48. Men'shchikov, I.E.; Shkolin, A.V.; Khozina, E.V.; et al. Thermodynamics of adsorbed methane storage systems based on peat-derived activated carbons. *Nanomaterials*, 2020, 10, 1379.

49. Men'shchikov, I.; Shkolin, A., Khozina, E.; et al. Peculiarities of thermodynamic behaviors of xenon adsorption on the activated carbon prepared from silicon carbide. *Nanomaterials.* 2021, 11(4), 971–994.

50. Rahman, K.A.; Loh, W.S.; Yanagi, H.; et al. Experimental adsorption isotherm of methane onto activated carbon at sub- and supercritical temperatures. *J. Chem. Eng. Data.* 2010, 55, 4961–4967.

51. Policicchio, A.; MacCallini, E.; Agostino, R.G.; et al. G. Higher methane storage at low pressure and room temperature in new easily scalable large-scale production activated carbon for static and vehicular applications. *Fuel.* 2013, 104, 813–821.

52. Policicchio, A.; Filosa, R.; Abate, S.; et al. Activated carbon and metal organic framework as adsorbent for low-pressure methane storage applications: an overview. *J. Porous Mater.* 2017, 24. 905–922.

53. Reljic, S.; Cuadrado-Collados, C.; Farrando-Perez, J.; et al. Carbon-based monoliths with improved thermal and mechanical properties for methane storage. *Fuel.* 2022, 324(Part C), 124753–124779.

54. Gor, G.Y.; Huber, P.; Bernstein, N. Adsorption-induced deformation of nanoporous materials: a review, *Appl. Phys. Rev.* 2017, 4, 011303

55. Shkolin, A.V.; Men'shchikov, I.E.; Khozina, E.V.; et al. Deformation of microporous carbon adsorbent Sorbonorit-4 during methane adsorption. *Chem. Eng. Data.* 2022, 67, 1699–1714.

56. Coudert, F-X.; Fuchs, A.H.; Neimark, A.V. Adsorption deformation of microporous composites. *J. Chem. Soc. Dalton. Trans.* 2016, 45, 4136–4140.

57. Fomkin, A.A. Adsorption of gases, vapors and liquids by microporous adsorbents. *Adsorption.* 2005, 11, 425–436.

58. Shkolin, A.V.; Fomkin, A.A. Thermodynamics of methane adsorption on the microporous carbon adsorbent ACC. *Russ. Chem. Bull.* 2008, 57, 1799–1805.

59. Tolmachev, A.M. Adsorption of gases, vapors, and solutions: I. Thermodynamics of adsorption. *Prot. Met. Phys. Chem. Phys. Surf.* 2010, 46, 955–963.

60. Shkolin, A.V.; Fomkin, A.A.; Tsivadze, A.Yu.; et al. Experimental study and numerical modeling: methane adsorption in microporous carbon adsorbent over the subcritical and supercritical temperature regions. *Prot. Met. Phys. Chem. Phys. Surf.* 2016, 52, 955–963.

61. Bakaev, V.A. *Molecular theory of physical adsorption. Doctoral (Phys.-Math.) Dissertation.* Moscow State University: Moscow, 1989.

62. Novikova, S.I. *Teplovoe rasshirenie tverdykh tel (Heat Expansion of Solids).* Nauka: Moscow, USSR, 1974, p. 293 (in Russian).

7 Utilization of Plastic Waste in Designing Tiles for Societal Usage

A Step Towards a Circular Economy

S.K. Dhawan, Ridham Dhawan and Vikas Garg

7.1 INTRODUCTION

If there is one type of municipal solid waste that has become ubiquitous in India and most developing countries, and largely seen along the shores and waterways of many developed countries, it is plastic waste. Much of it is not recycled, and ends up in landfills or as litter on land, in waterways and the ocean as shown in Figure 7.1. In particular, plastic carry bags are the biggest contributors of littered waste and every year, millions of plastic bags end up in the environment vis-a-vis soil, water bodies, water courses, etc., and it takes an average of one thousand years to decompose completely.

Piles of plastic in myriad forms lying by the roads, in the sewer, and in city dumps are a common sight in India. The metro cities are major culprits with Delhi producing 690 tons a day, Chennai 429 tons, Kolkata 426 tons and Mumbai producing 408 tons. The Central Pollution Control Board told the court that the 'total plastic waste which is collected and recycled in the country is estimated to be 9,205 tons per day (approximately 60 per cent of total plastic waste) and 6,137 tons remain uncollected and littered'. A survey conducted in 60 major cities found that 15,342.46 tons of plastic waste was generated every day, amounting to 56 lakh tons a year. Plastic waste has certainly become a threat to our existence. It has adversely affected the flora and fauna, and has also become a cause of further deteriorating environmental conditions.

The tons of plastic waste is collected and dumped on sites making surroundings look and smell awful. Society has to bear the consequences of vast utilization of these non-biodegradable plastics. Recycling of plastics is the only solution to this problem, the technology for converting plastic waste into tiles can divert the majority of plastic waste from landfill sites to commercial markets, thereby improving the environmental conditions of society and the quality of life. But the Supreme Court of India has reprimanded big cities for their poor waste management. Disposal of plastic waste is a major problem. It is non-biodegradable and it mainly consists of low-density polyethylene plastic bags, bottles, etc. Burning of these waste plastic bags causes environmental pollution. Our research on 'Utilization of waste plastic to tiles designing', offers a simple and novel process for the production of tiles from waste plastic bags and bottles. Thus, this technology not only provides a sustainable living for the people who are collecting them from the garbage but also converts waste into a useful product and saves the environment. These tiles can be used in buildings of toilets and rooms for the general public for societal benefits. Some specific tests like flammability test, water absorption, and mechanical strength of the tiles have been carried out as per ASTM standards.

Plastics due to their versatile nature are being widely employed in human lives. With the increasing use of plastics in different commercial applications they make up a fundamental part of our everyday lives [1]. Different types of plastics are widely used are thermoplastics and thermo-setting plastics.

DOI: 10.1201/9781003327646-7

FIGURE 7.1   Waste plastic bags have become a major source of pollution.

Among these, thermoplastics can be easily and cheaply moulded and re-moulded to different usable forms [2]. Thus, thermoplastics have a wide market globally due to its low processing cost, durability, strength and chemical resistance [3]. But the very factors which popularize the use of these plastics also make them a threat for the existence of life on Earth. Chemically, these plastics are non-biodegradable and can add to ground and water pollution; these factors make their disposal a topic of major concern [4, 5].

The technological advancements around the globe have led to an increased welfare level along with increased consumption per capita. This hike in consumption rate has gradually increased the quantum of waste plastic in municipal solid waste (MSW) causing widespread littering on landscape [6, 7]. Over 78% of this total plastic waste generated per year corresponds to thermoplastics mainly consisting of poly-olefins, which mainly consist of low-density polyethylene (LDPE), high-density polyethylene (HDPE) and polypropylene (PP), and the remaining 22% belongs to thermosetting polymer like epoxy resins, poly-urethane, etc. [8].

The major difference between two broad classes of plastics lies in the manner they respond to temperature changes. Thermoplastics melt and soften with an increase in temperature, whereas the thermosets polymerize and gets stiffer on increasing the temperature. Thermoplastics can further be differentiated based on their chemical structure and morphology. Thermoplastic polymers can show both amorphous and crystalline morphologies. The thermoplastics with randomly arranged polymeric chains, like polystyrene (PS) and polyvinyl chloride (PVC), exhibit amorphous structure, whereas polymers with ordered molecular arrangement form a crystalline structure with varying degrees of crystallinity that governs the physical and mechanical properties of the polymer structure. Another class of plastics that has emerged in the past decade is engineering plastics, which have found its application as a substitute to metals in small devices and structures with adequate physical and mechanical properties. They have many advantages like low processing cost, resistance to different chemicals and corrosion, low-density and ease of process ability. Although plastics do not have any superior properties in comparison to other materials but the favourable balance of different properties and relative ease of getting moulded into complex shapes make them popular in comparison to other engineered materials.

Among the different types of plastics available, thermoplastics are considered suitable for the fabrication of composites as they can undergo repeated melting processes whereas the thermosetting plastics can undergo only a single processing cycle and become irreversible solids post initial increase in temperature. Also, thermoplastics blend with different fillers with a lot of ease in extruders followed by moulding (injection or extrusion) of the composite in order to obtain the final product.

About 32% of the thermoplastic waste being generated is composed of low-density polyethylene (LDPE), which is commonly used for packaging and non-packaging film and sheet applications [9]. The different polymer products like plastic bags, carry-out bags, garbage bags, industrial sheeting, agricultural and commercial films, toys, house-wares and pipes that have a wide commercial market and day-to-day use are all composed of LDPE [10–12]. Thus, disposal of waste plastic is a major challenge for the local governing authorities looking into solid waste management (SWM) and sanitation [13]. In the absence of integrated solid waste management practices, the plastic waste generated is neither collected nor disposed of in a proper manner, causing littering of waste and choking of the sewage systems, having harmful impacts on public health and the surrounding environment [14].

Waste plastic persists for a long time as the natural degradation requires more than 500 years, thus making it the most visible constituent of waste dumps and landfills [5, 15, 16]. The challenge of disposing of plastics in municipal wastes has received little public attention until recently. The material has simply been buried or burned. But the rising costs of landfills, and of transporting wastes to other states, reduced the availability of space and concern about the health implications of incinerating some materials, have aroused the need for waste plastic utilization or recycling in an environmentally sound manner [9, 17, 18].

Recycling of waste is the best strategy for solid waste management, but it can also be seen as one current example of implementing the concept of industrial ecology, whereas in a natural ecosystem there are no wastes but only products. Recycling of plastics is one method by which one can reduce its negative impact on the environment and prevent depletion of resources and can therefore decrease energy and material usage per unit of output and so yield improved eco-efficiency [19–21]. In recent years, researchers have focused their attention on finding suitable methods for recycling and reutilization of plastic waste material in an economically and environmentally viable manner. Many workable technologies have been developed for recycling purpose, Butler et al. studied the current state-of-the-art processed and gave a review on the commercial-scale conversion of plastic waste to oil products [22]. The researchers have also reported the method for reutilization of waste plastic into new products by breaking the long polymeric chain by pyrolysis [23, 24]. Scientists have also developed techniques for utilization of plastic waste as a fine aggregate in a concrete mixture for obtaining a product of high compressive and flexural strength along with mechanical resistance comparable to conventional concrete products [25–27]. Much work has been done on building flexible road pavements by incorporating waste plastic into the matrix in road construction, bitumen plays the important role of the binder, but it has poor water resistance, so researchers and engineers have developed a technique to modify this bitumen by coating it with waste plastic to improve its properties. The innovation helps in strengthening the construction of a road thereby increasing its service life [28]. Also research has been carried out to use plastic waste as a carbon precursor for synthesizing carbon nano-tubes (CNTs) [29, 30].

According to the current scenario, the plastic recycling industry rather than solving the problem of plastic disposal is creating more problems and with the influx of plastic waste import it is getting aggravated. Thus, recycling of waste plastic in the most environmentally sound manner and by making its biggest drawback an advantage can be achieved my moulding it into tiles for building structures. The new product is environment friendly and its development may go a long way in solving the problem of handling plastic waste. Technological advancements and innovations in our society are highly extensive in nature and the different methodologies involved adversely affects the environmental conditions. With a slight increase in population there is an uncontrolled pumping of raw materials for developing both primary and secondary products, energy and for the urbanization of the environment. All this leads to harmful emissions and increased waste generation. Since we lack abundance of mineral resources there is a need for sustainable usage of the limited resources available. The resources required for fulfilling the demand of our growing population are limited and these include food, drinking water, fuel along with material required construction of buildings and development of infrastructure, therefore industry needs to ensure sustainable development for the welfare of people. Thus, there is a significant need for finding an alternative to conventional raw materials for production of tiles and other construction materials. In the present era, the major challenge lies in significantly reducing the energy usage while increasing the effectiveness of the material processing techniques. The present era demands sustainable construction, which has fostered research in the direction of alternative resources made of the newer materials, such as composites derived from renewable resources, which are obtained either directly from nature or from the waste of different industries, thereby encouraging the implementation of the recycling process. The awareness about these problems let research be focused on finding 'green chemistry' solutions for maximizing the efficiency and minimizing the waste generation. Thus, it involves the utilization of recyclable materials from different conventional industries by cleaner and non-polluting technologies to ultimately promote the preservation and protection of environment for future generations. Since cement and concrete are most widely used construction materials worldwide, the new approach must comprise of a change in the design of structures by introducing different alternatives like waste plastic as a substitute to concrete for making tiles.

As the modern construction for both commercial and residential buildings are subjected to diverse objectives and constraints, the plastic-based building materials can be effectively used as a replacement to conventional building materials, such as bricks, cement, concrete, wood, metal and glass. The plastic composite materials work in complimentary fashion with traditional ones to improve their performance by different and innovative applications in order to satisfy the needs of modern building construction industries. One of the major advantages of using plastic composite materials in the building materials is their low density and ability to be formed into complex shapes with great ease. For example, a thermoset like polyurethane foam has found application in building insulation due to the inherent nature of plastics of poor conductors of heat, a thermoplastic like polycarbonate which due to its transparent structure is being used for glazing, also wood plastic composite (WPC) is being widely used for railing and decking purpose. Plastics are hydrophobic in nature; thus, they can account for an attractive raw material for fabrication of composites with thermal and moisture barriers. Thus, plastics due to their wide spectrum of properties, extremely high durability, low maintenance, ease of processing and wide availability can be used for both structural and non-structural applications in the building and construction industry.

7.1.1 WHY IT IS IMPORTANT TO RECYCLE PLASTIC?

- Before we move to why it is important to recycle plastic first recall that what is plastic waste recycling. Plastic waste recycling can be defined as a process that converts the plastic scrap into a useful product or we can say that plastic waste recycling is a technique which helps in the minimization of plastic waste.
- The present world in which we are living is surrounded with plastic waste and this waste has a huge negative impact on the natural environment. It the plastic waste is present in oceans then it is polluting oceans if it is present in land, it is polluting land. Several animals die when this plastic enters into their food chain.
- Therefore, recycling of plastic waste provides us with a huge benefit in cleaning the environment, saving the life of humans and animals and also as a business opportunity.

7.1.2 WHY WASTE PLASTIC TO TILES TECHNOLOGY?

1. Disposal of plastic waste is a major problem. It is non-biodegradable and it mainly consists of low-density polyethylene plastic bags and other plastic scrap.

2. 500 trillion bags are consumed every year in the whole world. And this we are just throwing away, which is the main cause of plastic pollution.
3. Burning of these waste plastic bags causes environmental pollution. Burning 1 kilo of plastic releases 3 kilos of carbon dioxide, a gas that contributes to global warming.
4. Plastic waste going to a landfill site may take 500–1000 years to completely degrade.
5. The main objective of our research is to utilize waste plastic scrap for designing tiles in the building of structures and rooms for the general public for societal benefits. The various issues like mechanical strength, flame retardancy, water permeability, UV-protection from sunlight, acid and alkali resistance and antistatic response are the novelty of the technology.

7.1.3 BENEFITS OF TECHNOLOGY

The technology offers a simple and novel process for the production of tiles from waste plastic bags and bottles. Recycling of waste plastic bags and bottles into decorative coloured tiles creates a durable material from waste. Thus, this technology not only provides a sustainable living for the people who are collecting them from the garbage but also converts waste into a useful product and saves the environment.

Plastic products have made life easier for people living throughout the world. However, plastics also have numerous disadvantages. Plastic products after their use are thrown away in the environment. Usually, plastic bags, which are carelessly disposed of by consumers, are becoming an environmental nuisance. Plastic degradation in nature is very slow and a piece of plastic may last for several hundred years. Disposed plastic bags and bottles increase the volume of waste dumped or burned in dumpsites and landfills. Cows, stray animals, etc., eat them and choke, marine animals die, and drainage systems are clogged. The bags and bottles accumulate in ugly heaps alongside roads. Water pools in them, bringing mosquitoes and disease. Burning of these waste plastic bags causes environmental pollution. This technology provides a solution for the solid waste management problem and promotes a waste-to-usable technology programme, a much-needed impetus to India's recycling industry.

These tiles can be used in the building of toilets and rooms for the general public for societal benefits. Some specific tests like flammability test, water absorption and mechanical strength of the tiles have been carried out as per ASTM standards. Other parameters, like, environmental stability, resistance against strong acids and strong bases are also successfully tested.

7.1.4 MARKET POTENTIAL

Plastic recycling is growing in India and the whole world and the market is huge. It is essential to save the recyclable waste material from going to the waste processing and disposal sites and using up landfill space. Salvaging it at source for recycling could make profitable use of such material. This will save national resource and also save the cost and efforts to dispose of such waste. Currently, a very small percentage of waste plastics are recycled whereas the potential is much higher. As plastic consumption is expected to grow in coming years, the market for recycled plastic products is huge.

7.2 TYPE AND PROPERTIES OF FILLERS ADDED IN THE PLASTIC MATRIX

7.2.1 FLY ASH

Fly ash, the fine particulate material, is a type of industrial waste and its disposal is currently a major environmental concern worldwide. It is the waste product generated by the coal combustion process at pulverized coal-based thermal plants. It is produced when coal is pulverized in the presence of air in the combustion chamber of the boiler where it undergoes immediate ignition, releasing lots of heat and producing a molten mineral residue. The heat generated is extracted from the boiler and the molten mineral residue along with other flue gases is allowed to cool and form the ash. The hardened coarse ash particles, termed as 'bottom ash', settles at the bottom of the combustion chamber whereas the light weight pulverized fuel ash or the 'fly ash' can be extracted from the flue gases by any suitable particulate emission control method like electrostatic precipitation or cyclone separation. In the process, about 70–90% of fly ash is released while the coarse-grained bottom ash removed is only 10–30%. The mixture of fly ash and bottom ash in any significant proportion when mixed with a large amount of water to form a slurry to be deposited in ponds and lagoons where water eventually gets drained. This deposited ash is referred to as 'pond ash'. When bottom ash or fly ash or a mixture of them in any random proportion is conveyed or deposited in a dry manner then it is referred to as 'mound ash'. Fly ash, the major constituent of coal combustion residue, is an environmental pollutant having a complex heterogeneous mixture of amorphous and crystalline phases, which in turns makes it a potential resource material. The different physical, chemical and mineralogical properties of fly ash being generated in coal thermal power plants depend on the type of coal (bituminous or sub-bituminous), rate of combustion and conditions, emission control devices used and handling methods.

7.2.2 Physical Properties

Fly ash, a finely divided powder consists of spherical particles of size ranging between 10 to 100 microns. Figure 7.2 is the scanning electron micrograph (SEM) depicting the spherical surface morphology of the fly ah particles. The finely divided spherical particles help in improving the fluidity and workability of the fly ash as resource material.

Fly ash particles exhibit different colours from tan to dark grey, as shown in Figure 7.2, based on its chemical and mineral composition, but the colour is consistent for a particular thermal power plant and the coal source. The particles with high lime content are tan or light in colour, whereas particles with high iron content appear to be brownish in colour. The dark grey color of fly ash particles is due to the large content of unburnt carbon or organic material. The physical properties are summarized below in Table 7.1.

7.2.3 Chemical Properties

The chemical composition of fly ash plays a significant role in determining the chemical properties of the coal combustion residue. The chemical analysis of fly ash by x-ray fluorescence and other

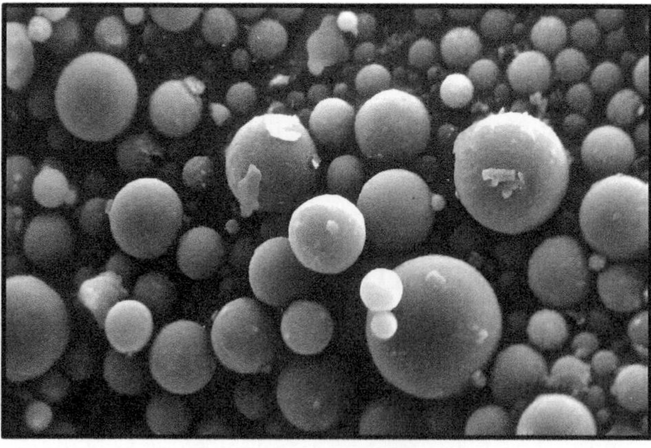

FIGURE 7.2   SEM image of fly ash particles at 2000× magnification.

TABLE 7.1
Physical Properties of Fly Ash

Structure	fine, powdery particles
Shape	spherical, either solid or hollow
Nature	mostly glassy (amorphous)
Particle size	less than a 0.075 mm or No. 200 sieve
Density	1.5–2.5 mg/m^3
Specific gravity	2.1 to 3.0
Specific surface area	170 to 1000 m^2 /kg
Colour	tan to grey to black (colour intensity varies from light to dark with increased amount of carbon content)

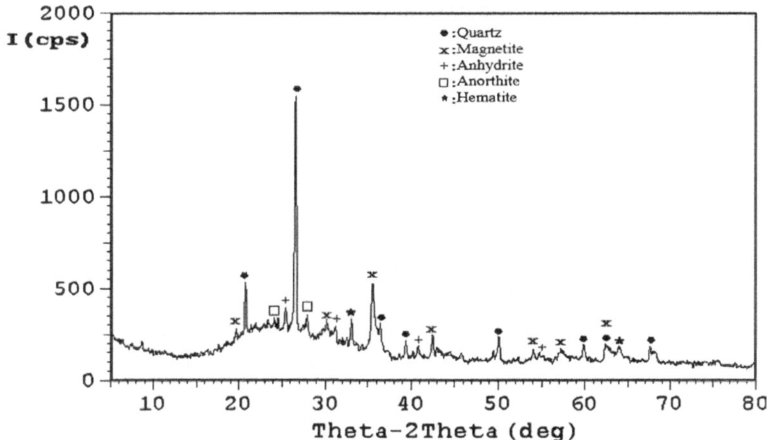

FIGURE 7.3   XRD of fly ash.

spectrometric techniques states that fly ash is basically a ferro-aluminosilicate material with the main elements aluminum (Al), silicon (Si), calcium (Ca), magnesium (Mg), iron (Fe) and sodium (Na) being present in their oxide form. In addition to this, fly ash also contains some toxic metals like lead (Pb), arsenic (As), mercury (Hg), chromium (Cr), nickel (Ni) and zinc (Zn). The table gives the weight percentage of different constituents present in the fly ash.

The x-ray diffraction (XRD) analysis of fly ash, as shown in Figure 7.3, quantifies the presence of different crystallite phases like quartz, magnetite, hematite, mullite, anorthite and other minerals.

7.2.4   DISPOSAL AND UTILIZATION OF FLY ASH

The fly ash is generally collected and found littered around the industrial plants and from there it is mixed with water and discharged into landfills or fly ash settling ponds. Fly ash is even stored in large quantities as waste heaps where if contaminated it can pose a serious environmental threat as it is a major source of inorganic pollution. As fly ash contains metal constituents, which originate due to the composition of coal uses and the way it undergoes the combustion process, in a reasonable amount, they can easily leach out in the environment during storage and can have deleterious effects on the flora and fauna. Since its disposal has such adverse effects on the environmental conditions and human health, its utilization in different ways can be an effective alternative and step towards solid waste management. In many developing countries, fly ash is being widely used in the construction industry. Its use in manufacturing of cement concrete is gaining momentum in countries like the UK, USA, India, etc., as fly ash, on being added to the mixture of cement concrete, improves its properties and lowers the cost of production. The incorporation of fly ash benefits the cement concrete in many ways as it improves the long-term strength, increased workability, lowers the heat of hydration, reduces the water permeability and alkali aggregate reactions and provides resistance to sulphate attacks. All these benefits come from the chemical reaction involved in the mixing process of fly ash and the cement concrete. During the setting and hydration process, concrete liberates lime, which then reacts with the fly ash present in the mixture to form additional cement like material to further improve the properties of the mix. Due to its ability to impart some exceptional properties to the existing materials, it has found application in ceramics, agriculture, metallurgical and environment-related fields. Common areas of use are cement, concrete, ready-mixed concrete, cement or lime-based fly ash bricks and blocks for walling prefabricated building elements, land reclamation, soil stabilization, road constructions, embankments, landfills, etc. Non-engineering applications are in agriculture, plant nutrients, ceramics, neutralizing soil acidity, metal extraction, etc.

7.2.5 RICE HUSK

With growing environmental consciousness at all levels of society, the pollution and health hazards especially associated with the agricultural industries, are coming under intense scrutiny from environmentalists and the governments. These agricultural wastes are mostly the by-products of oil and coal burning by-products, slag, rice husk ash, bagasse, fly ash, cement dust, stone crusher dust, marble dust, brick dust, sewer sludge, glass, tires, etc. This material is available in abundance and is discarded in about millions of tons every year. Such waste materials pose a serious environmental threat and lead to air pollution. They pose environmental problems like air pollution and leaching of hazardous and toxic chemicals (arsenic, beryllium, boron, cadmium, chromium, chromium (VI), cobalt, lead, manganese, mercury, molybdenum, selenium, strontium, thallium, and vanadium, along with dioxins and polycyclic aromatic hydrocarbon compounds, etc.) when dumped in landfills, quarries, rivers and oceans. Consequently, air and water pollution have been inextricably linked to environmental problems and climate change. Increasing concern for environmental protection, energy conservation with minimal impact on economy have been motivating researchers to look for safer alternatives to put these waste products to use. Rice is an important staple food for approximately half of the world population and more than 70 countries mainly China, India, Indonesia produce rice. The quantum of global production of paddy is close to 650 million tons per annum. The production of rice is dominated by Asia, where rice is the only food crop that can be grown during the rainy season in the waterlogged tropical areas. Asia generates over 90% of world rice production. Together, China and India accounted for over half of the world's rice supply. In India, Tamil Nadu is the third ranking state in the production of paddy after Andhra Pradesh and West Bengal. Paddy production is nearly 7 million tons in Tamil Nadu. Paddy, on average, consists of about 72% of rice, 5–8% of bran, and 20–22% of husk.

Globally, the annual output of rice paddy is approximately 600 million tons, of which 95% is produced by 20 countries. Assuming the ratio of a husk to paddy was approximately 20% and the ratio of an ash to husk was approximately 18%, this also means approximately 22 million tons of RH is produced and about 6.4 million tons of RHA is generated in China alone. So large amounts of RHA become refuse, and may not be available at all, and it is also a big problem to get rid of it. Rice husk is one of the most widely available agricultural wastes in many rice producing countries of the world. Rice hulls (or rice husks) are the hard-protecting coverings of grains of rice and removed from rice seed as a by-product during the milling process. Rice husk is used as a value-added raw material for different purposes. The annual global paddy production estimates for the year 2017 are

FIGURE 7.4   Rice husk to RHA.

678 million tons, which implies 149.16 million tons of rice husk production that year from which 37 million of tons of RHA can be obtained by heat treating the rice husk under controlled temperature conditions.

Rice milling generates a by-product known as husk. This surrounds the paddy grain. During the milling of paddy about 78% of weight is received as rice, broken rice and bran. The rest 22% of the weight of paddy is received as husk. This husk is used as fuel in the rice mills to generate steam for the parboiling process. This husk contains about 75% organic volatile matter, which burns up and the balance 25% of the weight of this husk is converted into ash during the firing process, which is known as rice husk ash (RHA) as shown in Figure 7.4. Rice husk was burnt approximately 48 hours under an uncontrolled combustion process. The burning temperature was within the range of 600 to 850 degrees. The ash obtained was ground in a ball mill for 30 minutes and its colour was seen as grey. This RHA in turn contains around 85%–90% amorphous silica. So, for every 1000 kg of paddy milled, about 220 kg (22%) of husk is produced, and when this husk is burnt in the boilers, about 55 kg (25%) of RHA is generated.

In the absence of its utilization, this huge quantity of RHA goes to waste and becomes a great threat to the environment, causing damage to the land and the surrounding areas in which it is dumped. In many countries, farmers dispose of rice straw by random burning in the open fields, which tends to be the most economical method of disposal. This practice not only generates smoke but also breathable dust that contains crystalline silica and other hazardous substances. This leads to many environmental problems related to human health and safety and diseases related to the lungs and eyes are very common in these areas. If most of the RHA were to be used in concrete, it would not only get rid of dumping RHA but would also have decreased the CO_2 emissions to the atmosphere by bringing down the cement production. Burning the husk under controlled temperature below 800°C can produce ash with silica mainly in amorphous form. High-value applications and current research investigations such as the use of RHA in manufacturing of silica gels, silicon chip, synthesis of activated carbon and silica, production of light weight construction materials and insulation, catalysts, zeolites, ingredients for lithium-ion batteries, graphene, energy storage/capacitor, carbon capture, and in drug delivery vehicles are presented. Use of RHA in potential future applications is also discussed. It is suggested that the amorphous silica rich RHA could become a potential resource of low-cost precursor to produce value-added silica-based materials for practical applications.

7.2.6   Classification and Composition of RHA

Rice husk constitutes of about 35% cellulose, 25% hemi-cellulose, 20% lignin, 17% ash (94% silica) and about 3% by weight of moisture. Since it contains 75% organic volatile matter, the balance 25% of the weight is converted into ash, known as rice husk ash (RHA), during the burning

TABLE 7.2
Chemical Composition and Physical Properties of RHAs

RHA Sample	1	2	3	4	5	6	7
SiO_2 (%)	92.40	94.60	87.86	91.71	86.98	87.3	87.2
Al_2O_3 (%)	0.30	0.3	0.68	0.36	0.84	0.15	0.15
Fe_2O_3 (%)	0.40	0.3	0.93	0.90	0.73	0.16	0.16
CaO (%)	0.70	0.40	1.30	0.86	1.40	0.55	0.55
MgO (%)	0.30	0.30	0.35	0.31	0.57	0.35	0.35
Na_2O (%)	0.07	0.20	0.12	0.12	2.46	1.12	1.12
K_2O (%)	2.31	1.80	–	3.13	5.14	8.55	8.55
Loss on ignition	2.31	1.80	–	3.13	5.14	8.55	8.55
Specific gravity (g/cm³)	2.10	2.05	–	–	2.10	2.06	2.06
Fineness: passing 45 lm (%)	–	98.20	–	–	–	99.00	99.00
Mean particle size (lm)	7.40	7.15	–	0.15	7.40	–	–

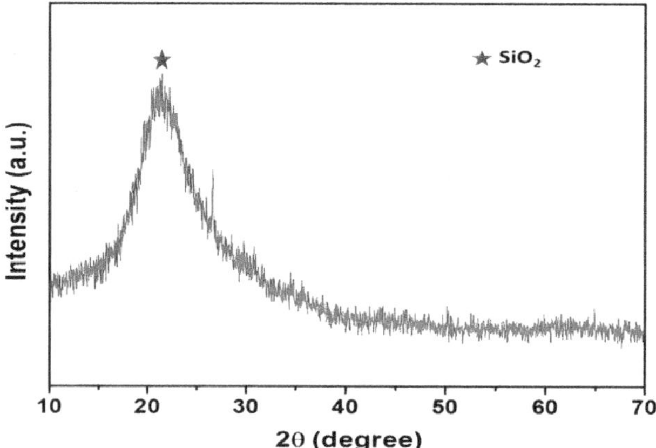

FIGURE 7.5 XRD pattern of RHA.

process. This RHA in turn contains around 85–95% of silica, which is mostly in an amorphous state but depends on the burning temperature and time. The chemical composition and physical properties of rice husk ash collected from seven different sources is given in Table 7.2.

7.2.7 APPLICATIONS OF RICE HUSK

Utilization of rice husk could solve the disposal problem and reduce the cost of waste treatment and also keep the environment free from pollution. Rice husk and its ash are used directly for manufacturing and synthesizing new materials. It is used as a fuel, fertilizer, and it is also used in the preparation of activated carbon. Rice husk acts as an adsorbent for heavy metal removal from waste water. Easy availability and low price of rice husk in rice-producing countries is an extra benefit towards the use of this material. RHA contains around 85–90% of silica, which is mostly in an amorphous state and it is a highly reactive pozzolanic material in the production of concrete due to its high silica content and high surface area and this can be confirmed by the x-ray diffraction pattern of RHA (Figure 7.5), which clearly shows the characteristic peaks of SiO_2. It has been reported that the optimum quantity of RHA can increase the mechanical properties of concrete.

Addition of RHA can enhance the strength and reduce the water absorption of concrete. To produce environment friendly and durable concrete products incorporation of RHA as a partial replacement of cement in concrete has gained importance. RHA acts as a supplementary binder in the concrete. The rice husk ash is a neutral green supplementary material that has applications in small to large scale. It can be used for waterproofing. It is also used as the admixture to make the concrete resistant against chemical penetration. There is a growing demand for fine amorphous silica in the production of high strength, low permeability concrete, for use in bridges, marine enviornments, nuclear power plants, etc. This market is currently filled by fumes or micro silica, being imported from Norway, China and also from Burma. Due to a limited supply of silica fumes in India and the demand being high, the price of silica fumes has risen to as much as US\$ 500/ton in India. From RHA, we can make organic micro-silica/amorphous silica, with silica content of above 89%, in a very small particle size of less than 35 microns, silica can find application for high-performance concrete. All the above applications make RHA a great filler for fabricating waste plastic composite tiles. This filler due to its high silica content will help in improving the physical and mechanical properties of the composite tiles thus fabricated. In addition, it will also help in reducing the flammability of the waste plastic composite tiles.

7.3　METHODOLOGY AND CHARACTERIZATION

The process of fabrication for different types of tiles from waste plastic first requires the selection of the type of thermoplastic to be used as the matrix for reinforcing any of the above fillers. The selection of the plastic to be reinforced might appear to be difficult but requires a general awareness about the behaviour of a particular type of plastic as a group and its characteristics as an individual plastic. The initial and utmost important step in designing the composite is defining its purpose and end application of the proposed product and to identify and study the service environment. The next step lies in assessing the suitability of different plastics based on their characteristics for most engineering components. A wide variety of plastic waste is generated, but our aim is to put to use different types of easily recyclable thermoplastics, as they have easy processing and do not add toxic pollutants to the environment, they include low-density polyethylene, high-density polyethylene and polypropylene. The choice of waste plastic to be reinforced is greatly influenced by the nature of application, cost of production and the service environment. Mostly for reinforcing the plastic waste with fillers to modify its physical, chemical and mechanical properties, conventional processing techniques of thermoplastics to produce a moulded article is used. The applications that require semi-finished products like sheets and rods for fabrication of structures, conventional methods like welding or machining can be used. Many applications require finished products that are quite complex in structure and shape like producing a pipe by extrusion or a telephone housing by injection moulding, these are produced in single operation, which involves continuous processing stages of heating, moulding and cooling or a repeated cycle of events. Since, there is a wide range of processing techniques available for plastics, it is important to have a basic understanding of the design process and processing methods as the wrong selection can further limit the choice of moulding method.

Plastic waste composite tiles were prepared by simple extrusion followed by a compression moulding technique. Waste plastic bags were collected, segregated and cleaned for utilization. After cleaning and drying they were shredded into smaller pieces and mixed with different filler materials as per the formulated composition. After this, the mixture is extruded in the form of wires using a single screw extruder. The wires were cut into pellets for further application. The pellets thus obtained were heated in an oven for about 30 minutes at 150°–180°C and then moulded into the tile of desired shape using compression or injection moulding technique. The schematic representation for fabrication of composite tiles is given in Figure 7.6.

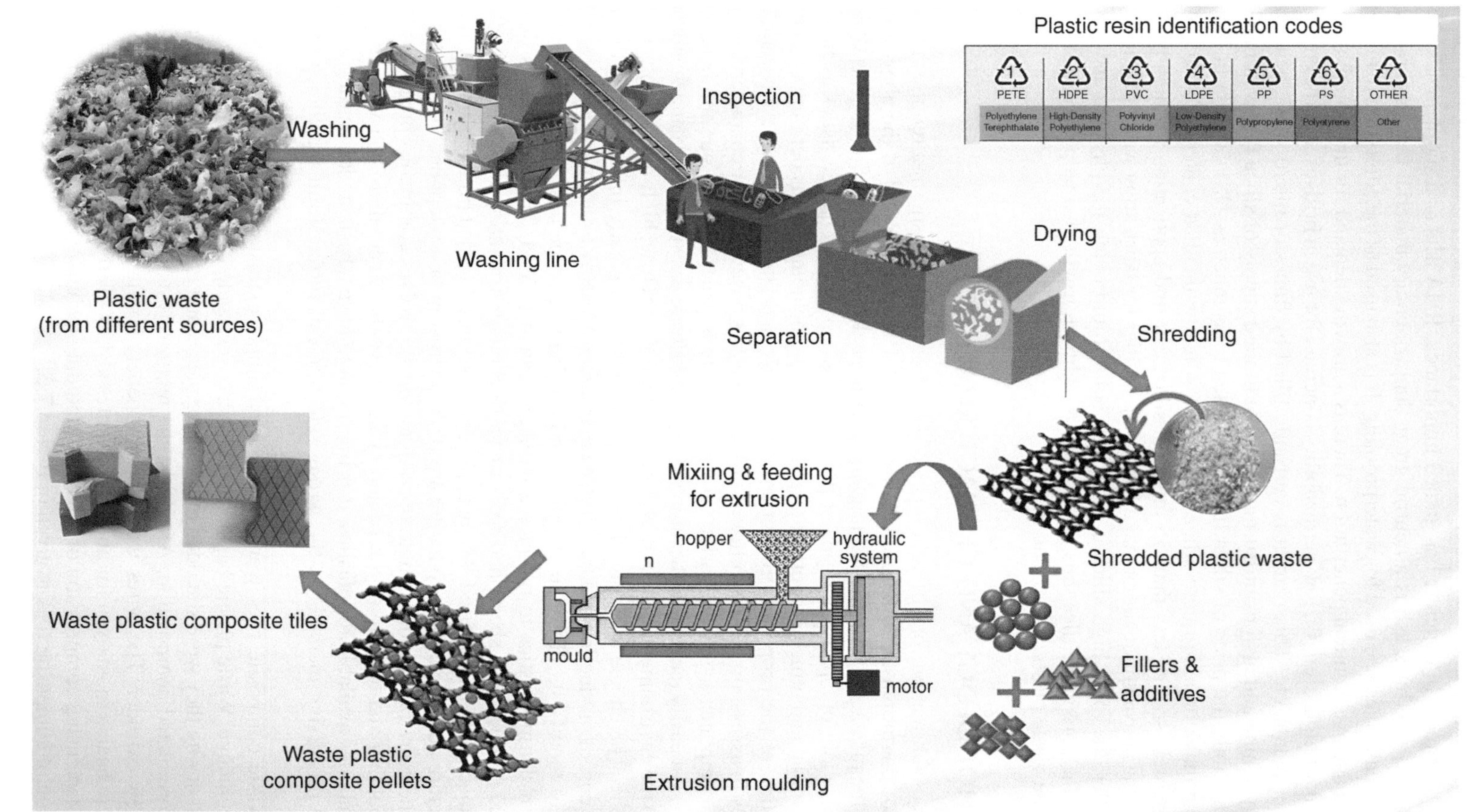

FIGURE 7.6   Schematic representation for fabrication of composite tiles.

7.3.1 Characterization of Waste Plastic Composite Tiles

7.3.1.1 Gas Analyzer Test

The fabrication of tiles involves melting the plastic waste at 140–150°C for mixing and moulding purpose. The evolution of gases during this process is a major issue of concern. The number of different gases released during the melting process of plastic waste and extrusion was measured using flue gas analyzer as shown in Figure 7.7. The flue gas analyzer was connected and installed to the chimney above the set up where the plastic tile was subjected to a temperature of 160–180°C as shown in Figure 7.8. The plastic composite was melted in open atmosphere and the gases evolved were analyzed using a standard gas analyzer under ambient temperature conditions.

FIGURE 7.7   Set up for flue gas analysis.

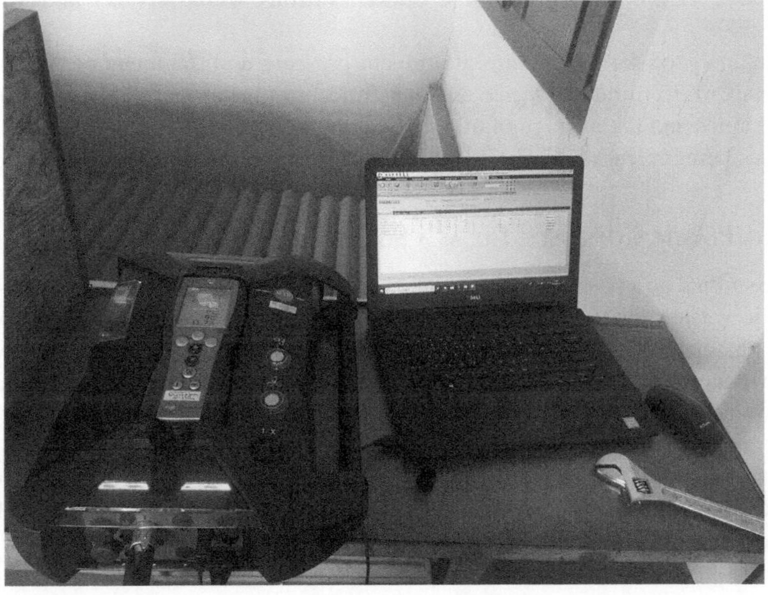

FIGURE 7.8   Flue gas analyzer.

TABLE 7.3
Amount of Gases Evolved from Composite Tiles

Gases	Amount Evolved
CO_2	0.5 %
CO	5 ppm
NOx	No evolution
SOx	No evolution
VOCs	No evolution

FIGURE 7.9   Sound transmission loss measurement in the reverberation chamber.

The amount of various gases released from the composite tile like carbon dioxide (CO2), carbon monoxide (CO), sulphur dioxide (SOx), nitrous oxide (NOx) and volatile organic compounds

VOCs were analyzed for 30 minutes at ambient pressure and the mean of all gases were taken at equal intervals of 30 minutes to give average concentration of gases evolved. The process was repeated three times and the average of all three values was taken to obtain the exact concentration of the flue gases being released. The average of these values is tabulated in Table 7.3.

7.3.2   Waste Plastic Composite Tile as a Sound Absorbing Material

The tiles were studied as a probable acoustical material. In this study, the sound–transmission properties of plastic waste tiles were investigated in the frequency region of 100–4000 Hz using the reverberation chamber method as shown in Figure 7.9 and absorption properties were measured using the impedance tube method. Based on the results, it appears that recycled plastic materials can be used to manufacture value-added acoustical panels and it was also found that plastic panels provide moderately superior acoustic properties as shown in Figure 7.10.

The study revealed that waste plastic tiles can also be used as a barrier for sound waves and can be explored for its usage in acoustics panels. Even a sound transmission loss of 35 dB was obtained having a noise reduction coefficient (NRC) of the order of 0.1.

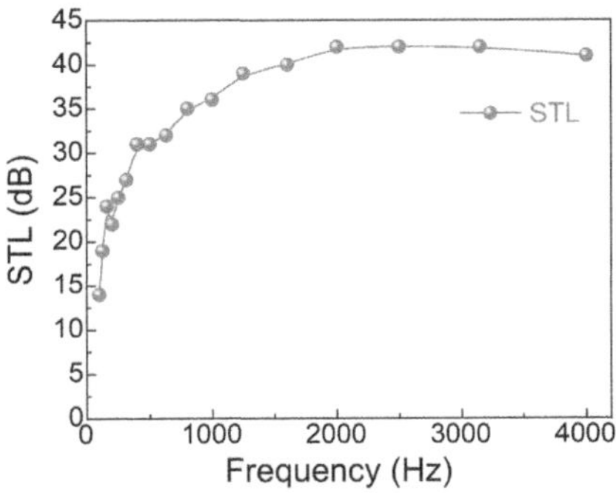

FIGURE 7.10 Acoustics characteristics of waste plastic tiles.

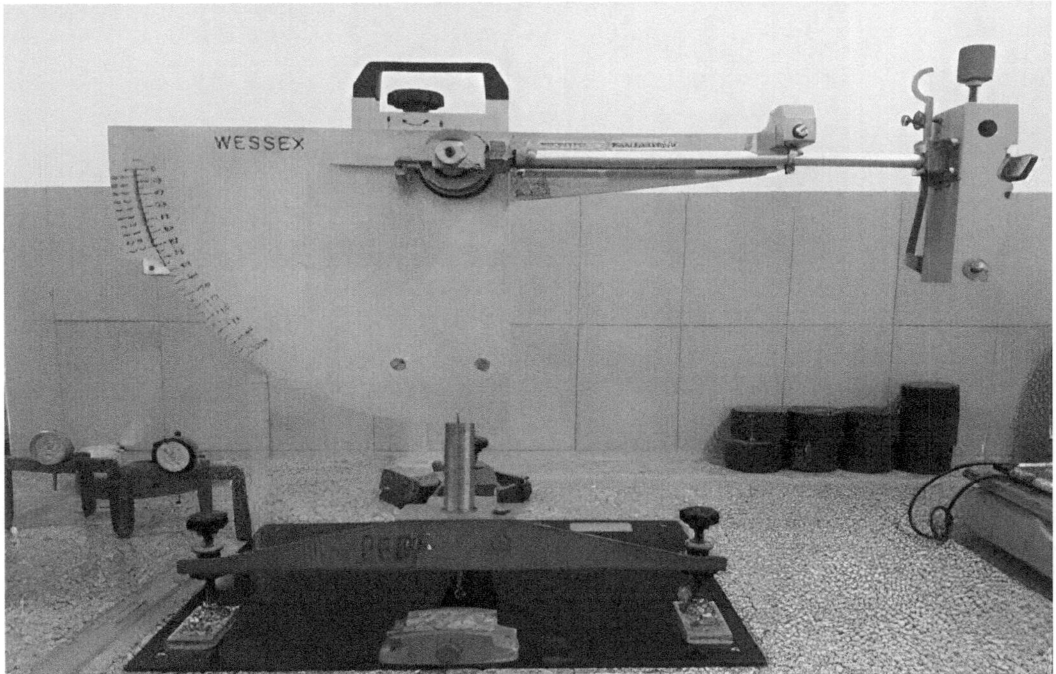

FIGURE 7.11 British Pendulum Tester.

7.3.3 BRITISH PENDULUM TEST

The paver tiles fabricated under this innovation can be put to use at external walkways, parks, footpaths, elevator lobbies and pedestrian crossings. For this purpose, there skid resistance was analyzed using national standard method British Pendulum Skid Resistance Tester (ASTM E303-93) as shown in Figure 7.11. The surface frictional properties of these tiles were studied in dry and wet conditions. Also, the effect of temperature change was analyzed. The greater the friction between the rubber slider and the surface, the more the skid resistance and the greater the value of the British Pendulum Number (BPN).

The samples of the composite tiles were tested for both dry and wet conditions. The pendulum with rubber slider perpendicular to the surface of the tile is allowed to swing over the surface and the value of BPN was recorded as shown in Figure 7.12.

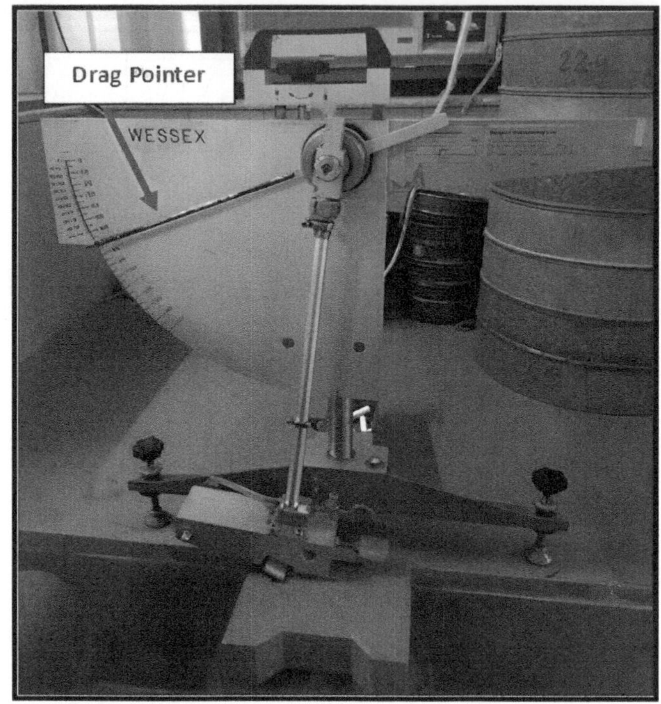

FIGURE 7.12   Skid resistance of composite tile by the British Pendulum Tester.

TABLE 7.4
Mean Values of the British Pendulum Test of
Paver Tiles under Dry and Wet Conditions

Tile	BPN (Dry)	BPN (Wet)
Paver tile (20 mm)	65	39
Paver Tile (60 mm)	80	40

The pendulum was allowed to swing three times to obtain three values of BPN and then the average was calculated to get the mean value. The same process is repeated for wet conditions with a precaution that the surface of the composite tile is wetted each time before taking the values. The mean values of BPN for dry and wet conditions of the composite tiles are given in Table 7.4.

Even the interlock floor tiles, pavement tiles and roof tiles have been made using these waste plastic scrap as shown in Figure 7.13.

7.4   APPLICATIONS

These decorative coloured tiles can be used in a variety of applications. The application of these tiles in day-to-day life includes the following:

- Making structures of rooms or toilets
- Deck floors
- Interlock tiles for pavements

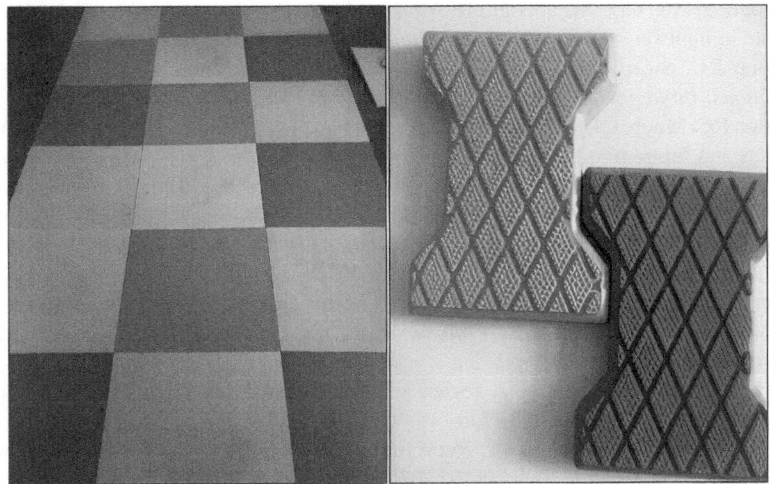

FIGURE 7.13 Interlock floor tiles and paver tiles made from waste plastic.

- Roof tiles
- Railings
- Fences
- Landscaping, horticulture and wall decoration
- Building exteriors, like, wall cladding and siding
- Park benches
- Moulding and trim
- Window and door frames
- Indoor and garden furniture
- Waterproof, durable floor tiles and floor coverings
- Colourful bricks and blocks to divide up rooms or outdoor areas.

7.5 CONCLUSION

The fabrication of composite tiles has been suggested as a newer approach for reutilization of waste plastic material in our day-to-day life and industry in an environmentally friendly and economical way. They can be used to build structures, which will be light weight, resistant to corrosion, chemically resistant, low cost of production, increased service life and most importantly put into use what is the menace for society -– plastic waste. Also, these plastic waste composite tiles can be a suitable replacement to traditional cement tiles as they are unbreakable, have improved mechanical strength and have a better service life.

ACKNOWLEDGEMENT

We thank Vikas Lifecare R&D team in getting the things done at an early pace and the Chairman, Vikas Group to allow us to publish the results.

REFERENCES

[1] Andrady AL, Neal MA. Applications and societal benefits of plastics. *Philosophical Transactions of the Royal Society B: Biological Sciences.* 2009;364(1526):1977–84.

[2] Akçaözoğlu S, Atiş CD, Akçaözoğlu K. An investigation on the use of shredded waste PET bottles as aggregate in lightweight concrete. *Waste Management*. 2010;30(2):285–90.

[3] Thompson RC, Swan SH, Moore CJ, vom Saal FS. Introduction: Our plastic age. *Philosophical Transactions: Biological Sciences*. 2009;364(1526):1973–6.

[4] Thompson RC, Moore CJ, vom Saal FS, Swan SH. Plastics, the environment and human health: Current consensus and future trends. *Philosophical Transactions of the Royal Society B: Biological Sciences*. 2009;364(1526):2153–66.

[5] Ryan PG, Moore CJ, van Franeker JA, Moloney CL. Monitoring the abundance of plastic debris in the marine environment. *Philosophical Transactions of the Royal Society B: Biological Sciences*. 2009;364(1526):1999–2012.

[6] Shaxson L. Structuring policy problems for plastics, the environment and human health: Reflections from the UK. *Philosophical Transactions of the Royal Society B: Biological Sciences*. 2009;364(1526):2141–51.

[7] Shent H, Pugh RJ, Forssberg E. A review of plastics waste recycling and the flotation of plastics. *Resources, Conservation and Recycling*. 1999;25(2):85–109.

[8] Al-Salem SM, Lettieri P, Baeyens J. Recycling and recovery routes of plastic solid waste (PSW): A review. *Waste Management*. 2009;29(10):2625–43.

[9] Masuda T, Kushino T, Matsuda T, Mukai SR, Hashimoto K, Yoshida S-i. Chemical recycling of mixture of waste plastics using a new reactor system with stirred heat medium particles in steam atmosphere. *Chemical Engineering Journal*. 2001;82(1):173–81.

[10] Angyal A, Miskolczi N, Bartha L. Petrochemical feedstock by thermal cracking of plastic waste. *Journal of Analytical and Applied Pyrolysis*. 2007;79(1):409–14.

[11] Shah AA, Hasan F, Hameed A, Ahmed S. Biological degradation of plastics: A comprehensive review. *Biotechnology Advances*. 2008;26(3):246–65.

[12] Pramila R, Ramesh KV. Biodegradation of low density polyethylene (LDPE) by fungi isolated from municipal landfill area. J. Microbiol. Biotechnol. Res. 2011;1(4):131–6.

[13] Guerrero LA, Maas G, Hogland W. Solid waste management challenges for cities in developing countries. *Waste Management*. 2013;33(1):220–32.

[14] Lithner D, Larsson Å, Dave G. Environmental and health hazard ranking and assessment of plastic polymers based on chemical composition. *Science of The Total Environment*. 2011;409(18):3309–24.

[15] Browne MA, Dissanayake A, Galloway TS, Lowe DM, Thompson RC. Ingested Microscopic Plastic Translocates to the Circulatory System of the Mussel, Mytilus edulis (L.). *Environmental Science & Technology*. 2008;42(13):5026–31.

[16] Oehlmann J, Schulte-Oehlmann U, Kloas W, Jagnytsch O, Lutz I, Kusk KO, et al. A critical analysis of the biological impacts of plasticizers on wildlife. *Philosophical Transactions of the Royal Society B: Biological Sciences*. 2009;364(1526):2047–62.

[17] Zia KM, Bhatti HN, Ahmad Bhatti I. Methods for polyurethane and polyurethane composites, recycling and recovery: A review. *Reactive and Functional Polymers*. 2007;67(8):675–92.

[18] Jambeck JR, Geyer R, Wilcox C, Siegler TR, Perryman M, Andrady A, et al. Plastic waste inputs from land into the ocean. *Science*. 2015;347(6223):768–71.

[19] Hamad K, Kaseem M, Deri F. Recycling of waste from polymer materials: An overview of the recent works. *Polymer Degradation and Stability*. 2013;98(12):2801–12.

[20] Singh N, Hui D, Singh R, Ahuja IPS, Feo L, Fraternali F. Recycling of plastic solid waste: A state of art review and future applications. *Composites Part B: Engineering*. 2017;115:409–22.

[21] Hopewell J, Dvorak R, Kosior E. Plastics recycling: Challenges and opportunities. *Philosophical Transactions of the Royal Society B: Biological Sciences*. 2009;364(1526):2115–26.

[22] Butler E, Devlin G, McDonnell K. Waste polyolefins to liquid fuels via pyrolysis: Review of commercial state-of-the-art and recent laboratory research. *Waste and Biomass Valorization*. 2011;2(3):227–55.

[23] Ray R, Thorpe RB. A comparison of gasification with pyrolysis for the recycling of plastic containing wastes. *International Journal of Chemical Reactor Engineering*. 2007 Oct 26; 5(1).

[24] Sharma BK, Moser BR, Vermillion KE, Doll KM, Rajagopalan N. Production, characterization and fuel properties of alternative diesel fuel from pyrolysis of waste plastic grocery bags. *Fuel Processing Technology*. 2014;122:79–90.

[25] Saikia N, de Brito J. Use of plastic waste as aggregate in cement mortar and concrete preparation: A review. *Construction and Building Materials*. 2012;34:385–401.

[26] Ismail ZZ, Al-Hashmi EA. Use of waste plastic in concrete mixture as aggregate replacement. *Waste Management*. 2008;28(11):2041–7.

[27] Bhogayata AC, Arora NK. Impact strength, permeability and chemical resistance of concrete reinforced with metalized plastic waste fibers. *Construction and Building Materials*. 2018;161:254–66.

[28] Huang Y, Bird RN, Heidrich O. A review of the use of recycled solid waste materials in asphalt pavements. *Resources, Conservation and Recycling*. 2007;52(1):58–73.

[29] Bazargan A, McKay G. A review – Synthesis of carbon nanotubes from plastic wastes. *Chemical Engineering Journal*. 2012;195–196:377–91.

[30] Gong J, Yao K, Liu J, Wen X, Chen X, Jiang Z, et al. Catalytic conversion of linear low density polyethylene into carbon nanomaterials under the combined catalysis of Ni_2O_3 and poly(vinyl chloride). *Chemical Engineering Journal*. 2013;215–216:339–47.

8 Synthesis of Geopolymer Materials Based on Non-Ferrous Metallurgy Slag and Fly Ash Using Mechanical Activation

A.M. Kalinkin, E.V. Kalinkina and E.A. Kruglyak

8.1 INTRODUCTION

Every year, a huge amount of metallurgical slags is produced in the world as a waste of processing various ores, amounting to hundreds of millions of tons [1]. The generation of electricity from coal combustion produces a similar amount of coal fly ash (FA), with less than 30% recycled [2]. The disposed slags and FA have become a matter of serious environmental concern. The most rational way to utilize FA and slags is to produce composite binders based on traditional Portland cement, as well as the relatively new types of cements, such as alkali-activated binders [3–6]. Alkali-activated binders or geopolymers are a promising class of materials that are prepared as a result of the interaction of natural and synthetic aluminosilicates with alkaline agents (for example, sodium hydroxide solution or water glass). Geopolymers are often regarded as a subclass of alkali-activated materials prepared by the reaction of low-calcium aluminosilicates, e.g., Class F FA, with the alkaline agent. With their energy saving, environmentally friendly processing, and high performance, geopolymers are gaining attention in civil engineering as a promising replacement for traditional Portland cement. Geopolymer materials can be used not only in construction but also as matrices for the immobilization of toxic waste, adsorbents and refractories [7–9].

To increase the reactivity of the aluminosilicate raw material and, consequently the geopolymer performance, mechanical activation (MA) is an effective tool [10, 11]. Mechanical treatment of solids in mills leads not only to a decrease in the particle size (increase in the specific surface area) but also results in the generation of structural defects, which is a very important factor in increasing the reactivity [12–15]. Another approach that is used to improve the performance of the geopolymer binder systems is the blending of aluminosilicate-rich materials, such as FA or calcined clays, with limestone, dolomite or other carbonate containing mineral additives [16–19]. In this chapter, both the influence of MA of aluminosilicate raw materials in various atmospheres (air, CO_2) and blending with carbonate additives on binding properties of the slag and fly ash are considered.

8.2 GEOPOLYMERS BASED ON NON-FERROUS METALLURGICAL SLAGS

8.2.1 Cu–Ni and Zn Slags Mechanically Activated in Air and in CO_2 Atmosphere

Compared to blast-furnace slag, non-ferrous metallurgy slags are characterized by a low content of calcium and a high content of iron [1]. As a result, the hydraulic activity of non-ferrous metallurgy

DOI: 10.1201/9781003327646-8

TABLE 8.1
The Chemical Composition of Slags, wt.%

Slag	SiO_2	Al_2O_3	Fe_2O_3	FeO	CaO	MgO	Na_2O	S	Co	Ni	Cu	ZnO	PbO
Cu–Ni slag	36.87	5.44	2.47	31.08	2.11	11.92	1.18	0.76	0.10	0.24	0.16	–	–
Zn slag	18.08	8.17	34.28	–	17.91	1.93	0.68	1.41	–	–	–	9.21	1.22

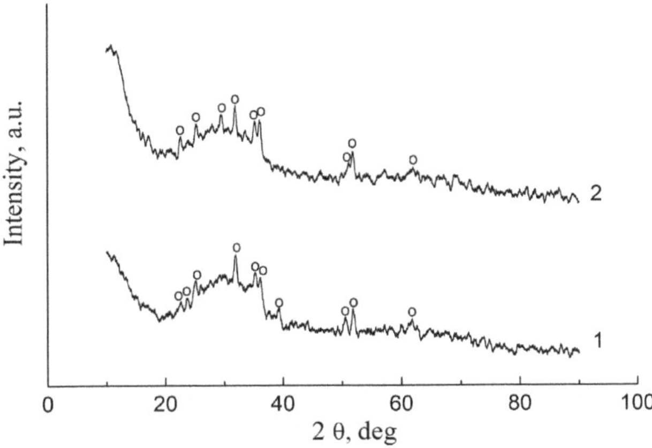

FIGURE 8.1 XRD patterns of Cu–Ni slag mechanically activated in CO_2 (1) and in air (2); o – olivine reflections. (Reproduced from [11] with permission from Elsevier.)

slags is less than that of blast-furnace slags. This section presents the results of studies on the effect of MA of non-ferrous metallurgy slags (copper-nickel and zinc slag) in air and in an atmosphere of carbon dioxide on their geopolymerization. The use of CO_2 as an MA medium is due to the following reasons. It is known that the reactivity of the surface layers of powder particles can change markedly during mechanical treatment in mills. An important factor affecting the reactivity of the powder surface is the gas medium in which MA is carried out [14, 15]. It was revealed previously that a prolonged grinding of Ca,Mg-containing silicates is accompanied by absorption of large amounts (>10 wt %) of atmospheric CO_2 by these minerals [20]. The carbonization effect is enhanced in the case of MA in a purely CO_2 atmosphere [21].

The research objects were granulated Cu–Ni slag from the Pechenganickel plant (Murmansk region), and zinc slag from Hindustan Zinc Ltd (Chittorgarh, Rajasthan, India). The chemical compositions of slags are shown in Table 8.1.

MA of the slags was carried out in an AGO-2 laboratory planetary mill in air and in CO_2 ($P(CO_2)$= 1 atm). Geopolymers were prepared by mixing a mechanically activated slag with an alkaline agent (liquid glass or NaOH solution). The fresh mortars were cured under relative humidity of 95±5% at ambient temperature of 20±2°C. The MA procedure and sample preparation are described in detail in [11, 22].

According to the analysis data, the content of CO_2 in the raw Cu–Ni slag was 0.015%. For samples of this slag after MA in air and in carbon dioxide atmosphere, the CO_2 content was 0.12% and 0.81%, respectively.

The XRD data of Cu–Ni slag after 10 min of MA in air and CO_2 (centrifugal factor – 20 g) are presented in Figure 8.1. The x-ray diffraction patterns for both samples are similar: against the background of an amorphous halo, they contain reflections of skeletal olivine crystals.

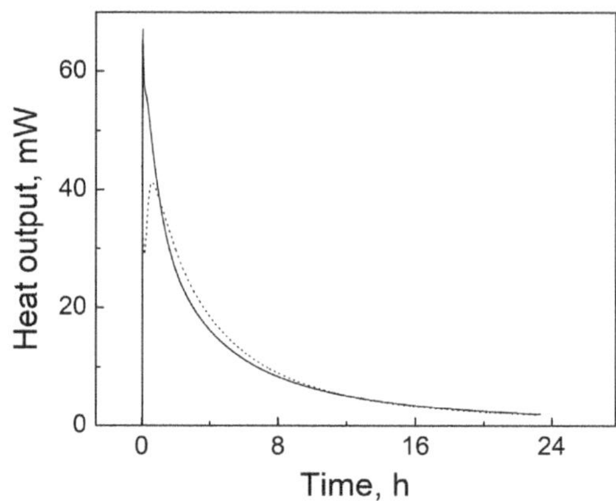

FIGURE 8.2 Heat release rate during the initial stages of geopolymerization of the Cu–Ni slag mechanically activated in air (dashed line) and in CO_2 (solid line) using liquid glass with a modulus of 1.5 as an alkaline reagent. (Reproduced from [11] with permission from Elsevier.)

The results of a calorimetric study of the hydration of samples of mechanically activated Cu–Ni slag, mixed with liquid glass solutions with a modulus of 1.5, are shown in Figure 8.2. The measurements were carried out using a TAM Air isothermal conduction calorimeter at 27°C, the mass of geopolymer samples was 7 g [11]. From the data presented in Figure 8.2, it follows that the geopolymerization of Cu–Ni slag mechanically activated in CO_2 starts earlier and proceeds more intensively compared to the slag after MA in air.

Higher reactivity of mechanochemically carbonized Cu–Ni slag with respect to the alkaline agent leads to an increase in the strength of geopolymers. Samples of geopolymers were prepared by mixing the mechanically activated slag and liquid glass with a modulus of 1.5. The consumption of the alkaline activator was 3 wt. % Na_2O in the composition of liquid glass in relation to the mass of slag. For the Cu–Ni slag after MA in CO_2, the compressive strength of geopolymers at the age of 1, 7 and 28 days was 54, 77 and 94 MPa, respectively. The strength for slag after MA in air was 51, 75 and 81 MPa, respectively.

Similar results were obtained for the geopolymerization of Zn slag mechanically activated in air and CO_2 for 3 min (centrifugal factor – 40 g). According to XRD data, the Zn slag contains a significant proportion of the amorphous phase, and the main crystalline phase is wustite FeO (Figure 8.3).

The CO_2 content of the raw Zn slag, Zn slag after MA in air and in carbon dioxide atmosphere was 0.102%, 0.145%, and 0.469%, respectively. The rate of heat release during geopolymerization of Zn slag, mechanically activated in CO_2 and mixed with 6 M NaOH solution, is noticeably higher than for the slag after MA in air (Figure 8.4). The calorimetric data were obtained using a TAM Air calorimeter at 27°C. Samples were prepared by mixing 7 g of slag with 3.5 ml of sodium hydroxide solution [22].

As in the case of the Cu–Ni slag, the MA of the Zn slag in carbon dioxide resulted in an increase in the binding properties of the geopolymer. The compressive strength of geopolymers based on Zn slag after MA in air and in CO_2 at the age of 1, 7 28 days was 11, 57, 74 MPa and 23, 59, 88 MPa, respectively.

For the Cu–Ni slag, we studied also the effect of the MA atmosphere on its hydration without the use of chemical activators. Samples were prepared using the slag mechanically activated in air and CO_2 for 10 min (centrifugal factor – 40 g) and distilled water. The water-to-slag ratio was 0.23. The experimental details are reported in [23]. As expected, without the use of an alkaline activator, slag hydration, according to calorimetric measurements (Figure 8.5), proceeds much more slowly compared to hydration in the geopolymer composition (Figure 8.2). The measurements were carried out using a TAM III isothermal calorimeter at 25°C for 100 days. At the same time, the observed

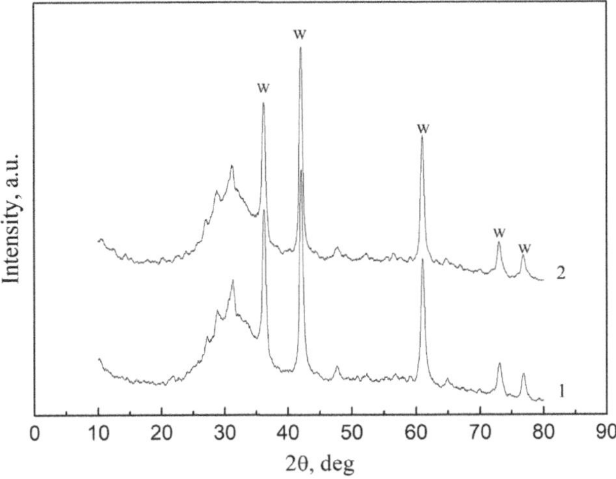

FIGURE 8.3   XRD patterns of Zn slag samples mechanically activated in CO_2 (1) and in air (2); w – wustite reflections. (Reproduced from [31] with permission from Chemistry for Sustainable Development.)

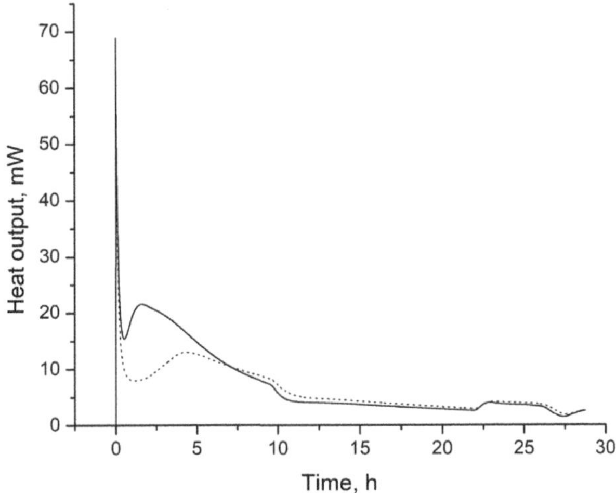

FIGURE 8.4   Heat release rate during the initial stages of geopolymerization of the Zn slag mechanically activated in air (dashed line) and in CO_2 (solid line) using 6 M NaOH solution as an alkaline reagent. (Reproduced from [31] with permission from Chemistry for Sustainable Development.)

trend is clearly manifested: MA in CO_2 significantly accelerates the slag hydration, since the main exothermal peak of the heat release rate curve in this case appears approximately 20 days earlier than for slag after MA in air (Figure 8.5).

To study the binding properties, the Cu–Ni slag after MA in air or in CO_2 was mixed with water to prepare paste. For the slag mechanically activated in air and in carbon dioxide atmosphere, the water-to-slag ratio was 0.230 and 0.241, respectively. The pastes were moulded to prepare specimens 1.41×1.41×1.41 cm and then were cured at 22 ± 2°C in tightly closed container. Compressive strength of the specimens prepared using the Cu–Ni slag mechanically activated in air or in CO_2 without an alkaline agent is shown in Figure 8.6. The Cu–Ni slag mechanically activated in air does not show binding properties. Compressive strength of the slag after MA in CO_2 cured for 7 days is equal to that of the slag mechanically activated in air. However, when the main hydration

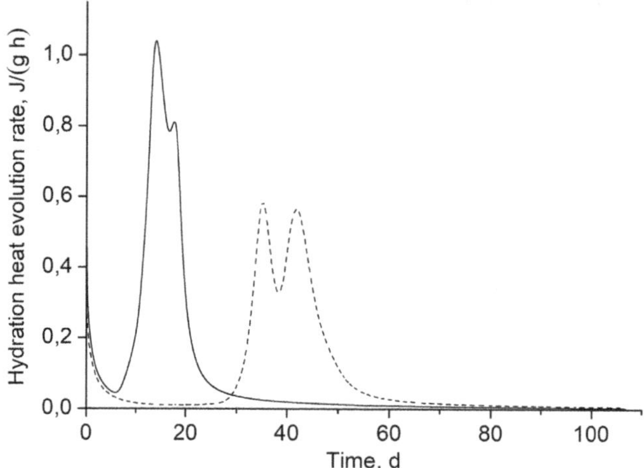

FIGURE 8.5   Heat release for the Cu–Ni slag after MA in air (dashed line) and in CO_2 (solid line) during hydration without using an alkaline reagent. (Reproduced from [23] with permission from Springer Nature.)

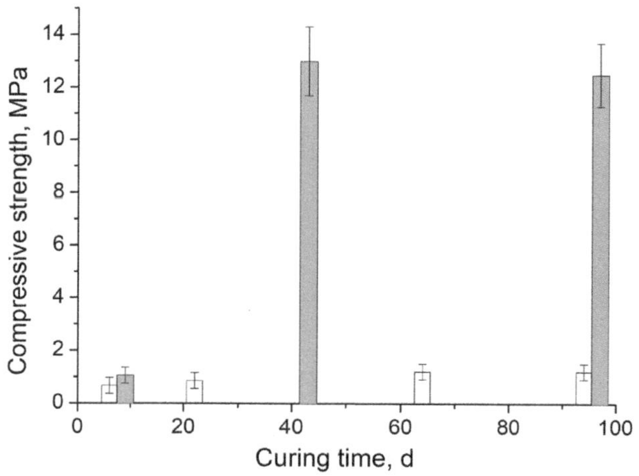

FIGURE 8.6   Compressive strength of the specimens prepared from the slag after MA in air (open columns) and in CO_2 (grey columns) without an alkaline agent as a function of curing time.

stage for the slag after MA in CO_2 is completed (ca. 40 days, Figure 8.5), the compressive strength increases up to 12–14 MPa (Figure 8.6).

Thus, preliminary MA of the Cu–Ni and Zn slag in CO_2 atmosphere significantly increases their hydraulic activity compared to MA in air.

8.2.2   Mechanically Activated Cu–Ni Slag Blended with Carbonates

We also investigated the possibility of replacing MA of Cu–Ni slag in an atmosphere of gaseous CO_2 on its solid phase carbonization by blending it with carbonate minerals at the stage of mechanical treatment in planetary mill. It was assumed, that joint MA of the slag and carbonate additives, as in the case of gaseous CO_2, will lead to the formation of phases that increase the reactivity of the slag particles with respect to alkaline agent. As carbonate additives, we used analytical grade

TABLE 8.2

Effect of Carbonate Additives on the Compressive Strength of the Mechanically Activated Cu–Ni Slag–Carbonate Blends

wt % CO_2 in the Additive	Compressive Strength, MPa								
	$MgCO_3$			$CaCO_3$			Natural Calcite		
	7*	28*	180*	7*	28*	180*	7*	28*	180*
0.25	2.0	9.5	10.6	1.4	8.0	8.5	4.5	12.1	15.1
0.5	10.5	12.3	17.2	10.9	23.8	24.1	9.3	22.1	22.3
1.0	3.7	6.0	7.2	2.2	2.2	3.1	3.1	4.2	4.9
2.0	2.0	2.6	3.2	1.5	2.8	2.1	1.2	1.8	1.9
3.0	1.1	1.0	1.7	1.7	4.0	3.0	1.4	1.5	1.9
	Natural dolomite			$SrCO_3$			$BaCO_3$		
0.25	1.5	4.9	9.4	9.3	14.3	19.6	13.2	15.7	18.8
0.5	7.9	13.7	17.9	9.9	10.5	14.0	10.5	8.9	13.7
1.0	12.4	20.2	19.0	7.6	12.0	12.8	0.9	3.1	3.6
2.0	9.6	17.5	20.0	5.3	5.4	5.1	2.4	0.6	0.6
3.0	12.9	6.2	14.6	2.4	2.5	2.7	1.0	0.6	0.6

Source: Reproduced from [24] with permission from Springer Nature.

*   Curing time, days

calcium carbonate ($CaCO_3$), magnesium carbonate ($MgCO_3$), strontium carbonate ($SrCO_3$), barium carbonate ($BaCO_3$), and natural minerals: calcite (Kovdor deposit, Murmansk region, Russia) and dolomite (Titan deposit, Murmansk region, Russia). MA of the slag blended with the carbonates was carried out in the AGO-2 laboratory centrifugal planetary mill in air for 330 s [24].

To study the binding properties, the mechanically activated blends were mixed with water to prepare specimens 1.41 × 1.41 × 1.41 cm, which hardened under humid conditions at a temperature of 20–22°C. The effect of nature and content of the carbonates on the compressive strength is shown in Table 8.2. For comparison, Table 8.3 presents the strength for the similarly prepared binders based on the slag mechanically activated in air and CO_2 atmosphere with no carbonate additives.

Please note that the data presented in Tables 8.2 and 8.3 refer to binders prepared without alkaline agent. It is seen that small amounts of carbonate additives (within 1 wt % in terms of CO_2) increase the strength (Table 8.2). There is an optimal amount of carbonate additives that provides the highest strength of the mechanically activated blend. A decrease or increase in this amount leads to a decrease in strength. The optimal amount is 0.5–1.0% CO_2 relative to the weight of the slag for the synthetic and natural Ca and Mg carbonates and 0.25% CO_2 for $SrCO_3$ and $BaCO_3$. The strength of the samples based on the slag blended with carbonates is 10–30 times larger than that of the additive-free slag samples mechanically activated in air and is only slightly lower than the strength of the mechanically activated slag carbonized through CO_2 absorption from the gas phase (Table 8.3).

We suggest that the increase in the reactivity of the slag after preliminary MA in CO_2 is due to carbon dioxide mechanosorption by silicates [20]. Under the effect of MA, CO_2 is absorbed by outer layers of slag particles to form distorted carbonate groups, which have a characteristic split band of asymmetric stretching vibrations in their IR spectrum in the range 1400–1550 cm^{-1} [11]. The mechanochemically carbonized slag surface more actively reacts with alkaline agent or water

TABLE 8.3
Effect of Mechanical Activation Atmosphere on the Compressive Strength of the Cu–Ni Slag

Mechanical Activation Atmosphere	Compressive Strength, MPa		
	7*	28*	180*
air	0.6	0.7	1.87
CO_2	13.0	15.6	23.0

Source: Reproduced from [24] with permission from Springer Nature.

*   Curing time, days

in comparison with the slag mechanically activated in air. In simplified form, the mechanochemical reaction between CO_2 and the olivine-based slag can be represented as

$$MgFeSiO_4 + 2CO_2 = MgCO_3 + FeCO_3 + SiO_2.$$

Strictly speaking, mechanosorption of carbon dioxide by the slag does not lead to the formation of individual carbonates or silica but is accompanied by the formation of an amorphous carbonate–silicate phase [20], which displays enhanced reactivity.

To understand the changes that are produced by mechanical activation of the slag blended with carbonate additives, we measured IR spectra of mechanically activated slag–calcite mixtures (solid-state carbonization) and, for comparison, IR spectra of the slag mechanically activated in a CO_2 atmosphere (gas-phase carbonization) (Figure 8.7).

It is seen that, with increasing MA time, the single carbonate peak of the slag–calcite mixture gradually broadens and eventually splits (Figure 8.7, spectra 1–3), and its shape and position approach those of the analogous doublet in the spectrum of the slag activated in a CO_2 atmosphere (spectrum 4). Analysis for CO_2 indicates that the CO_2 content of the slag–calcite mixture mechanically activated for 330 s is essentially equal to the sum of the initial amount of CO_2 in the slag and the amount added to the mixture in the form of calcite. According to IR spectroscopy data, this means that the joint milling of the slag and carbonate additive leads to calcite decomposition according to the reaction $CaCO_3 = CaO + CO_2$, accompanied by mechanosorption of the released CO_2 by the slag particles in the form of distorted groups. The IR spectrum of the resultant activation product is similar to that of the slag mechanically activated in a CO_2 atmosphere.

8.3   GEOPOLYMERS BASED ON FLY ASH

In this section, geopolymerization of mechanically activated FA blended with natural calcite and dolomite is considered. The research object was Class F FA collected from Apatity thermal power plant (Murmansk Region, Russia). The main crystalline phases in the FA were quartz and mullite, hematite and magnetite were present in small amounts. Natural calcite (Kovdor massif, Murmansk Region, Russia) and dolomite (Titan deposit, Murmansk Region, Russia) were used as carbonate additives. Augite (<2%) and a minor amount of feldspar were present in calcite as admixtures. The main admixtures in dolomite were quartz (4–5 %) and a minor amount of calcite. Table 8.4 presents the chemical composition of the FA, calcite and dolomite.

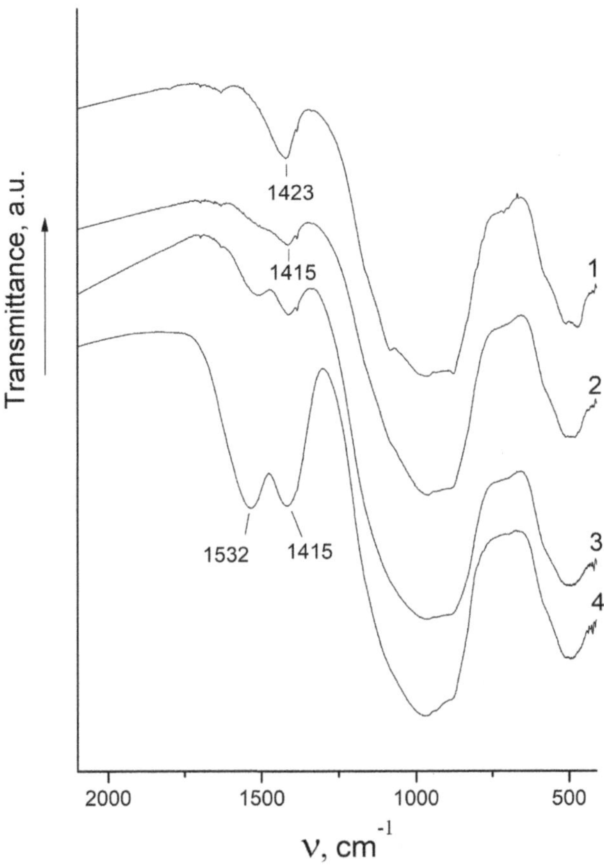

FIGURE 8.7 IR spectra of the (slag + $CaCO_3$) mixture (calcite content 1 wt % in terms of CO_2) after MA for 30 s (1), 330 s (2), and 900 s (3) in air; the slag mechanically activated in a CO_2 atmosphere (4.7 wt % CO_2) (4). (Reproduced from [24] with permission from Springer Nature.)

TABLE 8.4
Chemical Composition of the FA, Calcite and Dolomite

Material	SiO_2	Al_2O_3	Fe_2O_3	FeO	CaO	MgO	SO_3	Na_2O	K_2O	C	P_2O_5	TiO_2	L.O.I.
FA	56.26	18.39	8.58	0.69	2.14	2.60	0.18	4.04	1.32	0.88	0.32	1.13	2.28
calcite	0.24	0.47	0.67	–	52.1	1.44	0.15	1.76	0.56	–	0.05	0.05	43.0
dolomite	4.76	0.18	0.23	–	30.66	21.96	0.01	0.12	0.09	–	–	–	44.35

MA of the blends was carried out in an AGO-2 laboratory planetary mill (centrifugal factor – 40 g). The mechanically activated blends were mixed with 8.3 M NaOH solution to prepare pastes. The mass ratio of Na_2O in sodium hydroxide solution to the mechanically activated blends in the paste was equal to 0.06. Prepared specimens were cured in a relative humidity of 95 ± 5% at 22 ± 2°C before the compressive strength testing. The experimental details are described in [25, 26].

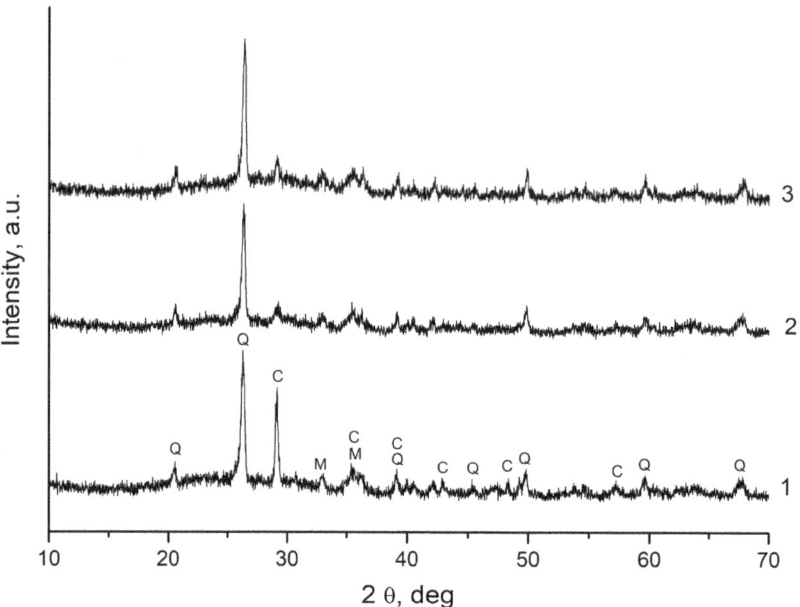

FIGURE 8.8 XRD patterns of the (95 % FA + 5 % calcite) blend mechanically activated for 30 s (1) and 180 s (2), and of the geopolymer prepared using the blend milled for 180 s after 28 d of curing (3). Phases marked are as follows: Q – quartz, M – mullite, C – calcite.

8.3.1 MECHANICALLY ACTIVATED FA BLENDED WITH CALCITE

The x-ray diffraction patterns of the (95% FA + 5% calcite) blend mechanically activated for 30 s and 180 s are shown in Figure 8.8. From a comparison of the peak intensities it follows that, the structure disordering of calcite caused by MA is noticeably more pronounced than that of quartz and mullite present in the FA. This can be explained by differences in the hardness of minerals: for quartz, mullite and calcite, the hardness on the Mohs scale is 7, 6.3–7.5 and 3, respectively.

As a typical example of a geopolymer based on a blend containing 5% calcium carbonate or less, Figure 8.8 shows an x-ray diffraction pattern of a geopolymer based on a (95% FA + 5% calcite) blend, mechanically activated for 180 s, at the age of 28 days (curve 3). In the x-ray diffraction pattern of the geopolymer, compared to the blend (Figure 8.8, curve 2), no new peaks are observed, and their intensities do not change. Therefore, the geopolymerization product is x-ray amorphous. This result is consistent with the data of previous studies, according to which the interaction of FA with alkaline agents at temperatures close to room temperature does not lead to the appearance of crystalline newly formed phases [27, 28]. The main reaction product of the mechanically activated (FA + calcite) blends was sodium containing aluminosilicate hydrate gel (N-A-S-H gel). Its formation is confirmed by the appearance of a halo in the $2\theta = 25–35°$ region (Figure 8.8, curve 3).

Figures 8.9 and 8.10 show the compressive strengths of geopolymers based on the mechanically activated (FA + calcite) blends at the age of 7 and 28 days, respectively.

From the data shown in Figures 8.9 and 8.10, it follows that the strength increases with an increase in the proportion of calcite in the blend. The optimal time of MA compositions is 180 s. The greatest positive effect of the calcite additive is observed in the early stages of hardening. The strength of a geopolymer based on a (90% FA + 10% calcite) blend cured for 7 days is 8.0, 3.5 and 2.9 times higher than that of a geopolymer based on 100% FA at MA times of 30, 180 and 400 s, respectively.

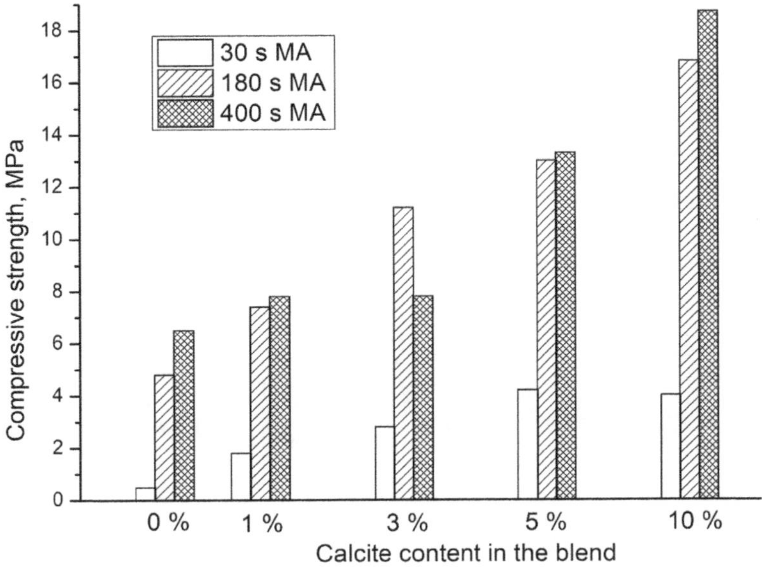

FIGURE 8.9 Effect of calcite content and MA time on the 7 d compressive strength of geopolymers based on the (FA + calcite) blends.

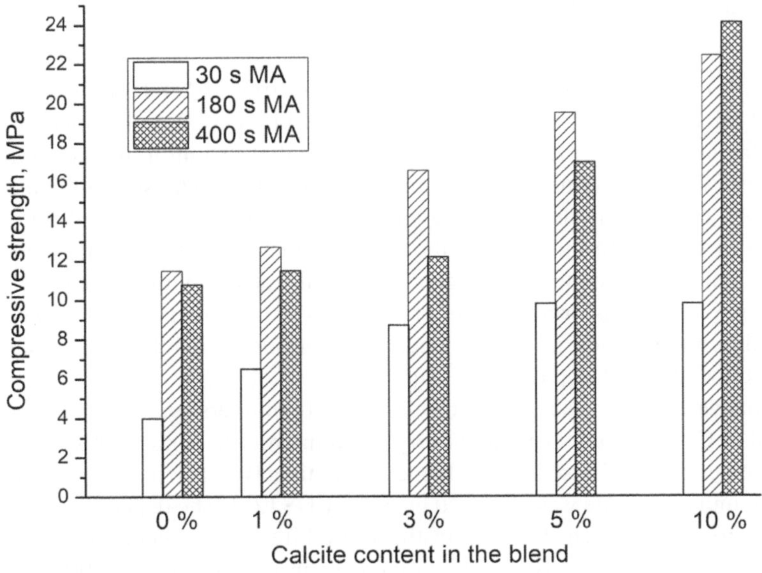

FIGURE 8.10 Effect of calcite content and MA time on the 28 d compressive strength of geopolymers based on the (FA + calcite) blends.

At the age of 28 days, the corresponding strength is 2.5, 1.9 and 2.2 times higher. An increase in the MA time (more than 400 s) and a calcite content (more than 10%) is undesirable, since in this case the workability of the geopolymer paste is noticeably reduced and the growth of the geopolymer strength significantly slows down.

The XRD patterns of a geopolymer synthesized using the (90% FA + 10% calcite) blend display distinctive features in comparison with the geopolymers based on blends with a lower content of

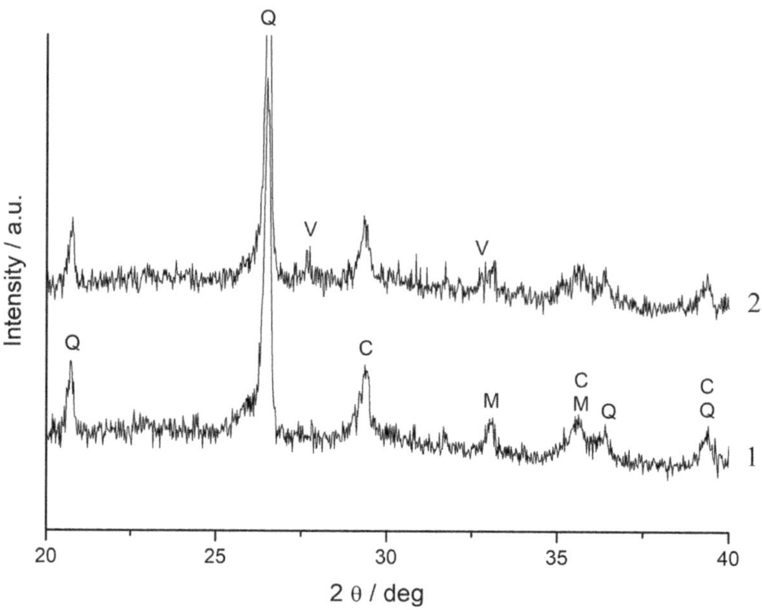

FIGURE 8.11   XRD patterns of the (90 % FA + 10 % calcite) blend mechanically activated for 180 s (1) and of the geopolymer prepared using this blend after 28 d of curing (2). Phases marked are as follows: Q – quartz, M – mullite, C – calcite, V – vaterite.

calcite. In the XRD pattern of the geopolymer based on the (90% FA + 10% calcite) blend, mechanically activated for 180 s, at the age of 28 days (Figure 8.11, curve 2) in comparison with the XRD pattern of the blend (Figure 8.11, curve 1) peaks appear that belong to the newly formed phase – vaterite which is a $CaCO_3$ polymorph. It should be noted that the intensity of the main reflection of calcite ($2\theta=29.4°$) in the x-ray diffraction pattern of the geopolymer (Figure 8.11, curve 2) somewhat decreased in relation to this peak in the x-ray diffraction pattern of the blend (Figure 8.11, curve 1). This confirms the conclusion regarding partial transformation of calcite into vaterite in this geopolymer. Such a transformation can proceed through the recrystallization of calcite, which is the most stable modification of $CaCO_3$ under the considered conditions, in NaOH solution in accordance with the Ostwald step rule [29, 30].

It is likely that vaterite can also form in the case of geopolymers synthesized using mixtures with a lower content of $CaCO_3$, but its amount is very small in order to confirm this by XRD analysis. In addition, for the geopolymer based on the blend containing 10% calcite, the partial transformation of calcite to calcium hydroxide was observed by SEM/EDS [25]. The data obtained make it possible to explain the above-mentioned decrease in the workability of the geopolymer paste at elevated contents of calcium carbonate (>10%) and an increase in the MA time (>400 s). In this case, the rate and degree of transformation of calcite into calcium hydroxide should increase, which leads to a decrease in the water-to-solid ratio due to the consumption of water for the formation of $Ca(OH)_2$ and, consequently, to the decrease in the paste workability.

The increase in the strength of the geopolymer due to the addition of calcite to the FA can be explained not only by the well-known microfiller effect. Calcite particles, as well as newly formed vaterite and calcium hydroxide, are the centers of the formation of the N-A-S-H gel – the main binding phase of the geopolymer. The use of MA increases the reactivity of the fly ash with respect to the alkaline agent, and also leads to an increase in calcite surface area, which accelerates the gel formation.

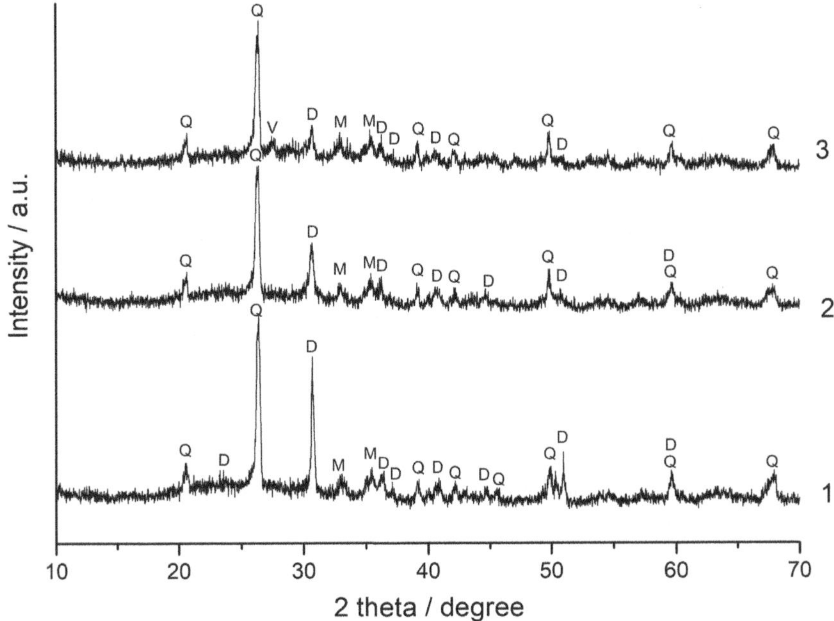

FIGURE 8.12 XRD patterns of the (90% FA + 10% dolomite) blend after MA for 30 s (1), 180 s (2), and of the geopolymer synthesized using the blend mechanically activated for 180 s cured for 360 d. Phases marked are as follows: Q – quartz, M – mullite, D – dolomite, V – vaterite.

8.3.2 MECHANICALLY ACTIVATED FA BLENDED WITH DOLOMITE

Figure 8.12 shows representative XRD patterns of the (90% FA + 10% dolomite) blend mechanically activated for 30 s and 180 s. As in the case of calcite containing blends (Figure 8.8), from a comparison of the peak intensities it is seen that structural disordering caused by MA is much more pronounced for the carbonate mineral (dolomite) than for mullite and quartz. Obviously, this is due to the difference in the hardness of the minerals. Mohs hardness for quartz, mullite and dolomite is 7.0, 6.3–7.5 and 3.5–4.0, respectively.

The compressive strengths of geopolymers based on the (FA + dolomite) blends at the age of 7 and 28 days are shown in Figures 8.13 and 8.14, respectively. Blending the FA with dolomite generally leads to an increase in strength with an increase in the proportion of the added carbonate mineral.

Comparing additions of calcite or dolomite to FA, it should be noted that calcite is preferable in terms of the geopolymer strength. For the (FA + calcite) blends containing 1, 3, 5 and 10% calcite mechanically activated for 180 s, the geopolymer strength at the age of 7 days was 7.4, 11.2, 13.0 and 16.8 MPa, respectively (Figure 8.9). For the (FA + dolomite) blends, the corresponding strength was 4.8, 4.9, 4.7 and 11.2 MPa, respectively (Figure 8.13). With an increase in the curing time, the difference in strength between geopolymers prepared using calcite and dolomite decreases. For the (FA + calcite) blends containing 1, 3, 5 and 10% calcite mechanically activated for 180 s, the geopolymer strength after 28 days of curing was 12.7, 16.6, 19.5 and 22.4 MPa, respectively (Figure 8.10). For the (FA + dolomite) blends, the corresponding strength was 12.0, 12.5, 13.8 and 18.1 MPa, respectively (Figure 8.14).

Similar results were obtained for geopolymers based on metakaolin with additions of 20% calcite or 20% dolomite and water glass as an alkaline agent (without the use of MA raw materials) [16]. The increased strength of geopolymers due to the addition of calcite to metakaolin compared to the

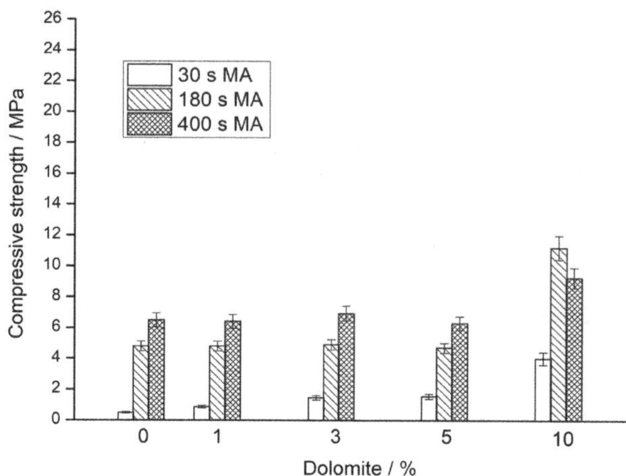

FIGURE 8.13 Effect of dolomite content and MA time on the 7 d compressive strength of geopolymers based on the (FA + dolomite) blends.

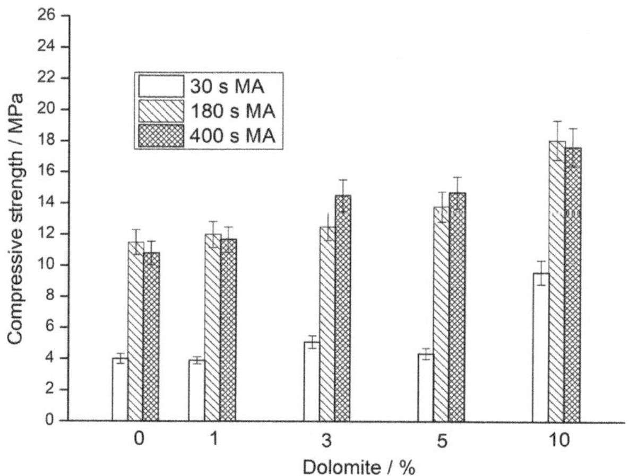

FIGURE 8.14 Effect of dolomite content and MA time on the 28 d compressive strength of geopolymers based on the (FA + dolomite) blends.

addition of dolomite was explained by the higher level of dissolved calcium, which affects the structure of the gel and the strength of the geopolymer matrix. In addition, the strength of the composite geopolymer depends on the contact strength of the geopolymer matrix with particles of unreacted carbonate minerals. Therefore, differences in strength can also be determined by differences in the surface properties of calcite and dolomite.

8.4 CONCLUSIONS

The Cu–Ni slag and Zn slag are suitable raw materials for the synthesis of geopolymers. MA of the Cu–Ni slag and Zn slag in CO_2 atmosphere was accompanied by chemisorption of carbon dioxide by the outer layers of the slag particles. This can be used as a tool to improve the reactivity of the slag and geopolymer performance. According to the calorimetric measurements, the carbonized layer of the non-ferrous slag particles after MA in CO_2 showed higher reactivity with respect to the alkaline agents as compared to the slag mechanically activated in air. For the geopolymers based on Cu–Ni slag prepared using liquid glass and geopolymers based on Zn slag prepared using NaOH solution, the compressive strength was higher in the case of the slag mechanically activated in CO_2 compared to the corresponding geopolymers based on the slag mechanically activated in air.

The reactivity of the Cu–Ni slag can be improved also through the MA of the slag with small additions of carbonate minerals. The mechanically activated Cu–Ni slag blended with carbonates of alkaline-earth metals, hardened without an alkaline agent. There was an optimum amount of carbonate additive that provides the highest strength of the mechanically activated blend.

The addition of carbonates (calcite, dolomite) and MA of the blends is also a promising way to increase the efficiency of geopolymers based on FA. From the point of view of the geopolymer strength, the addition of calcite is preferable to dolomite, although the difference in strength decreases with increasing curing time.

In the presented results, a planetary mill was used as a mechanical activator. Future research should be directed towards achieving similar results using other types of mills (e.g., vibration mills and attrition mills) currently used in the industry for large scale production.

ACKNOWLEDGMENTS

This study was partially funded by RFBR, project number 20-03-00486.

REFERENCES

1. Piatak N.M., Parsons M.B., Seal R.R. Characteristics and environmental aspects of slag: A review. *Appl. Geochem.* 2015. V. 57. P. 236–266.
2. Yao Z.T., Ji X.S., Sarker P.K., Tang J.H., Ge L.Q., Xia M.S., Xi Y.Q. A comprehensive review on the applications of coal fly ash. *Earth-Science Rev.* 2015. V. 141. P. 105–121.
3. Provis J.L. Alkali-activated materials. *Cem. Concr. Res.* 2018. V. 114. P. 40–48.
4. Davidovits J. Geopolymers: Ceramic-like inorganic polymers. *J. Ceram. Sci. Technol.* 2017. V. 8. P. 335–349.
5. Krivenko P. Why alkaline activation – 60 years of the theory and practice of alkali-activated materials. *J. Ceram. Sci. Technol.* 2017. V. 8. P. 323–333.
6. Kalombe R.M., Ojumu V.T., Eze C.P., et al. Fly ash-based geopolymer building materials for green and sustainable development. *Materials* 2020. V. 13. P. 5699.
7. Singh N.B., Middendorf B. Geopolymers as an alternative to Portland cement: An overview. *Constr. Build. Mater.* 2020 V. 237. P. 117455.
8. Luukkonen T., Heponiemi A., Runtti H., Pesonen J., Yliniemi J., Lassi U. Application of alkali-activated materials for water and wastewater treatment: A review. *Rev. Environ. Sci. Biotechnol.* 2019. V. 18. P. 271–297.
9. Shehata N., Sayed E.T., Abdelkareem M.A. Recent progress in environmentally friendly geopolymers: A review. *Sci. Total. Environ.* 2021. V. 762. 143166.
10. Kumar S., Kumar R. Mechanical activation of fly ash: Effect on reaction, structure and properties of resulting geopolymer. *Ceram. Int.* 2011. V. 37. P. 533–541.
11. Kalinkin A.M., Kumar S., Gurevich B.I., Alex T.C., Kalinkina E.V., Tyukavkina V.V., Kalinnikov V.T., Kumar R. Geopolymerisation behavior of Cu-Ni slag mechanically activated in air and in CO_2 atmosphere. *Int. J. Mineral Process.* 2012. V. 112–113. P. 101–106.
12. Lapshin O.V., Boldyreva E.V., Boldyrev V.V. Role of mixing and milling in mechanochemical synthesis (Review). *Russ. J. Inorg. Chem.* 2021. V. 66. P. 433–453.
13. Baláž P., Achimovicová M., Baláž M., Billik P. et al. Hallmarks of mechanochemistry: From nanoparticles to technology. *Chem. Soc. Rev.* 2013. V. 42. P. 7571–7637.
14. Avvakumov E.G. *Mechanical methods for activation of chemical process.* Nauka: Novosibirsk, Russia, 1986. P. 305. (In Russian)
15. Heinicke G. *Tribochemistry.* Akademie-Verlag: Berlin, 1984. P. 495.

16. Yip C., Provis J., Lukey G., van Deventer J. Carbonate mineral addition to metakaolin-based geopolymers. *Cem. Concr. Compos.* 2008. V. 30. P. 979–985.

17. Cwirzen A., Provis J.L., Penttala V., Habermehl-Cwirzen K. The effect of limestone on sodium hydroxide-activated metakaolin-based geopolymers. *Constr. Build. Mater.* 2014. V. 66. P. 53–62.

18. Perez-Cortes P., Escalante-Garcia J.I. Alkali activated metakaolin with high limestone contents - Statistical modeling of strength and environmental and cost analyses. *Cem. Concr. Compos.* 2020. V. 106. P. 103450.

19. Bayiha B.N., Billong N., Yamb E., Kaze R.C., Nzengwa R.E. Effect of limestone dosages on some properties of geopolymer from thermally activated halloysite. *Constr. Build. Mater.* 2019. V. 217. P. 28–35.

20. Kalinkina E.V., Kalinkin A.M., Forsling W., Makarov V.N. Sorption of atmospheric carbon dioxide and structural changes of Ca and Mg silicate minerals during grinding. I. Diopside. *Int. J. Miner. Process.* 2001. V. 61. P. 273–299.

21. Kalinkin A.M., Nevedomskii V.N., Kalinkina E.V. Nanostructure of $CaMgSi_2O_6$ (diopside) and $CaTiO_3$ (perovskite) mechanically activated in carbon dioxide. *Inorganic Mater.* 2008. V. 44. P. 639–644.

22. Alex T.C., Kalinkin A.M., Nath S.K., Gurevich B.I., Kalinkina E.V., Tyukavkina V.V., Kumar S. Utilization of zinc slag through geopolymerization: Influence of milling atmosphere. *Int. J. Miner. Process.* 2013. V. 123. P. 102–107.

23. Kalinkin A.M., Gurevich B.I., Myshenkov, M.S., Kalinkina E.V., Zvereva I.A. A calorimetric study of hydration of magnesia-ferriferous slag mechanically activated in air and in CO_2 atmosphere. *J. Therm. Anal. Calorim.* 2018. V. 134. P. 165–171.

24. Kalinkina E.V., Gurevich B.I., Kalinkin A.M., Mazukhina S.I., Tykavkina V.V., Zalkind O.A. Binding properties of ferromagnesian slags after mechanical activation with alkaline-earth carbonates. *Inorganic Materials.* 2014. V. 50. P. 1179–1184.

25. Kalinkin A.M., Gurevich B.I., Myshenkov M.S., Chislov M.V., Kalinkina E.V. et al. Synthesis of fly ash-based geopolymers: Effect of calcite addition and mechanical activation. *Minerals* 2020. V. 10. P. 827.

26. Kalinkin A.M., Gurevich B.I., Kalinkina E.V., Chislov M.V., Zvereva I.A. Geopolymers based on mechanically activated fly ash blended with dolomite. *Minerals* 2021. V. 11. P. 700.

27. Kumar S., Mucsi G., Kristály F., Pekker P. Mechanical activation of fly ash and its influence on micro and nano-structural behaviour of resulting geopolymers. *Adv. Powder Technol.* 2017. V. 28. P. 805–813.

28. Lee W.K.W., Deventer J.S.J.V. Use of infrared spectroscopy to study geopolymerization of heterogeneous amorphous aluminosilicates. *Langmuir* 2003. V. 19. P. 8726–8734.

29. Bernal S.A., Provis J.L., Walkley B., San Nicolas R. et al. Gel nanostructure in alkali-activated binders based on slag and fly ash, and effects of accelerated carbonation. *Cem. Concr. Res.* 2013. V. 53. P. 127–144.

30. Navrotsky A. Energetic clues to pathways to biomineralization: Precursors, clusters, and nanoparticles. *Proc. Natl. Acad. Sci. USA* 2004. V. 101. P. 12096–12101.

31. Kalinkin A.M., Alex T.C., Nath S.K., Gurevich B.I., Kalinkina E.V., Tyukavkina V.V., Kumar S. Geopolymers based on mechanically activated non-ferrous slags. *Chem. Sustain. Develop.* 2013. V. 21. P. 593–601.

9 Waste to Wealth
Upcycling of Solid Waste Plastics

*Mayank Pathak, Bhashkar Singh Bohra and
Nanda Gopal Sahoo*

9.1 INTRODUCTION

The exponential and unorthodox expansion in universal development has resulted to an exponential increase in the generation of various types of solid wastes. Plastics, tyres, food, animal dung, woody biomass, and their combinations, make up a major portion of the world's solid waste. Plastic gets a significant and growing share of around 10 wt% of these solid wastes generated per year because of its large-scale manufacturing and environmental effect [1]. Plastic is the kind of polymer that is flexible, malleable and durable and due to this they can be used in numerous applications to make human life more comfortable and convenient. It has been employed in almost every industry, including packaging, construction, textiles, consumer goods, housing, transportation, electrical and electronics, and industrial machinery. Plastics' adaptability in various industries has drawn the attention of today's globe due to its remarkable strength-to-weight ratio, durability, cheap cost, minimal maintenance and corrosion resistance, making it an economically appealing alternative for future applications. As suggested by the review, global plastic output hit a new high of 280 million tonnes, with China accounting for 23% of the total, followed by the European Union (21%), the United States (16%) and Japan (5%) in 2011 [2]. As a result, plastics' output is expected to double by 2035 and quadruple by 2050, having increased more than 20-fold between 1950 and 2018 as suggested by statistics provided by statista.com [3]. Plastic, the magical material, has a flipside, it does not degrade naturally. Usually commodity plastics (~99%) originating from petrochemicals (i.e., synthetic plastic) are non-biodegradable in nature or degrade at a very slow rate [4]. The main obstacle for plastic degradation is the high molecular weight of the plastic. The microorganism cannot grasp the higher molecular weight compounds [5]. The continuous improvement in plastic properties with the help of some additives, such as plasticizers, stabilizers, cross-linking agents, etc., according to our requirement, make its degradation very difficult [6]. The alarming situation originating from plastic is due to its non-biodegradable nature, short lifetime uses and inappropriate dumping, such as throwing it in water streams from running car windows, or dumping filled garbage containers on land, etc., which is of great concern and contributes to environmental, economic and waste management problems [7].

Furthermore, human use of plastics in the linear pattern of 'take, make, use and discard' is the principal cause of natural resource depletion, waste creation, environmental degradation, climate change and adverse effects on living organisms due to which large trajectories of plastic waste can be seen everywhere. The hazardous chemicals and non-biodegradable nature of waste plastics pose a dilemma for our environment, humans, ecosystems and natural resources. Managing plastic waste through landfill and incineration not only destroys the resources but also generates subsidiary problems, for example, it may be turned into micro and nano-plastics that are detrimental to aquatic

DOI: 10.1201/9781003327646-9

animals when a little portion of it reaches rivers, lakes, and seas, while plastic garbage produces heat and carbon dioxide when it is buried in landfills or burnt, conclusively disturbs the ecosystem. As a result, there is presently no safe means to dispose of plastic waste, which results in significant environmental damage during the manufacture, usage and disposal processes.

Only 9% of plastic garbage is recycled, and most of the recovered plastic is used to make less recyclable and low-value items like plant pots and garden furniture [8, 9]. For instance, 'primary recycling', 'secondary recycling' (mechanical recycling), 'tertiary recycling' (chemical recycling) and 'quaternary recycling' are the four main types of plastic recycling processes now available. Plastics are sorted, cleaned, crushed, melted and remoulded in both primary and secondary recycling processes [10]. However, as a result of cross-linking reactions caused by thermal degradation, trace acids, and immiscibility with other plastics, products made this way frequently suffer from deteriorated properties (e.g., discolouration or loss of strength caused by decreases in molar mass). This approach of producing low-value products is also known as 'downcycling'. In quaternary recycling, plastics that have been discarded are burnt to create heat for electricity generation because they have a high heating value of 20 to 40 MJ kg^{-1}, which is equivalent to crude oil. This method is still not adopted by many countries due to emission of toxic gases and low energy efficiency [11]. Another method is tertiary recycling, which is also called chemical recycling in which plastics can be converted into original monomers or new chemical moieties by the means of glycolysis and pyrolysis [12, 13].

Furthermore, a new strategy 'upcycling' has evolved compared to traditional recycling, under which plastics can be converted into fuels, chemicals and valuable carbon nanomaterials [14]. Upcycling of waste plastics into fuels takes heat energy which can be compensated by reuse of the by-product while upcycling into carbon nanomaterials is considered as the novel procedure to manage waste plastics. In the process of conversion of waste plastics into carbon nanomaterials, the final product can be used in various field of applications including adsorbents, water purification, energy storage systems, etc. Although, disposal methods such as landfilling, incineration and recycling are not sufficient to overcome this problem, a lot of research work has been accomplished by researchers using advanced techniques to convert waste plastics into value-added products. Here, in this chapter we discuss plastic wastes and their types, as well as their recycling and upcycling methods.

9.2 CLASSIFICATION OF PLASTICS

Most plastics are the synthetic polymer formed by addition or condensation polymerization. Before upcycling or recycling the plastic waste, we need to understand the plastics and their classifications. Most of the plastics are classified by their thermal properties, broadly categorizing them into two groups – thermoplastics and thermosetting plastics, which are further discussed [15].

9.2.1 THERMOPLASTIC

A thermoplastic is a plastic (polymer) in which the molecules are held together by weak secondary bonding forces that are softened when exposed to heat [16] and do not undergo chemical change in their composition as shown in Figure 9.1. So, when the temperature is increased, bonds will break and then occurs the phase change leading to convert the plastic from the solid state to the liquid state. Due to this sort of melting, it can be moulded again and again. Additionally, in the case where heating bonds are broken, this leads to plastic'sstrength and hardness loosening, while on cooling, solidification takes place which helps the plastic to regain its original properties, such as original strength and hardness. Conclusively, plastics that show the behaviour of altering the shape, are known as thermoplastic polymers. They are also considered as non-biodegradable plastics. Common thermoplastics range from 20,000 to 500,000 in molecular mass. Some examples of thermoplastic are areylic ABS,

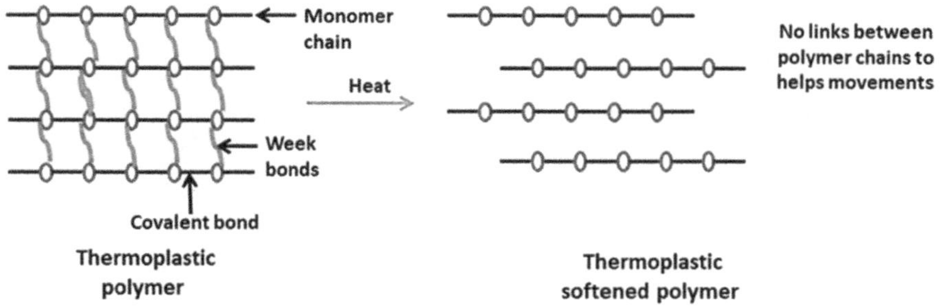

FIGURE 9.1   Thermal behaviour of thermoplastic polymer.

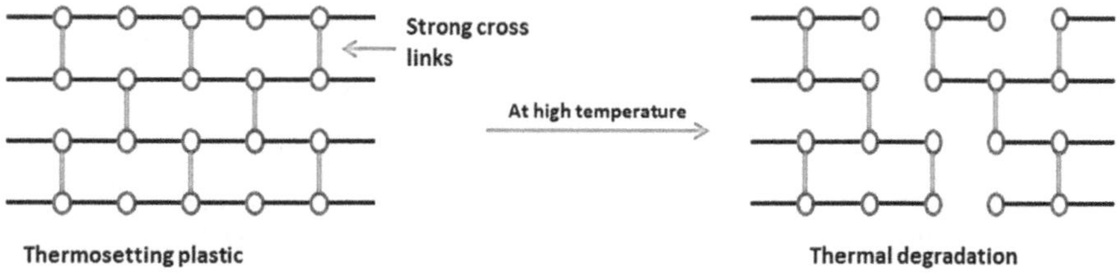

FIGURE 9.2   Thermal behaviour of thermosetting plastic.

polytetrafluoroethylenes (PTFE), polyacrylonitrile (PAN), polystyrene (PS), polyethylene (PE) and polyvinyl chloride (PVC) [15].

9.2.2   Thermosetting Plastic

These are the type of plastics in which the molecules are joint together by strong covalent bonds and have extremely cross-linked structures that become hardened or solidified when exposed to heat and undergo chemical change in their composition [16]. The behaviour of thermosetting plastic at high temperature is shown schematically in Figure 9.2. It can be said that the process cannot be reversed, and physical shape cannot be altered. In the case of elevating the temperature of thermosetting plastic, the bonds will break and automatically the plastic will get degraded, or it will get decomposed. The reason behind the hardness and strength of thermosetting polymers is a three-dimensional cross-linked structure of the polymer resins due to which it retains its properties during the moulding process, which leads to the non-recyclable nature of this type of plastic. These plastics are usually stronger, harder and more brittle than thermoplastic and cannot be reclaimed from waste. These resins are insoluble in almost all inorganic solvents. Some common examples of thermosetting plastics are menthol, epoxy resin, phenol formaldehyde, urea and Bakelite, etc. [15].

The thermoplastic polymer contains two dimensions and no cross-link between the bond or its adjacent molecules, while thermosetting plastic consists of a three-dimension cross-link leading to higher thermosetting plastic strength than the thermoplastic, as shown in Figure 9.3. Thermoplastics are typically manufactured by the process known as addition polymerization whereas thermosetting plastic is manufactured by the process known as condensation polymerization. In the case of addition polymerization, directly the addition between the two adjacent molecules takes place and there is no formation of water or any other substituent molecule, while in the case of condensation polymerization there is a formation of water.

Both types of plastics are wildly used in our day-to-day life starting from a pen to a polythene bag. Use of plastics has become essential in all sectors of the economy. At present, mostly 50 to 60%

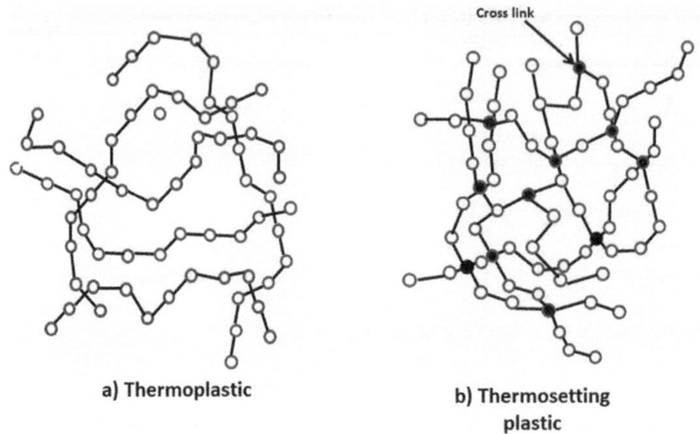

FIGURE 9.3   Structure of thermoplastic polymer and thermosetting plastic.

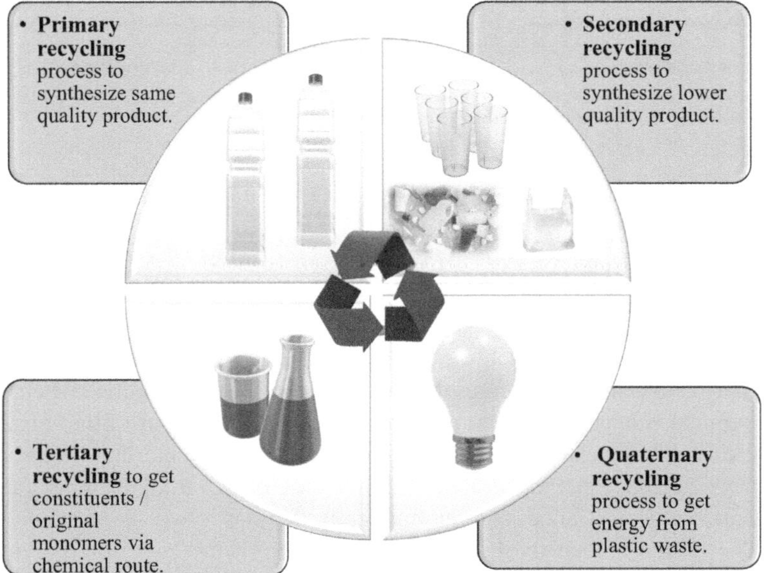

FIGURE 9.4   Various kinds of plastics and their respective applications with symbols.

of plastics are used for single use disposable applications, such as packing materials like carry bags, packing of food, etc., and agricultural tools, and 20 to 25% of plastics are used for infrastructure applications, such as cable coating pipes, and the remaining part of plastics are used for consumer applications with intermediate lifespan, such as in furniture, transports, textile, etc. [17]. For a clear view of all kinds of plastics, and their respective applications in numerous fields, see Figure 9.4.

This figure indicates that we are surrounded by various kinds of plastics in our daily routine life. But these plastics are toxic in nature as they create pollution in the environment from the start of the manufacturing process to their usage and finally from their improper disposal due to the release of harmful chemicals in the environment such as vinyl chloride, dioxins, benzene, plasticizers, phthalates, bisphenol, formaldehyde, etc. These harmful chemicals are passed to us through air, soil and water. For instance, polyethylene (PE) is used in shopping bags, toy, bottles, etc., and are carcinogenic in nature. Phthalates present in emulsion ink, toys, footwear are creating hormonal disturbance carcinogenic, reducing sperm counts, causing infertility and reducing immunity. Polyvinyl alcohol (PVA) is used in packaging, vinyl siding, utility items and in cosmetics, creating birth, genetic problems, skin disease, vision problems, deafness, digestion problem, liver-related problems and carcinogenic. PVA produces dioxin (potent synthetic chemical) causing cancer, reducing immunity and reproductively. Bisphenol A is a monomeric building block for polycarbonate plastic. During polymerization of polycarbonate, some bisphenol A remains unbound. This unbound bisphenol is released from food and drink containers to food and drink at elevated temperature. The presence of these chemicals in our body disturb the natural hormonal messaging system, which causes cancer, reduces immune system and triggers obesity and diabetes [18, 19].

Plastic waste is also disposed in oceans. In oceans, these plastics are broken down into small fragments (in micron level) due to the weathering effect. This plastic debris causes physical harm to marine mammals through ingestion or entanglement, for example, sea turtles, sea birds, fish and other mammals ingesting plastic debris by mistake due to considering the debris as food. Ingestion can lead to internal injuries and infection, malnutrition in mammals when plastic debris is collected in the stomach. Plastic debris release toxic chemicals, which cause death and reproductively failure in mammals, such as fish and shellfish. Entanglement in the marine system causes suffocation drowning, starvation and vulnerability to predators [20, 21]. Due to these reasons, proper treatment of waste plastics is needed, which drives the requirement for waste plastics to be recycled and upcycled into value-added products instead of using landfills, incineration and organic disposal; this will not only solve the global problem of waste plastics but will also provide a path turning waste to wealth.

9.3 RECYCLING OF PLASTIC WASTE

Plastic waste management is a dramatically increasing issue worldwide, which cannot be resolved by technologies, such as landfills, etc. Such technologies, which are consuming a huge amount of energy, generating secondary wastes and destroying the environment along with human health, should not be felicitated. Recycling techniques, which are not creating secondary waste and environmental issues along with no consumption of a huge amount of energy, should be adopted and considered due to their sustainability. Recycling technology mainly comprises four strategies including primary, secondary, tertiary and quaternary recycling, which are discussed under the upcoming subheads.

9.3.1 PRIMARY RECYCLING

When waste plastic, which is uncontaminated, can be reused directly by conversion into new products whilst pertaining its original properties, the process is referred to as primary recycling, also known by the term close-loop recycling. It is only performed by manufacturers whenrecycling the plastic waste generated after completing various industrial tasks [22]. In this process, plastic waste is shredded, crushed and milled for the proper mixing with additives followed by systematic cleaning to purify and remove unwanted impurities [23]. Thermoplastic polymers like PE, PET and PP can be recycled through this process as this crushed and purified plastic can be converted into a new product only by melting followed by the introduction of injection moulding or heat pressing processes, etc. [22, 23]. The manufacturing of PET bottles from the mixture of discarded PET and virgin PET polymers is an example of such a technique. Recently, some researchers from KU Leuven used this technique to convert the old and discarded back covers of TVs into new back covers [24]. But due to selective collection and use of uncontaminated plastics, this technique is not adopted by most of the recyclers.

9.3.2 SECONDARY RECYCLING

In contrast to primary recycling, secondary recycling includes separation and purification of contaminants of food waste, additives, dyes and another polymer matrix from the original polymer. Sometimes this process is called 'downcycling' as in this process, mechanical properties may get reduced due to the contamination of the polymer with another polymer matrix and due to the presence of trace amounts of water and acids leading to chain scissions. For example, sometimes PET impurities can be seen in the PVC, which reduce its properties. The problem of the presence of water and acid can be resolved by the excessive drying along with the addition of stabilizers while the problem of contamination can be resolved by various techniques including Ft-ir spectroscopy, optical colour camera, laser sorting [17] and XRD techniques [25]. Some examples of secondary recycling are the use of car shredder residue to make composites for some new car components and recycling into a new form from discarded PU foam by means of crushing and remoulding. As per the discussion, it can also be seen that sometimes secondary recycling techniques may become complex and so much more expensive, which further leads to tertiary recycling.

9.3.3 TERTIARY RECYCLING

It is often called chemical recycling because in this process the long chain of polymers in plastic are transformed into tiny molecules using chemical routes. Some chemical routes adopted under chemical recycling are glycolysis, pyrolysis, hydrolysis and gasification [22, 26]. Using these strategies, the waste plastic polymer may get converted into their original monomers or novel chemical feedstocks due to which this process is also referred to as feedstock recycling. For example, depolymerization can take place in the case of PLA, PU and PET polymers using this technique and these monomers can again be used for the manufacturing of an original polymer [27]. However, techniques like pyrolysis require a lot of heat energy in the absence of oxygen due to which this technique becomes extensively energy consuming and non-selective in nature.

9.3.4 QUATERNARY RECYCLING

Sometimes referred to as incineration, quaternary recycling is a popular process, which gives rise to heat for electricity generation by burning the plastic waste in order to recover the energy due to the high heating value of waste plastics equivalent to crude oil [28]. This technique is mainly adopted when there is heavily mixed and contaminated plastics including packages of medical supplies and toxic goods. The problems like landfilling and lack of fossil-fuel-based energy production leading to environment issues can be aborted using this technique. However, the energy recovery efficiency is low in this case. In addition, the probability of leaking of hazardous and toxic gases into the environment is very high in this technique [11]. Due to the limitation of all above mentioned recycling processes, another term upcycling came into emergence, which is discussed in the upcoming sections. Collectively, all the categories of recycling techniques are shown in Figure 9.5.

9.4 UPCYCLING

Waste conversion into wealth is the basis of the upcycling technique. In the upcycling process, plastic is converted into a better quality and higher value product. Upcycling helps in plastic waste management by converting plastic into value-added products such as chemicals for monomer feedstock, fuels, carbon nanomaterials, etc., due to the presence of carbon content in a significant amount [29]. For example, polyethylene contains 85.6%, polystyrene 92.2%, and polypropylene contains 85.6% carbon content [2]. Hence it can be used as a carbon source for carbon-based value-added products, for example, activated carbon, carbon nanotubes, graphene, carbon fibers, etc. The non-wetting nature of plastic makes it useable in road construction. Water penetration across the bitumen

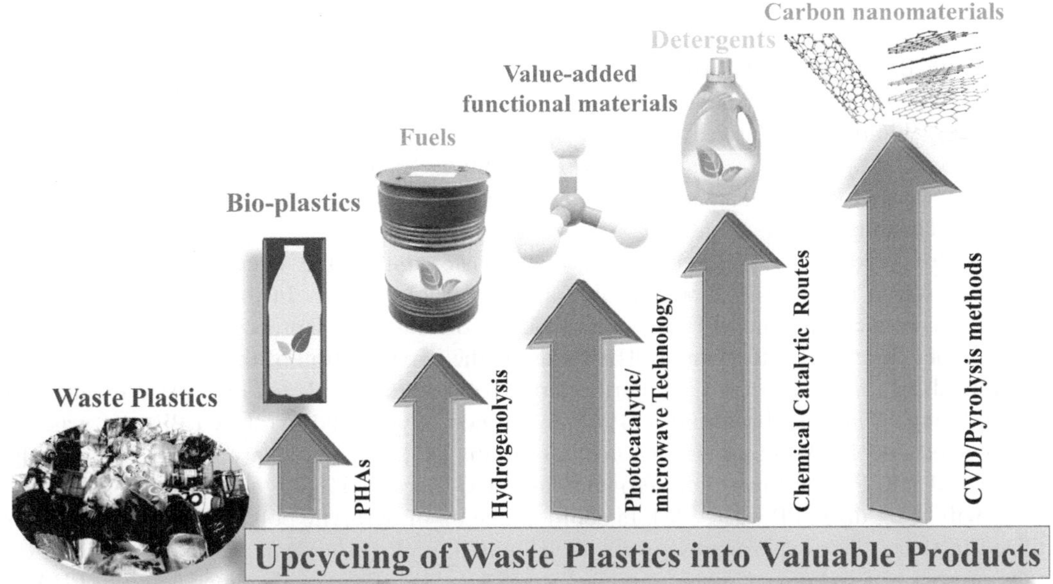

FIGURE 9.5   Plastic waste recycling using numerous techniques.

film makes potholes in roads, however, plastic coating over the aggregates prevent the penetration of water due to the non-wetting nature of plastic. Polymers also have higher softening temperature, which reduce the stripping of aggregates during summer season [29]. For a better understanding, various upcycling techniques are discussed below, separately.

9.4.1 Upcycling into Bioplastics

Biodegradable plastics are decomposed by living organisms. These plastics are nearly equivalent to the synthetic petrochemical-based plastics in all aspects with an additional quality of being able to decompose naturally into safe bio-products in a short period of time as compared to conventional plastic [18]. The main advantage of these polymers over conventional polymers is that they are totally manufactured by renewable biomass such as vegetables, tree leaf, food waste, animal shells and by microorganism rather than fossil fuels, which are very cheap, and this bio-waste does not remain as waste instead it is converted into useful items. Non-biodegradable plastic waste can easily be upcycled into biodegradable bioplastics via bio-upcycling technology. In such a technique, waste plastics are first pyrolyzed or biologically degraded into monomers and then the depolymerized/degraded monomers are used as the feedstock in the bio-upcycling technique to produce value-added biodegradable plastics [18]. There are two main approaches for the development of biodegradable polymer: the production of a synthetic polymer with an enhanced degradability nature without reducing the material properties, such as addition of functional moieties or bio-component, which helps in the deterioration of the polymer chain via post-polymerization treatments [29, 30]. The main purpose behind this is to enhance the microbial attack on the polymer chain. The degradation rate for conventional plastic will become much better when producing block copolymer with hydrolyzable polymeric molecules such as starch, ethylene glycol, lactic acid, caprolactone [31]. New synthetic polyester-based biodegradable polymers are frequently used in the in-packaging industry, paper coating and garbage bags. The polyester blended with polylactic acid and starch-based thermoplastic are degraded into carbon dioxide and water in the environment when exposed to a microorganism and in compost disposal. Polylactic acid is a synthetic polymer which is completely mineralized and frequently used in the medical field since 1990 [29]. Unfortunately, these polymers show a less durable nature than synthetic polymers in some cases. Hence, the second approach is based on the development of plastic and totally depends on biological molecules, such as starch, proteins, cellulose acetate, lactic acid, caprolactone and polysaccharides [29, 31]. The mechanical properties of biologically derived polymers have been increased by the addition of plasticizers, nanomaterials or by controlling the production condition. Examples of bioplastics or sustainable polymers are poly(hydroxyalkanoates) (PHAs), made by many bacteria and used in packaging, biomedical and coating application [32].

Polyhydroxyalkanoates (PHAs) are the main basic unit in the manufacturing process of bioplastics which is a sort of biodegradable and renewable plastic. In this regard, waste polyethylene was upcycled into mcl-PHA through chemo-biotechnological process by some researchers, in which they microbially fermented PHA by using the waxy portion of the pyrolyzed PE as the carbon feedstock in the process [33]. Furthermore, oxygenated pyrolyzed PE wax fermented into PHA utilizing *C. necator* bacterium [34] while the non-oxygenated pyrolyzed PE wax was also bio-upcycled into PHA by utilizing *C. necator* strains [35]. Waste polypropylene (PP) can also be upcycled into PHA by microbial fermentation of *C. necator* H16 in which additional carbon source will be oxidized PP waste [36]. In addition, the oil of pyrolytic polystyrene (PS) waste was utilized in shake flask experiments for the development of *Pseudomonas putida* CA-3 and PHA in the presence of sodium ammonium phosphate (SAP) [37]. PS was also upcycled into PHA by some researchers using oxidatively pyrolyzed PS as a carbon source in fermenting the *C. necator* strain followed by the biosynthesis of PHA for generation of bioplastics [38]. Furthermore, these techniques of upcycling into bioplastics not only offers the renewable plastics an alternative to petroleum-based plastics, but also it reduces environmental issues as PHAs are biodegradable, providing a path towards a circular economy instead of a linear economy. But there are some obstacles which prevent the use of biodegradable polymers: the cost of biodegradable polymers is higher than petrochemical-based polymers. Another drawback is limited availability of monomers, their processability, cost, performance limitation and rate of degradation. Bioplastics such as starch and cellulose are not favourable for mechanical recycling as well as for incineration because their calorific value is very low. The next obstacle is the absence of a suitable infrastructure for the development of biodegradable polymers such as the higher cost of the manufacturing technology of biodegradable polymers. Hence further research is required to make the biodegradable polymers' manufacturing cost more effective.

9.4.2 Upcycling into Fuels, Waxes and Value-Added Products

Due to the fact that plastics along with fuels and waxes have a high carbon content, they are easily interconvertible via chemical methods. The hydrogenolysis method can be used for selective depolymerization of polymers like plastics into liquid fuels and waxes via C–C bond cleavage along with aromatization. In this regard, some scientists converted polyethylene into liquid alkylaromatics using a platinum-supported catalyst without the presence of any solvent and hydrogen [39]. The *in situ* synthesis of hydrogen via the dehydroaromatization process of shorter hydrocarbon chains, which results in the creation of aromatic compounds, is an important step. The hydrogenolysis process, which requires hydrogen as a reactant, can be used to create these chains from larger hydrocarbon chains. Furthermore, some researchers used a bifunctional metal/acid catalyst-based zeolite catalytic system for producing a mixture of gasoline, diesel, and jet fuel range hydrocarbons from low-density polyethylene via mild hydrocracking [40]. Mostly, platinum is used to crack the carbon–carbon bonds as its hydrogenation activity is very high. However, some researchers found that Ru can also be used for hydrogenation to facilitate the inner carbon–carbon bond cleavage to generate shorter chain alkanes [41, 42]. In this regard, the Ru/Nb_2O_5 catalyst was fabricated to break C–C and C–O bonds in various types of waste plastics, including PET, PS, etc., via hydrogenolysis leading to the conversion of mixed plastics into arenes [43, 44]. Researchers used another method called alkane metathesis for breaking C–C bonds via catalytically dehydrogenating PE and light alkanes to generate two novel olefins, which then again hydrogenated back to synthesize saturated hydrocarbons [45].

In addition, some researchers prepared oil from waste plastic. His team took 1 ton of plastic waste after washing and drying processing. This plastic mixture was pyrolyzed in the presence of catalyst (aluminum silicates nearly 2.5% of the plastic mass) at (603–723) K. Three products were obtained: pyrolysis oil (60–70%), gas (15–20%) and carbon black (20–30%). Oil mixture contains a different ratio of material 40% oil + 60% petrol, 10% oil + 90% diesel, 30% oil + 70% diesel and 50% oil + 50% diesel and various characteristic confirms the presence of pure fuel oil [46]. As per a few researchers, PE waxes could also be easily utilized to get oils and most of the waste plastics including PET bottles, films and bags can be upcycled into liquid fuels and wax products using a cross alkanes metathesis system (CAMS) [47].

9.4.3 Upcycling into Value-Added Functional Materials

Plastic waste can also be utilized to obtain value-added function materials, hydrogen, and chemical feedstocks via numerous emerging technologies. In this regard, some researchers upcycled waste plastics into hydrogen and fine chemicals using CdS/CDO_x QDs and nickel phosphide supported on cyanamide-functionalized carbon nitride using ambient-temperature photo-reforming [48, 49]. Another ambient temperature photocatalytic process was introduced, which was also non-toxic in nature, to produce high-value carboxylic acid fuels from waste plastics [50, 51]. To extract more than 97% of hydrogen in mixed plastics, some scientists used the microwave technique with an inexpensive iron-based catalyzt to convert the mechanically pulverize plastic waste [52]. This microwave technology can also be used in the oxidative process to upcycle high density polyethylene into valuable chemicals like succinic, glutaric, adipic acids, longer dicarboxylic acids and propionic acids, which all are also water soluble and can further be transformed into plasticizers for processing of polylactic acid (PLA) [53].

Along with C–C bond cleavage, the addition of new functionalities during the bond cleavage is desirable to get high value materials. For that, post polymerization functionalization can be a possible approach that catalytically incorporates a functional group along the polymer backbone via CH activation and does not break the C–C chain. In this regard, some researchers upcycled polyethylene into a powerful adhesive using selective CH functionalization with polyfluorinated ruthenium porphyrin as the catalyst [54]. The use of this technique to upcycle polyethylene via CH functionalization may be limited in its application due to cost ineffectiveness.

9.4.4 Upcycling into Detergents

Waste plastics not only can be upcycled into carbon nanomaterials, fuels, waxes and functional materials but can also be converted into more valuable chemical products like detergents. In this regard, researchers proposed a catalytic process in which waste plastics can be converted into valuable chemical building blocks for manufacturing daily life house holding products [39]. For instance, polyethylene waste can be converted into long chain alkyl aromatics. Furthermore, these long chain alkyl aromatics can be chemically sulfonated to synthesize biodegradable surfactants at an ambient temperature range, which additionally produces a little amount of light gas such as methane as by-product [55]. Therefore, many more methods including plastic–solvent process combinations can be used to deal with real-world plastic waste, but a single type of plastic will have the solution of a specific chemical conversion process for upcycling [56]. After the realization of such techniques, non-fossil-based plastics may become more attractive in daily life products and can be reintroduced again into a high value product. These future developments can give a path towards a commercially feasible upcycling process and then the plastics will not be considered as waste anymore, instead they will be seen as a valuable raw material, which can then drive wealth from waste.

9.4.5 Upcycling into Carbon Nanomaterials

The recycling techniques are somehow not environment friendly and economically feasible. Furthermore, recycling takes a lot of time, money and cannot convert waste plastics into products which can generate wealth including the cost of conversion. Recently, the scientific community started focusing on converting waste plastics into carbon nanomaterials, which can further be used in numerous applications. The market cost of carbon nanomaterials is very high which can compensate the cost of the whole conversion process. This process not only resolves the problem of waste plastics but also generates high value products. The number of processes including the CVD process and pyrolysis process can be utilized to convert waste plastics into valuable carbon nanomaterials, such as graphene, carbon nanotube and others.

9.4.5.1 Graphene

Graphene is a two-dimensional nanomaterial which is made up of sp^2 hybridized carbon atoms, arranged in a hexagonal array. Graphene is getting attention day by day due to its tremendous properties like high surface area, high mechanical strength and high electrical conductivity. These tremendous properties allow the utilization of graphene in various fields, including water purification, energy storage and biomedicals. Many methods have been developed for the synthesis of graphene using highly pure graphite (HPG) as a precursor material which enhances the cost of manufacturing. In solution to this, many researchers synthesized graphene using waste plastics instead of HPG due to the high content of carbon in plastics. For instance, graphene was prepared from waste plastics using CVD technique at a temperature of 1050°C under the inert atmosphere of Ar/N_2 using Cu foil as the catalytic support [57]. Other researchers synthesized graphene from waste plastics on a bulk scale using a two-step pyrolysis method. In the first step, they chopped, washed, dried and mixed the plastics with the catalyst and pyrolyzed it on the lower temperature around 400°C and pyrolyze it again on higher temperatures of around 750°C under the atmosphere of N_2 in the second step [58]. Furthermore, the graphene flakes were synthesized from polypropylene waste using organically modified montmorillonite (OMMT) clay as a catalyst [59]. Many other methods like flash synthesis, flash Joule heating were also utilized to upcycle the waste plastics into graphene without the use of any catalyst [60, 61].

9.4.5.2 Carbon Nanotube (CNT)

When a graphene sheet is rolled up in a particular dimension, carbon nanotube is obtained in a cylindrical form with both open ends. The carbon atoms in carbon nanotube are also sp^2 hybridized with a hexagonal array. At present, CNT is a valuable carbon nanomaterial due to its high electrical, mechanical and thermal properties along with high specific surface area. Many methods of CNT synthesis using various precursors are already developed by researchers including laser ablation, arc discharge, CVD, etc. However, the high-cost process and lower yield have invoked scientists to find an alternative way to synthesize CNT using cheaper precursors. To overcome this hurdle, researchers have been focusing on waste plastic as a precursor material to synthesize CNT. When two or more than two concentric CNTs come into existence, they then are collectively referred to as multi-walled carbon nanotubes (MWCNTs). In this order, researchers successfully converted high- and low-density plastics into MWCNTs using high-temperature exfoliation [62]. MWCNTs were also synthesized using polypropylene (PP) as a precursor via a one-step CVD method at different temperature ranges from 600 to 700°C in inert environment of Ar/H_2 for 1 hour [63]. Other researchers converted polypropylene into CNTs using different mechanisms in which a nickel-based catalyst was used [64]. A systematic pyrolysis approach can be used to get CNTs from waste plastics, which has been adopted by some scientists. They converted waste plastics into CNTs using a three-step pyrolysis approach in which ferrocene metal sandwich was used as a catalyst [65]. Similarly, another two-step pyrolysis approach was used by other researchers to decompose polypropylene into CNTs and hydrogen using a HZSM-5 zeolite in the presence of a nickel-based catalyst [66]. CNTs were also synthesized from waste polythene using a nickel-based catalyst through the pyrolysis method in inert atmosphere of argon, which was further used in organic dye absorption applications [67].

9.4.5.3 Some Other Valuable Carbon Nanomaterials

Except graphene and carbon nanotubes, there are many other carbon nanomaterials, which are highly valuable due to their extraordinary electrical, mechanical and optical properties in the fields of sensing, electronics, conductivity, biochemical sensing and energy storage. In this regard, many scientists have successfully synthesized other valuable carbon nanomaterials using different precursors. But as the plastic problem is the main concern of today's scientific community, researchers are focusing on utilizing waste plastics as precursor materials for the synthesis of such valuable carbon nanomaterials except graphene and carbon tubes. In this regard, polyethylene terephthalate (PET) waste was utilized to synthesize the carbon nanostructure from PET waste using ferrocene as a catalyst at 800°C for 100 minutes [68]. Another carbon nanostructure from polyethylene waste was prepared by researchers via a catalytic pyrolysis approach using a MgO supported bimetallic catalyst at 400°C [69]. Some other carbon nanomaterials like carbon dots were also successfully synthesized from upcycling the plastic waste via normal heating in which polypropylene was used as a precursor at different temperature ranges and it was shown that thus prepared carbon dots can be used in photocatalyst, bioimaging as well as in sensing applications [52]. Some researchers also developed fluorescent carbon dots via upcycling of waste plastics in which they used plastic waste such as waste bottles, used cups and polyethylene bags to transform them into fluorescent carbon dots [70]. Along with some new nanomaterial like solid carbon spheres, hollow carbon spheres and nitrogen doped hollow carbon spheres synthesized from polypropylene waste via the two-stage CVD method at different temperatures [71]. All methods to upcycle waste plastics are displayed in Figure 9.6, collectively.

9.5 CONCLUSION AND FUTURE PERSPECTIVE

A future without plastic is unthinkable as plastic plays an inevitable role in the daily life of mankind to promote the living standards but indiscriminate use of plastic and its totally inert behaviour to all environmental factors, leads to plastic waste generation all over the world. Waste plastic produces so many harmful effects such as the generation of toxic pollutants, spoiling the natural behaviour of soil, water, and air, creating a disturbance to the ecosystem, and originating many diseases in humans and mammals. This inert material is not degraded naturally in the environment or takes nearly 400 years or more to biodegrade, leading towards the solution of recycling/degrading waste plastics through numerous approaches, but these recycling techniques are not efficient as they are time consuming, cost-ineffective and even generate secondary wastes. In order to solve these problems, researchers are interested in converting the various kinds of plastic wastes into value-added products, which is the present need to convert waste to wealth, leading to a greener atmosphere. In this regard, this chapter provides information about various kinds of plastics, as well as their recycling and their upcycling techniques converting them into numerous value-added products. This chapter also gives information about possible solutions for upcycling of waste plastics by converting them into bioplastics, fuels, waxes, value-added functional materials, detergents, carbon nanomaterials as the economic value of fuels, waxes, value-added functional materials, detergents, carbon nanomaterials like graphene and CNT from waste plastics, is much higher than traditional recycling. This chapter possibly provides a direction for not only solving the global problem of

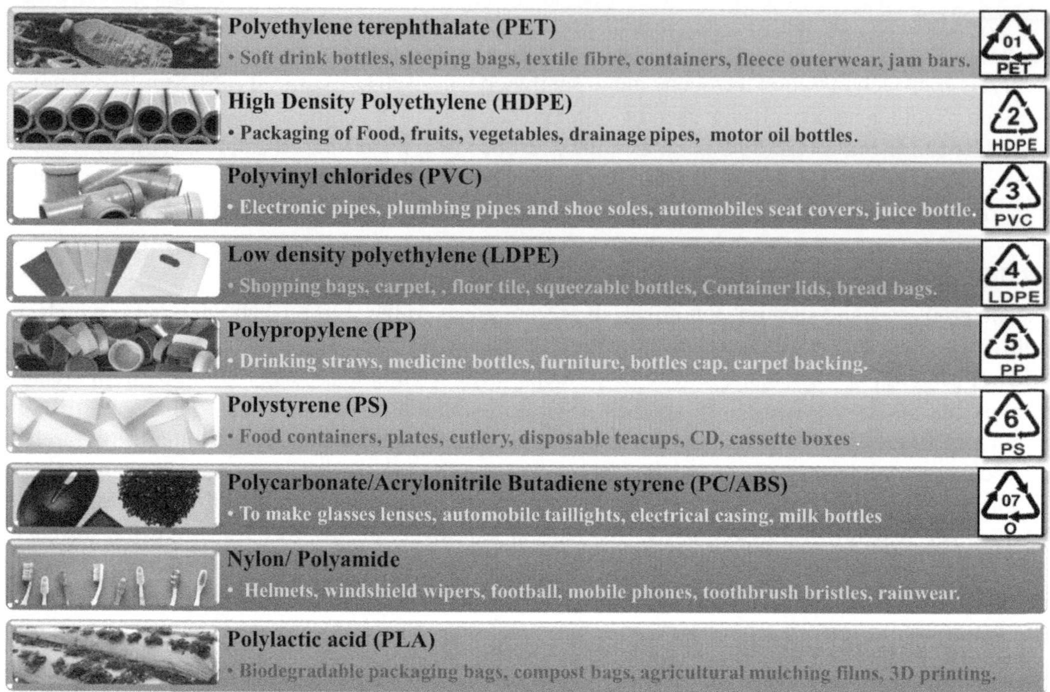

FIGURE 9.6 Various methods to upcycle the waste plastics.

plastic waste but also provides a way to generate wealth from the waste, leading to greener energy and a circular economy. In conclusion, the techniques mentioned for upcycling of waste plastics may become cheaper, feasible and greener in the upcoming decades as researchers' interest is increasing day by day to solve the global issue, leading to the development of various other future technologies to generate wealth via the upcycling of waste plastics.

ACKNOWLEDGEMENTS

The authors acknowledge DST INSPIRE Division (Ref. No.-IF180347), New Delhi, India for giving the necessary financial funds for work. The author(s) is highly grateful to the National Mission on Himalayan Studies (NMHS), Almora, India (Ref. No.-GBPNI/NMHS-2019-20/MG) for the financial assistance.

REFERENCES

[1] G. Celik, R.M. Kennedy, R.A. Hackler, M. Ferrandon, A. Tennakoon, S. Patnaik, A.M. LaPointe, S.C. Ammal, A. Heyden, F.A. Perras, Upcycling single-use polyethylene into high-quality liquid products, *ACS Central Science* 5(11) (2019) 1795–1803.

[2] C. Zhuo, Y.A. Levendis, Upcycling waste plastics into carbon nanomaterials: a review, *Journal of Applied Polymer Science* 131(4) (2014) 39931.

[3] M. Garside, Global plastic production statistic, 2019. www.statista.com/statistics/282732/global-production-of-plasticssince-1950/.

[4] S. Doetsch-Kidder, *Social change and intersectional activism: The spirit of social movement*, Springer 2012.

[5] A.A. Shah, F. Hasan, A. Hameed, S. Ahmed, Biological degradation of plastics: a comprehensive review, *Biotechnology Advances* 26(3) (2008) 246–265.

[6] B. Singh, N. Sharma, Mechanistic implications of plastic degradation, *Polymer Degradation Stability* 93(3) (2008) 561–584.

[7] S.K. Kale, A.G. Deshmukh, M.S. Dudhare, V.B. Patil, Microbial degradation of plastic: a review, *Journal of Biochemical Technology* 6(2) (2015) 952–961.

[8] A.L. Brooks, S. Wang, J.R. Jambeck, The Chinese import ban and its impact on global plastic waste trade, *Science Advances* 4(6) (2018) eaat0131.

[9] B.C. Gibb, *Plastics are forever*, Nature Publishing Group, 2019.

[10] A. Rahimi, J.M. García, Chemical recycling of waste plastics for new materials production, *Nature Reviews Chemistry* 1(6) (2017) 1–11.

[11] A.S. Burange, M.B. Gawande, F.L. Lam, R.V. Jayaram, R. Luque, Heterogeneously catalyzed strategies for the deconstruction of high density polyethylene: plastic waste valorisation to fuels, *Green Chemistry* 17(1) (2015) 146–156.

[12] S.C. Kosloski-Oh, Z.A. Wood, Y. Manjarrez, J.P. de Los Rios, M.E. Fieser, Catalytic methods for chemical recycling or upcycling of commercial polymers, *Journal of Biochemical Technology* 8(4) (2021) 1084–1129.

[13] G.W. Coates, Y.D. Getzler, Chemical recycling to monomer for an ideal, circular polymer economy, *Nature Reviews Materials* 5(7) (2020) 501–516.

[14] Y.J. Sohn, H.T. Kim, K.A. Baritugo, S.Y. Jo, H.M. Song, S.Y. Park, S.K. Park, J. Pyo, H.G. Cha, H.J.B.J. Kim, Recent advances in sustainable plastic upcycling and biopolymers, *Biotechnology Journal* 15(6) (2020) 1900489.

[15] M. Alauddin, I. Choudhury, M. El Baradie, M. Hashmi, Plastics and their machining: a review, *Journal of Materials Processing Technology* 54(1–4) (1995) 40–46.

[16] J.G. Speight, *Handbook of industrial hydrocarbon processes*, Gulf Professional Publishing 2019.

[17] J. Hopewell, R. Dvorak, E. Kosior, Plastics recycling: challenges and opportunities, *Philosophical Transactions of the Royal Society B: Biological Sciences* 364(1526) (2009) 2115–2126.

[18] V. Koushal, R. Sharma, M. Sharma, R. Sharma, V. Sharma, Plastics: issues challenges and remediation, *International Journal of Waste Resources* 4(1) (2014) 1–6.

[19] P. Pavani, T.R. Rajeswari, Impact of plastics on environmental pollution, *Journal of Chemical Pharmaceutical Sciences* 3 (2014) 87–93.

[20] R.E. Hester, Marine pollution and human health, ed. R.E. Hester, R.M. Harrison, R. Harrison, and R. Hester, *The Royal Society of Chemistry* (2011) P011–P012.

[21] R. Proshad, T. Kormoker, M.S. Islam, M.A. Haque, M.M. Rahman, M.M.R. Mithu, Toxic effects of plastic on human health and environment: A consequences of health risk assessment in Bangladesh, *International Journal of Health* 6(1) (2018) 1–5.

[22] S. Al-Salem, P. Lettieri, J. Baeyens, The valorization of plastic solid waste (PSW) by primary to quaternary routes: from re-use to energy and chemicals, *Progress in Energy Combustion Science* 36(1) (2010) 103–129.

[23] V. Goodship, Plastic recycling, *Science Progress* 90(4) (2007) 245–268.

[24] J.R. Peeters, P. Vanegas, T. Devoldere, W. Dewulf, J.R. Duflou, Closed loop recycling of plastic housing for flat screen TVs, 2012 electronics goes Green 2012+, *IEEE* (2012) 1–6.

[25] I. Arvanitoyannis, L. Bosnea, Recycling of polymeric materials used for food packaging: current status and perspectives, *Food Reviews International* 17(3) (2001) 291–346.

[26] A. Brems, J. Baeyens, R. Dewil, Recycling and recovery of post-consumer plastic solid waste in a European context, *Thermal Science* 16(3) (2012) 669–685.

[27] J.H. Clark, J.A. Alonso, J.A. Villalba, J. Aguado, D.P. Serrano, D. Serrano, *Feedstock recycling of plastic wastes*, *Royal Society of Chemistry*, (1999).

[28] A.J. Martín, C. Mondelli, S.D. Jaydev, J. Pérez-Ramírez, Catalytic processing of plastic waste on the rise, *Chem* 7(6) (2021) 1487–1533.

[29] S.P. Davila, L.G. Rodríguez, S. Chiussi, J. Serra, P. González, How to sterilize polylactic acid based medical devices?, *Polymers* 13(13) (2021) 2115.

[30] J.A. Glaser, *Biological degradation of polymers in the environment*, IntechOpen, 2019.

[31] M.A. Russo, C. O'Sullivan, B. Rounsefell, P.J. Halley, R. Truss, W.P. Clarke, The anaerobic degradability of thermoplastic starch: polyvinyl alcohol blends: Potential biodegradable food packaging materials, *Bioresource Technology* 100(5) (2009) 1705–1710.

[32] M. Nitschke, S.G. Costa, J. Contiero, Rhamnolipids and PHAs: recent reports on Pseudomonas-derived molecules of increasing industrial interest, *Process Biochemistry* 46(3) (2011) 621–630.

[33] M.W. Guzik, S.T. Kenny, G.F. Duane, E. Casey, T. Woods, R.P. Babu, J. Nikodinovic-Runic, M. Murray, K.E. O'Connor, Conversion of post consumer polyethylene to the biodegradable polymer polyhydroxyalkanoate, *Applied Microbiology Biotechnology* 98(9) (2014) 4223–4232.

[34] I. Radecka, V. Irorere, G. Jiang, D. Hill, C. Williams, G. Adamus, M. Kwiecień, A.A. Marek, J. Zawadiak, B. Johnston, Oxidized polyethylene wax as a potential carbon source for PHA production, *Materials* 9(5) (2016) 367.

[35] B. Johnston, G. Jiang, D. Hill, G. Adamus, I. Kwiecień, M. Zięba, W. Sikorska, M. Green, M. Kowalczuk, I. Radecka, The molecular level characterization of biodegradable polymers originated from polyethylene using non-oxygenated polyethylene wax as a carbon source for polyhydroxyalkanoate production, *Bioengineering* 4(3) (2017) 73.

[36] B. Johnston, I. Radecka, E. Chiellini, D. Barsi, V.I. Ilieva, W. Sikorska, M. Musioł, M. Zięba, P. Chaber, A.A. Marek, Mass spectrometry reveals molecular structure of polyhydroxyalkanoates attained by bioconversion of oxidized polypropylene waste fragments, *Polymers* 11(10) (2019) 1580.

[37] P.G. Ward, M. Goff, M. Donner, W. Kaminsky, K.E. O'Connor, A two step chemo-biotechnological conversion of polystyrene to a biodegradable thermoplastic, *Environmental Science Technology* 40(7) (2006) 2433–2437.

[38] B. Johnston, I. Radecka, D. Hill, E. Chiellini, V.I. Ilieva, W. Sikorska, M. Musioł, M. Zięba, A.A. Marek, D. Keddie, The microbial production of polyhydroxyalkanoates from waste polystyrene fragments attained using oxidative degradation, *Polymers* 10(9) (2018) 957.

[39] F. Zhang, M. Zeng, R.D. Yappert, J. Sun, Y.-H. Lee, A.M. LaPointe, B. Peters, M.M. Abu-Omar, S.L. Scott, Polyethylene upcycling to long-chain alkylaromatics by tandem hydrogenolysis/aromatization, *Science* 370(6515) (2020) 437–441.

[40] S. Liu, P.A. Kots, B.C. Vance, A. Danielson, D.G. Vlachos, Plastic waste to fuels by hydrocracking at mild conditions, *Science Advances* 7(17) (2021) eabf8283.

[41] Y. Nakaji, M. Tamura, S. Miyaoka, S. Kumagai, M. Tanji, Y. Nakagawa, T. Yoshioka, K. Tomishige, Low-temperature catalytic upgrading of waste polyolefinic plastics into liquid fuels and waxes, *Applied Catalysis B: Environmental* 285 (2021) 119805.

[42] J.E. Rorrer, G.T. Beckham, Y. Román-Leshkov, Conversion of polyolefin waste to liquid alkanes with ru-based catalysts under mild conditions, *JACS Au* 1(1) (2020) 8–12.

[43] S. Lu, Y. Jing, B. Feng, Y. Guo, X. Liu, Y. Wang, H2-free plastic conversion: Converting PET back to BTX by unlocking hidden hydrogen, *ChemSusChem* 14(19) (2021) 4242–4250.

[44] Y. Jing, Y. Wang, S. Furukawa, J. Xia, C. Sun, M.J. Hülsey, H. Wang, Y. Guo, X. Liu, N. Yan, Towards the circular economy: converting aromatic plastic waste back to arenes over a Ru/Nb2O5 catalyst, *Angewandte Chemie International Edition* 60(10) (2021) 5527–5535.

[45] X. Jia, C. Qin, T. Friedberger, Z. Guan, Z. Huang, Efficient and selective degradation of polyethylenes into liquid fuels and waxes under mild conditions, *Science Advance* 2(6) (2016).e1501591.

[46] L.N. Rao, J. Jayanthi, D. Kamalakar, Conversion of waste plastics into alternative fuel, *International Journal of Engineering Sciences & Research Technology* (2015).

[47] L.D. Ellis, S.V. Orski, G.A. Kenlaw, A.G. Norman, K.L. Beers, Y. Román-Leshkov, G.T. Beckham, Engineering, Tandem heterogeneous catalysis for polyethylene depolymerization via an olefin-intermediate process, *ACS Sustainable Chemistry* 9(2) (2021) 623–628.

[48] T. Uekert, H. Kasap, E. Reisner, Photoreforming of nonrecyclable plastic waste over a carbon nitride/nickel phosphide catalyst, *Journal of the American Chemical Society* 141(38) (2019) 15201–15210.

[49] T. Uekert, M.F. Kuehnel, D.W. Wakerley, E. Reisner, Plastic waste as a feedstock for solar-driven H 2 generation, *Energy Environmental Science* 11(10) (2018) 2853–2857.

[50] S. Gazi, M. Đokić, K.F. Chin, P.R. Ng, H.S. Soo, Visible light–driven cascade carbon–carbon bond scission for organic transformations and plastics recycling, *Advanced Science* 6(24) (2019) 1902020.

[51] X. Jiao, K. Zheng, Q. Chen, X. Li, Y. Li, W. Shao, J. Xu, J. Zhu, Y. Pan, Y. Sun, Photocatalytic conversion of waste plastics into C2 fuels under simulated natural environment conditions, *Angewandte Chemie International Edition* 59(36) (2020) 15497–15501.

[52] M.P. Aji, A.L. Wati, A. Priyanto, J. Karunawan, B.W. Nuryadin, E. Wibowo, P. Marwoto, Polymer carbon dots from plastics waste upcycling, *Environmental Nanotechnology, Monitoring & Management* 9 (2018) 136–140.

[53] E. Bäckström, K. Odelius, M. Hakkarainen, Trash to treasure: microwave-assisted conversion of polyethylene to functional chemicals, *Industrial Engineering Chemistry Research* 56(50) (2017) 14814–14821.

[54] L. Chen, K.G. Malollari, A. Uliana, D. Sanchez, P.B. Messersmith, J.F. Hartwig, Selective, catalytic oxidations of C–H bonds in polyethylenes produce functional materials with enhanced adhesion, *Chem* 7(1) (2021) 137–145.

[55] X. Li, W. Xing, S. Zhuo, J. Zhou, F. Li, S.-Z. Qiao, G.-Q. Lu, Preparation of capacitor's electrode from sunflower seed shell, *Bioresource Technology* 102(2) (2011) 1118–1123.

[56] I. Vollmer, M.J. Jenks, M.C. Roelands, R.J. White, T. van Harmelen, P. de Wild, G.P. van Der Laan, F. Meirer, J.T. Keurentjes, B.M. Weckhuysen, Beyond mechanical recycling: giving new life to plastic waste, *Angewandte Chemie International Edition* 59(36) (2020) 15402–15423.

[57] G. Ruan, Z. Sun, Z. Peng, J.M. Tour, Growth of graphene from food, insects, and waste, *ACS Nano* 5(9) (2011) 7601–7607.

[58] S. Pandey, M. Karakoti, S. Dhali, N. Karki, B. SanthiBhushan, C. Tewari, S. Rana, A. Srivastava, A.B. Melkani, N.G. Sahoo, Bulk synthesis of graphene nanosheets from plastic waste: an invincible method of solid waste management for better tomorrow, *Waste Management* 88 (2019) 48–55.

[59] J. Gong, J. Liu, X. Wen, Z. Jiang, X. Chen, E. Mijowska, T. Tang, Upcycling waste polypropylene into graphene flakes on organically modified montmorillonite, *Industrial Engineering Chemistry Research* 53(11) (2014) 4173–4181.

[60] W.A. Algozeeb, P.E. Savas, D.X. Luong, W. Chen, C. Kittrell, M. Bhat, R. Shahsavari, J.M. Tour, Flash graphene from plastic waste, *ACS Nano* 14(11) (2020) 15595–15604.

[61] K.M. Wyss, J.L. Beckham, W. Chen, D.X. Luong, P. Hundi, S. Raghuraman, R. Shahsavari, J.M. Tour, Converting plastic waste pyrolysis ash into flash graphene, *Carbon* 174 (2021) 430–438.

[62] V.G. Pol, P. Thiyagarajan, Remediating plastic waste into carbon nanotubes, *Journal of Environmental Monitoring* 12(2) (2010) 455–459.

[63] N. Mishra, G. Das, A. Ansaldo, A. Genovese, M. Malerba, M. Povia, D. Ricci, E. Di Fabrizio, E. Di Zitti, M. Sharon, Pyrolysis of waste polypropylene for the synthesis of carbon nanotubes, *Journal of Analytical Applied Pyrolysis* 94 (2012) 91–98.

[64] Z. Jiang, R. Song, W. Bi, J. Lu, T. Tang, Polypropylene as a carbon source for the synthesis of multi-walled carbon nanotubes via catalytic combustion, *Carbon* 45(2) (2007) 449–458.

[65] Z. Yang, Q. Zhang, G. Luo, J.-Q. Huang, M.-Q. Zhao, F. Wei, Coupled process of plastics pyrolysis and chemical vapor deposition for controllable synthesis of vertically aligned carbon nanotube arrays, *Applied Physics A* 100(2) (2010) 533–540.

[66] J. Liu, Z. Jiang, H. Yu, T. Tang, Catalytic pyrolysis of polypropylene to synthesize carbon nanotubes and hydrogen through a two-stage process, *Polymer Degradation Stability* 96(10) (2011) 1711–1719.

[67] Y. Zheng, H. Zhang, S. Ge, J. Song, J. Wang, S. Zhang, Synthesis of carbon nanotube arrays with high aspect ratio via Ni-catalyzed pyrolysis of waste polyethylene, *Nanomaterials* 8(7) (2018) 556.

[68] N.A. El Essawy, S.M. Ali, H.A. Farag, A.H. Konsowa, M. Elnouby, H.A. Hamad, Green synthesis of graphene from recycled PET bottle wastes for use in the adsorption of dyes in aqueous solution, *Ecotoxicology Environmental Safety* 145 (2017) 57–68.

[69] A.A. Aboul-Enein, A.E. Awadallah, Production of nanostructured carbon materials using Fe–Mo/MgO catalysts via mild catalytic pyrolysis of polyethylene waste, *Chemical Engineering Journal* 354 (2018) 802–816.

[70] S. Chaudhary, M. Kumari, P. Chauhan, G.R. Chaudhary, Upcycling of plastic waste into fluorescent carbon dots: an environmentally viable transformation to biocompatible C-dots with potential prospective in analytical applications, *Waste Management* 120 (2021) 675–686.

[71] P.K. Tripathi, S. Durbach, N.J. Coville, CVD Synthesis of solid, hollow, and nitrogen-doped hollow carbon spheres from polypropylene waste materials, *Applied Sciences* 9(12) (2019) 2451.

10 Recycling of Waste Heat Energy from Engine Exhausts

O.V. Lozhkina and V.N. Lozhkin

10.1 INTRODUCTION

Air pollution still remains a big problem in urbanized territories across the world. Many research studies reveal diesel heavy-duty vehicles (HDVs, including trucks and buses) to be the major contributors to harmful $PM_{2.5}$, NO_x, CO and other pollutants and greenhouse gas emissions [1–5].

Wang with coauthors [1], Toronto (Canada), using a multiple linear regression model (taking into consideration parameters, such as the share of HDVs in road traffic, vehicle speed, air temperature and road-side concentrations of NO, NO_2, CO and black carbon (BC)) found the difference between emission levels in the road-side environment was more correlated with the number of HDVs rather than the number of cars. Expectedly, NO_x and BC levels were more correlated with the share of HDVs than the CO level. Liu et al. [2], during their research near major roads in Detroit, MI, USA, to better understand the relationship between vehicle emissions and road-side air pollution, evaluated three traffic density-based indices (major-road density (MRD), all-traffic density (ATD) and heavy traffic density (HTD)) and revealed the black carbon concentration was highly correlated with the volume of HDVs. According to the report [3] of the Centre for Science and Environment, Delhi, India, Delhi's vehicles account for 62% of PM emission and 68% of NO_x emission, while transit HDVs entering the city account for 30% of PM emission and 22% of NO_x from total road transport emissions. Jin et al., in their report "Air quality and health impacts of heavy-duty vehicles in G20 economies" [4], confirm HDVs are the main contributors to exhaust emissions responsible for 86% of on-road diesel NO_x and 78% of on-road diesel BC. We have also revealed in our previous research [5, 6] that HDVs are responsible for 69% of NO_x emissions and 78% of PM emissions in St. Petersburg.

To reduce the negative effects of exhaust emissions from diesel HDVs, many countries implement strict emission standards for new vehicles [4]. Euro VI-equivalent standards are now in force in Canada, the European Union, India, Japan, South Korea, Turkey, the United Kingdom, and the United States. The Russian Federation has not yet adopted the Euro VI-equivalent standard. The key technologies to meet these standards are a modern diesel oxidative catalyst (DOC) to reduce the emission of CO and hydrocarbons (HC), diesel particulate filter (DPF) to reduce the emission of PM and selective reduction catalyst (SRC) to reduce NO_x emission [7–9].

The DOC, DPF and SRC are operating properly in hot conditions. The results of many studies reveal vehicle emissions are significantly higher at the cold start and warm-up period [10–12]. This is mainly due to incomplete combustion and catalyst inefficiency at a temperature below 300°C. The duration of the warm-up period depends considerably on the ambient temperature and initial temperature of the engine and catalytic convertor [13–15].

A vehicle's engine and catalytic convertor operates in urban traffic, as a rule, in transient conditions, accompanied by a continuous change in time of torque and crankshaft speed due to acceleration and breaking [16, 17]. Increased emissions of pollutants and insufficient temperature of

DOI: 10.1201/9781003327646-10

the catalyst are typical problems of urban driving cycles. Starting the internal combustion engine at a low ambient temperature is also an urgent problem of diesel vehicles' operation in winter.

It is well known that more than 60% of the energy produced in an internal combustion engine is wasted to the ambient air in the form of heat [18–25], even though it could be reused to facilitate the engine and catalyst start, to maintain the catalyst efficiency in transient driving conditions, to improve the efficiency of the engine, contributing thus to fuel economy and pollutants and GHG emissions' reduction.

The goal of the present work was to develop and test an on-board phase-change heat-storage device (PCHSD) capturing waste heat energy from diesel engine exhausts. It helps to facilitate the engine start at low ambient temperature and increase the catalyst efficiency of urban driving.

10.2　WASTE HEAT RECOVERY FROM DIESEL ENGINE

10.2.1　Main Pathways of Internal Combustion Engine Heat Loss in Heavy-Duty Diesel Engine – Potential Sources for Energy Recovery

Waste heat recovery (WHR) is the use of thermal waste energy to accomplish useful work. The goal of WHR is to generate additional functioning. Higher-quality heat sources allow a large share of the waste heat to turn into work [18–27].

About 25–42% of the total fuel energy produced is utilized to generate useful work in modern on-road heavy-duty vehicles while the rest is wasted in the form of heat and friction [21, 22]. Figure 10.1 illustrates the main sources of internal combustion engine heat loss in a heavy-duty diesel engine [18].

Heat waste is mainly due to heat losses with exhaust and coolant streams into the ambient air. In an internal combustion engine, heat is directly wasted with hot exhaust gas, and it is indirectly lost through a heat exchanger from the engine coolant radiator, from the exhaust gas recirculation and charge air cooler rejects. All these waste pathways are potentials for heat recovery.

To evaluate the potential for waste heat recovery, an energy and exergy thermodynamic analysis is needed. Figure 10.2 presents such an example [18, 21].

Exhaust gas and EGR are in priority for heat recovery as they have high temperature (Figures 10.1 and 10.2). Exergy analysis is a key tool to estimate potential efficiency for heat recovery. The comparative analysis of the data indicates that 61.9–64.7% of heat losses from the EGR cooler may be available as exergy. It is followed by exhaust gas emitted from selective reduction catalyst: 54.0–54.7%

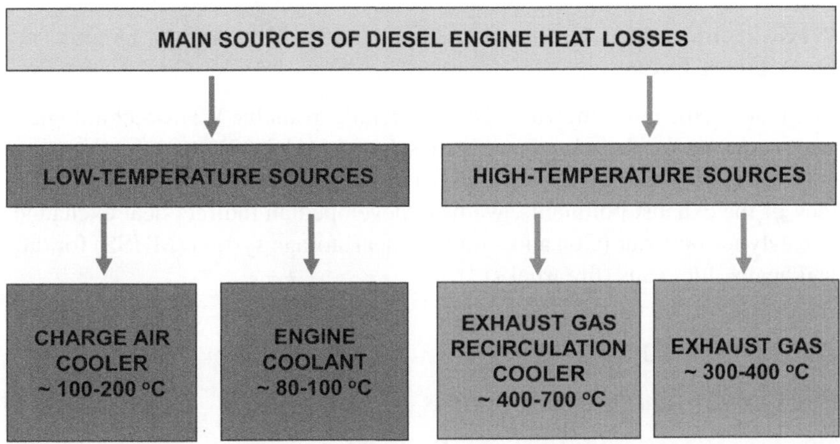

FIGURE 10.1　Main sources of internal combustion engine heat loss in a heavy-duty diesel engine.

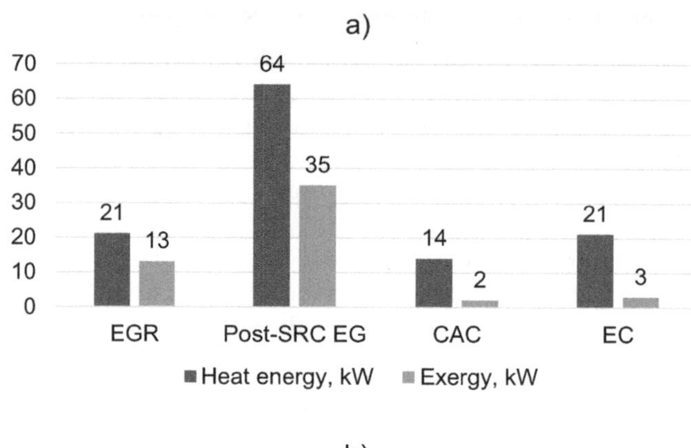

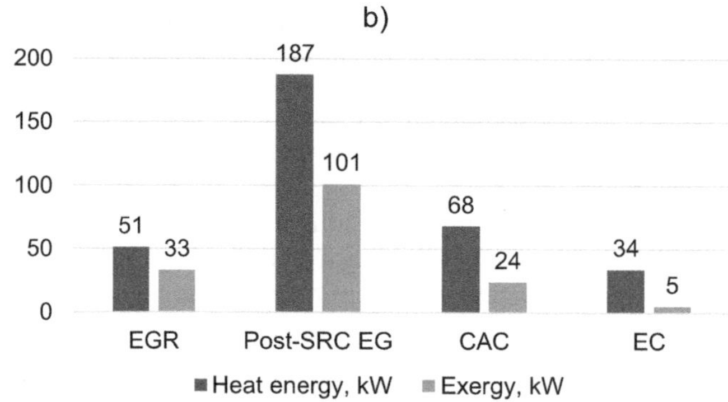

FIGURE 10.2 Energy and exergy of waste heat sources for two diesel engines assuming a heat rejection temperature of 36°C for engine power of 136 kW (a) and 348 kW (b).

EGR – exhaust gas recirculation; post-SRC EG – post selective reduction catalyst exhaust gas; CAC – charge air cooler; EC – engine coolant.

may be available as exergy. Charge air cooler and engine coolant are expectedly less promising for heat recovery because of considerably lower temperature. The exergy may account, respectively, for 14.3–35.3 and 14.3–14.7%.

10.2.2 WHR-TECHNOLOGIES USED FOR THE CAPTURE OF HEAT WASTE ENERGY FROM HEAVY DUTY DIESEL ENGINE

Table 10.1 summarizes the most known and commercially available WHR-technologies used for the capture of heat waste energy from heavy-duty diesel engine [18, 27].

To facilitate the engine starting at low ambient temperature and increasing the catalytic conversion efficiency of the exhaust pollutants, we have developed an indirect heat exchanger – a device combining a catalytic converter (CC) and a mineral heat storage system (MHSS) for diesel-powered city buses and heavy-duty vans (fire trucks).

10.3 MATERIALS AND METHODS

10.3.1 PCM-BASED THERMAL ACCUMULATORS

There are different ways to accumulate heat: (1) sensible heat storage; (2) storage of latent heat of phase change; (3) chemical storage of heat. The second one is realized through the phase change,

TABLE 10.1
Waste Heat Recovery Techniques from Diesel Engines [18, 27]

WHR-Techniques	Modo Operandis	Availability
Heat exchanger	Direct heat exchange	Commercially used to heat the driver's cab or saloon
Direct electrical conversion devices (thermoelectric generator (TEG), piezo-electric devices, thermophotovoltaics (TPV)	Direct conversion of waste heat to electricity via the Seebeck effect using solid-state devices such as, for example, TEG for electricity generation	Commercially used for seat heating
Thermodynamic cycles (Rankine cycle (RC), organic Rankine cycle (ORC), Kalina cycle, Brayton cycle, Stirling cycle)	RC and ORC are mostly in use for the heat recovery through a working fluid (water in the case of high-temperature waste heat and an organic fluid having a lower boiling temperature for the recovering of low and medium temperature waste heat) to drive a turbine to produce mechanical or electrical energy	Widely used for diesel marine engines, still few applications for HDV diesel engines

like, melting–crystallization, sublimation or evaporation–condensation. Thermal accumulators (TAs) based on the phase-change materials (PCMs) undergoing a reversible melting–crystallization phase transition are currently of the greatest practical interest in the automotive industry, building industry, medicine, etc. [28–33]. PCMs are able to store latent heat energy for a long time and then give it to the consumer when needed. Their main advantages are as follows: (1) providing a high density of stored energy in a narrow temperature range; (2) providing a quasi-constant temperature of the coolant at the outlet of TA; (3) relatively low pressure in the PCM.

Since PCMs make it possible to store large amounts of heat in a relatively small heat-storage volume, they are promising for the recovery of waste heat from diesel HDV exhaust taking into account the requirements for the size and weight of an additional device to be placed on the vehicle board. Crystal hydrates of salts and bases, organic substances, salts and bases, various mixtures of these substances can be used as PCMs [28–33].

When choosing a PCM and designing a heat exchanger for the heat recovery from diesel engine exhausts, parameters such as weighted average flow rate and temperature of exhausts should be considered.

10.3.2 Design of Heat Exchanger

Figure 10.3 illustrates a constructive scheme of phase-change heat-storage device (PCHSD); Figure 10.4 shows a PCHSD installed in the engine cooling system of a city bus to facilitate starting at low temperatures and maintain thermal stability [32–33].

PCHSD consists of a body (1), a removable cover (2), which has an inlet (3) and outlet (4), into which the inlet pipe (5) and the outlet pipe (6) are installed. A heat exchanger consisting of coaxially arranged 'annular' cylindrical capsules (7) with annular gaps (8) for the passage of a liquid coolant or exhaust gas. Cylindrical 'annular' capsules (7) and annular gaps (8) have the same radial dimensions (thickness). Removable cover (2) is bolted (10) to the ring (9) welded to the body. The accumulation of heat is carried out due to the PCM located in sealed cylindrical 'annular' capsules (7). The conversion between the heat and the potential energy within the mineral material involves a phase change. An additional pump, switched on by an on-board electric storage battery, pumps the liquid antifreeze coolant in a closed loop. In the process of heat exchange between the liquid heat carrier and the PCM, the latter undergoes a reversible phase change from a liquid state to a solid

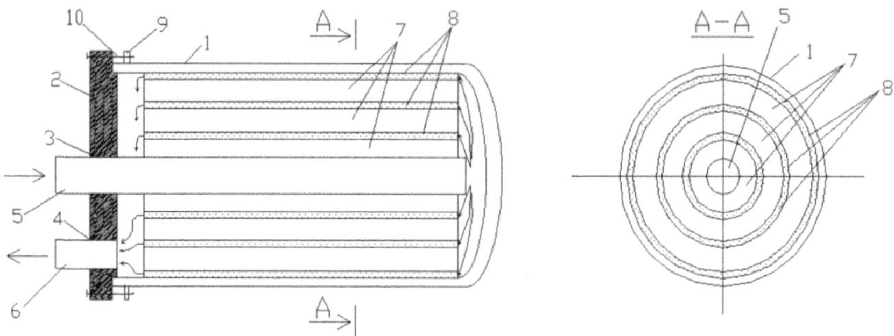

FIGURE 10.3   Phase-change heat-storage device.

1 is the body; 2 is the removable cover; 3, 4 are the inlet and outlet openings in the removable cover, respectively; 5, 6 are the inlet and outlet pipes, respectively; 7 are the cylindric coaxially located capsules filled with phase-change heat-storage materials; 8 are the annular gaps; 9 is the ring; 10 is the bolted connection.

FIGURE 10.4   Phase-change heat-storage device installed in the engine cooling system of a city bus.

state, and gives the latent heat of crystallization to the liquid heat carrier. The liquid carrier transfers the heat to the internal combustion engine through the cavity of its cooling jacket.

10.3.3 Design of Catalytic Converter–Heat Exchanger Device

We also have developed a system combining a catalytic converter and a phase-change heat storage device (CC-PCHSD) – Figure 10.5 [33].

A catalytic convertor and a phase-change heat-storage device are located in a cylindrical body (1). A PCHSD includes an inlet pipe (2), a diffuser (3), a heat exchanger consisting of coaxially located annular capsules (4) filled with PCM. Annular gaps (5) are formed between 'annular' capsules (4). Annular capsules (4) and annular gaps (5) have the same radial dimensions (thickness). There is a layer of thermal insulation (6) between the cylindrical body (1) and the heat exchanger. The catalytic converter consists of a block reactor (7), a blind expansion cavity (8), and an outlet pipe (9).

The device is installed in the exhaust system of a diesel engine instead of the noise muffler. PCM allows storing the thermal energy of the exhaust gas when the engine operates at full capacity and to deliver the heat back to the catalytic converter when the engine runs at idling mode or after its stop. The reverse heat delivery allows maintaining the optimum temperature for the effective catalytic conversion of exhaust pollutants. PCM may be a binary system NaCl-NaOH with a phase-change temperature of 314°C. Such a device may be used at ambient temperature up to minus 40°C.

10.4 RESULTS AND DISCUSSION

10.4.1 Estimation of Efficiency of Capsule PCHSD Engine Preheater

An integrated calculating method for the engine pre-start heating with the capsule PCHSD (Figures 10.3 and 10.4) was developed and described in detail by Professor V.V. Shulgin [32]. It includes a method for the design calculation and a method for the verification. The first one is to determine the main design parameters of the PCHSD and auxiliary elements. The purpose of the second one is to determine the temperatures of the internal combustion engine, PCHSD, exhaust gas and coolant involved in the heat exchange, as well as to estimate the charging time of the PCHSD and the hydraulic resistances.

Several PCHSDs were proposed using $Ba(OH)_2 \cdot 8H_2O$ ($T_{phase\ change}$ = 351 K), KNO_3-$LiNO_3$ ($T_{phase\ change}$ = 406 K) or $NaCH_3COO \cdot 3H_2O$ ($T_{phase\ change}$ = 331 K) as PCMs. Such PCMs may be used at ambient temperature up to –40°C.

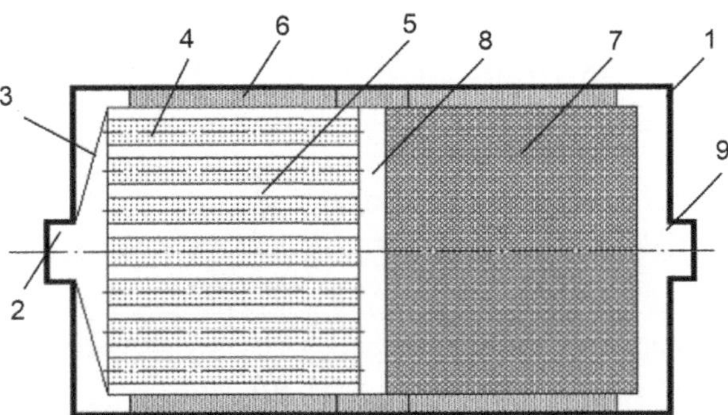

FIGURE 10.5 Design of a catalytic converter combined with a phase-change heat-storage device.

1 is the cylindric body; 2 is the inlet branch pipe; 3 is the diffuser and heat exchanger; 4 are the central cylindrical and coaxially capsules; 5 are the annular gaps; 6 is the insulation layer; 7 is the catalytic convertor; 8 is the blind expansion cavity; 9 is the branch pipe for gas discharge to the air.

The Ba(OH)$_2$·8H$_2$O-based PCHSD allows, for example, at ambient temperature 253 K, to accumulate the wasted heat from the coolant or exhaust gas for 1.14–1.43 hours, to store the energy required for starting engine heating for 17 hours, and to warm up the KAMAZ-7408.10 diesel engine to an average temperature of 312–315 K in 540–630 seconds. At the same time, the antifreeze located in the notch space of the diesel engine is heated from 253 to 328–332 K, and the maximum thermal power output of the PCHSD is 57 kW.

Laboratory and field experiments of the Ba(OH)$_2$·8H$_2$O-based PCHSD were carried out on a LiAZ-5256 city bus equipped with a KamAZ-7408.10 diesel engine [32].

Laboratory determined energy characteristics of the PCHSD are as follows: heat accumulated during charging of PCHSD, Q_{charge} = 14124 kJ; heat lost during 24 h, Q_{lost} = 665 + 42 kJ; average power of dissipation in the environment q_{dissip} = 10.5+0.7 Wt; heat released during discharge, $Q_{discharge}$ = 11424 + 345 kJ; average discharge power $q_{discharge}$ = 11.3+0.5 Wt.

On-board field experiments showed the PCHSD charging is non-stationary and continues for 60–120 minutes. Heat can be stored for 11–14 hours at an ambient temperature from –2 to –20°C. The release of accumulated heat, as well as its accumulation, is also non-stationary. The average rate of increase in the antifreeze temperature reaches very high values during the first 30 s from the start: at the outlet of the PCHSD – up to 1.38°C/s, and at the outlet of the thermostatic box of the engine – up to 0.993°C/s. This indicates the PCHSD delivers a large starting thermal power. Then (in the next 150), the average rate of change in the temperature of antifreeze decreases from 0.143–0.193 to 0°C/s. Subsequently, with an increase in time from 180 to 330–390 s, the antifreeze temperature reaches 33–37°C at all observed points and practically does not change [32].

The performed field tests have shown that the PCHSD is efficient, easy to manage and reliable in use.

10.4.2　MODEL OF CC-PCHSD

We developed a mathematical model of a CC-PCHSD basing on the thermal balance approach to describe the heat exchange and the classical Arrhenius kinetics approach to describe the catalytic transformation of pollutants (CO and HC):

$$\theta_{wall} = \frac{2\Omega}{N\eta_{disch} + 2(\Omega+1)}, \tag{10.1}$$

$$\Omega = 1/2\left\{\left[\left(N\cdot\eta_{disch} + 2\right)^2 + 8t\right]^{1/2} - \left(N\cdot\eta_{disch} + 2\right)\right\}, \tag{10.2}$$

$$\bar{\theta} = \frac{2 + N\cdot\eta_{disch}\cdot\theta_{wall}}{2 + N\cdot\eta_{disch}}, \tag{10.3}$$

$$\theta_{outlet} = \frac{2 - N\cdot\eta_{disch}\left(1 - 2\theta_{wall}\right)}{2 + N\cdot\eta_{disch}}, \text{ where} \tag{10.4}$$

θ_{wall} is the dimensionless temperature of the capsule surface, $\bar{\theta}$ is the average dimensionless temperature of exhaust gas in the PCHSD, θ_{outlet} is the outlet (from the PCHSD) dimensionless temperature

TABLE 10.2
Basic Thermophysical Properties of NaOH

Parameter	Value
Melting–crystallization phase change temperature, K	572
Specific heat of phase change, kJ/kg	393
Solid phase density, kg/m³	2130
Liquid phase density, kg/m³	1780
thermal conductivity coefficient of the solid phase, W/(m·K)	1.8

of exhaust gas, Ω is the dimensionless thickness of solidified heat storage substance, N is the heat transfer number, η_{disch} is the discharge energy efficiency, t is the dimensionless time.

The crystallization at pure heat conductivity is determined by three dimensionless parameters – dimensionless time t, dimensionless heat transfer number N and dimensionless discharge energy efficiency η.

Catalytic conversion of pollutants depends on the temperature in the catalyst according to the approximation:

$$\widetilde{T}_s = \widetilde{T}_{\text{inlet}} + \frac{q_0}{c_{gV}} \cdot \left[\left(C_{\text{initial}} - C_{\text{out}} \right) + \left(C_g - C_s \right) \cdot \left(\frac{D}{a_g} \right)^{\frac{2}{3}} \right], \text{ where} \tag{10.5}$$

$\widetilde{T}_s$ is the temperature of the solid phase in the catalyst, K; $\widetilde{T}_{\text{inlet}}$ is the inlet temperature of the gas flow, K; q_0 is the thermal effect of catalytic reaction, J/kg; c_{gV} is the specific heat capacity of the gas, J/(m³K); C_{initial} is the inlet concentration of pollutant, kg/m³; C_{out} is the exhaust concentration of pollutant, kg/m³; C_g is the concentration of pollutant in the gas phase (gas flow), kg/m³; C_s is the concentration of pollutant on the catalyst surface, kg/m³; D is the diffusion coefficient, m²/s, a_g is the thermal diffusivity coefficient, m²/s.

We calculated the characteristics of CC-PCHSD for a LiAZ-5256 city bus with the KamAZ-7408.10 diesel engine using the proposed approach.

We assumed that the distances from the axis of the central capsule to the cylindrical surfaces of the 'annular' capsules (r_i) are as so: $r_1 = 20$ mm, $r_2 = 25$ mm, $r_3 = 35$ mm, $r_4 = 40$ mm, $r_5 = 50$ mm, $r_6 = 55$ mm, $r_7 = 65$ mm, $r_8 = 70$ mm, $r_9 = 80$ mm, $r_{10} = 85$ mm, $r_{11} = 95$ mm, $r_{12} = 100$ mm.

We also assumed the PCHSD is supposed to heat exhaust gas of KamAZ-7408.10 diesel engine operating at idling mode from the $T_{\text{inlet}} = 373$ K to the $T_{\text{outlet}} = 553$ K. Sodium hydroxide NaOH, characteristics are given in Table 10.2, can be used as PCM for this purpose.

Taking into account the assumptions described above, we have determined the characteristics of the CC-PCHSD (at $N = 2.15$ and $\eta = 0.9$):

- the average dimensionless temperature of exhaust gas in the PCHSD $\bar{\theta}$ is 0.547;
- the outlet (from the PCHSD) dimensionless temperature of exhaust gas θ_{outlet} is 0.0955;
- the dimensionless thickness of solidified heat storage substance Ω is 0.173;
- the catalyst efficiency at the idle and cold start may be increased from 4 to 74% for hydrocarbons and from 10 to 86% for carbon monoxide by maintaining the temperature at the proper level using the phase-change storage device instead of the noise muffler.

10.5　CONCLUSION

About 25–42% of the energy produced by diesel HDV internal combustion engines is used efficiently, and the rest is wasted in the ambient air in the form of heat. A promising way to increase their efficiency is heat recovery of the exhaust gas and coolant water. This contributes to fuel economy and pollutants and GHG emissions' reduction.

A phase-change heat-storage device (PCHSD) was developed with the aim to facilitate engine starting at low ambient temperature and a system combining a catalytic converter and a phase-change heat-storage device (CC-PCHSD) aimed at the improving of the catalyst efficiency operating in a transient mode in urban conditions.

Heat recovery helps to start the engine and heat the vehicle interior at low ambient temperature, improving the efficiency of pollutants' catalytic conversion in transient modes of engine and catalyst operation, storing waste heat, and preventing the engine and exhaust system from overheating.

ACKNOWLEDGEMENT

This chapter is dedicated to the memory of our colleague Professor Vassiliy Valentinovich Shulgin, who made a significant contribution to the development of the theory and application of technologies for the recycling of waste heat from diesel engine exhausts in the Russian Federation.

REFERENCES

1. Wang, J.M., Jeong, Ch.-H., Hilker, N., et al. Near-road air pollutant measurements: Accounting for inter-site variability using emission factors. *Environmental Science & Technology*. *52* (16), 9495 (2018).

2. Liu, S.V., Chen F.L., Xue, J. Evaluation of traffic density parameters as an indicator of vehicle emission-related near-road air pollution: A case study with NEXUS measurement data on black carbon. *International Journal of Environmental Research and Public Health*. *14* (12), 1581 (2017).

3. Chattopadhyaya, V., Roychowdhury, A., Nasim, U. *TRUCKS: Heavy-duty Pollution and Action*. Centre for Science and Environment, New Delhi (2016). DOI:10.13140/RG.2. 2.33002.11204.

4. Jin, L., Braun, C., Miller, J., Buysse, C. Air quality and health impacts of heavy-duty vehicles in G20 economies. (2021). Available at: https://theicct.org/wp-content/uploads/2021/12/g20-hdv-impacts-jul2021_0.pdf.

5. Lozhkina, O.V., Lozhkin, V.N. Estimation of road transport related air pollution in Saint Petersburg using European and Russian calculation models. *Transportation Research Part D: Transport*. *36*, 178–189 (2015).

6. Lozhkin, V., Gavkalyuk, B., Lozhkina, O., Evtukov, S., Ginzburg, G. Monitoring of extreme air pollution on ring roads with PM2.5 soot particles considering their chemical composition (case study of Saint Petersburg). *Transportation Research Procedia*. *50*, 381–388 (2020).

7. Russell, A., Epling, W.S. Diesel oxidation catalysts. *Catalysis Reviews - Science and Engineering*. *53* (4), 337–423 (2011).

8. Cho, S.M. Properly apply selective catalytic reduction for NOx removal. *Chemical Engineering Progress*. *90*, 39–45 (1994).

9. Hirata, K., Masaki, N., Ueno, H., Akagawa, H. Development of urea-SCR system for a heavy-duty commercial vehicles. *SAE Technical Paper*. *1*, 1860 (2005).

10. Ramanathan, K., West, D. H., Balakotaiah, V. Optimal design of catalytic converters for minimizing cold-start emissions. *Catalysis Today*. *98* (3), 357–373 (2004).

11. Ye, S., Yap, H., Kolaczkowski, S.T., Robinson, K., Lukyanov, D. Catalyst 'light-off' experiments on a diesel oxidation catalyst connected to a diesel engine—Methodology and techniques. *Chemical Engineering Research and Design*. *90*, 834–845 (2012).

12. Matthaios, V.N., Kramer, L.J, Sommariva, R., Pope, F.D., Bloss, W.J. Investigation of vehicle cold start primary NO_2 emissions inferred from ambient monitoring data in the UK and their implications for urban air quality. *Atmospheric Environment*. *199*, 402–414 (2019).

13. Lozhkin, V., Lozhkina, O., Dobromirov V. A study of air pollution by exhaust gases from cars in well courtyards of Saint Petersburg. *Transportation Research Procedia. 36*, 453–458 (2018).

14. Weilenmann, M., Favez, J.-Y., Alvarez, R. Cold-start emissions of modern passenger cars at different low ambient temperatures and their evolution over vehicle legislation categories. *Atmospheric Environment. 43*, 2419–2429 (2009).

15. Weilenmann, M. Regulated and nonregulated diesel and gasoline cold-start emissions at different temperatures. *Atmospheric Environment. 39*, 2433–2441 (2005).

16. Lozhkina, O.V., Lozhkin, V.N. Estimation of nitrogen oxides emissions from petrol and diesel passenger cars by means of on-board monitoring: effect of vehicle speed, vehicle technology, engine type on emission rates. *Transportation Research Part D: Transport and Environment. 47*, 251–264 (2016).

17. Methods for calculating the dispersion of emissions of harmful (polluting) substances in the air. Available at: http://docs.cntd.ru/document/456074826 (2017). [In Russian]

18. Jääskeläinen, H. Waste Heat Recovery. DieselNet Technology Guide. Available at: https://dieselnet.com/tech/engine_whr.php#:~:text=Waste%20heat%20from%20the%20EGR,system%20more%20significant%20%5B3711%5D%20

19. Baidya, D. 2019. Diesel exhaust heat recovery: a study on combined heat and power generation strategy for energy-efficient remote mining in Canada. Thesis. p. 137 DOI:10.14288/1.0378291

20. Avaritsioti, E. Environmental and economic benefits of car exhaust heat recovery. *Transportation Research Procedia. 14*, 1003–1021 (2016).

21. Pradhan, S., Thiruvengadam, A., Thiruvengadam, P., Besch, M. et al., Investigating the potential of waste heat recovery as a pathway for heavy-duty exhaust aftertreatment thermal management. SAE Technical Paper 2015-01-1606 (2015).

22. Mohd. Quasim Khan, S. Malarmannan, G. Manikandaraja. Power generation from waste heat of vehicle exhaust using thermoelectric generator: A review. *IOP Conference Series: Materials Science and Engineering. 402*, 012174 (2018).

23. Rijpkema, Jelmer. Waste heat recovery in heavy duty diesel engines. thermodynamic potential of Rankine and flash cycles for low and high temperature heat sources. Thesis for the degree of licentiate of engineering in thermo- and fluid dynamics. Chalmers University of Technology. 28 p. (2018).

24. Hossain, S.N., Bari, S. Waste heat recovery from exhaust of a diesel generator set using organic fluids. *Procedia Engineering. 90*, 439–444 (2014).

25. Valencia, G., Fontalvo, A., Cárdenas, Yu. Duarte, J., Isaza, C. Energy and exergy analysis of different exhaust waste heat recovery systems for natural gas engine based on ORC. *Energies. 12*, 2378 (2019).

26. Hassan, M.A., Samanta, R. Heat exchanger assisted exhaust heat recovery with thermoelectric generator in heavy vehicles. *Energy Technology. 9* (7), 2100037 (2021).

27. Varshil, P., Deshmukh, D. A comprehensive review of waste heat recovery from a diesel engine using organic rankine cycle. *Energy Reports. 7,* 3951–3970 (2021).

28. Yazdani, M.R., Laitinen, A., Helaakoski, V., Farnas, L.K., Kukko, K., KariSaari, K., Vuorinen, V. Efficient storage and recovery of waste heat by phase change material embedded within additively manufactured grid heat exchangers. *International Journal of Heat and Mass Transfer. 181*, 121846 (2021).

29. Kumar, N., Gupta, S.K. and Sharma, V.K. Application of phase change material for thermal energy storage: An overview of recent advances. *Materials Today: Proceedings.* (2021). *44*, 368–375. DOI:10.1016/j.matpr.2020.09.745

30. Yang, T., King, W.P., and Miljkovic, N. Phase change material-based thermal energy storage. *Cell Reports Physical Science. 2*, 100540 (2021).

31. Sun, K., Kou, Y., et al. The design of phase change materials with carbon aerogel composites for multi-responsive thermal energy capture and storage. *Journal of Materials Chemistry A. 2*, 1213–1220 (2021). Available at: https://pubs.rsc.org/en/content/articlelanding/2021/ta/d0ta09035b

32. Shulgin, V.V. Theory and practice of using phase transition thermal accumulators in motor vehicles. Thesis for the degree of Doctor of Technical Sciences, St. Petersburg, p. 501 (2004). [In Russian].

33. Lozhkin, V., Lozhkina, O. Catalytic converter with storage device of exhaust gas heat for city bus. *Transportation Research Procedia. 20*, 412–417 (2017).

11 Use of Conductive Material in the Anaerobic Digestion for Improving Process Performance

A Review

Satchidananda Mishra, Sagarika Panigrahi and Debapriya Kar

11.1 INTRODUCTION

Historically, fossil fuels have played a significant role in shaping the global and economic landscapes of the world as we know it today. However, their positives are considerably outnumbered by their negatives, such as the emittance of greenhouse gases, which directly results in global warming, a terrifying real-world consequence. It is therefore the need of the hour to look into alternativee sources of fuel such as biofuels, which can be produced using renewable materials available locally. Biogas is one such biofuel, and one of the many industrially recognized processes that can be utilized to produce biogas is called anaerobic digestion (AD) [1]. As illustrated in Figure 11.1, AD is recognized as a technology that transforms waste materials into valuable resources and is appropriate for treating sewage sludge and wastewater, agricultural waste and organic waste. It offers the advantages of energy production, biofertilizer, wastewater treatment and waste reduction.

The AD process begins with the breaking down of organic matter and ends with the production of carbon dioxide and methane through a series of reactions that are metabolically supported by various microorganisms involved in the reaction medium. It is an established practice that utilizes an environment devoid of any oxygen to 'manage' waste material. Simply put, this process causes the breakdown of organic matter without oxygen. It has been used to treat a variety of wastes including brewery wastewater, food waste, fruit and vegetable waste, household waste, and agricultural waste. However, despite being highly advantageous, AD is not widely favoured as a source of biofuel production because its mechanism is complex. It is subject to slight changes in environmental conditions and reactants. Compared to other chemical processes, AD has also been observed to have a long starting-up time and an equally long retention time – thus causing the entire process to be dreadfully slow. This slowness has been attributed to the type of microbial communities involved [2, 3]. The final stage of a biological process is a result of the many shorter steps carried out in a coordinated fashion by microorganisms. This means that the extent of electronic exchange between different microorganisms involved in the process contributes to its overall kinetics. From this direct correspondence, it can be inferred that a fast and efficient mechanism for facilitating the transfer of electrons would accelerate the reaction. Electron transfer between microbial communities involved in the reaction, or interspecies electron transfer, is essential in AD for the oxidation of complex organic matter and the reduction of carbon dioxide into methane. This transfer relies

DOI: 10.1201/9781003327646-11

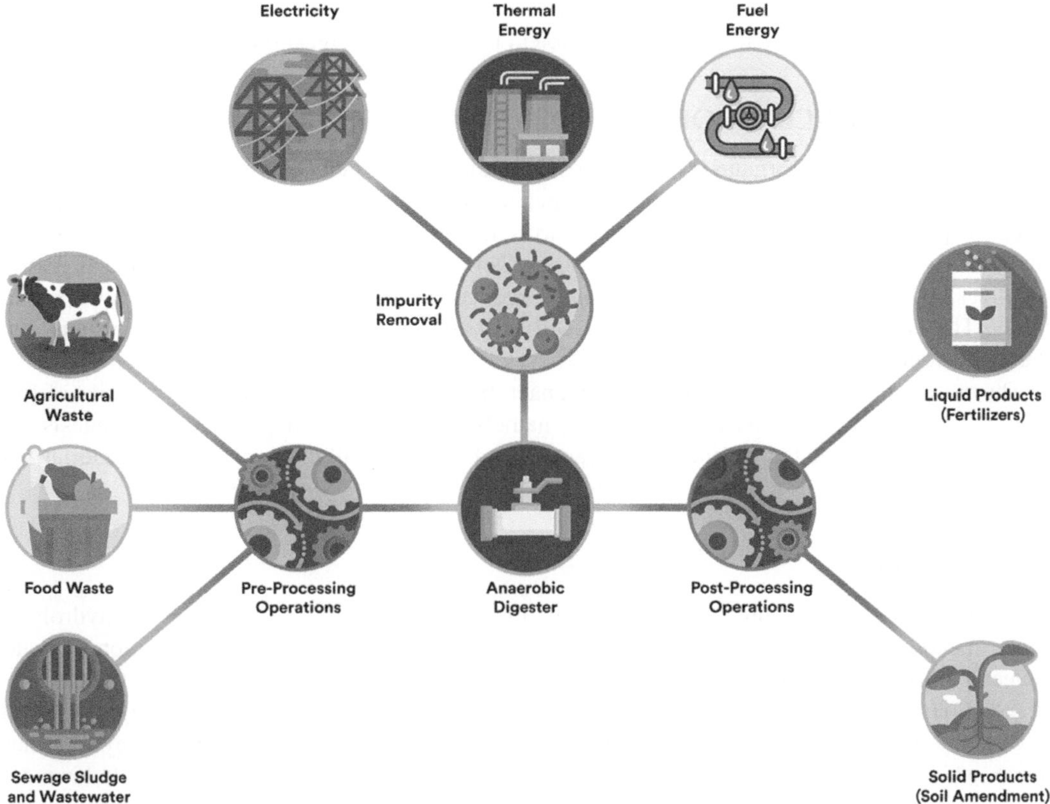

FIGURE 11.1 Components of an anaerobic system.

on the transactions between hydrogen or formate-based electron carriers, and a disturbance in this cross-feeding syntrophic partnership can result in the accumulation of byproducts like fatty acids and hydrogen, both of which can hamper AD to a great extent. To overcome these setbacks, a new mechanism called direct interspecies electron transfer (DIET) has been looked into and evaluated by several researchers around the world due to the surge in popularity of AD, which has led to the adoption of a variety of additives to benefit from several positive synergies, including improved digestate quality and methane output [4, 5].

DIET involves an electron exchange between the two parties, wherein, the syntroph utilizes multi-heme cytochromes and a network of pili to oxidize a reduced food substrate and provide electrons to a methanogen. The methanogen then converts carbon dioxide to methane using electrons from the syntroph's oxidative metabolism. DIET is used to speed up the kinetics of methanogenesis processes to promote the syntrophic interaction of microbes and increase CH_4 production. These anaerobic microbes transform organic materials into more stable forms and generate high-energy biogas mainly consisting of methane and carbon dioxide. Without the need for oxygen, bacteria break down organic materials to create intermediary molecules like sugar, hydrogen and acetic acid, which are ultimately transformed into biogas. To enhance digestion, these biological processes and the ensuing chemical reactions must be optimized [6]. This paper summarizes and reviews the recent literature on the conductive material-mediated AD process, first revisiting the foundational theory of anaerobic digestion, then examining recent developments and findings that demonstrate the suitability of conductive materials as promoters for anaerobic digestion. This paper summarizes and reviews the recent literature on the conductive material-mediated AD process, first revisiting the foundational theory of anaerobic digestion, then examining recent developments and findings that demonstrate the suitability of conductive materials as promoters for anaerobic digestion.

11.2 BASIC PRINCIPLES OF ANAEROBIC DIGESTION

In traditional AD, microorganisms decompose organic matter under oxygen-free conditions into methane-containing biogas. This process is generally seen to occur naturally in wetlands like marshes, ponds and swamps, as well as in artificially produced such as sewage digesters and landfills. By converting a portion of the volatile solids' fraction to biogas, anaerobic digestion stabilizes organic matter in wastewater solids, reduces pathogens and odours and lowers total solids. The end-products of this process usually contain ammonia-nitrogen-based nutrients and stabilized solids [7, 8].

The conversion of biomass into biogas by anaerobic decomposition, is aided by a series of four sequential steps as illustrated in Figure 11.2, namely hydrolysis, acidogenesis, acetogenesis and methanogenesis, propelled by the distinct activity of metabolically interconnected bacterial and archaeal groupings.

11.2.1 HYDROLYSIS

The first stage in the breaking down of complex organic materials into simpler forms is hydrolysis, which means the breaking of chemical bonds by the addition of water. Upon reaction of water molecules with the ions present in the medium, a change in the pH and subsequently, the cleavage of the O–H bond occurs. Multiple microorganisms secrete different hydrolyzing enzymes during hydrolysis, including nuclease, which breaks down nucleic acid into purines and pyrimidines, lipase, which breaks down lipids into fatty acids, and cellulase and amylase, which break down cellulose into sugar and alcohol. The primary role of these enzymes is to break down larger molecules into smaller pieces that microorganisms can take into their cells and use for energy and nutrition. Bacterial colonization, with hydrolytic bacteria coating solid surfaces, is the first stage of hydrolytic processes. Enzymes produced by bacteria on a particle's surface can be utilized by both hydrolytic as well as other types of bacteria. In the second stage, bacteria on the particle's surface continue to degrade it at a steady rate. The rate of decomposition during the hydrolysis stage is determined by the type of substrate, with lignin, cellulose and hemicellulose decomposing more slowly than proteins. The hydrolysis stage generates monomers, which are consumed by various facultative and obligate anaerobic bacteria in the following stage of digestion. It is a slow stage that can affect the overall digestion process. Hydrolysis can be accelerated by pre-treating the polymer such that penetration by microorganisms or extracellular enzymes is arrested. However, because of the high cost of pre-treatment, the technology is not yet commercially viable [7].

11.2.2 ACIDOGENESIS

During acidogenesis, also referred to as the anaerobic oxidation phase or fermentation stage, microorganisms transform the hydrolysis stage product into intermediary compounds like volatile fatty acids (VFAs), alcohols, organic acids, carbon dioxide (CO_2) and hydrogen gas (H_2), which act as substrates for the anaerobic digestion of methanogenesis archaea. These intermediaries are necessary for methanogenic bacteria to produce methane (CH_4), the main component of biogas. The success of this partnership depends on the amount of hydrogen in the system as well as partial pressure. Methane-producing microorganisms and anaerobic oxidation bacteria must collaborate to produce methane because this stage produces hydrogen, increasing the hydrogen partial pressure, and methanogenic bacteria use hydrogen, lowering the partial pressure within the system. Acidogenesis is the fastest anaerobic digestion intermediate. In this step, volatile fatty acids (VFAs)

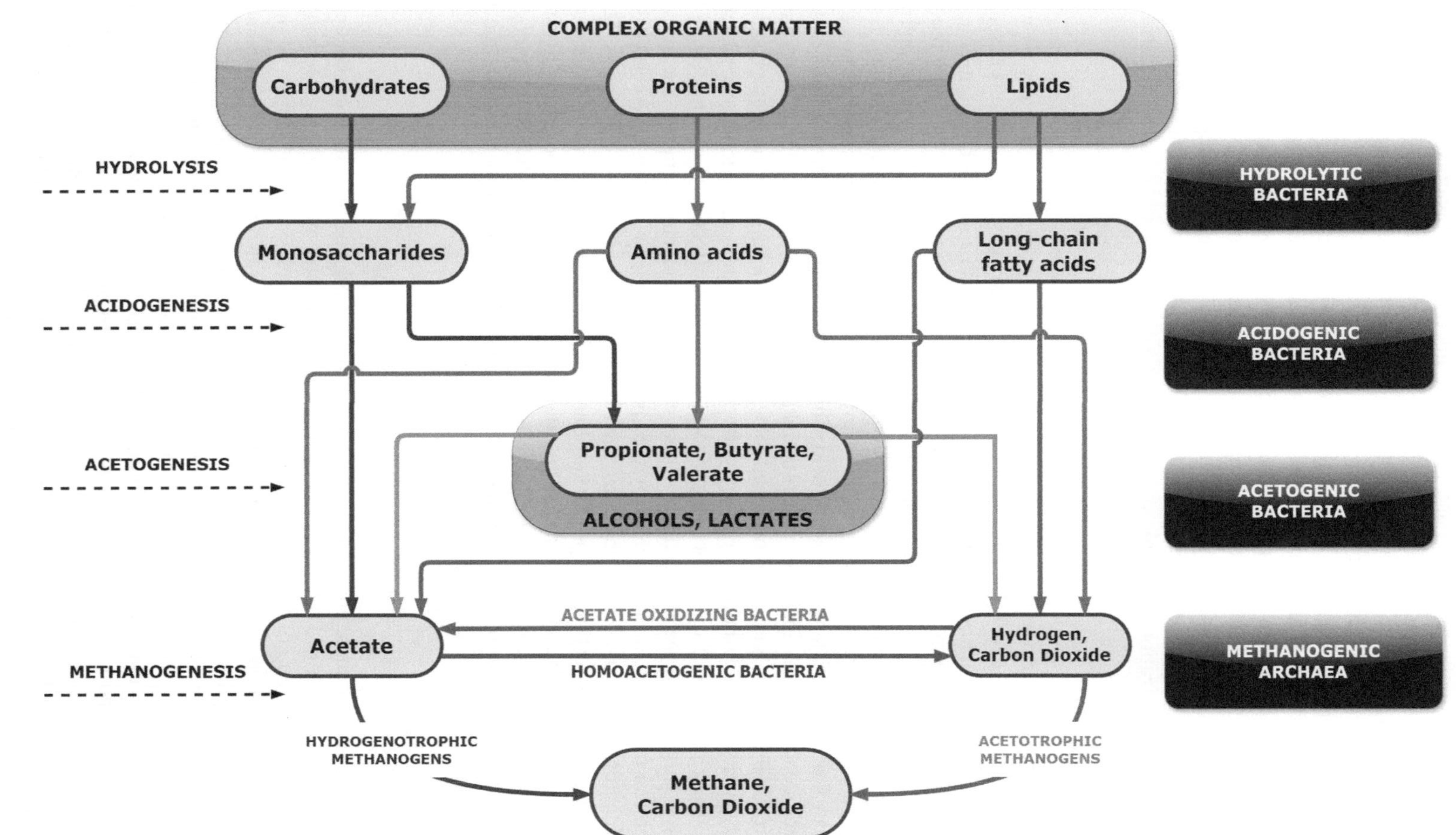

FIGURE 11.2	Mechanism of anaerobic digestion.

are produced more effectively at a pH more than 5, whereas ethanol is formed at a pH less than 5 and pH less than 4 ends the reaction phase. Anaerobic microbes find energy in amino acids and peptides from protein degradation. Functional suppression of acidogenic bacteria at higher loading rates inversely affects acidification yield and OLR. Increasing the operating temperature enhances hydrolysis and acidogenesis but inhibits acetogenesis. When heated to encourage hydrolysis and acidogenesis, VFA production may exceed the slower-reacting acetogenic and methanogenic bacteria's capacity, raising pH and inhibiting AD acetogenesis and methanogenesis [9, 10]. During acidogenesis, co-digestion with multiple feedstocks enhances the population of digester bacteria, nutritional balance, decomposition of organic matter, and production of volatile fatty acids (VFAs). By enhancing the system's resistance to acidity, co-digestion promotes acidogenic bacteria and stabilizes pH. However, for co-digestion to be most effective, precise feedstock selection, blending ratios and process optimization are required.

11.2.3 ACETOGENESIS

Acetogenesis is the third anaerobic metabolic stage in which acetogenic bacteria carry on with the degradation of previously generated acidogenesis products. Acetogens require a lower concentration of hydrogen in the system. To achieve this, they often establish a syntrophic arrangement with methanogens that consume hydrogen, which increases their functionality by reducing the partial pressure due to hydrogen [11–13]. Acetogens cannot directly convert ethanol to methane and carbon dioxide; instead, they must first convert the ethanol to acetic acid, which results in the release of molecular hydrogen. Acetogens are incapable of converting ethanol to methane and carbon dioxide directly. They first convert ethanol into acetic acid. This results in the release of molecular hydrogen. As hydrogen concentrations increase, electron sinks accumulate in the liquid. These cannot be utilized directly by methanogens and thus need to be broken down further. The acetogenesis stage is thermodynamically viable because of the syntrophic relationship, in which some bacteria are fed on the waste products of other microorganisms. Since H_2 is effectively a proton (H^+) plus an extra electron, this interspecies H_2 transfer is synonymous with an electron transfer. The overall rate of digestion can be considerably impacted by the speed of this electron transfer. Acetogenic bacteria are encouraged by high carbon substrates, such as proteins, lipids and carbohydrates. This can be accomplished by co-digestion and proper substrate mixing, which equally distributes nutrients. Acetogenic bacteria growth can be enhanced by extending the retention period of the digester and adding trace metals and micronutrients, such as iron, cobalt, nickel and molybdenum. Acidogenesis can influence the efficiency of biogas production. Acetogenic bacteria further degrade the VFAs, resulting in CH_3COOH, CO_2 and H_2 [14–17].

11.2.4 METHANOGENESIS

Methanogenesis is the ultimate phase of the AD process, where organic molecules are decomposed and biogas is largely generated by autotrophic and hydrogenotrophic methanogens. Methanogenic bacteria are a specific group of microorganisms that play a key role in the generation of methane during this process. Archaea-associated methanogenic microorganisms generate methane as a metabolic byproduct in the absence of oxygen, while methanogenic bacteria convert intermediate compounds into methane and carbon dioxide during the methanogenic stage. The main source of methane produced during digestion is the transformation of acetic acid by heterotrophic methane bacteria, despite the fact that only a limited number of bacteria have this capability. As a result of the conversion of H_2 and CO_2 into CH_4 in this reaction, entropy is reduced and heat is released, which is in contrast to acetate-oxidation, where entropy increases in anaerobic cultures. Methane can be produced by either cleaving acetic acid molecules or by reducing carbon dioxide. The most typical metabolic route is hydrotrophic methanogenesis. In Acetoclastic methanogenesis, acetate is transformed into methane without any intermediate reactions – it is the direct conversion of acetate to CH_4, wherein the microbes involved grow at a slower rate and are hypersensitive to environmental changes. The conversion of acetate and hydrogen into carbon dioxide and methane is known as hydrogenotrophic methanogenesis, and the bacterias that perform this conversion are purely anaerobic. Hydrogenotrophic methanogens can withstand environmental changes to a larger extent as compared to acetoclastic methanogens. Another classification divides bacteria involved in this stage into two categories: acetophilic and hydrogenophilic. It has been shown that methanogenesis is the rate-limiting step in the anaerobic process, and it can be improved by using mixed cultures. Methane output may be limited by high hydraulic loading rates that wash away digester methanogens. Methanogenesis can be halted by feedstock containing high concentrations of ammonia, sulfide, heavy metals and organic acids via blocking the activity of methanogens. Methanogens require enough retention time to engage the substrate and create methane, as they grow more slowly than acidogens.

11.3   EFFECT OF CONDUCTIVE MATERIALS ON ANAEROBIC DIGESTION

Through interspecies electron transfer (IET), bacteria convert CO_2, H_2 and organic molecules into CH_4 during anaerobic digestion. For IET to function, a balance between bacteria that produce and use hydrogen must be maintained via a syntrophic relationship between hydrogen and formate electron carriers. Anaerobic digestion can be impacted by any imbalances in the balance between two bacteria that help each other out. Hydrogenotrophic methanogens are the most common type of methanogens to use hydrogen as an electron donor and acquire energy from CO_2 to CH_4 reduction. If bacteria consuming hydrogen and formate in IET failed to perform scavenging, propionate and butyrate would build up and hinder anaerobic decomposition.

Kato et al. [18] suggested that nonbiological conductive materials might also be able to mediate cell-to-cell electron transport. Adding conductive material to anaerobic reactors helps to increase the effectiveness of methane production, while also making the process more stable. This results in higher methane production rates, shorter lag times before methane formation, and more efficient removal of organic waste. The addition of conductive materials can improve the performance of anaerobic digestion via the direct interspecies electron transfer (DIET) mechanism. The conductive substance enhances anaerobic electron transfer and microbial activity, serves as a beneficial matrix for immobilizing microbial cells, and improves both methane output and the quality of biogas. The CM facilitates direct intercellular electron transfer (DIET) through a conduction-based mechanism, where electrons are transferred from electron-donating cells to electron-accepting cells. Conductive materials can substitute for pili and other cell components that are engaged in external electron transport [19, 20]. Research has demonstrated that bacteria can improve the direct flow of electrons between different species by utilizing components made of conductive materials, rather than relying on intercellular contact. The metabolic function of microbial groups relies on the presence of conductive materials or direct physical contact between cells, which enables direct interspecies electron transfer. This eliminates the requirement for electron transfer media such as e-pili or OmcS. The addition of conductive materials may require less energy to satisfy the growth of syntrophic partners and metabolize more substrates because the bacteria in DIET did not need to highly express the critical genes encoding e-pili and OmcS. Conductive materials that have been demonstrated to enhance the speed and efficiency of DIET-mediated anaerobic digestion. Table 11.1 presents a comprehensive overview of the relative benefits associated with anaerobic digestion, anaerobic digestion facilitated by DIET, and anaerobic digestion facilitated by conductive material-mediated DIET. Conductive materials such as GAC, magnetite, and charcoal, among others, have been recognized for their ability to facilitate electron transmission, despite not being living components. These materials are classified into two categories: (1) conventional conductive materials and (2) novel conductive materials.

TABLE 11.1
Table of Comparison of Advatanges of Anaerobic Digestion, Anaerobic Digestion Facilitated by DIET and Anaerobic Digestion Facilitated by Conductive Material Mediated DIET

Advantages oF AD	Advantages of AD Facilitated by DIET	Advantages of AD Facilitated by CM Mediated DIET
Green Benefits	**Efficiency Enhancement**	**Specific Role of Conductive Materials**
• Quenching of green house gases	• Increase in rate of reaction and organic waste digestion efficiency.	• Provide high surface area
• Reduction of chemical waste	• Increase in efficiency of overall anaerobic digestion process.	• Higher number of active sites.
• Inhibition of air / water / soil pollution		• Higher reactivity and specificity.
• Maintains global carbon cycle		• Encourages unrestricted mobility and dispersibility in the AD medium.
		• Lessens the negative impact of ammonia produced during AD.
Energy Restoration	**Lesser Complex Pathway**	
• Biogas Regeneration	• Shuttle mechanism involving Hydrogen / Formate circumvented.	
• No carbon footprint		
Recovery of Nutrients	**Natural Applications**	**Reaction Mechanics and Yield**
• High nutrient retention in organic manure.	• Economical for solid waste management.	• Shortening of lag period and startup time.
• Reduces use of synthetic fertilizers	• Prevents digesters from going bad from organic overloading.	• Considerable improvement in yield of Methane.
• Restores soil fertility	• Nurtures the biogeochemical processes of terrestrial wetlands.	• Stability enhancement of biochemical system.

AD = Anaerobic Digestion; **DIET** = Direct Interspecies Electron Transfer; **CM** = Conductive Material

11.3.1　Conventional Materials

11.3.1.1　Iron-Oxides

It is a conductive material with low environmental malignancy, accessibility, strong magnetism and higher conductivity. The role that iron oxides play in biological and geological reactions is significant. The conductivity of iron oxides changes the kinetics of methanogenesis by shortening the lag phase, resulting in higher yields. In tests to understand the DIET mechanism, different types of iron oxide particles can be exploited because they have varying conductivities. This was initially done using soil bacteria, and it was discovered that iron oxide nanoparticles can support the respiratory systems of soil microbial cultures. It has been demonstrated that in enriched microbial communities from rice paddy soils, the rate of methanogenesis may be predicted using the conductivity of iron oxides. In this work, the effects of various iron oxides on the methanogenesis of acetate and ethanol were examined. It was found that (semi)conductive iron oxides could accelerate the pace of methanogenesis and reduce lag time. According to the study, the inclusion of magnetite, hematite or ferrihydrite can enhance methanogenesis. This is due to the development of an electrical syntrophic connection between Geobacter and Methanosarcina via semi-conductive iron oxides [21, 22]. A previous study found that there are specific bacterial-archaeal associations in digesters near geothermal active regions. These associations are thought to be due to the physicochemical properties of the water in that region. Inorganic elements, like iron, may act as an accelerator for improved electron transport. Iron-reducing bacteria, which are known to aid in the breakdown of complicated substrates, may be strengthened by iron oxides. Magnetite has been demonstrated to be a conductive material in syntrophic associations in both defined and mixed microbial consortiums. The effect of the addition of magnetite on methane production and organic degradation was studied over a prolonged period and it was found that the reactor with added magnetite performed significantly better in terms of both methane production and organic degradation.

11.3.1.2   Activated Carbon

Studies have shown that the cultures of *Geobacter metallireducens* and *Methanosarcina barkeri* can create more methane from ethanol when activated carbon is added. Sulfide was absorbed by the activated carbon, which can impede methanogenesis. The benefits of using GAC (granulated activated carbon) as an anaerobic digestion stimulator are immense: GAC has a large surface area for accommodating microbial adhesions and can help with electron transfer. Adding GAC to a biomass aggregate can help enrich exoelectrogens and hydrogenotrophic methanogens. GAC addition to co-cultures also decreases the lag phase, which aids in DIET. In reactors supplemented with GAC and carbon cloth, VFA concentrations were significantly lower, which accelerated the breakdown of OFMSW. Through a decrease in the lag phase and an increase in methane output, carbon-based conductive materials like GAC can be employed to support DIET. There is not much research that demonstrates how adding GAC affects diverse wastes. They all stated that conductive GAC accelerated the rate of digestion of hypothetical DIET participants, leading to the assertion that conductive GAC can be used as an effective electron transfer mediator in real wastewater streams. Although biochar and activated carbon's capacity to support a DIET has recently received attention, the effects of related microbial growth and soluble chemical absorption must be ignored. Via the DIET mechanism, the use of conductive materials can improve AD performance. The impact of DIET on AD performance is supported by the enhanced methane production brought on by the addition of activated carbon and electroactive microorganisms. It was discovered that *Sporanaerobacter* and *Methanosarcina* were concentrated on the carbon surface of carbon fabric and granular activated carbon, both of which were known to encourage the synthesis of methane [23].

11.3.1.3   Carbon Cloth, Fibres, Felt

Due to their conductive attributes, carbon cloth and felt are often used as electrode materials in bioelectrochemical systems. Dang et al. [24] found that using carbon cloth as a conducting material in a methane production reactor increased methane production, COD removal, and VS removal significantly. They came to the conclusion that even at very high VFA concentrations, these conductive materials are still efficient. Another study found that the DIET process, which was mediated by carbon cloth, significantly influenced the enrichment of the microbial population in the solid waste leachate during AD. Chen et al. [25] found that adding carbon cloth to cultures of two types of bacteria (*G. metallireducens* and *G. sulfurreducens*) increased their metabolism. This was not seen when carbon cloth was added to cultures of another type of bacteria (*M. barkeri*). They hypothesized that the carbon cloth's electrical conductivity was to blame for the elevated metabolism. Carbon fibres have significant chemical and physical properties because of their regular hexagonal rings, which are parallel to the fibre's long axis. Carbon fibre surfaces are easily capable of forming biofilms, which enhances AD performance and lengthens the time biomass is retained inside reactors. Carbon fibres may be employed as DIET promoters, according to recent research. Carbon cloth and carbon felt help to alleviate the inhibiting condition caused by acidic impact and high hydrogen partial pressure in anaerobic digesters, increasing DIET. The researchers argued that in the presence of carbon fibre in a difficult environment, such as an acidic pH, the main electron transport pathways transition from IHT to DIET. Numerous studies over the years have focused on anaerobic digestion systems with a variety of substrates that effectively used conductive materials like carbon cloth [26] and magnetite to mediate DIET [27].

11.3.2 Novel Materials

In numerous microbial electrochemical systems, the new carbon-based nanomaterials, such as graphene and single- or multiwalled carbon nanotubes, have been used as electrodes. They have been used to simulate DIET in AD and offer a wide range of physicochemical features such as large surface area, high electrical conductivity, and strong resistance against corrosion [28]. Despite having one-thousandth of the conductivity of activated carbon, the boosting effect of biochar is nearly identical to that of actived carbon. The methanogenic DIET mechanism can be triggered by any conductivity larger than a specific threshold. DIET may be intimately related to the surface redox groups of conductive carbon compounds like quinone or quinone hydrogen. Improvements to DIET for methane generation are being made using carbon-based compounds such as biochar, graphene and GAC. These materials have a high electrical conductivity, are biocompatible, chemically stable, lightweight and are inexpensive. Due to its excellent conductivity, biocompatibility and catalytic reactive sites, GAC in particular has been employed in batch and continuous type reactors with a variety of substrates, including ethanol, butyrate, propionate, acetate and waste-activated sludge.

11.3.2.1 Iron-Based Nanoparticles

Recently, iron-based nanoparticles such as magnetite and hematite have been exploited for biosensors, bio-separation, and targeted drug delivery. Through the formation of syntrophic partnerships between bacteria and archaea, these substances promote the DIET process. These nanoparticles' modes of operation might be distinct from those of conductive materials made of carbon. Nanoparticles that combine with biomass are a hindrance to the DIET process. By quickening the DIET process, magnetite nanoparticles enhance syntrophic methanogenesis. Because of direct suppression by Fe(III), reduction of archaea by Fe(III), substrate competition, mass transfer restrictions, and other negative effects on anaerobes, the anaerobic digestion process can be constrained by adding conductive materials. The presence of magnetite has been found to inhibit the expression of the wild-type OmcS gene, as well as promote the formation of DIET. However, magnetite cannot replace conductive carbon materials. This was discovered in a co-culture system where the addition of magnetite did not promote DIET in an e-pili deficient strain. Magnetite enables OmcS-deficient strains to realize extracellular electron transfer. Furthermore, transmission electron microscopy has revealed that, like OmcS, magnetite is distributed between cells and along the e-pili. These findings indicated that magnetite, rather than e-pili, played the role of OmcS. In contrast to conductive carbon compounds, magnetite has a stimulating impact on microbial cells. This is most probably because conductive carbon and magnetite have different sizes and structural compositions. Magnetite particles typically range in size from 20 to 50 nm, which is smaller than microbial cells, but conductive carbon materials are noticeably larger and provide a greater surface area for microbial attachment [29]. The presence of extracellular electron transfer capacities in bacteria can also be enhanced by conductive iron materials, according to numerous studies, which may be significant to DIET.

11.3.2.2 Biochar

Biochar is a type of black carbon that is made up of minerals, amorphous and graphitic carbon, and mutable organic molecules. It is studied as a conductive material because of its graphitic structure. The high internal surface area of biochar is due to its porous structure, which also explains how soluble organic matter, gases, nutrients and redox-active compounds like quinones can diffuse inside. It is carbon-rich and biostable in nature, produced by the thermal transformation of biomass that is used as a soil amendment to improve soil quality. Heavy metals, antibiotics, pentachlorophenol, atrazine and tylosin are just a few of the harmful chemicals in water that can be treated using biochar, a less expensive carbon-based conductive material. Biochar encourages methanogenesis by promoting a DIET. Because it may absorb potential inhibitors, it can provide matrices for microbial adhesion and attachment and reduce inhibition in anaerobic digesters. Biochar encourages the production of biogas, which aids in the maintenance of the VFA balance in the digester. The syntrophic conversion of alcohols and VFAS to methane can be promoted by the conductivity of biochar in a co-culture of *G. metallireducens* and *Methanosarcina barkeri*. This makes it an excellent organic fertilizer that can help improve nutrient retention in digestate. Despite having a conductivity a thousand times lower than GAC, biochar can cause DIET. The use of activated carbon and biochar in engineered systems has been researched in a variety of settings. According to one of these researches, biochar may be essential to the process of converting ethanol into methane because it is the only substance that allows a methanogenic coculture to use ethanol to produce methane [30]. The main goals of using biochar and activated carbon in anaerobic digestion have been to remove inhibitors through adsorption and to encourage the growth of attached bacteria.

11.3.2.3 Others: Graphene, Nanotubes

Graphene's antimicrobial properties could potentially inhibit the growth of microorganisms in anaerobic environments. This is significant because graphene has unique features, such as high mechanical strength, extremely high electrical conductivity, and large surface area, which could improve AD performance. The effect of graphene on the growth of microorganisms in anaerobic environments has been studied recently. It was found that graphene has the potential to improve the performance of anaerobic digestion due to its unique properties. Additionally, adding nano-sized graphene to the reactor significantly increased methane production rates and activity. Single-walled or multi-walled carbon nanotubes (SWCNT, MWCNT), graphene, and other new carbon-based non-conventional carbonaceous nanomaterials have all been used as electrodes in a variety of microbial electrochemical systems. These CMs have been employed to promote DIET in AD because of their many physicochemical properties, which include great corrosion resistance, strong electrical conductivity, and a large surface area. Lin et al. [31] found that the optimal methane production rate and yield from ethanol requires a concentration of 1 g/L graphene, and also found that a much smaller amount of graphene is required for DIET stimulation than activated carbon. Due to graphene's greater conductivity and larger specific surface area, this is most likely the case. A study found that in continuous operation at room temperature, the methane yield at a graphene dosage of 30 mg/L was 14.3% higher than the control, but the methane yield at a dosage of 120 mg/L showed a minor inhibitory effect [32].

11.4 NEGATIVE EFFECTS OF CONDUCTIVE MATERIALS-MEDIATED AD

The application of conductive materials (CMs) in anaerobic digestion (AD) has a number of drawbacks. Direct interspecies electron transfer (DIET) and methanogenesis have been reported to be enhanced by carbon-based CMs, whereas microbial methanogenesis can be hindered by metal-based CMs such as Ag and Mg nanoparticles ferrihydrite [33]. The effectiveness of CMs in AD depends on a number of physicochemical characteristics, including conductivity, pore size, shape and surface area. These characteristics also affect variables like electrical conductivity, medium redox potential, and volatile fatty acid (VFA) concentration.

Despite progress, there is still a significant information gap regarding the fundamental mechanisms driving the interaction between CMs and microbial consortia in AD environments. Research has revealed that while the application of DIET has the potential to increase methane production from leachate from municipal solid waste (MSW), the syntrophic metabolism of alcohols and volatile fatty acids (VFAs) may not always be improved by the addition of magnetite and granular activated carbon (GAC) [34]. Adding iron-based CMs could increase the danger of corrosion, which could

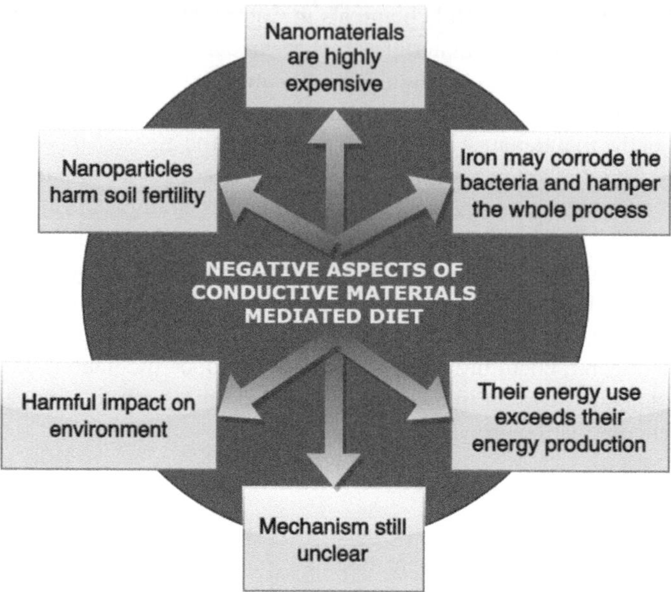

FIGURE 11.3 Negative effects of conductive material-mediated DIET.

jeopardize process efficiency overall and bacterial integrity [35]. Moreover, the incorporation of conductive nanoparticles may have an adverse effect on soil fertility, hence giving rise to apprehensions regarding the wider environmental consequences of CM-mediated AD [36]. A higher mass transfer resistance between bacteria and substrates may be the cause of the potential inhibitory effect of excess conductive material, as carbon nanotubes at a dose of 5 g/L were demonstrated to minimize methane production in contrast to dosages of 0.5 and 1 g/L [37]. Although the use of nanomaterials has great promise for improving electron transport and methanogenesis, their inherent high cost presents substantial economic obstacles. Since the energy required for the manufacture of many CMs is greater than the energy produced by improved methanogenesis, there are questions about the sustainability of CM-mediated AD. These complex challenges, as highlighted in Figure 11.3, underline the importance of conducting thorough research to understand and minimize the harmful effects linked to CM-mediated AD systems, emphasizing the need for a better grasp of CM-mediated AD processes to create more efficient and eco-friendly methods.

11.5 CONCLUSION AND FUTURE SCOPE

It has been shown that introducing conductive materials to anaerobic digestion systems can significantly reduce the lag time, even in the presence of high hydrogen partial pressure, high ammonia nitrogen concentration and sulfur inhibition. Anaerobic digestion systems are designed to break down organic nitrogen-containing materials without the use of oxygen. However, this method can result in the production of large amounts of ammonia, a poisonous chemical that can prevent microbial activity and cell growth. The addition of biochar and magnetite to an anaerobic digestion system can increase methane yield in an ammonia inhibition environment. Anaerobic digestion's efficiency is decreased by hydrogen sulfide, a consequence of sulfate reduction, which can considerably impede methanogen activity. But numerous studies have demonstrated that DIET mediated by conductive materials is likely to improve methane production in anaerobic digestion systems handling organic wastes with high sulphate contents. It is well known that DIET outperforms IHT and that setting up DIET mediated by conductive materials can boost methane output. Graphite, magnetite, biochar and carbon cloth are examples of conductive materials where advances in methane output have been reported and verified. By encouraging the production of more pure methane, accelerating and optimizing electron transport between microbial populations, decreasing lag times, raising COD removal efficiency, and raising methane production rates, conductive materials have the potential to improve anaerobic digestion efficiency. It can also improve the process stability and lessen the harm that temperature swings, acidic working environments, high H_2 partial pressure, and enhanced syntrophic metabolism to speed up the reaction pace of the biochemical digestion process. Even though CMs can speed up the process, several variables can hinder the incorporation of CMs in AD, including mass transfer restrictions, substrate competition, methanogen decrease and microbial toxicity. The impact of various conductive materials on the acceleration of anaerobic digestion and operational costs is influenced by factors such as cost, distinctive properties, and interactions with diverse substrates. The limitations associated with the incorporation of a single conductive material can be mitigated through the utilization of a combination of conductive materials, composites comprising diverse CMs, or CMs coupled with other additive materials. Future research should focus on integrating the appropriate conductive material based on the qualities of the substrate, which necessitates an in-depth understanding of the fundamental physics and chemistry of electron transport facilitated by CM and interactions with microorganisms. Understanding the impact of the physicochemical properties of CM, including surface area, porosity and conductivity, on methanogenesis processes is of utmost importance, which can encourage to use wastes from industries to be used as additive/conductive material. Multidisciplinary research must advance in order to minimize negative effects and maximize the additional benefits of CMs in AD systems, as well as address concerns about the soil fertility of the digestate, corrosion of the digestor and piping, and the economic viability of CM. To guarantee maximum process efficiency and environmental sustainability, certain actions are required. By solving current problems and seizing new opportunities, CM integration has the potential to revolutionize AD practices, promote sustainable bioenergy production, eradicate waste, aid in the fight against climate change and boost energy security. It requires further research and development to fully utilize CMs' potential to improve waste reduction and waste-to-energy conversion.

REFERENCES

[1] Panagiotopoulos IA, Lignos GD, Bakker RR, Koukios EG. Effect of low severity dilute-acid pretreatment of barley straw and decreased enzyme loading hydrolysis on the production of fermentable substrates and the release of inhibitory compounds. *J Clean Prod* 2012;32:45–51. https://doi.org/https://doi.org/10.1016/j.jclepro.2012.03.019.

[2] Zahoor, Wang W, Tan X, Guo Y, Zhang B, Chen X, et al. Mild Urea/KOH pretreatment to enhance enzymatic hydrolysis of corn stover with liquid waste recovery for plant growth. *J Clean Prod* 2021;284:125392. https://doi.org/https://doi.org/10.1016/j.jclepro.2020.125392.

[3] Cai Y, Zheng Z, Schäfer F, Stinner W, Yuan X, Wang H, et al. A review about pretreatment of lignocellulosic biomass in anaerobic digestion: Achievement and challenge in Germany and China. *J Clean Prod* 2021;299:126885. https://doi.org/https://doi.org/10.1016/j.jclepro.2021.126885.

[4] Panigrahi S, Sharma HB, Dubey BK. Integrated bioprocessing of urban organic wastes by anaerobic digestion coupled with hydrothermal carbonization for value added bio-carbon and bio-product recovery: A concept of circular economy. *Sustain. Environ. Geotech.* 2020: 47–54.

[5] Panigrahi S, Dubey BK. A critical review on operating parameters and strategies to improve the biogas yield from anaerobic digestion of organic fraction of municipal solid waste. *Renew Energy* 2019;143:779–97.

[6] Panigrahi S, Dubey BK. Electrochemical pretreatment of yard waste to improve biogas production: Understanding the mechanism of delignification, and energy balance. *Bioresour Technol* 2019;292:121958. https://doi.org/10.1016/j.biortech.2019.121958.

[7] Panigrahi S, Tiwari BR, Brar SK, Dubey BK. Thermo-Chemo-Sonic pretreatment of lignocellulosic waste: Evaluating anaerobic biodegradability and environmental impacts. *Bioresour Technol* 2022;361:127675.

[8] Panigrahi S, Sharma HB, Dubey BK. Overcoming yard waste recalcitrance through four different liquid hot water pretreatment techniques – Structural evolution, biogas production and energy balance. *Biomass Bioenergy* 2019;127: 105268. https://doi.org/10.1016/j.biombioe.2019.105268.

[9] Hashemi B, Sarker S, Lamb JJ, Lien KM. Yield improvements in anaerobic digestion of lignocellulosic feedstocks. *J Clean Prod* 2021;288:125447. https://doi.org/https://doi.org/10.1016/j.jclepro.2020.125447.

[10] Liu M, Wei Y, Leng X. Improving biogas production using additives in anaerobic digestion: A review. *J Clean Prod* 2021;297:126666. https://doi.org/https://doi.org/10.1016/j.jclepro.2021.126666.

[11] Möller K, Müller T. Effects of anaerobic digestion on digestate nutrient availability and crop growth: A review. *Eng Life Sci* 2012;12:242–57.

[12] Kumar P, Barrett DM, Delwiche MJ, Stroeve P. Methods for pretreatment of lignocellulosic biomass for efficient hydrolysis and biofuel production. *Ind Eng Chem Res* 2009;48:3713–29.

[13] Mu L, Zhang L, Zhu K, Ma J, Ifran M, Li A. Anaerobic co-digestion of sewage sludge, food waste and yard waste: Synergistic enhancement on process stability and biogas production. *Sci Total Environ* 2020;704:135429.

[14] Xu N, Liu S, Xin F, Zhou J, Jia H, Xu J, & Dong W. Biomethane production from lignocellulose: Biomass recalcitrance and its impacts on anaerobic digestion. *Front Bioeng Biotechnol* 2019;7:191. https://doi.org/10.3389/fbioe.2019.00191.

[15] Aljerf L. Méthode de synthèse économique et environnementale de la cardolite alkylée. *Am J Innov Res Appl Sci* 2016;3:415–22.

[16] Sun XF, Xu F, Sun RC, Fowler P, Baird MS. Characteristics of degraded cellulose obtained from steam-exploded wheat straw. *Carbohydr Res* 2005;340:97–106.

[17] Merk S, Blume A, Riederer M. Phase behaviour and crystallinity of plant cuticular waxes studied by Fourier transform infrared spectroscopy. *Planta* 1997;204:44–53.

[18] Kato S, Hashimoto K, Watanabe K. Methanogenesis facilitated by electric syntrophy via (semi) conductive iron-oxide minerals. *Environ Microbiol* 2012;14:1646–54.

[19] Lu JS, Chang JS, Lee DJ. Adding carbon-based materials on anaerobic digestion performance: A mini-review. *Bioresour Technol* 2020;300:122696. https://doi.org/10.1016/j.biortech.2019.122696.

[20] Park JH, Kang HJ, Park KH, Park HD. Direct interspecies electron transfer via conductive materials: A perspective for anaerobic digestion applications. *Bioresour Technol* 2018;254:300–11. https://doi.org/10.1016/j.biortech.2018.01.095.

[21] Yin Q, Wu G. Advances in direct interspecies electron transfer and conductive materials: Electron flux, organic degradation and microbial interaction. *Biotechnol Adv* 2019;37:107443. https://doi.org/10.1016/j.biotechadv.2019.107443.

[22] Wang L-Y, Nevin KP, Woodard TL, Mu B-Z, Lovley DR. Expanding the diet for DIET: Electron donors supporting direct interspecies electron transfer (DIET) in defined co-cultures. *Front Microbiol* 2016;7:236.

[23] Jensen MB, Strübing D, de Jonge N, Nielsen JL, Ottosen LDM, Koch K, et al. Stick or leave – Pushing methanogens to biofilm formation for ex situ biomethanation. *Bioresour Technol* 2019;291:121784. https://doi.org/10.1016/J.BIORTECH.2019.121784.

[24] Dang Y, Sun D, Woodard TL, Wang LY, Nevin KP, Holmes DE. Stimulation of the anaerobic digestion of the dry organic fraction of municipal solid waste (OFMSW) with carbon-based conductive materials. *Bioresour Technol* 2017;238:30–8. https://doi.org/10.1016/j.biortech.2017.04.021.

[25] Chen S, Rotaru AE, Liu F, Philips J, Woodard TL, Nevin KP, et al. Carbon cloth stimulates direct interspecies electron transfer in syntrophic co-cultures. *Bioresour Technol* 2014;173:82–6. https://doi.org/10.1016/j.biortech.2014.09.009.

[26] Azizi SMM, Zakaria BS, Haffiez N, Kumar A, Dhar BR. Pilot-scale investigation of conductive carbon cloth amendment for enhancing high-solids anaerobic digestion and mitigating antibiotic resistance. *Bioresour Technol* 2023;385:129411.

[27] You G, Wang C, Wang P, Chen J, Gao Y, Li Y, et al. Long-term transformation of nanoscale zero-valent iron explains its biological effects in anaerobic digestion: From ferroptosis-like death to magnetite-enhanced direct electron transfer networks. *Water Res* 2023;241:120115.

[28] Zhang W, Xu Y, Ma J, He J, Ye H, Song J, et al. Efficient, robust, and flame-retardant electrothermal coatings based on a polyhedral oligomeric silsesquioxane-functionalized graphene/multiwalled carbon nanotube hybrid with a dually cross-linking structure. *ACS Appl Mater Interfaces* 2023;15:4430–40.

[29] Dastjerdi R, Montazer M. A review on the application of inorganic nano-structured materials in the modification of textiles: focus on anti-microbial properties. *Colloids Surfaces B Biointerfaces* 2010;79:5–18.

[30] Yuan H-Y, Ding L-J, Zama EF, Liu P-P, Hozzein WN, Zhu Y-G. Biochar modulates methanogenesis through electron syntrophy of microorganisms with ethanol as a substrate. *Environ Sci Technol* 2018;52:12198–207.

[31] Lin R, Cheng J, Zhang J, Zhou J, Cen K, Murphy JD. Boosting biomethane yield and production rate with graphene: The potential of direct interspecies electron transfer in anaerobic digestion. *Bioresour Technol* 2017;239:345–52.

[32] Qiu L, Deng YF, Wang F, Davaritouchaee M, Yao YQ. A review on biochar-mediated anaerobic digestion with enhanced methane recovery. *Renew Sustain Energy Rev* 2019;115:109373. https://doi.org/10.1016/j.rser.2019.109373.

[33] Wu Y, Wang S, Liang D, Li N. Conductive materials in anaerobic digestion: From mechanism to application. *Bioresour Technol* 2020;298:122403.

[34] Nabi M, Liang H, Cheng L, Yang W, Gao D. A comprehensive review on the use of conductive materials to improve anaerobic digestion: Focusing on landfill leachate treatment. *J Environ Manage* 2022;309:114540.

[35] Liu Yiwei, Li Xiang, Wu Shaohua, Tan Zhao, & Yang Chunping. Enhancing anaerobic digestion process with addition of conductive materials. *Chemosphere* 2021;278:130449.

[36] Maurice N. Impact of nanoparticles on soil microbes for enhancing soil fertility and productivity. *Nanotechnol. Sustain. Agric.* 2023: 75–127.

[37] Hao Y, Wang Y, Ma C, White JC, Zhao Z, Duan C, et al. Carbon nanomaterials induce residue degradation and increase methane production from livestock manure in an anaerobic digestion system. *J Clean Prod* 2019;240:118257.

12 Self-Propagating High-Temperature Synthesis (SHS) Technology for the Disposal of Radioactive Waste

M.I. Alymov and T.V. Barinova

12.1 INTRODUCTION

The development of nuclear energy is currently the most promising way of providing real economic growth. In this regard, it is particularly important to overcome environmental problems of nuclear energy and specifically the problems of the safe management of radioactive waste and its disposal. Due to the serious biological hazard, the greatest attention is paid to issues related to the development of methods for management of high-level radioactive waste (HLW) where up to 99% of spent heat-generating elements is concentrated. Therefore, it is HLW that poses the greatest radioecological danger. Nowadays, methods of converting waste into calcinates or a mixture of oxides with subsequent curing in a glass-like matrix are developed. Only borosilicate and aluminophosphate glasses as a matrix find practical use.

However, the glasses have relatively low chemical resistance especially at high temperatures, which leads to their devitrification and the release of dangerous radionuclides into the biosphere.

Therefore, in parallel with the improvement of glass materials for the long-term disposal of HLW, various types of rock-forming mineral-based ceramics showing high chemical, thermal and radiation resistance are developed. They include titanate and zirconate pyrochlores (rare earth elements (REEs), An, Ca, etc.)$_2$(Ti, Zr)$_2$O$_7$, titanate and aluminate perovskites (Ca, REEs, An, etc.)(Ti, Al) O$_3$, and Y–Al garnets (Y, Ln, An, Ca, etc.)$_3$Al$_5$O$_{12}$, in which HLW actinoids can be isomorphically incorporated. These materials are prepared by hot isostatic pressing, cold pressing and sintering, induction melting in a cold crucible.

The above-mentioned methods require complex and expensive technological equipment, which, after completion of operation, becomes high active waste of the second order, requiring disposal. An alternative way is a self-propagating high-temperature synthesis (SHS) process. The suggested SHS method [1] meets modern requirements for promising energy and resource-saving technologies. The main SHS parameters are the following: initiation time of 0.05–2 s; temperature in the combustion wave of up to 3500 K; burning velocity of up to 150 mm/s; heating rate of 10^3–10^6 K/s. SHS equipment is easy to manufacture and maintain. SHS has the following advantages: (i) high process rates that excludes significant loss of volatile components, (ii) reducing conditions of synthesis, which ensures the transfer of highly charged actinide ions into lower valence states compatible with the lattices of mineral concentrator phases of radionuclide, and (iii) thermodynamic stability of SHS-obtained matrix materials and their higher chemical resistance. The disadvantage is the high porosity of combustion products because of high synthesis temperatures and short duration of the melt crystallization process in the combustion wave. In order to obtain high-density matrices with a given size, we used SHS compaction including combustion followed by pressing of a hot sample

DOI: 10.1201/9781003327646-12

into cylindrical high-density matrices having homogeneous structure, composition and properties throughout the sample [2–4]. Synthesis and compacting of ceramics was carried out in moulds of ISMAN design using a hydraulic press with automatic control. The mixture loaded in a mould was preliminarily compacted at room temperature. The initiation of the SHS process is carried out by local heating using an electric coil. Further, the process is distributed spontaneously along the sample in the form of combustion wave.

Bank sand was used as a medium through which pressure was passed to the combusting composition with simultaneous removal of adsorbed gases escaping during combustion. Moreover, it was a fairly effective heat insulator.

It is known that the quality of the products synthesized in such a way is determined by the optimal combination of a number of technological parameters of SHS compacting: prepressing and pressing pressure, delay time, and time of holding under pressure. The most sensitive of them is the delay time, which is directly related to the mixture composition.

A mineral-like matrix is formed by combustion from green mixture consisting of components – Al or Ti powders as reducing agents and Fe_2O_3, MoO_3, CuO, MnO_2 as oxidizing agents – ensuring the redox process with a high exothermic effect and structure-forming additives (for example, Al_2O_3, SiO_2, ZrO_2, Y_2O_3) in addition to HRW (mechanical mixtures of oxides). Note that the structure-forming oxides as diluents reduce the intensity of the process. When selecting the charge composition, the requirement for formation of a minimum volume of gas phase during combustion must be adhered to.

Combustion products are represented by an oxide composite material containing a matrix based on analogues of minerals of rocks containing isomorphic inclusions of HLW, and metal particles reduced from oxides. Such particles improve mechanical and thermophysical characteristics. The chemical resistance of matrices was evaluated by rates of leaching of HLW simulator elements in a distilled water [5].

12.2 SHS-ASSISTED IMMOBILIZATION OF RADIOACTIVE WASTE: MODEL EXPERIMENTS

12.2.1 IMMOBILIZATION OF THE ENTIRE SPECTRUM: SrO, Cs$_2$O, ACTINIDES, REEs AND CORROSION PRODUCTS

The model HLW compositions are the following (wt %): 9.6 Cs_2O (CsOH), 6.6 SrO, 20.9 Fe_2O_3, 3.3 NiO, 2.0 Cr_2O_3, 0.2 MnO, 10.3 ZrO_2, 10.2 MoO_3, 0.5 MgO, 12.0 La_2O_3, 0.2 B_2O_3, 11.1 Ce_2O_3, 6.3 Pr_2O_3, 3.9 Sm_2O_3, and 2.9 Y_2O_3.

Immobilization of HLW containing nuclides with different half-lives (from 30 years for Cs and Sr to several thousand years for actinides and some fission products) was carried out by including elements of HLW in matrices based on $CaTiO_3$ perovskite and $CaZrTi_2O_7$ zirconolite. Cs cannot be included in the $CaTiO_3$ and $CaZrTi_2O_7$ structures due to large size of its ion. Therefore, aluminum and silicon oxides were introduced into the green mixture to form aluminosilicate compounds containing Cs [2, 3, 6–8]. At combustion temperatures (T_c) of 1770–1870 K, cylindrical dense matrices (up to 700 g in weight) consisting of $CaTiO_3$, $CaZrTi_2O_7$, Cs-containing aluminosilicate phase based on Al_2O_3, SiO_2, CaO and TiO_2, and iron-based phase were obtained (in case of the introduction of 10 wt % of HLW). In order to prevent corrosion of reduced iron, NiO and Cr_2O_3 were additionally introduced into the starting mixture by decreasing the proportion of Fe_2O_3. Control experiments have shown that the magnitude of Cs losses depends on the type of oxidizer [Ca(NO$_3$)$_2$ or F2O3] and on its amount in a charge. The percentage of immobilized Cs grows with increasing amount of Ca(NO3)2 in green mixtures. At [Ca(NO3)2] = 8 wt %, the latter attained a value of 80%. For higher amounts of added Ca(NO3)2, we failed to obtain monolith ceramic blocks. In the case of Fe2O3 as an oxidizer, the loss of Cs was found to attain a value of 100%, although the burned samples represented dense ceramics. In order to ensure the formation of dense blocks and minimize the loss of Cs during the immobilization process, in further experiments we used a mixture of Fe2O3 and Ca(NO3)2 taken in proportions that ensured the completeness of reaction.

Despite the short residence times of HLW components in combustion wave, the immobilization process is accompanied by significant losses of Cs and non-uniform distribution of remaining Cs in the ceramic block: Cs content in the center was significantly less than that at the periphery. To reduce Cs losses, T_c was reduced to 1520 K by introducing TiO_2, Al_2O_3 and SiO_2 into the green mixture in ratios for formation of $CsAlSi_2O_6$. In this case, the content of Cs in the central part and at the periphery of the synthesized matrix was the same (0.51 and 0.50 wt %, respectively). Water resistance tests showed also the same leaching rates of Cs for both regions.

The product obtained at 1520 K was found to consist of $CaTiO_3$, $CaZrTi_2O_7$, TiO_2 and crystalline Cs-containing alumosilicate phase ($CsAlSi_2O_6$ type).

Chemical binding of Cs to $CsAlSi_2O_6$ completely excluded its loss by transition into gas phase during synthesis.

The leaching rate of Cs (R) in ceramic synthesized at T_c = 1520 K was found to be $(0.5–2.4) \cdot 10^{-6}$ g/(cm^2·day) [5] that was higher by an order than that in matrix obtained at T_c = 1770–1870 K ($R = (2.3–6.0) \cdot 10^{-7}$ g/(cm^2·day) [4]. The leaching rate of Sr $R = 5.6 \cdot 10^{-6}$ g/(cm^2·day) for matrix synthesized at lower T_c was also higher than that in case of 1770–1870 K ($R = (1.7–5.2) \cdot 10^{-7}$ g/(cm^2·day). Note that an open porosity in ceramics prepared at T_c = 1520 K and 1770–1870 K was 3–5% and 0.2–0.4%, respectively.

12.2.2 IMMOBILIZATION OF HLW GRAPHITE

Currently, uranium–graphite reactors are decommissioned in some countries thus resulting in a great amount of radioactive waste (irradiated graphite). The main radionuclide polluting graphite is the biologically hazardous ^{14}C (half-life is 5730 years). In addition, there are spent nuclear fuel fragments and spills. The high activity of spills is determined by U, Pu, Am, and Cm actinides included in their composition. Irradiated graphite containing spills forms graphite waste HLW. Immobilization of graphite HLW is a complex task having no satisfactory solution yet.

The SHS reaction proposed for the immobilization of graphite HLW can be represented by the following scheme:

$$3TiO_2 + 4Al + 3C = 3TiC + 2Al_2O_3 \tag{12.1}$$

This reaction shows two chemical stages occurring simultaneously during the combustion:

$$3TiO_2 + 4Al = 3Ti + 2Al_2O_3 \tag{12.2}$$

$$3Ti + 3C = 3TiC \tag{12.3}$$

The proposed technology is characterized by a high degree of binding of carbon ^{14}C into $Ti^{14}C$ compound [9].

The composition of model graphite HLW consisted of Y_2O_3, La_2O_3, $CsNO_3$, SrO_2, ZrO_2, CaO, Fe_2O_3, NiO and Cr_2O_3 in equal quantities.

Experiments showed that the heat release in Eq. (12.1) is not sufficient for conducting SHS compaction, which requires complete combustion of starting billet [10]. The adiabatic T_c for Eq. (12.1) (composition 1) was 2343 K, which insignificantly exceeds the melting point of aluminum oxide (T_m = 2320 K). In this case, measured T_c was 1860 K. The frontal combustion of the pressed billet was characterized by underburning; heat release did not allow to compensate technological heat losses.

In order to increase heat release and to obtain dense ceramics, 11.0 and 12.4 wt % high-energy thermite additive (Fe_2O_3 (57.3), Cr_2O_3 (12.0), NiO (5.4), Al (25.3)) was added into composition *1* – this is composition 2. This made it possible to compensate for the lack of heat in the system due to high heat release in the reaction:

$$Fe_2O_3 + 2Al = 2Fe + Al_2O_3 \tag{12.4}$$

Introduction of 12 wt % thermite additive was found to increase T_c to 1960 K. Large particles of incompletely burned graphite are found in the combustion products of the investigated mixtures, especially in mixture 1. This allowed one to trace the stages of its solid-phase reaction with titanium. The penetration of iron, titanium and other impurities into graphite grains is found. This may be explained by the peculiarities of its crystalline structure. The practical conclusion is made that it is desirable to use finely dispersed single-size graphite powder for the preparation of a mixture.

Aside from increasing combustion temperatures, there were changes in microstructure and phase composition of combustion products [11]. It is confirmed experimentally that, under combustion of a composite that does not contain a thermite additive, formation of titanium carbide occurs by the solid phase pattern; upon addition of the thermite additive $Fe_2O_3(Cr_2O_3, NiO) + Al$, titanium carbide crystallizes from the iron-based melt. At complete combustion of composites that contain and do not contain a thermite additive, no particular difference in the impurity distribution in the combustion products was found. In the case of the composite containing no thermite additive, impurities of Fe, Ni, Cr and Si are associated with carbide regions; impurities of Sr, La and Y are associated with oxide regions; and the Ca impurity is present in both groups. In the case of the composite containing a thermite additive, impurities of Fe, Ni and Cr are included in the metal phase composition, and impurities of Ca, Sr, La and Y are associated with the Al_2O_3 formations.

The behaviour of cesium in the synthesis of combustion products depends on the thermal conditions of the combustion. In the composite with a thermite additive (the highest combustion temperature), cesium is not found. In the part of these samples where the combustion process stopped at the initial stage (the minimum temperature), an appreciable amount of cesium is present. In the composite with a thermite additive of 11.0%, cesium in the form of inclusions is concentrated in the TiO_2 grains; in the samples with a thermite additive of 12.4%, the cesium content is lower; it is situated mainly at the periphery of the TiO_2 grains, although it is also observed in the composition of the grains. In addition, a small amount of cesium was found in completely combusted samples of the composite containing no thermite additive (transient temperature of combustion). In this case, it is associated with oxide inclusions and it is also observed inside residual graphite particles [11]. The combustion of composition 1 resulted in the solid-phase synthesis of main final products (Al_2O_3 and TiC). The formed product had high porosity and low strength. During the combustion of composition 2, the synthesis and crystallization of TiC occurred in the iron melt, while Al_2O_3 crystallized from its own melt. As a result, pore-free (open porosity of about 1%) and high-strength (compression strength $\sigma_c \leq 0.9$ GPa) ceramic blocks were obtained. In both cases, the synthesis was accompanied by large losses of Cs and as a result, Cs did not bind to aluminosilicate phase. To solve the problem of cesium fixation, the underburned parts of the SHS-compacted product were investigated. In these studies, TiO_2 was used as an ore concentrate with porous structure. It was found that the unburned graphite particles of samples derived from composition 1 contained a small amount of Cs. It is not inconceivable that during cooling of combustion products, graphite was intercalated with Cs thus forming stable C_nCs at $n = 8$ and 60 [12].

Completely burned samples synthesized from composition 2 with 12.4% thermite additive contained no Cs. However, in case of introduction of 11% thermite additive, the underburned parts where the combustion process had stopped at the initial stage showed Cs inclusions in the pores of TiO_2 grains. Upon gradual heating of TiO_2/Cs ($CsNO_3$) system to 1040 K [13], Cs was shown to

TABLE 12.1
Heat release in the studied compositions

No.	Composition	Q_T, MJ/kg
1	TiO_2, Al, C	2.9
2	TiO_2, Al, C, HLW*	3.0
3	TiO_2, Al, C, Fe_2O_3**, HLW	3.7
4	TiO_2, Al, C, $CsNO_3$***	3.6
5	TiO_2, Al, $CsNO_3$****	3.0
6	Fe_2O_3, Al	4.5–5.3*****

*   2.8 wt % oxides modeling HLW.
**   Additive in excess of 100% to composition 1.
***   Additive in excess of 100% to stoichiometric composition of Eq. (12.1).
**** Additive in excess of 100% to stoichiometric composition: $2TiO_2 + 4Al + N_2 + O_2$ [13].
***** Depending on the raw material composition.

interact with titanium dioxide thus forming $Cs_2Ti_6O_{13}$. Presumably, atmospheric oxygen took part in this interaction:

$$2Cs + 6TiO_2 + 1/2O_2 = Cs_2Ti_6O_{13} \tag{12.5}$$

At temperatures above 1040 K, double oxide starts to dissociate. To optimize the synthesis of dense and strength ceramic matrix materials, it is important to know the thermal parameter of the synthesis. In [14], the heat release for preparing $TiC–Al_2O_3$ was experimentally measured using a BKS-3 combustion bomb calorimeter (Table 12.1) [14].

It follows from the data in the table that the mixtures studied are markedly inferior in heat release level to the classic iron oxide–aluminum thermite (mixture 6). Comparison of mixtures 1, 2 and 3 allows us to quantitatively evaluate the increase in heat release on the addition of 12.4 wt % iron oxide–aluminum thermite to the TiO2–Al–C combustion system. The results lead us to conclude that heat release at a level of 3 MJ per kilogram of the starting mixture is near the exothermicity limit. The addition of cesium nitrate (instead of the oxides that model RAW in mixture 2) to mixture 4 markedly increases the heat release. Comparison of mixtures 4 and 5 indicates that the presence of carbon increases the heat release level by almost 40%, which is due to the highly exothermic reaction $Ti + C = TiC$.

12.2.3 Immobilisation of the Fraction Actinid/Zirconium/REEs and Corrosion Products

Fractionation of HLW leads to the formation of fraction of actinide/Zr/rare earth elements enriched with actinides. Another source of such waste is operations for the conversion of weapons-grade plutonium into U-Pu oxide fuel. The most studied phase for the immobilization of this HLW fraction is $Y_2Ti_2O_7$ pyrochlore.

The composition of model HLW (wt %) includes 25.0 CeO_2, 46.1 La_2O_3, 4.6 Gd_2O_3, 19.6 ZrO_2, 3.8 MnO_2, 0.9 Fe_2O_3. $Y_2Ti_2O_7$ pyrochlore has high isomorphic capacity with respect to actinides but does not have the necessary radiation resistance [15]. The radiation and chemical stability of pyrochlore increases when titanium atoms are replaced with zirconium [16]. To introduce Zr into the pyrochlore structure, it was proposed to use total charge compositions for obtaining $Y_2Ti_2O_7$ and $CaZrTi_2O_7$ according to Eqs. (12.6) and (12.7) [17]:

$$(2-x)Ti+xTiO_2 + Y_2O_3+2(2-x)/3MoO_3 \rightarrow Y_2Ti_2O_7 + 2(2-x)/3Mo, \qquad (12.6)$$

$$(2-x)Ti+xTiO_2+ZrO_2+CaO+2(2-x)/3MoO_3 \rightarrow CaZrTi_2O_7+2(2-x)/3Mo, \qquad (12.7)$$

Experiments showed that the composition containing $Y_2Ti_2O_7$ and $CaZrTi_2O_7$ phases in a ratio of 2:1 is optimal for preparing dense matrices with pyrochlore structure. The synthesized ceramics had a cast structure consisting of fused coarse crystalline fragments of $Y_2(Ti,Zr)_2O_7$ and Mo precipitates.

Pyrochlore crystals had a ring structure. The outer region of crystals is enriched with Zr, here 25% of titanium atoms were replaced with Zr, while about 20% of titanium atoms were replaced in the center. Microscopic studies revealed a high concentration of La and Ce in the phase based on titanium and lanthanum oxides (possibly lanthanum titanate). Primary studies of the water resistance of the synthesized ceramics showed that the leach rates of Y, La and Ce at 20°C on the 10th day were less than 10^{-8} g cm^{-2} day^{-1}, and the content of Ca and Ti in the solutions was below the detection limit. The open porosity of the ceramic varies from 2.1 to 4.3%, and the compression strength is no less than 230 MPa. The microhardness of the pyrochlore phase ranges from 840 to 1144 kg·mm^{-2}, exceeding that of natural pyrochlore (557–663 kg·mm^{-2}).

When using MoO_3 as an oxidizer, it was revealed the following disadvantages: (i) during synthesis, there were significant losses of Mo (up to 30 wt %) in the gas phase due to sublimation of MoO_3, which contributes to stoichiometry violation and elevated gas release during SHS process; (ii) water resistance tests of matrices showed the transition of Mo into a solution that increases the leaching area.

To avoid the above-mentioned disadvantages, the syntheses were carried out using Fe_2O_3 as an oxidizer [18–22]. The syntheses were carried out without the use of hot pressing.

Thermodynamic analysis of $Ti–ZrO_2–CaO–Y_2O_3–Me_nO_m$ system (Me = Mo, Fe, Ni, Cr, Mn, Cu) showed the possibility and prospects of using iron oxide as an oxidizer for the synthesis of matrices based on pyrochlore and zirconolite [18]. The optimal green composition – 2.5–5.0 CaO; 22.5–35.0 Y_2O_3; 7.5–22.5 ZrO_2; 45.0–60.0 (40% Ti + 60% Fe_2O_3) (wt %) – for preparing matrices based on pyrochlore enriched with Zr was selected. The adiabatic temperature of this composition was 2100–2400 K. The measured combustion temperature was 1700–1970 K, which was lower than the calculated one. However, in this case, ceramic based on the pyrochlore phase was formed.

Variation in the starting composition made it possible to obtain a matrix material consisting of two phases: titanate pyrochlore $Y_2Ti_2O_7$ containing elements of HLW and metallic iron. The substitution of titanium atoms with zirconium in the pyrochlore lattice in this case was 26 at % [19].

In SHS process, the density of charge billets affects the burning velocity and combustion temperature. An increase in the density from 1.1 to 2.7 g/cm^3 was found to favour a decrease in T_c from 1820–1870 K to 1720 K [20]. At the same time, the pyrochlore lattice parameter increased insignificantly from 10.153 to 10.165 Å and the fraction of secondary phases (perovskites, ilmenites) in the matrix decreased. The change in the unit cell parameter indicates the modification of the crystal composition. At the density of above 2.7 g/cm^3, the stationary combustion mode transferred into to self-oscillating one; as a result, the layered combustion product was formed.

Pressing of the sample, which was carried out immediately after SHS, led to no change in the phase composition; however, there was a decrease in its porosity.

The amount of HLW introduced into the green mixture was found to influence the combustion temperature, phase composition of matrix, and elemental composition of pyrochlore crystals. The ceramic containing 10 wt % HLW consisted of pyrochlore $Y_2Ti_2O_7$ and metallic Fe [21]. The unit cell parameter (a) for synthesized pyrochlore was 10.194 Å, which corresponds to pyrochlore $Y_2(Ti_{0.7}Zr_{0.3})_2O_7$. Microscopic studies showed that the substitution of titanium atoms with zirconium is 28 at %. The total content of lanthanides in separated pyrochlore crystals ranged from 0.9

to 1.6 at %. Metallic iron precipitated up to 2 μm in size was uniformly distributed. For 20% HLW, the pyrochlore unit cell parameter was 10.166 Å that indicates a lower percentage of titanium substitution with zirconium (about 20 at %). In this case, the matrix was represented by pyrochlore, $La_{0.66}TiO_3$ lanthanum titanate, zirconium dioxide, $FeTiO_3$ ilmenite and reduced iron. Lanthanum and cerium were concentrated in the $La_{0.66}TiO_3$ phase. Note that with increasing HLW content, the porosity of combustion products increased, which is obviously caused by a decrease in the combustion temperature from 1870 to 1570 K.

It was not possible to significantly reduce the porosity of matrices by the synthesis carrying out in a SHS reactor under pressure of argon or compressed air of 1.6 and 3.2 MPa [22]. It was noted that as the gas pressure increases, the yield of crystalline phases of pyrochlore and perovskite increases and amount of zirconolite decreases. In addition, there was a decrease in the calculated unit cell parameters of pyrochlore $Y_2(Zr_{0.3}Ti_{0.7})_2O_7$, which is apparently due to the smaller amount of zirconium embedded in the structure.

12.3　COMBINED USE OF SHS AND HOT PRESSING: FORCED SHS COMPACTION OF LARGE-SIZED CERAMIC BLOCKS

SHS compaction technology for manufacturing large-sized (weighing up to 1.75 kg) ceramic carbide–oxide matrices containing nuclide simulators in metal containers was developed [4]. The proposed routine had no thermal insulation of containers and removal of free container part (ceramic block fills almost completely the volume of the container). The green mixture contained up to 15% of composite iron–aluminum termite in addition to oxides of iron, nickel and chromium. In the carbide–oxide matrix, there were mineral-like phases – ilmenite and aluminum–yttrium garnet – for the immobilization of HLW elements. Synthesized ceramic matrix materials were characterized by high density (open porosity of about 1%) and high compressive strength ($\sigma_c \leq 1$ GPa). Figure 12.1 shows the overall view of ceramic matrix synthesized in the steel container dedicated for burial.

12.4　CONCLUSIONS

Matrix materials based on pyrochlore, zirconolite, perovskite, garnet, and pollucite, titanium carbide and aluminum oxide were prepared by SHS process. Elements imitating HLW radionuclides

FIGURE 12.1　Overall view of ceramic matrix containing HLW imitators in a steel container.

are included isomorphically into the crystal lattices of synthetic analogues of rock-forming minerals and do not form independent phases in the matrix material. Dense ceramic was fabricated by SHS compaction. The high chemical stability of matrices and the high degree of binding of HLW elements were shown. Combined use of SHS and pressing for manufacturing large-sized matrix ceramics containing HLW elements in metal containers was proposed. The proposed economic and environmental-friendly method for HLW immobilization can become the basis of the technological process used in the final stage of reprocessing of spent nuclear fuel.

REFERENCES

1. *Merzhanov A. G.* Russ. Chem. Rev. 2003. Vol. 72. P. 289.
2. *Barinova T. V., Borovinskaya I. P., Ratnikov V. I., and Ignat'eva T. I.* Radiochemistry. 2008. Vol. 50. No 3. P. 316–320.
3. *Barinova T. V., Borovinskaya I. P., Ratnikov V. I. and Ignat'eva T. I.* Radiochemistry. 2008. Vol. 50. No 3. P. 321–323.
4. *Barinova T. V., Borovinskaya I. P., Ratnikov V. I., Ignat'eva T. I., and Belikova A. F.* Radiochemistry. 2013. Vol. 55. No 6. P. 539–543.
5. *ISO, Long-Term* Leach Testing of Solidified Radioactive Waste Forms. ISO-6961, International Standard Organization, Switzerland, 1982, p. 1.
6. *Barinova T. V., Borovinskaya I. P., Ratnikov V. I., Ignatieva T. I., and Zakorzhevsky V. V.* J. Self-Prop. High-Temp. Synth. 1998. Vol. 7. No 1. P. 129–135.
7. *Barinova T. V., Borovinskaya I. P., Ratnikov V. I., Zakorzhevsky V. V. Ignatieva T. I.* Int. J. Self-Prop. High-Temp. Synth. 2001. Vol. 10. No 1. P. 77–82.
8. *Barinova T. V., Borovinskaya I. P., Ratnikov V. I., Ignatieva T. I., and Zakorzhevsky V. V.* Key Eng. Mater. 2002. Vol. 217. P. 193–200.
9. *Karlina O. K., Klimov V. L., Ojovan M. I., Pavlova G. Yu., and Dmitriev S. A., Yurchenko A. Yu.* J. Nucl. Mater. 2005. Vol. 345. P. 84–85.
10. *Kobyakov V. P., Barinova T. V., and Ratnikov V. I.* Inorg. Mater. 2009. Vol. 45. No 2. P. 211–219.
11. *Kobyakov V. P., Barinova T. V., and Ratnikov V. I., Borovinskaya I. P.* Inorg. Mater. 2009. Vol. 45. No 6. P. 694–701.
12. *Kobyakov V. P., Barinova T. V., and Sichinava M. A.* Inorg. Mater. 2011. Vol. 47. No 3. P. 290–295.
13. *Kobyakov V. P., and Barinova T. V.* Int. J. Self-Prop. High-Temp. Synth. 2011. Vol. 20. No 3. P. 159–163.
14. *Kobyakov V. P., Barinova T. V., Mashkinov L.B., and Sichinava M. A. Inorg. Mater.* 2012. Vol. 48. No 5. P. 549–551.
15. *Volkov Yu. F, Tomilin S. V., Lukinykh A. N., Lizin A. A., Elesin A. A., Yakovenko A. G., Spiryakov V. I., Konovalov V. I., Chistyakov V. M., Bychkov A.V., and Jardine L. J.* Radiochemistry. 2004. Vol. 46. P. 358–364.
16. *Laverov N. P., Yudintsev S. V., Stefanovsky S. V., and Ewing R. Ch.* Dokl. Earth. Sc. 2012. Vol. 443. No 6. P. 526–531.
17. *Barinova T. V., Borovinskaya I. P., Ratnikov V. I., Ignat'eva T. I., Belikova A.F., Sachkova N. V., and Khomenko N. Yu.* Int. J. Self-Prop. High-Temp. Synth. 2011. Vol. 20. No 2. P. 67–71.
18. *Barinova T. V., Podbolotov K. B., Borovinskaya I. P., and Shchukin A. S.* Radiochemistry. 2014. Vol. 56. No 5. P. 554–559.
19. *Podbolotov K. B., Barinova T. V., Khina B. B., and Velichko M. V.* Int. J. Self-Prop. High-Temp. Synth. 2019. Vol. 28. No 4. P. 239–244.
20. *Barinova T. V., Borovinskaya I. P., Barinov V. Yu., Kovalev I. D., and Podbolotov K. B.* Int. J. Self-Prop. High-Temp. Synth. 2017. Vol. 26. No 2. P. 119–123.
21. *Barinova T. V., Borovinskaya I. P., Barinov V. Yu., Kovalev I. D., and Shchukin A.S.* Int. J. Self-Prop. High-Temp. Synth. 2017. Vol. 26. No 1. P. 54–59.
22. *Podbolotov K.B., and Barinova T.V.* Glass Ceram. 2019. Vol. 75. No 11–12. P. 451–456.

13 Conversion of Carbon Dioxide to Fuel, Feed and High Value Chemicals

Hema Jha, Upakrishnam S.V. Sanskrit and Brajesh Kumar Dubey

13.1 INTRODUCTION

Climate modelling predicts the impact of increased carbon dioxide (CO_2) concentration to be irreversible and lasting for the next 1000 years if current emissions are reduced to zero, though this modelling does not consider artificial carbon capture and sequestration (Solomon et al., 2009). Intergovernmental Panel on Climate Change (IPCC) predicts a temperature rise of 1.1–6.4°C, and a mean sea level rise of 0.18–0.59 m in this century, and emphasizes the use of carbon capture and storage (CCS) to withstand this change (Blamey et al., 2010). In general, CCS technologies can capture the CO_2 from emission sources and the atmosphere, and deliver it to geological sites or in the deep sea for permanent storage, alongside, depleted oil and gas fields are often porous and have been recognized as a potential sequestration site (Li et al., 2016; Xiaoding & Moulijn, 1996). Instead, carbon capture and utilization (CCU) are defined as the valorization of CO_2 after capture, labelling waste CO_2 as a resource. Technologies for CCU have been widely explored in academia and industry, considering CO_2 as benign raw material and its potential of replacing oil and gas in many synthetic applications (Alper & Yuksel Orhan, 2017; Truong & Mishra, 2021). Figure 13.1 indicates a flowchart for carbon capture, storage and the utilization of atmospheric CO_2.

The advantages of CCU are multifold, which include a reduction in greenhouse gas (GHG) emissions, a cutback dependence on petroleum-based products, storing renewable energy, creating green jobs, and establishing a circular carbon economy. Various routes for CO_2 utilization and conversion are known including solvent extraction, enhanced oil recovery (EOR), electrocatalytic conversion, mineral carbonation, polymerization and copolymerization, algae cultivation, and synthesis of fine chemicals. Due to thermodynamic stability of CO_2 molecules, any of this conversion process demands energy, which can again lead to CO_2 emissions. Hence, it becomes important to assess the potential conversion routes, which will require lower energy, generates a product that has significant market value, and replaces a product from the market, which is relatively less environmentally friendly. Combinedly, carbon capture, storage and utilization (CCUS) are discussed worldwide, for a long time, among policymakers, entrepreneurs and researchers; however deployment of CCUS projects is obstructed by high capital cost, technical performance, uncertainty in policies and regulations (Li et al., 2016). In the report published by IPCC, the main criticisms for CO_2 capture and usage in industrial applications are the duration of carbon storage in the product and the net GHG emission to remain positive (IPCC, 2014). These obstacles keep the CO_2 valorization technologies under continuous scrutiny to prove themselves as a true decarbonization scheme; thus the literature is full of investigation to explore the green conversion routes and exploit CO_2 as an abundant resource. Along this line, taking CO_2 as a C1 source of carbon, conversion into a thermoplastic polymer (poly methyl acrylate) containing 65% by weight CO_2 was made possible, which

DOI: 10.1201/9781003327646-13

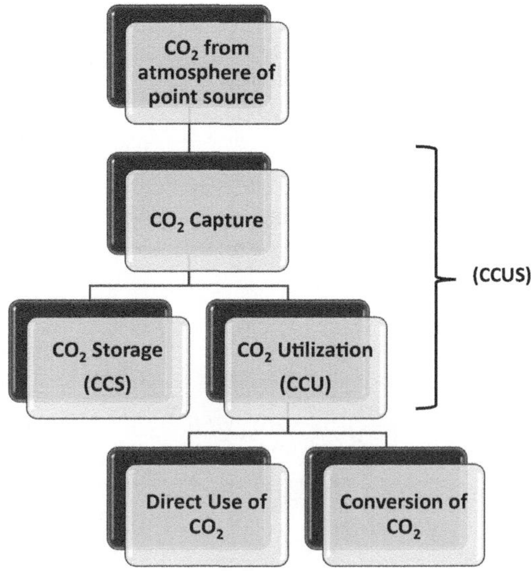

FIGURE 13.1 Flow of carbon capture, storage and utilization scheme.

can be recycled repeatedly (Styring et al., 2014). Another potential route at present is 'Power-to-X' technology, where renewable energy can be stored as chemical energy in dense carbon compounds, utilizing CO_2 as feedstock (Smith et al., 2019). The goal of net zero maintains the research in this domain exciting, while the successful demonstrations and full-scale implementations remain challenging.

The objective of this chapter is to provide an overview of the existing CO_2-derived product and their production methods, and also assess the economic viability and environmental impact of CO_2 conversion routes. To frame this discussion, a glance at the physical and chemical properties of CO_2 is presented first. This discussion is limited to evaluating the CCU technology for CO_2 valorization and does not elaborate on CCS technologies and their role in climate mitigation.

13.2 PHYSICAL AND CHEMICAL PROPERTIES OF CARBON DIOXIDE

The physical and chemical property of any compound determines the behaviour of the compound under ambient conditions or when subjected to any chemical change. A few chemical and physical properties of the CO_2 molecule is presented in Figure 13.2.

From basic chemistry, it is known that molecules vibrate having a periodic motion of one atom with respect to another atom within the molecule. There are three vibration modes in CO_2 molecules, which are symmetric stretch, anti-symmetric stretch and bend. The anti-symmetric stretch of the molecule is infrared active and hence the root cause for the CO_2 gas to be a GHG gas. There are other highly potent GHG gases like methane, nitrous oxide, etc., however, the relatively higher concentration of CO_2 in the atmosphere accounts for two thirds of the global warming effect coming only from CO_2.

The chemical reactivity of CO_2 depends upon the polarization of the carbon–oxygen bond having a partially positively charged carbon atom and a partially negatively charged oxygen atom. Theoretically, a nucleophile with lone pair of electrons, an electron-rich π-bond, or a carbon-metal σ-bond can target the electron-deficient carbon atom in CO_2. The ability of CO_2 to form coordination complexes with metals provides the basis for metal-induced reaction and catalysis in CO_2. At present, 13 geometries of CO_2-metal complexes are known. If the monometallic complex is formed by utilizing an oxygen atom, the carbon atom becomes more available for nucleophile attack

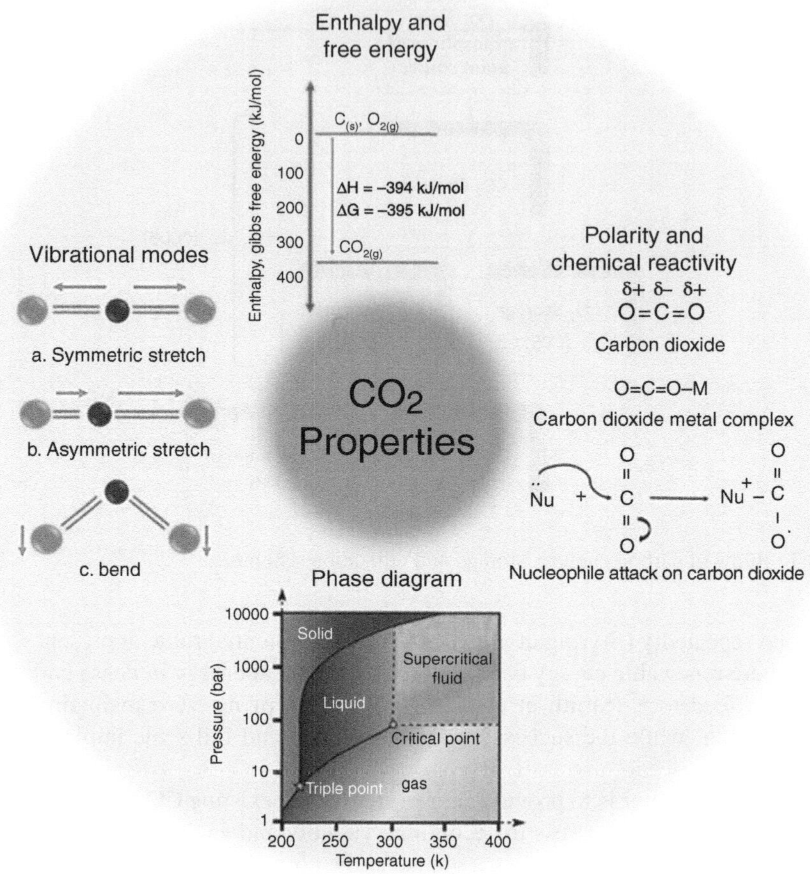

FIGURE 13.2 Physical and chemical properties of carbon dioxide.

without any change in geometry, while if the loosely held electron of metals bonds with the carbon atom in the complex compound, the geometry changes, and carbon atom cannot be attacked by the nucleophile.

The Gibbs energy change for the formation of CO_2 is highly negative (–395 kJ/mol), and the enthalpy of formation for CO_2 is also very high (–394 kJ/mol), hence the resulting molecule is very stable. Although the reaction with CO_2 can also be exothermic, some compounds react with CO_2 to give products having total enthalpy less than that of the reactants. For example, the reaction between ethylene oxide and CO_2 to produce ethylene carbonate is an exothermic reaction with a total energy release of 144 kJ/mol (North, 2015). In the literature, there are expressions for predicting the enthalpy, entropy and free energy for CO_2 and claimed to be overall more efficient than computational methods that rely on huge experimental and spectroscopic data (Wang et al., 2019).

At atmospheric pressure and a temperature of –78°C, CO_2 sublimes directly to gaseous phase from solid phase since the triple point of CO_2 is at a relatively higher pressure (5.8 bar) and temperature (–51°C). At temperature and pressure of 31°C and 73 bar the gas converts into a supercritical fluid, consequently, it expands to fill the entire space available like a gas while having a density like a liquid. Owing to its accessible critical point conditions, CO_2 is considered a green solvent in industries. Advanced flow characteristics and fluid properties such as solubility and miscibility of CO_2 with other solvents have also been investigated by microfluidics and nanofluidic to assess the behaviour of systems for some advanced applications, like EOR and sequestration of CO_2 in deep sea reservoirs (Bao et al., 2017). Likewise, other prospects for alternative energy and simultaneous CO_2 sequestration are proposed and include the replacement of CH_4 by CO_2 in natural gas hydrate fields because the hydrate-forming conditions for CO_2 are more thermodynamically stable (Sinehbaghizadeh et al., 2022). The authors suggested the clathrate hydrates formed by CO_2 can also be thought of as a next-generation raw material for industrial processing. Besides the fluid characteristic, solid phase properties are also accurately modelled from equations of state, which helps in understanding the fluid-solid phase thermodynamic correlations for CO_2 hydrate conversion (Span et al., 2013).

Apart from CO_2 conversion methods, which will be discussed in upcoming sections, CO_2 transport is recognized as an important part of CCUS, and requires sophisticated design of pipes and vessels. Fracture propagation control, friction, flow topology, heat transfer during depressurization and liquefication, and vapour-liquid equilibrium is being studied extensively for economic, efficient and safe transportation of CO_2 (Munkejord et al., 2016).

13.3 CARBON DIOXIDE AS FEEDSTOCK, FUEL AND HIGH-VALUE CHEMICALS

There is a wide range of applications of CO_2, either using it directly or through a conversion process to yield a final product. Before exploring the potential usage of CO_2, it is important to note that the direct usage and conversion could only be followed after the successful capture of CO_2 from the atmosphere or effluent gases from the industry. Methods for CO_2 capture include absorption, adsorption, membrane separation, cryogenics, ionic liquid, chemical looping, etc. (Li et al., 2016), CO_2 capture in mineral carbonation and algae production will be considered as a conversion process in this discussion. To establish an environmentally sustainable CO_2 capture technology, the application of activated carbon derived from biomass was reviewed by Maniarasu et al. (2022). The review provides an extensive list of diverse biomass-based biochar, which were investigated for CO_2 adsorption under varying process conditions, and it was concluded that at low temperatures and high pressure the adsorption of CO_2 is maximum.

The major direct use of CO_2 is in EOR, in which the supercritical CO_2 is injected in oil reservoirs to reduce the viscosity for the easy recovery of oil, alongside, part of injected CO_2 can also be sequestered in porous rocks (Desport & Selosse, 2022). The authors also highlighted that when used as a refrigerant, CO_2 can be assumed to be in an artificial sink where it is in a cyclic process away from the atmosphere. In its supercritical state, the compound serves as a suitable solvent for many chemical reactions, extraction, and separation processes (Li et al., 2016). The benign raw material characteristics of CO_2 include non-toxicity, non-flammability, and abundance. These properties enable its usage in the food and beverage industry, for example, in carbonated soft drinks (Li et al., 2016), decaffeination of coffee, winemaking, etc. (Desport & Selosse, 2022). Other direct uses of CO_2 include applications in fire extinguishing and safety applications, applications in the steel industry, and electronics manufacturing (Desport & Selosse, 2022; Xiaoding & Moulijn, 1996). Discussion on CO_2 capture and storage/direct use is limited to this point. Further, this chapter will delve deeper into understanding the potential conversion of CO_2 for its valorization, primarily through the routes outlined in Figure 13.3.

For the conversion of CO_2 into fuel, microalgae production is widely investigated and appreciated. Microalgae fix CO_2 by the natural photosynthesis process, hence posing marginal environmental pollution as compared to fossil fuel when the target application is to obtain fuel. The biofuel produced from algal biomass can directly be used in existing motor engines without design modification, further, algal biomass can also be converted to other products such as pigments, protein, etc. (Li et al., 2016). Many of the thermochemical techniques for biomass processing were reviewed in detail and deduced that economic challenges are roadblocks in their deployment as a sustainable technology (Mathimani et al., 2019). Other potential fuels which can be synthesized from CO_2 are

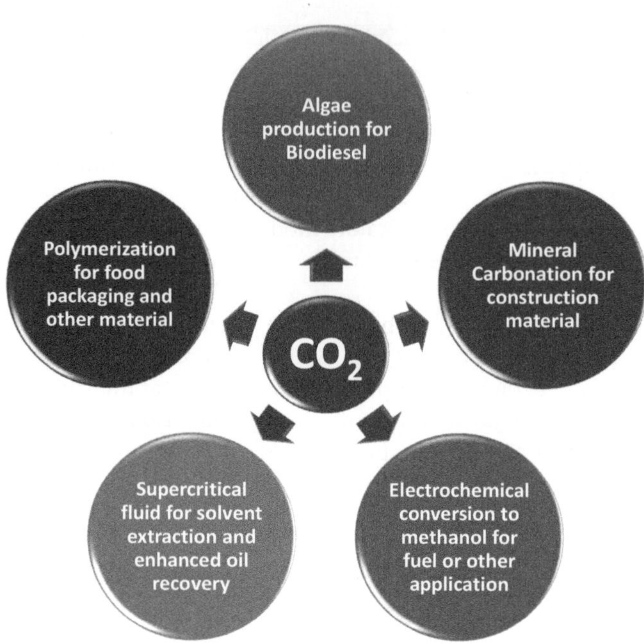

FIGURE 13.3   Potential CO_2 valorization routes.

methane, syngas and methanol. Methanol produced from CO_2 can be used in fuel cells and can be a green electricity source listing its name in 'Power-to-X' technology, wherein small molecules (H_2O, N_2, CO_2, etc.) are converted into dense molecules essentially having power source as renewable energy (e.g., solar, wind, geothermal, etc.) (Corma & Garcia, 2013; Smith et al., 2019). Along this line, Carbon Recycling International developed the first commercial plant in Iceland, which uses CO_2 emitted from volcano activity to produce methanol relying on geothermal energy as the power source (Desport & Selosse, 2022). Further, using methanol produced from this route, in a mixed substrate CO_2 conversion into microbial organic acids is thought to increase the overall carbon capture potential (Steiger et al., 2017).

Conversion to high-value chemicals requires CO_2 to react with hydrocarbons and nitrogen-containing compounds. A wide range of products can be synthesized from CO_2 including salicylic acid, alcohols, esters, lactones, carbamates, urethanes, urea derivatives, various copolymers, and polymers, polycarbonates, etc. (Venkata Narayana Reddy et al., 2022; Xiaoding & Moulijn, 1996). CO_2 can also be used as a weak acid in the neutralization process, for example, in water purification in swimming pools. When the conversion route does not include any catalyst, it is highlighted to be more sustainable and products including oxazolidinones, quinazoline-2,4(1H,3H)-diones, benzimidazoles, 1,6-dioxospiro/1,6-dioxaspiro derivatives, polymers, N-formamides, N-methylamines and carboxylic acids, etc., are identified to be synthesized without any catalyzt (Truong & Mishra, 2021).

The cementitious product synthesized from mineral carbonation of CO_2 and other raw material, preferably industrial waste, steel-making slag, or mining waste, can be used as a cement blend or in a concrete block in construction (Galina et al., 2019; Li et al., 2016). Authors suggest the huge potential to trap the CO_2 in these mineralized carbonates for a longer period than other carbon utilization methods. Also, this process serves the simultaneous purpose of dealing with heavy alkali metals that are present in this hazardous waste. Also, the CO_2 can directly be fed from the point source without purification in mineral carbonation, hence reducing the burden of energy-extensive capture technology (Desport & Selosse, 2022). The carbonation process is thermodynamically favourable but the kinetics is slow, hence extensive energy usage is associated with this route of CO_2 utilization (Alper & Yuksel Orhan, 2017).

Cyclic carbonates can also be produced by the reaction of epoxides and CO_2 using metal-organic frameworks as the heterogenous catalyzt for the reaction (Pal et al., 2020). The authors presented multiple schematics for conversion along with the appropriate catalyzt used and emphasized this route of conversion into fine chemicals could be a potential carbon fixation strategy upon the synthesis of improvised metal-organic frameworks. Synthesis of other fine chemicals (e.g., synthetic aspirin, phenyl salicylates) can be done by salicylic acid, which is produced by electrophilic substitution of CO_2 and phenolate (Alper & Yuksel Orhan, 2017). The author highlighted another fine chemical, carbon nanotube could be produced by thermal cracking of CO_2 with magnesium, however, it was emphasized that the overall contribution from fine chemicals remains limited in CO_2 utilization even if they are of high value.

Again, by the reaction of epoxides and CO_2 following another pathway, the reaction can result in polycarbonates, having cyclic carbonate as a competing product, which was tried to be minimized by using computational chemistry (Darensbourg & Yeung, 2014). These polycarbonates can be utilized in packaging materials or coating. The co-polymerization of CO_2 can lead to 50% of the mass contribution from CO_2 molecules in polymer, producing a biodegradable, biocompatible polymer with characteristics similar to common fossil-based thermoplastics (Muthuraj & Mekonnen, 2018). These plastics are suggested to be suitable for food packaging, however, due to low glass transition temperature, these CO_2 co-polymers are having lower strengths. Other polymers, including dimethyl carbonates, ethylene carbonates, polyols, etc., can also be synthesized by copolymerization and find application in a wide range of products that our current petroleum-based lifestyle exploits (Desport & Selosse, 2022).

CO_2 fixation in polyester, urea-formaldehyde resin, and melamine-formaldehyde resin was also spotlighted utilizing CO_2 as a C1 source (Alper & Yuksel Orhan, 2017). The authors concluded the product formation technologies into three categories: (1) mature technologies with industrial-scale production of urea, methanol, cyclic carbonate and salicylic acid, (2) emerging technologies with demonstration plants producing CO_2 based polymers, formic acid, dimethyl-ether, and dry reforming of CO_2, and, (3) applied research in lab or bench scale for isocyanates, organic carbonates, lactone, and carboxylic acid synthesis.

13.4 OVERVIEW OF CARBON DIOXIDE CONVERSION TECHNOLOGIES

The restricted reactivity of the molecule is due to very high enthalpy (+805 kJ/mol) of carbon–oxygen bond and the highest oxidation state (+4) of carbon. However, the activation energy for reactions can be lowered if an appropriate catalyzt is available and/or highly reactive chemicals are used to perform the conversion of CO_2 (Pal et al., 2020). CO_2 conversion employs multiple technologies including mineral carbonation, catalytic hydrogenation, reduction with complex metal hybrids, electrocatalytic, photocatalytic, and biological methods (Li et al., 2016). An overview of the conversion mechanism in some potential conversion technology is followed.

13.4.1 ELECTROCHEMICAL REDUCTION

In a typical electrochemical cell, electrodes are employed to avail electrical charge and transfer it to the electrolytic solution through a passage of ion exchange membranes, which separates the cathodic and anodic chambers. Electrodes are crucial to the electrochemical reaction, for example, using a gas diffusion electrode with a layer of the catalyzt increases the overall reduction. An example of a reduction reaction is presented in reaction (1). To reduce the activation energy required for the conversion process, the electrode catalyzt should be highly active and selective (Hori et al., 1994). Therefore, by optimizing the electrode material and its arrangement, the reactor's overall performance is maximized. Solid fuel cells, metal electrodes in aqueous electrolytes and the catalyzts for enhancing conversion in cells were reviewed (Lim et al., 2014). It was elucidated that solid fuel cells function at high temperatures for the solid electrolyte to become permeable for oxygen ions, consequentially CO and H_2 are the main products with good current density, while low-temperature reduction leads to a high applied voltage and low current densities. The electrochemical reduction can produce synthetic chemicals and fuels including CO, methanol, ethanol, ethylene and ammonia (Smith et al., 2019).

$$CO_2 + 6H^+_{(aq)} + 6e^- \rightarrow CH_3OH_{(aq)} + H_2O_{(l)} \tag{13.1}$$

13.4.2 Photocatalytic Reduction

Due to the availability and abundance of sunlight, the photoreduction of CO_2 is one of the most enticing processes for CO_2 conversion since it mimics the natural process of photosynthesis. The light-sensitive catalyst that transforms CO_2 into useful chemicals is known as a photocatalyst (Yaashikaa et al., 2019). A typical photoreduction electrode is made up of a semiconductor and several transition metals complexes, many of which serve as photocatalysts. Representation of photocatalysis can be seen in Figure 13.4.

Semiconductors absorb photons, causing excited electrons to move from a valence band to the conducting band. These excited electrons are subsequently sent to a photocatalyst complex, where they convert CO_2 to CO and other beneficial organic molecules like formic acid, methane, methanol, etc. In terms of energy efficiency and sustainability, this method is superior to other methods (e.g., electrochemical methods, thermal catalytic methods) (Saravanan et al., 2021).

13.4.3 Thermochemical Conversion

13.4.3.1 Reverse Water Gas Shift Reaction (RWGS)

In the RWGS reaction, CO_2 is treated with H_2 to produce CO and water, as shown in reaction (2), at an expense of total enthalpy of 41 kJ/mol. This method is quite famous for its CO_2 mitigation property as well as for obtaining water during the process at the same time (Saravanan et al., 2021). The CO produced by this reaction is often employed in many industries such as steel, pharmaceutical and biotechnology (Roy et al., 2018).

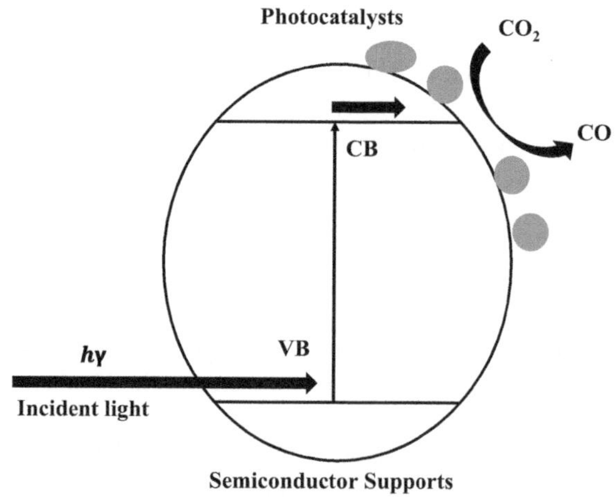

FIGURE 13.4   Photocatalytic reduction of carbon dioxide (Hu et al., 2013).

$$CO_2 + H_2 \rightarrow CO + H_2O \tag{13.2}$$

This reaction is endothermic and operates at high temperatures (550–575°C) and low pressure (0–5 bar) (Hu et al., 2013). With a minor change in temperature and pressure, the reaction can proceed in both forward and backward directions depending on its conditions (Agarwal et al., 2022; Kho et al., 2017).

13.4.3.2 Reforming Process

When methane and CO_2 react to produce synthesis gas, the process is known as dry reforming; if water is added, it is known as bi-reforming (Hu et al., 2013). The reaction involved in dry reforming is shown in reaction (13.3).

$$CH_4 + CO_2 \rightarrow 2CO + 2H_2 \tag{13.3}$$

The bi-reforming technique, which combines dry and steam methane reforming, can also be used to produce syngas. The reaction for bi-reforming is presented in reaction (13.4) and reaction (13.5).

$$CH_4 + H_2O \rightarrow CO + 3H_2 \tag{13.4}$$

$$CO_2 + CH_4 \rightarrow 2CO + 2H_2 \tag{13.5}$$

Another reforming procedure called tri-reforming combines the two reforming techniques of steam methane reforming and dry reforming, as well as partial oxidation of methane. The capacity of this technique to alter the CO to H_2 ratio in the product to produce a range of hydrocarbons makes it notable (Agarwal et al., 2022; Roy et al., 2018).

13.4.4 MINERAL CARBONATION

Mineral carbonation mimics the natural process known as rock weathering, in which an aqueous suspension of alkaline minerals is exposed to CO_2 to produce synthetic carbonates (Galina et al., 2019; Saravanan et al., 2021). One of the few thermodynamically advantageous reactions involving the low-energy molecule CO_2 is the conversion of CO_2 into solid mineral carbonates, which can occur at temperatures close to ambient temperature (Yaashikaa et al., 2019). An example of CO_2 mineralization reaction is shown below, in which CO_2 interacts with portlandite to produce calcite and water.

$$Ca(OH)_2 + CO_2 \rightarrow CaCO_3 + H_2O \tag{13.6}$$

The above reaction is exothermic ($\Delta H = -68$ kJ/mol) and thermodynamically favoured ($\Delta G = -74.61$ kJ/mol) and occurs at the ambient room conditions without any extra energy input (Saravanan et al., 2021). Dissolution-precipitation reactions are common in the carbonation process, which involves the dissolution of the elemental species from the reactant solid(s) and the solubilization of CO_2 into the liquid phase (such as water), followed by the precipitation of carbonate mineral solids from a supersaturated solution. Amorphous calcium carbonate, vaterite, aragonite and calcite are a few examples of the polymorphs of calcium carbonate that can be produced as a result of this process.

Mineral carbonation to produce recycled concrete aggregates was evaluated for its contribution to climate mitigation and was estimated to complement the geological storage for CO_2 capture (Hori et al., 1994). Aggregates and binding agents produced via mineral carbonation has the potential to replace existing synthetic and natural sources of these vital components of conventional construction materials. Because of the big size of the construction market all around the world, mineral carbonation is considered to be one of the best ways of CO_2 utilization (Sick et al., 2022). Also, this process supports the growing demand for 'green building' signifying the increase in environmental awareness globally.

13.4.5 BIOLOGICAL CONVERSION

The natural ability of the microorganisms to convert gaseous CO_2 into valuable products can be seen as an alternative to complementary chemical processes. Biologically, CO_2 is utilized in two ways, photosynthetically and non-photosynthetically. It is important to note that although the non-photosynthetic approach can produce higher biomass yields than the photosynthetic approach, they also suffer from poor life cycle analysis (Vieira et al., 2013).

13.4.5.1 Photosynthetic Approach

Algae can be considered self-replicating machines that transform CO_2 and light into goods with additional value (Hori et al., 1994). Biofuel produced is seen as an alternative to crude oil, besides it is observed that the investment in algal biofuel research increases when the crude oil price rises. Algal biomass produces high yields, i.e., 30 to 50 times more oil production than conventional crops (Saravanan et al., 2021). There are many benefits of algal cultivation, including the possibility of using saline water and non-arable land, intrinsic adaptability of biological systems to different feedstocks and operating conditions that make the system flexible. Furthermore, low CO_2 concentrations and contaminants coming from industrial CO_2 sources are tolerable by these systems. Green algae cultivation also results in various biofuels and coproducts produced from the process are (1) biodiesel, or fatty acid methyl ester, (regarded as a part of the mature industry), (2) nutritional protein for both human and animal consumption, although the restrictions on algae strain for human consumption are caused by food safety laws (Sick et al., 2022), (3) biogas, a low-value by-product, by anaerobic digestion of the waste generated in the biomass processing, which mainly comprises protein and carbohydrates.

Other microbes that follow photosynthetic pathways are cyanobacteria, e.g., acetogens. They are prokaryotic and are recognized as the planet's first known form of life. Cyanobacteria have a two to four times higher photosynthetic efficiency than plants and do not compete with food crops for land (Roy et al., 2018). These microbes can produce biofuels and other products (ethanol, butanol, fatty acids, limonene, etc.).

13.4.5.2 Non-Photosynthetic Approach

Due to the slow growth of photosynthetic microorganisms, non-photosynthetic microbes have become extremely relevant due to their wide diversity and relatively higher growth rates. This approach is observed in chemolithotrophs. Unlike photoautotrophs, chemolithotrophs are capable of assimilating CO_2 at high concentrations of CO_2 and ambient temperatures (less of an energy penalty). Hence, microorganism can turn the process into an effective and affordable method of bio-mitigation of CO_2 (Anand et al., 2020). This eliminates problems with photosynthetic production like cell shadowing and enables chemolithotrophs to carry out light-independent CO_2 fixation.

Also, by supplying microorganisms with electrons in a microbial electrosynthesis cell, bioelectrochemical systems (BES) create artificial photosynthesis and use metabolic pathways to convert CO_2 into small organic compounds. (Lovley & Nevin, 2013). At the cathode, microbes transform CO_2 into organic materials, which usually takes place in an anaerobic environment to stop oxygen reduction by available electrons and producing harmful by-products like H_2O_2. Figure 13.5 represents a typical bioelectrochemical cell for CO_2 conversion.

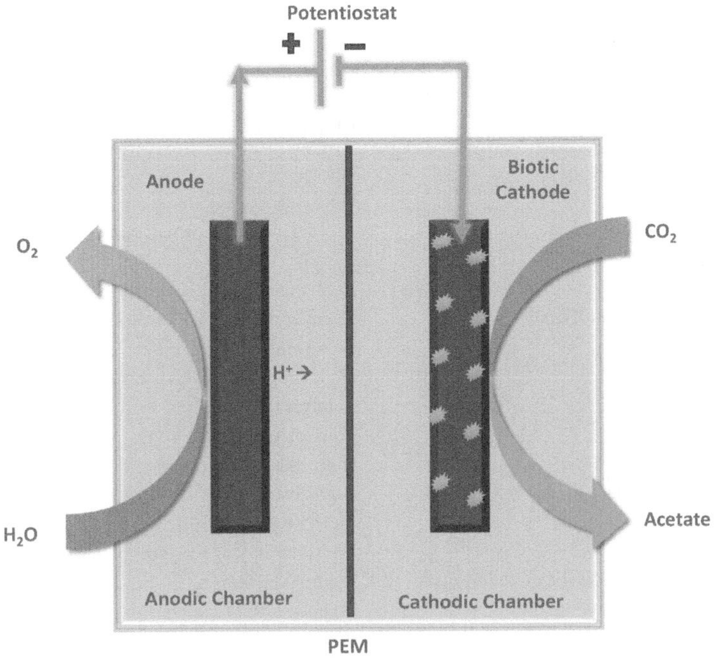

FIGURE 13.5   Bioelectrochemical system for CO_2 reduction to acetate.

13.5   ECONOMIC AND ENVIRONMENTAL SUSTAINABILITY ASSESSMENT

The European Commission considers the sequestration of CO_2 to be permanent if the product stores carbon for at least 1000 years, and long-term if it is trapped in the product for more than 100 years. Carbon utilization technologies can have advantages beyond reducing GHG emissions, such as improved energy and chemical feedstock security, less risk of environmental harm from feedstock extraction, and other advantages including fostering a circular carbon economy (Bruhn et al., 2016).

Mineral carbonation is paid enough attention to be evaluated and enlist itself as a promising CCUS technology. The main advantages of this process are its ability to utilize CO_2 in flue gas without prior purification, utilization of industrial and mining waste as raw material, and a product (solid carbonates for building materials) of considerable market value (Desport & Selosse, 2022). All these advantages make the process environmentally appealing and reduce the overall cost of the process, however, reaction kinetics need to be improvised for the process to be more economically attractive (Olajire, 2013). Carbon credit generated for a balance between CO_2 emission and CO_2 capture in this process can lead to large-scale implementations and make it economically viable (Galina et al., 2019).

In algal cultivation, the low efficiency of solar energy utilization and CO_2 capture lowers the annual flux, alongside, high operating and capital costs making the process financially taxing (Wilcox, 2012; Yaashikaa et al., 2019). Algal culturing can be a land- and water-intensive process as a result of the design of cultivation ponds or bioreactors to maximize sun exposure through huge volumes and surface area. Theoretically, it would take 25–37 acres of farming to absorb all the CO_2 from a power station that produces 10 kilotons of CO_2 every day (Hori et al., 1994). To synthesize goods with high value, it is possible to combine photosynthetic, fermentative and chemical processes, for instance, a novel combined algal processing technique that can extract several compounds from algal biomass is investigated by Michalak & Chojnacka (2015). The electrochemical reduction of

TABLE 13.1
Benefits and Challenges of CO_2 Conversion Technologies

Process	Benefits	Challenges	References
Electrolysis	Easy to scale up Highly selective No additional heat requirement Electrolyte recycling is possible	High cost Optimization of Faradaic efficiency, energy efficiency, and current density is required	Jhong et al., 2013; Saravanan et al., 2021
Photo-catalytic reduction	No requirement of additional energy Environment friendly	Product and catalyst separation Catalyzt requires optimization Low yield	Gao et al., 2020; Kumar et al., 2012
Thermal approaches	High yield	High temperature and pressure Catalyzt instability Low selectivity High cost	Galadima et al., 2011; Hu et al., 2013
Chemo-enzymatic approach	Environment friendly Highly selective	High cost Recycling of catalyzt Unavailability of ion-type co-factors	Shi et al., 2015
Biological conversion	Operates on ambient conditions Environment friendly	Product separation Maintenance of culture Genetic knockout Long time for reaction and processing	Appel et al., 2013

CO_2 as a 'Power-to-X' technology is also envisioned to be associated with a smooth energy transition from fossil fuel to renewable energy (e.g., solar, geothermal, wind). However, the process is still technically and economically challenging, and requires CO_2 based fuel cells and electrolyzes to be optimized (Lim et al., 2014; Smith et al., 2019).

Desport & Selosse (2022) have evaluated CCUS strategies for providing a net negative emission and concluded that the production of fuels from CO_2 can be carbon neutral at the very best, also this conversion route is combating economic challenges. The authors highlight that CO_2 conversion to urea should not be considered climate mitigating, as the sequestration is short term, alongside more potent GHG, nitrous oxides are also released. The less mature techniques to produce polycarbonates, cyclic carbonates and plastics have been envisioned as the most likely synthetic application after CO_2 conversion and a potential net negative CCU strategy, as it could sequester CO_2 for the long term by employing recycling as the end of life (EOL) management (Alper & Yuksel Orhan, 2017; Desport & Selosse, 2022). A list of CO_2 conversion technologies with its advantages and limitation is presented in Table 13.1.

13.5.1 Life Cycle Assessment of Carbon Dioxide Utilization

Although during the conversion process CO_2 is consumed, the process itself has energy demand, and therefore GHG emissions associated, which could be more or less than the converted CO_2 gas. (Bruhn et al., 2016; Von Der Assen et al., 2013). The process involved in creating a product from CO_2 conversion may increase the overall emission but if the CO_2-derived product is replacing another product in the market that has greater GHG emission in its life cycle, the overall emission will be reduced (Von Der Assen et al., 2013).

The LCA principle and frameworks are ascribed by the International Standards Organization, and they take into account the stages of a product's life cycle that involve the acquisition of raw materials, the conversion of raw materials into products, the transportation and distribution of those products to points of sale and/or distribution, their use, and EOL management. Hence life cycle analysis answers some important question (National Academies Press, 2019) like

- Is the net GHG emission negative when waste CO_2 is transformed into a product?
- Do various energy and environmental effects of carbon utilization methods have to be balanced?
- Which conversion product (building materials, plastics, biodiesel, etc.) results in the largest net reduction in GHG emissions?
- What effects do common CO_2 conversion materials, such as catalysts, energy-carrying substances like H_2, and solvents, have on the environment?

The application of LCA on CCU is increasing and has been summarized in several recent overviews and reviews, with some discussion of ongoing bottlenecks and research needs (Artz et al., 2018; Cuéllar-Franca & Azapagic, 2015; Rahman et al., 2017; Von Der Assen et al., 2014). One of the LCA investigations suggests that the environmental impact reduction for CO_2 conversion is maximum for formic acid production followed by production of CO and methanol (Sternberg et al., 2017). It is important to note that if non-renewable energy is used in conversion technologies, the energy demand and climate implications of the CO_2-based synthetic hydrocarbon fuel are greater than those of the existing fossil fuels. CO_2-based fuels can be slightly superior to current fossil fuels when powered by renewable energy (Van Der Giesen et al., 2014).

13.6 PERSPECTIVES AND CONCLUSIONS

The properties of the CO_2 molecule make it unique as a very stable molecule, nevertheless fitting suitable as feedstock for chemical conversion in numerous applications. Mineral carbonates as a construction material, fine chemicals, and plastics by polymerization and copolymerization, biodiesel from algal cultivation, and methanol from electrochemical reduction with renewable energy are attractive CO_2 conversion routes. The conversion of CO_2 to hydrocarbon fuels can be thermocatalytic, photocatalytic or electrocatalytic for which reviews on materials for conversion were done, and the need for developing an inexpensive, robust, and regenerable catalyzt was highlighted, alongside designing the reactors to accommodate catalyzt adsorption-regeneration steps to proceed towards commercial installations (Das et al., 2020; Vu et al., 2021). Several direct applications as a refrigerant and solvent in extraction and EOR label CO_2 as an important green material.

Furthermore, it is deduced that one technology implementation for CO_2 valorization cannot be envisioned with available knowledge, hence a portfolio of CCU technologies should be identified and deployed (Li et al., 2016). Along this line, Quadrelli et al. (2015) suggested the catalyzed hydrogenation of CO_2 as a key short-term technology, followed by electrochemical conversion as the mid-term solution, and finally, photoreduction and biological capture as long-term sustainable methods and have provided the most probable timescale for implementation of these technologies to establish a circular carbon economy. Additionally, the discovery of novel CO_2 utilization reactions and processes may result in the creation of new strategies that are less detrimental and more economic. Economic and environmental impact assessment of CCU technologies becomes challenging, owing to the complexity of the value chain, technology readiness levels, storage time of carbon, etc. To combat this challenge use of prospective energy models, and also the concept of CCU as negative emission technology (CCUNET) was introduced by Desport & Selosse, (2022). The emphasis of discussing each CCU, CCS, CCUS, and CCUNET separately is worth considering, and for each CCU technology investigation, scrutiny with LCA tools, and CCUNET strategy needs to be laid out (Desport & Selosse, 2022). The strengthening of research centres by the modern government and the responsibility of the scientific community to resolve this sustainability puzzle is essential (Venkata Narayana Reddy et al., 2022).

The field of CO_2 valorization for value addition, simultaneously mitigating climate change is advancing and demanding further investigations and implementation approaches to be in accordance, globally. Moving towards a circular economy and sustainable lifestyles, CO_2 must be considered a potential resource to replace existing petroleum-based products.

REFERENCES

Agarwal, A. S., Rode, E., Sridhar, N., & Hill, D. (2022). Conversion of CO_2 to value added chemicals: Opportunities and challenges. In *Handbook of Climate Change Mitigation and Adaptation*. 1585–1623. https://doi.org/10.1007/978-1-4614-6431-0

Alper, E., & Yuksel Orhan, O. (2017). CO_2 utilization: Developments in conversion processes. *Petroleum, 3*(1), 109–126. https://doi.org/10.1016/j.petlm.2016.11.003

Anand, A., Raghuvanshi, S., & Gupta, S. (2020). Trends in carbon dioxide (CO_2) fixation by microbial cultivations. *Current Sustainable/Renewable Energy Reports, 7*(2), 40–47. https://doi.org/10.1007/s40 518-020-00149-1

Appel, A. M., Bercaw, J. E., Bocarsly, A. B., Dobbek, H., Dubois, D. L., Dupuis, M., Ferry, J. G., Fujita, E., Hille, R., Kenis, P. J. A., Kerfeld, C. A., Morris, R. H., Peden, C. H. F., Portis, A. R., Ragsdale, S. W., Rauchfuss, T. B., Reek, J. N. H., Seefeldt, L. C., Thauer, R. K., & Waldrop, G. L. (2013). Frontiers, opportunities, and challenges in biochemical and chemical catalysis of CO2 fixation. *Chemical Reviews, 113*(8), 6621–6658. https://doi.org/10.1021/cr300463y

Artz, J., Müller, T. E., Thenert, K., Kleinekorte, J., Meys, R., Sternberg, A., Bardow, A., & Leitner, W. (2018). Sustainable conversion of carbon dioxide: An integrated review of catalysis and life cycle assessment. *Chemical Reviews, 118*(2), 434–504. https://doi.org/10.1021/acs.chemrev.7b00435

Bao, B., Riordon, J., Mostowfi, F., & Sinton, D. (2017). Microfluidic and nanofluidic phase behaviour characterization for industrial CO2, oil and gas. *Lab on a Chip, 17*(16), 2740–2759. https://doi.org/10.1039/c7lc00301c

Blamey, J., Anthony, E. J., Wang, J., & Fennell, P. S. (2010). The calcium looping cycle for large-scale CO2 capture. *Progress in Energy and Combustion Science, 36*(2), 260–279. https://doi.org/10.1016/j.pecs.2009.10.001

Bruhn, T., Naims, H., & Olfe-Kräutlein, B. (2016). Separating the debate on CO2 utilisation from carbon capture and storage. *Environmental Science and Policy, 60*, 38–43. https://doi.org/10.1016/j.env sci.2016.03.001

Corma, A., & Garcia, H. (2013). Photocatalytic reduction of CO2 for fuel production: Possibilities and challenges. *Journal of Catalysis, 308*, 168–175. https://doi.org/10.1016/J.JCAT.2013.06.008

Cuéllar-Franca, R. M., & Azapagic, A. (2015). Carbon capture, storage and utilisation technologies: A critical analysis and comparison of their life cycle environmental impacts. *Journal of CO2 Utilization, 9*, 82–102. https://doi.org/10.1016/j.jcou.2014.12.001

Darensbourg, D. J., & Yeung, A. D. (2014). A concise review of computational studies of the carbon dioxide-epoxide copolymerization reactions. *Polymer Chemistry, 5*(13), 3949–3962. https://doi.org/10.1039/c4p y00299g

Das, S., Pérez-Ramírez, J., Gong, J., Dewangan, N., Hidajat, K., Gates, B. C., & Kawi, S. (2020). Core-shell structured catalysts for thermocatalytic, photocatalytic, and electrocatalytic conversion of CO2. *Chemical Society Reviews, 49*(10), 2937–3004. https://doi.org/10.1039/c9cs00713j

Desport, L., & Selosse, S. (2022). Perspectives of CO2 utilization as a negative emission technology. *Sustainable Energy Technologies and Assessments, 53*(PC), 102623. https://doi.org/10.1016/j.seta.2022.102623

Galadima, A., Garba, Z. N., Ibrahim, B. M., Almustapha, M. N., Leke, L., & Adam, I. K. (2011). Biofuels production in Nigeria: The policy and public opinions. *Journal of Sustainable Development, 4*(4), 22–31. https://doi.org/10.5539/jsd.v4n4p22

Galina, N. R., Arce, G. L. A. F., & Ávila, I. (2019). Evolution of carbon capture and storage by mineral carbonation: Data analysis and relevance of the theme. *Minerals Engineering, 142*(August), 105879. https://doi.org/10.1016/j.mineng.2019.105879

Gao, Y., Qian, K., Xu, B., Li, Z., Zheng, J., Zhao, S., Ding, F., Sun, Y., & Xu, Z. (2020). Recent advances in visible-light-driven conversion of CO2 by photocatalysts into fuels or value-added chemicals. *Carbon Resources Conversion, 3*(November 2019), 46–59. https://doi.org/10.1016/j.crcon.2020.02.003

Hori, Y., Wakebe, H. H. I., Tsukamoto, T., & Koga, O. (1994). Electrocatalytic process of CO selectivity in electrochemical reduction of CO_2 at metal electrodes in aqueous media. *Electrochimica Acta, 39*(11–12), 1833–1839.

Hu, B., Guild, C., & Suib, S. L. (2013). Thermal, electrochemical, and photochemical conversion of CO2 to fuels and value-added products. *Journal of CO2 Utilization, 1*, 18–27. https://doi.org/10.1016/j.jcou.2013.03.004

Jhong, H. R. M., Ma, S., & Kenis, P. J. (2013). Electrochemical conversion of CO2 to useful chemicals: Current status, remaining challenges, and future opportunities. *Current Opinion in Chemical Engineering, 2*(2), 191–199. https://doi.org/10.1016/j.coche.2013.03.005

Kho, E. T., Tan, T. H., Lovell, E., Wong, R. J., Scott, J., & Amal, R. (2017). A review on photo-thermal catalytic conversion of carbon dioxide. *Green Energy and Environment, 2*(3), 204–217. https://doi.org/10.1016/j.gee.2017.06.003

Kumar, B., Llorente, M., Froehlich, J., Dang, T., Sathrum, A., & Kubiak, C. P. (2012). Photochemical and photoelectrochemical reduction of CO 2. *Annual Review of Physical Chemistry, 63*, 541–569. https://doi.org/10.1146/annurev-physchem-032511-143759

Li, P., Pan, S. Y., Pei, S., Lin, Y. J., & Chiang, P. C. (2016). Challenges and perspectives on carbon fixation and utilization technologies: An overview. *Aerosol and Air Quality Research, 16*(6), 1327–1344. https://doi.org/10.4209/aaqr.2015.12.0698

Lim, R. J., Xie, M., Sk, M. A., Lee, J. M., Fisher, A., Wang, X., & Lim, K. H. (2014). A review on the electrochemical reduction of CO_2 in fuel cells, metal electrodes and molecular catalysts. *Catalysis Today, 233*, 169–180. https://doi.org/10.1016/j.cattod.2013.11.037

Lovley, D. R., & Nevin, K. P. (2013). Electrobiocommodities: Powering microbial production of fuels and commodity chemicals from carbon dioxide with electricity. *Current Opinion in Biotechnology, 24*(3), 385–390. https://doi.org/10.1016/j.copbio.2013.02.012

Maniarasu, R., Rathore, S. K., & Murugan, S. (2023). Biomass-based activated carbon for CO_2 adsorption–A review. *Energy and Environment. 34*(5), 1674–1721. https://doi.org/10.1177/0958305X221093465

Mathimani, T., Baldinelli, A., Rajendran, K., Prabakar, D., Matheswaran, M., Pieter van Leeuwen, R., & Pugazhendhi, A. (2019). Review on cultivation and thermochemical conversion of microalgae to fuels and chemicals: Process evaluation and knowledge gaps. *Journal of Cleaner Production, 208*, 1053–1064. https://doi.org/10.1016/j.jclepro.2018.10.096

Michalak, I., & Chojnacka, K. (2015). Algae as production systems of bioactive compounds. *Engineering in Life Sciences, 15*(2), 160–176. https://doi.org/10.1002/elsc.201400191

Munkejord, S. T., Hammer, M., & Løvseth, S. W. (2016). CO_2 transport: Data and models — A review. *Applied Energy, 169*, 499–523. https://doi.org/10.1016/j.apenergy.2016.01.100

Muthuraj, R., & Mekonnen, T. (2018). Recent progress in carbon dioxide (CO_2) as feedstock for sustainable materials development: Co-polymers and polymer blends. *Polymer, 145*, 348–373. https://doi.org/10.1016/j.polymer.2018.04.078

National Academies of Sciences, Engineering, and Medicine. (2019). Gaseous Carbon Waste Streams Utilization: Status and Research Needs. Washington, DC: The National Academies Press. https://doi.org/10.17226/25232

North, M. (2015). What is CO_2? Thermodynamics, Basic Reactions and Physical Chemistry. In *Carbon Dioxide Utilisation: Closing the Carbon Cycle: First Edition*. Elsevier B.V. https://doi.org/10.1016/B978-0-444-62746-9.00001-3

Olajire, A. A. (2013). Valorization of greenhouse carbon dioxide emissions into value-added products by catalytic processes. *Journal of CO₂ Utilization, 3–4*, 74–92. https://doi.org/10.1016/j.jcou.2013.10.004

Pal, T. K., De, D., & Bharadwaj, P. K. (2020). Metal–organic frameworks for the chemical fixation of CO_2 into cyclic carbonates. *Coordination Chemistry Reviews, 408*, 213173. https://doi.org/10.1016/j.ccr.2019.213173

Quadrelli, E. A., Armstrong, K., & Styring, P. (2015). Potential CO_2 Utilisation Contributions to a More Carbon-Sober Future: A 2050 Vision. In *Carbon Dioxide Utilisation: Closing the Carbon Cycle: First Edition*. Elsevier B.V. https://doi.org/10.1016/B978-0-444-62746-9.00016-5

Rahman, F. A., Aziz, M. M. A., Saidur, R., Bakar, W. A. W. A., Hainin, M. R., Putrajaya, R., & Hassan, N. A. (2017). Pollution to solution: Capture and sequestration of carbon dioxide (CO_2) and its utilization as a renewable energy source for a sustainable future. *Renewable and Sustainable Energy Reviews*, *71*(January), 112–126. https://doi.org/10.1016/j.rser.2017.01.011

Roy, P. S., Song, J., Kim, K., Park, C. S., & Raju, A. S. K. (2018). CO_2 conversion to syngas through the steam-biogas reforming process. *Journal of CO2 Utilization*, *25*(February), 275–282. https://doi.org/10.1016/j.jcou.2018.04.013

Roy, S., Cherevotan, A., & Peter, S. C. (2018). Thermochemical CO_2 hydrogenation to single carbon products: Scientific and technological challenges. *ACS Energy Letters*, *3*(8), 1938–1966. https://doi.org/10.1021/acsenergylett.8b00740

IPCC. (2014). Climate change 2014: Mitigation of climate change; Working Group III contribution to the Fifth Assessment Report of the Intergovernmental Panel on Climate Change, Cambridge University Press, New York, NY.

Saravanan, A., Senthil kumar, P., Vo, D. V. N., Jeevanantham, S., Bhuvaneswari, V., Anantha Narayanan, V., Yaashikaa, P. R., Swetha, S., & Reshma, B. (2021). A comprehensive review on different approaches for CO_2 utilization and conversion pathways. *Chemical Engineering Science*, *236*, 116515. https://doi.org/10.1016/j.ces.2021.116515

Shi, J., Jiang, Y., Jiang, Z., Wang, X., Wang, X., Zhang, S., Han, P., & Yang, C. (2015). Enzymatic conversion of carbon dioxide. *Chemical Society Reviews*, *44*(17), 5981–6000. https://doi.org/10.1039/c5cs00182j

Sick, V., Stokes, G., & Mason, F. C. (2022). CO_2 utilization and market size projection for CO_2-treated construction materials. *Frontiers in Climate*, *4*(May), 1–5. https://doi.org/10.3389/fclim.2022.878756

Sinehbaghizadeh, S., Saptoro, A., & Mohammadi, A. H. (2022). CO_2 hydrate properties and applications: A state of the art. *Progress in Energy and Combustion Science*, *93*(July), 101026. https://doi.org/10.1016/j.pecs.2022.101026

Smith, W. A., Burdyny, T., Vermaas, D. A., & Geerlings, H. (2019). Pathways to industrial-scale fuel out of thin air from CO_2 electrolysis. *Joule*, *3*(8), 1822–1834. https://doi.org/10.1016/j.joule.2019.07.009

Solomon, S., Plattner, G. K., Knutti, R., & Friedlingstein, P. (2009). Irreversible climate change due to carbon dioxide emissions. *Proceedings of the National Academy of Sciences of the United States of America*, *106*(6), 1704–1709. https://doi.org/10.1073/pnas.0812721106

Span, R., Gernert, J., & Jäger, A. (2013). Accurate thermodynamic-property models for CO_2-rich mixtures. *Energy Procedia*, *37*, 2914–2922. https://doi.org/10.1016/j.egypro.2013.06.177

Steiger, M. G., Mattanovich, D., & Sauer, M. (2017). Microbial organic acid production as carbon dioxide sink. *FEMS Microbiology Letters*, *364*(21), 1–4. https://doi.org/10.1093/femsle/fnx212

Sternberg, A., Jens, C. M., & Bardow, A. (2017). Life cycle assessment of CO_2-based C1-chemicals. *Green Chemistry*, *19*(9), 2244–2259. https://doi.org/10.1039/c6gc02852g

Styring, P., Quadrelli, E. A., & Armstrong, K. (2014). Carbon Dioxide Utilisation: Closing the Carbon Cycle: First Edition. *Carbon Dioxide Utilisation: Closing the Carbon Cycle: First Edition*, 1–311. https://doi.org/10.1016/C2012-0-02814-1

Truong, C. C., & Mishra, D. K. (2021). Catalyst-free fixation of carbon dioxide into value-added chemicals: A review. *Environmental Chemistry Letters*, *19*(2), 911–940. https://doi.org/10.1007/s10311-020-01121-7

Van Der Giesen, C., Kleijn, R., & Kramer, G. J. (2014). Energy and climate impacts of producing synthetic hydrocarbon fuels from CO_2. *Environmental Science and Technology*, *48*(12), 7111–7121. https://doi.org/10.1021/es500191g

Venkata Narayana Reddy, B., Venkata Shiva Reddy, B., Suresh Kumar, N., & Chandra Babu Naidu, K. (2022). Materials for conversion of CO_2. *Biointerface Research in Applied Chemistry*, *12*(1), 486–497. https://doi.org/10.33263/BRIAC121.486497

Vieira, É. D., Da Graça Stupiello Andrietta, M., & Andrietta, S. R. (2013). Yeast biomass production: A new approach in glucose-limited feeding strategy. *Brazilian Journal of Microbiology*, *44*(2), 551–558. https://doi.org/10.1590/S1517-83822013000200035

Von Der Assen, N., Jung, J., & Bardow, A. (2013). Life-cycle assessment of carbon dioxide capture and utilization: Avoiding the pitfalls. *Energy and Environmental Science*, *6*(9), 2721–2734. https://doi.org/10.1039/c3ee41151f

Von Der Assen, N., Voll, P., Peters, M., & Bardow, A. (2014). Life cycle assessment of CO_2 capture and utilization: A tutorial review. *Chemical Society Reviews*, *43*(23), 7982–7994. https://doi.org/10.1039/c3cs60373c

Vu, T. T. N., Desgagnés, A., & Iliuta, M. C. (2021). Efficient approaches to overcome challenges in material development for conventional and intensified CO_2 catalytic hydrogenation to CO, methanol, and DME. *Applied Catalysis A: General, 617*(January), 118119 https://doi.org/10.1016/j.apcata.2021.118119

Wang, J., Jia, C. S., Li, C. J., Peng, X. L., Zhang, L. H., & Liu, J. Y. (2019). Thermodynamic properties for carbon dioxide. *ACS Omega, 4*(21), 19193–19198. https://doi.org/10.1021/acsomega.9b02488

Wilcox, J. (2012). *Carbon capture textbook.* Springer Science & Business Media.

Xiaoding, X., & Moulijn, J. A. (1996). Mitigation of CO_2 by chemical conversion: Plausible chemical reactions and promising products. *Energy & Fuels, 10*(2), 305–325.

Yaashikaa, P. R., Senthil Kumar, P., Varjani, S. J., & Saravanan, A. (2019). A review on photochemical, biochemical and electrochemical transformation of CO_2 into value-added products. *Journal of CO_2 Utilization, 33*(March), 131–147. https://doi.org/10.1016/j.jcou.2019.05.017

14 Agro-Residues' Waste to Wealth for a Circular Economy and Sustainable Development

Preeti Chaturvedi Bhargava, Neha Kamal,
Deepshi Chaurasia, Shivani Singh and Anuradha Singh

14.1 INTRODUCTION

The unprecedented developmental activities have set up constraints on the ecological health of the environment leading to serious climate change issues and health hazards and sustainable management of resources has become a necessity. Besides, the surge in global population is proportional to cause exhaustion and pollution of natural resources. Various measures havebeen taken in order to combat pollution and other related environmental complications. Agro-residues contribute to a substantial amount of waste generation that is usually burnt leading to greenhouse gas emissions, global warming, smog formation, and depletion of land and soil quality (Lohan et al., 2018). These residues being abundant can help economy in the form of sources of bioactive compounds, the sustainable management of the same can help in extracting benefits (in form of energy and other value-added products) (Pattanaik et al., 2019). Additionally, the exploitation of agro-residues as energy sources can lead to energy self-sufficiency and help promote circular economy. The concept of 'waste to wealth' fits perfectly for agro-residues as its sensible utilization can render benefits of an additional source of income, reduces dependency on fossil fuels and emission of greenhouse gases (GHGs), provides high resource efficiency without requiring any additional land, cost or input, helps in enhancing soil quality, carbon sequestration, wastewater treatment and assists in phytoremediation (Europe, 2019).

Keeping in view the potential benefits of agro-residues the current chapter is mainly focused towards a detailed description of this energy-rich biomass, composition and valorization. Insights about the different processing routes that can be undertaken to convert these residues into value-added products has also been discussed. The work also highlights the contribution of agro-residues in attaining circular economy, its techno-economic aspects, with the future perspectives and challenges faced in its conversion.

14.2 AGRO-RESIDUES: DEFINITION, CLASSIFICATION AND COMPOSITION

Agro-residues (or agricultural residues) basically refer to the solid or liquid remnants produced after harvesting and processing of fruits, vegetables and crops such as straw, stalks, leaves, root branches, twigs, trimmings correspond to primary biomass residue. On the other hand, secondary biomass agro-residues generated from processing of the main substrates (fruits/vegetables/crops) include husks, bagasse, molasses, sawdust, corncobs, pomace, etc. (Figure 14.1) (Singh nee' Nigam P 2009; Santana-Méridas et al., 2012; Mohammed et al., 2018).

DOI: 10.1201/9781003327646-14

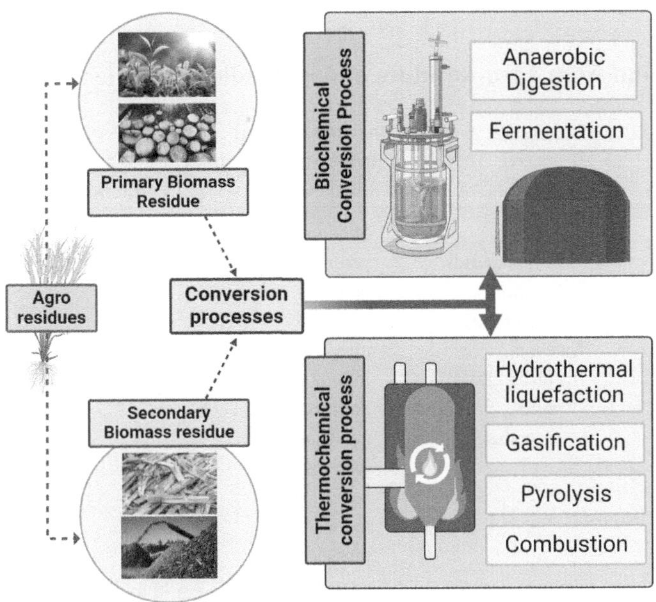

FIGURE 14.1 Agro-residues: its classification and processing methods.

These carbon-rich materials function as potent candidates for production of sustainable and renewable energy resource. The composition of agro-residues contains cellulose (40–50%), hemicellulose (25–35%) and lignin (15–20%) contributes to its complex rigid structure that is majorly explored for its sustainable utilization (Ginni et al., 2021). Owing to the recalcitrant property of these components, the agro-residues are primarily subjected to various pre-treatment methods to enhance its application to achieve the objective of waste to wealth. The different pre-treatment and processing methods are further explained in section 14.3. Generally, agro-residues are mainly employed either as fertilizers in fields or as fodder to feed the livestock whereas substantial quantities of these residues are burnt (i) to avoid the cost and effort behind its transport for judicious management, and/or (ii) to clear off the land for consecutive sowing of other crops (Wang et al., 2013; Yadav and Singh, 2011; Bhuvaneshwari et al., 2019; Shiet al., 2014). Burning of these residues causes detrimental effects on the environment such as deterioration of soil quality and increases in air pollution (Cassou, 2018; Tripathi et al. 2019). The methods for extraction of bioactive compound from agro-residues and its medicinal values are described in Table 14.1.

14.3 PROCESSING OF AGRO-RESIDUES: CONCEPT OF WASTE MINIMIZATION AND RECYCLING

14.3.1 Biochemical Conversion Process

The agro-residues exhibits favourable features for bioprocessing like biological compatibility, stereo uniformity and hydrophobicity, is essential to obtain value-added products such as fibres, fuels, chemicals, etc. The biochemical conversion process of agro-residues is mainly targeted towards disarrangement of hemicellulose using techniques like anaerobic digestion and fermentation to facilitate easy availability of cellulose by decreasing its crystallinity and polymerization ability (Baruah et al., 2018). The hemicellulose forms a network between cellulose and lignin by forming interlinkage between the components (Ginni et al., 2021).

Anaerobic digestion process involves action of microorganisms alone or in combination with enzymes in anaerobic conditions to achieve successful conversion of moist agro-residues into

TABLE 14.1
Bioactive Compound from Agro-Residues and its Medicinal Values

S. No.	Agro-Residues	Bioactive Compound	Pharmacological Attributes/Medicinal Value	Extraction Method	References
1.	Citrus peel	Limonene, Polyphenoids, limonoids, carotenoids	Higher antioxidant potential, cytotoxic potential	GC-MS	Rossi et al., 2020; Saini et al., 2019
2.	Grape seed	Catechin, epicatechin gallate, flavonoids, proanthocyanidin	Antioxidation, anti-inflammation, anticancer, neuroprotection, lowering lipid, bacteriostatic, reduced blood pressure	-	Chen et al., 2020
3.	Grape peel	Phenolic compound, Flavonoids	Antioxidant activity	Solid liquid extraction (CE), Ultrasound assisted extraction (UAE)	Ruales et al., 2017; Garcia et al., 2015
4.	Pineapple husk	Phenolic compound	Antioxidant activity	Microwave-assisted extraction (MAE)	Alias et al., 2017
5.	Acerola waste	Flavonoids and anthocyanins	Antioxidant activity	Ultrasound assisted extraction (UAE)	Rezende et al., 2017
6.	Papaya seed	Sulforaphane and phenolic compound	Antioxidant activity	High hydrostatic pressure (HHPE), Ultrasound(UAE), Conventional extraction (CE)	Briones-Labarco et al., 2015
7.	Carrot pulp	Phenolic compound	Antioxidant activity, antimicrobial and antimutagenic activity	Ultrasonic bath, HPLC-DAD-ESI-MS	Vodnar et al., 2017
8.	Cardamon, radish, turnip aerial and leaves	Phenolic compound	Antioxidant activity, antibacterial and hepatotoxic activity	HPLC-DAD/ESI-MS	Chihoub et al., 2019
9.	Corn (bran, germ, germ meal), wheat (bran, shorts, germ), rice (bran, germ, husk)	Total carotenoids, Total phenolic content	Antioxidant activity,	Folin-Ciocalteau reagent method	Smuda et al., 2018
10.	Defatted rice bran	Total monomeric phenolics, Total phenolic content	Antioxidant activity,	HPLC	Zhao et al., 2018
11.	Rice husk	Polyphenols, total phenolics, total flavonoids, caffeic acid, ferulic acid, p-hydroxybenzoic acid, p-coumaric acid	Antioxidant activity, cellular antioxidant activity, antiproliferative activity	HPLC chromatogram	Gao et al., 2018

TABLE 14.1 (Continued)
Bioactive Compound from Agro-Residues and its Medicinal Values

S. No.	Agro-Residues	Bioactive Compound	Pharmacological Attributes/Medicinal Value	Extraction Method	References
12.	Wheat bran	Total phenolic content, Total alkylresorcinols content	Arabinoxylans	Folin-Ciocalteau reagent method	Zhao et al., 2017
13.	Grape seed extract	Phenolic compound, epigallocatechin gallate, epicatechin, proanthocyanidins B_4, flavonols, phenolic acid, monomeric catechin	Highly antioxidants, anti-oxidative, anti-inflammatory activity, anti-cancer	2D-LC-MS, HPLC-FLD, HPLC	Kuhnert et al., 2015

products like biogas with high energy content and bioethanol, etc. Digestate, a by-product during the process, is utilized in agricultural fields as manure (Mohammed et al., 2018; Ramesh et al., 2021, Khoshnevisan et al., 2019). On the other side, fermentation is a popular conventional technique of bioethanol production from residues rich in sugar content. The method is further assisted by microbes like yeast and bacteria for production of alcohol and acid (Ramesh et al., 2019; Demirbas 2009).

14.3.2 THERMOCHEMICAL CONVERSION PROCESS

Unlike biochemical conversion process, the thermo chemical conversion operates at extreme reactions conditions involving high temperature and pressure with/without oxygen to break down the agro-wastes into simpler products to obtain heat energy, biofuels and biochemicals (Ullah et al., 2015). The method includes processes like combustion, pyrolysis, gasification and hydrothermal liquefaction. Owing to the low-bulk density of agro-residues, the residues are subjected to pre-treatment steps like drying and size reduction prior to degradation via thermochemical route. Combustion, pyrolysis and gasification are best suited for biomass with low moisture content, whereas hydrothermal liquefaction is found suitable for high moisture feedstocks (Ramesh et al., 2021).

Combustion is a traditional thermochemical method that encompasses direct burning of agro-residues in the presence of excessive air/oxygen to produce energy in the form of heat for power generation. The method is regulated by factors like biomass type, its composition and plant age (Ramesh et al., 2019). The biomass for combustion should have low ash content with a set up for ash removal during the combustion process. Pyrolysis involves thermo-chemical decomposition of organic feedstock at high temperatures ranging from 300–800°C in complete anaerobic conditions. The end products generated in the process arebiochar (solid), bio-oil (liquid) and biogas (gas). The proportion of these products formed depends on factors, such as type of substrate, heating rate, temperature and residence time (Chen et al., 2017; Mettu et al., 2020). Bio-oil, biogas and biochar serve as important sources of renewable energy and can offer environmental applications.

Gasification involves formation of gaseous biofuel through thermal cracking and partial combustion of agro-residues between 600–900°C in the presence of gasifying agent under regulated aerobic conditions (Hosseini et al. 2012; Shahbaz et al., 2016; Ramesh et al., 2021). The gasification process is induced by factors like feedstock type and size, its moisture content, the gasifying agent, temperature and catalyst (Mettu et al., 2020).The gaseous biofuel or the syngas produced (consisting of a mixture of CO, H2, CH4, CO2, N2 and light hydrocarbons) is explored as fuel in burners or gas engines for thermal applications and power generation. The syngas is also transformed into chemicals through catalytic processes (Ibarra-Gonzalez and Rong, 2019).

Another thermochemical conversion, hydrothermal liquefaction functions by converting agro-residues into liquid fuels at high temperatures (523–647K) and pressure (4–22 MPa) in an environment to degrade the recalcitrant biopolymeric structure into liquid component, the biocrude (Elliott et al., 2015; He et al., 2018; Zhang et al., 2018). The formation of biocrude involves hydrolysis, degradation, decarboxylation, repolymerization, etc. (Leng et al., 2020). The process is applied for biomass rich in moisture and offers advantages such as high-energy recovery rate and the biocrude formed can be used in place of conventional fossil fuels (Hao et al., 2021).

14.4 VALORIZATION OF AGRO-RESIDUES: SUSTAINABLE PRODUCTION OF VALUE-ADDED PRODUCTS

14.4.1 ENZYMES

The catalytic potential of enzymes has gained special attention in pharma, oil chemical industries (hydrolysis of fat and oil, bio-detergent production), food products (modified cheese, flavour enhancement), agriculture (Salim et al., 2017; Savino et al., 2021). Agro-residues such as sugarcane bagasse, corncob, rice bran, wheat bran, wheat straw has been widely investigated for the production of enzymes (Ravindran et al., 2018; Bharathiraja et al., 2017). Studies have reported significant production of enzymes like proteases, laccases, lipases, amylases through microbial assisted submerged fermentation (SmF) or solid-state fermentation (SSF) (Sadh et al., 2018) (Table 14.2).

SSF is preferred more than SmF for the production of enzymes from agro-residues due to features like high yield, low energy requirement and simpler extraction process to obtain enzymes (de Castro et al., 2015; Abu Yazid et al., 2017). Optimal production of pectinases has been reported

TABLE 14.2
Types of Agro-Residues, Enzyme, Techniques and its Application in Various Fields

S. No.	Enzyme	Agro-Residues	Application	Techniques	References
1.	Cellulose	Groundnut shells	Biofuel production	SSF	Kakde and Aithal, 2020
		Ragi husk	Cellulose hydrolysis	SSF	Ishchi and Sibi, 2019
		Soybean husk	Lignin hydrolysis	SSF	Salazar et al., 2019
2.	Pectinase	Orange peel	Food and beverage industries	SSF	Kaur and Gupta, 2017
		Wheat bran	-	SSF	Sethi et al., 2016
3.	Laccase	Rice straw	Dye decolourization	SSF	Ali et al., 2020
		Wheat straw, rice husk	Waste reduction	SSF	Chhaya and Gupte, 2019
		Millet husk	Pesticides decontamination	SSF	Srinivasan et al., 2019
4.	Lipase	Mango waste	Economical production	SmF	Pereira et al., 2019
		Rice bran	Biofuel generation	SSF	Putri et al., 2020
5.	Protease	Orange peels	Production for white soft cheese	SSF	Wehaidy et al., 2020
		Pulse four, vegetable pills	Detergent	SSF	Thakrar et al., 2020
6.	Xylanase	Peach palm	Food industry	SSF	Carvalho et al., 2018

from orange peel using fungus *Aspergillus niger*through SSF (Mahmoodi et al., 2019). Similarly, apple pomace has also been reported for the production of pectinases using a co-culture of *Bacillus species* through SmF (Kuvvet et al., 2019). Different agro-residues like sugarcane bagasse, orange peel waste and tobacco have been found as a potent carbon source for the production of cellulase enzyme by different microbial species (Srivastava et al., 2017; Salomão et al., 2019; Buntić et al., 2019).

14.4.2 BIOFUELS

The requirement as a renewable fuel source has become a necessity to meet its unstoppable demand in present and future scenarios. Biofuel with its unique features such as being environment friendly, energy efficient, low cost, renewability, etc., provides the ultimate solution to meet the growing energy requirement. The lignocellulosic content of agro-waste has been vastly studied for generation of high-energy biofuels such as bioethanol, biodiesel, biogas, etc. After undergoing transformation via a biochemical or thermochemical route, the feedstocks are processed for production of biofuels. In a study by Chintagunta et al. (2017), the holocellulose content (60–80%) of the pineapple leaves post-harvesting, was exploited for bioethanol production through saccharification and fermentation using yeast yielding 7.12% v/v bioethanol. Rice bran has been used for bioethanol production through fermentation by yeast under optimized conditions (*Saccharomyces cerevisiae* MTCC 4780) (Agrawal et al., 2019). Similarly in a study by Onuguh et al. (2022) the potential of starch containing agro-wastes viz. cassava peels, yam peels and potato peel was compared for bioethanol production. Biodiesel, the alkyl esters of fatty acids, are mainly produced from agro-wastes rich in residual oil. The bio-oil in spent coffee powders may be involved for the production of biodiesel through transesterification (Sakuragi et al., 2016). In another study, groundnut shell has been investigated for biodiesel production through enzyme-mediated transesterification (Udeh et al., 2018). Various other agro-residues like switch grass, rice husk, rice and wheat straw, jatropha oil, castor oil, grasses, etc., have also been reported as potential candidates for biodiesel generation (Ginni et al., 2021). Another popular biofuel, biogas mainly produced from anaerobic digestion is a blended product consisting of CH4, CO2, H2S, moisture and siloxanes (Sindhu et al., 2019). The production cost of biogas is highly economical as compared to other fuels (Vasudevan et al., 2020). Spoilt fruit wastes and corncob was used as feedstock for the production of biogas with less corrosive composites after undergoing pre-treatment and optimization (Oladejo et al., 2022).

14.4.3 BIOCHAR

The biochar obtained as result of pyrolysis is rich in carbon and is a highly stable product. A conversion of biomass into biochar utilizes the complete potential of the former as an energy provider and offers sustainable management (Chaurasia et al., 2022). The role of biochar inagriculture to increase fertility and media for waste water reclamation has been studied extensibility. Ahmad et al. (2022) reported biochar prepared from sewage sludge augmented with soil impacted soil nutrients, microbial abundance and plant growth. The addition of biochar not only improves the soil quality but also promoted the growth of beneficial rhizospheric bacteria. The properties of biochar like large surface area, presence of functional groups and its porous structure makes it an ideal adsorbent of different organic compounds- dyes (Ahmad et al., 2020; Khan et al., 2020), antibiotics, etc., furthermore, the biochar may be modified to enhance its performance efficiency (Shukla et al., 2021). Owing to the fruitful benefits of biochar such as carbon sequestration, removal of organic/inorganic pollutants, waste water treatment, production of biofuels, soil quality enhancement, etc., it can be successfully implemented as an approach towards attaining the waste to wealth concept (Neogi et al., 2022; Khan et al., 2021) (Table 14.3).

TABLE 14.3
Types of Agro-Residues, Biochar and its Application

S. No.	Agro-Residues Derived Biochar	Techniques	Application	References
1.	Corn cob, coffee husk	XRD, BET, SEM, XRF	As(V) and Pb(II) removal	Cruz et al., 2020
2.	Acacia (*Acacia auriculiformis*)	BET, SEM,	Congo red dye removal	Nguyen et al., 2021a
3.	Wattle bark, mimosa, coffee husk	FTIR, SEM, BET	Methyl orange adsorption	Nguyen et al., 2021b
4.	Walnut shells	BET	Ni(II) ions absorbent	Georgieva et al., 2020
5.	Raw jujube seeds	BET	Zn(II), Pb(II)	Gayathri et al., 2021
6.	Rice husk	FTIR, SEM-EDX	Methylene blue dye removal	Ahmad et al., 2020
7.	Rice husk	FTIR, SEM, SEM-EDX	Congo red dye removal	Khan et al., 2020
8.	Wheat straw	-	Enhance soil pH	Huang et al., 2018
9.	Alfalfa hays	FTIR, XRD	Removal of tetracycline	Jang and Kan, 2019
11.	Rice husk	GC-MS	Bio-oil production	Cai et al., 2018
12.	Seaweed biomass	SEM, TEM, XRD, XPS, FTIR	Removal of cationic dye	Zhou et al., 2022

14.4.4 BIOFERTILIZERS

Biofertilizers contains live or latent cells of microorganisms that solubilizes phosphates, fixes nitrogen, and when applied to plants or a seed, increases its growth (Kumar et al., 2022). The agro-residues are used as a source of nutrition by microorganisms, producing commercially important chemicals and compost, which can be used as biofertilizers (Bibi et al., 2022). Although bio-fertilizers cannot completely overcome the chemical fertilizers, butcan support a more sustainable agricultural system (Okur et al., 2018). Agro-waste, such as crop residues, stalks, animal faeces, roots, husks, and shells, are used as biofertilizers to improve food production. When applied to plants, bio-fertilizers include living microorganisms that enhance development by giving primary nutrients. While carrying out their normal operations, microorganisms stimulate nitrogen fixation, phosphorous solubilization, and the formation of growth-promoting compounds. They all are critical for agricultural plant growth (Lamina et al., 2022). Biofertilizers are inexpensive, and are classified according to their function. These include *Rhizobium, Azospirillum, Cyanobacteria, Azolla* and *Phosphate solubilizing bacteria (PSB)*. These biofertilizers are beneficial and do not cause any environmental risk as with chemical fertilizers. The main limitation to biofertilizers is the lack of awareness among farmers who are content with traditional agricultural operations and refuse to change their methods (Bhattacharjee et al., 2014).

14.4.4.1 Classification of Biofertilizers

Biofertilizers are made from various microorganisms and their interactions with agricultural plants. Based on their nature and purpose, they can be classified into several categories.

> *Rhizobium*: *Rhizobium* is a bacterium that colonizes legume roots and fixes atmospheric nitrogen in a symbiotic relationship. It has various physiology and, shapes from free-living to nodular bacteroids. They are the most efficient biofertilizer for fixing of nitrogen (Rohela et al., 2022).
>
> *Azotobacter*: It has free nitrogen-fixing capabilities and potential in agriculture. *Azotobacter* species are gram-negative that live in an open environment. It can produce vitamins and plant hormones like thiamine and riboflavin and aid in nitrogen fixation (Kumar et al., 2022).

Azospirillum: *Azospirillum lipoferum* and *Azospirillum brasilense* are common graminaceous plant intercellular inhabitants in the soil, rhizosphere, and root tissue. Graminaceous plants have a symbiotic connection with growth-promoters, nitrogen fixation, disease resistance, and drought tolerance.

Cyanobacteria: Rice agriculture has benefited from the use of cyanobacteria (blue-green algae). It was formerly widely promoted as a rice crop biofertilizer, but it is no longer attracting the attention of rice growers across India. Under the optimum conditions, the advantages of algalization might be as high as 20–30 kg N/ha (Shamina et al., 2022).

Azolla: *Azolla* is increasingly used for sustainable animal feed production, despite its traditional usage as a bio-fertilizer for wetland paddy (due to its capacity to fix nitrogen). Proteins, vital amino acids, vitamins, and minerals occur in *Azolla*. It increases rice yield by 30–60 kg N/ha (Prabakaran et al., 2022).

Phosphate solubilizing microorganisms (PSM): Phosphate solubilizing bacteria acts as a biofertilizer, converting phosphate into accessible soluble forms. It has been highlighted that the use of PSM in agriculture is an environmentally friendly strategy that can increase agricultural productivity and soil fertility. PSM protects plants from numerous diseases, promotes plant development, and aids in nitrogen fixation (Sharma et al., 2021).

Plant growth-promoting rhizobacteria (PGPR): These are rhizobacteria that help plants by promoting development traits (Kloepper et al., 1980). Studies have shown that PGPR-based biocontrol agents have a high potential for plant development and control plant diseases (Myo et al., 2019), especially under stressed growing situations (Lyu et al., 2019).*Bacillus* and *Pseudomonas* species produce phytohormones or growth regulators that induce crops to have more fine roots, increasing the absorptive surface for water and nutrient uptake (Bajracharya 2019).

14.4.4.2 Bioactive Compounds

The agro-processing business produces a lot of waste in the form of peels, kernels and pulp that causes environmental pollution by its disposal in outdoor land or municipal landfills (Sadh et al., 2018), that ideally can be used to recover phytochemicals/bioactive substances in food, cosmetics and pharmaceutical application (Lemes et al., 2022). Due to environmental concerns, these industries are increasingly obliged to find an alternative use for agro-residues. On the other hand, extracting bioactive compounds is an attractive and cost-effective approach (Pai et al., 2022). Bioactive compounds are secondary metabolites (antibiotics, mycotoxins alkaloids, food-grade colours, phenolic compounds, and plant growth hormones) (Singh et al., 2022). In recent years, several food bioactive components have been developed as pharmaceutical components (capsules, pills, gels, liquors, solutions powders and granules, etc.) containing extracts or phytochemicals that have been linked to a physiological function, either directly or indirectly (Sasidharan et al., 2011) and are termed as 'nutraceuticals'. Nutraceuticals like carotenoid lycopene, alliaceae (onion, garlic) extracts contain sulphur compounds, glucosinolate extracts; phytosterol, polyphenols like lignans, proanthocyanidins, anthocyanins, stilbenes, flavonols, hydroxycinnamates, coumarins, ellagic acid (EA), ellagitannins (ETs) and isoflavones are some of the most common phytochemicals found in nutraceuticals (Maurya et al.,2022; Rahman et al., 2022).

14.4.4.3 Organic Acids

The organic acid term is generally used in the food, beverage, cosmetics, and pharmaceutical sectors, known preservatives to control deterioration and extend shelf life (Sharma et al., 2008). Citric acid, lactic acid, acetic and gluconic acid are the common organic acids. One of the essential organic acids is citric acid, used in the food sector due to its acidic flavour and for functions like acidulation, chelation, emulsification, preservation, taste improvement and plasticizer, etc. Valorization of food and waste management for the production of organic acid has been reported by Ventorino et al. (2017).

Lactic acid (LA) is a commercially helpful hydroxycarboxylic acid used in the cosmetic, food, chemical and pharmaceutical industries, etc. LA has recently gained economic interest in manufacturing polylactic acid, an environmentally benign biodegradable polymer (Hofvendahl et al., 2000). Lignocellulosic biomass, have been promoted as cost-effective substrates for LA synthesis. Lactic acid-producing heterotrophic bacteria are required for the microbial generation of LA and a variety of growth nutrients such as vitamins, nucleotides, amino acids, fatty acids, peptides, etc. (Moldes et al., 2000).

Gluconic acid (GA) and its intermediates are primarily acidity regulators utilized as food, medicinal, dairy and hygiene additives. It is commonly found in non-alcoholic drinks, pickles, meat-based goods as a food preservative; as cleaning agents for food-processing units. GA is formed when microorganisms oxidize glucose. *A. niger* is employed in the industrial synthesis of gluconic acid from glucose. GA is prepared by degrading agro-residue into glucose and then converting it to gluconic acid using enzymes. Using agro-waste for gluconic acid production has a lot of potential (Gu et al., 2022). Similarly,acetic acid is formed by oxidizing ethanol through acetic acid bacteria, majorly deployed in vinegar preparation. For the commercial production, two techniques (1) enzymatic and (2) nonenzymatic bioconversion have been employed. Mostly yeast *S. cerevisiae* and *Zymonas mobilis*, ferment sugar into ethanol and transform it into acetic acid by *Gluconacetobacter sp.* and *Acetobacter sp.* Grape pomace, sugarcane molasses, fruit wastes and other sugar-rich agricultural residues can be used to produce ethanol at a low cost. Currently, employing a fed-batch to do both stages simultaneously is considered a higher cost–benefit (Koul et al., 2022). The valorization of agro-residues into value-added products such as enzymes, biofuels, organic acids, biochar, biofertilizers, bioactive compounds and its typesare listed in Figure 14.2.

14.5 ROLE OF AGRO-RESIDUES IN THE CIRCULAR ECONOMY

Circular economy blends bio-economy with emphasis on bio-products and bioenergy that has a balanced ideology between a productive economy and a safe environment (Tseng et al., 2021).5R values favouring the circular economy agenda are reuse, recycling, reduction, repurpose and recovery (Kirchherr et al., 2017). Iqbal and Kang (2021) argued that a Waste-to-Energy (WTE) supply chain could recycle the bio-waste for profit. The WTE supply chain is closely associated with circular economy and needs to be managed effectively from production to consumption. The practical implementation of eight task forces, including quality control (QC), R&D, economic instruments, information network, technical assistance, public–private partnership (PPP), and internal cooperation, have argued to optimize the WTE supply chain and CEC (Pan et al., 2015). The increasing energy demands and climate change issues make WTE supply chain management crucial. Therefore, a circular economy ensures that the WTE supply chain is effectively applied in agriculture and agro-based industries.

Agricultural waste such as palm tree waste was the largest contributor of agricultural waste in Asia, especially in Malaysia and Indonesia, with palm production in these two countries accounting for 84% of global palm production being exported worldwide (Ritchie and Roser, 2021). Residue recycling from the palm oil industry had restored the confidence in transforming agro-based waste into energy management, thereby supporting the 'UN sustainable development goals of energy needs'. Thus, the agro-based industry can generate new energy sources to resolve the issue of clean energy demand, and waste management systems along with reducing greenhouse gas emissions (Figure 14.3).

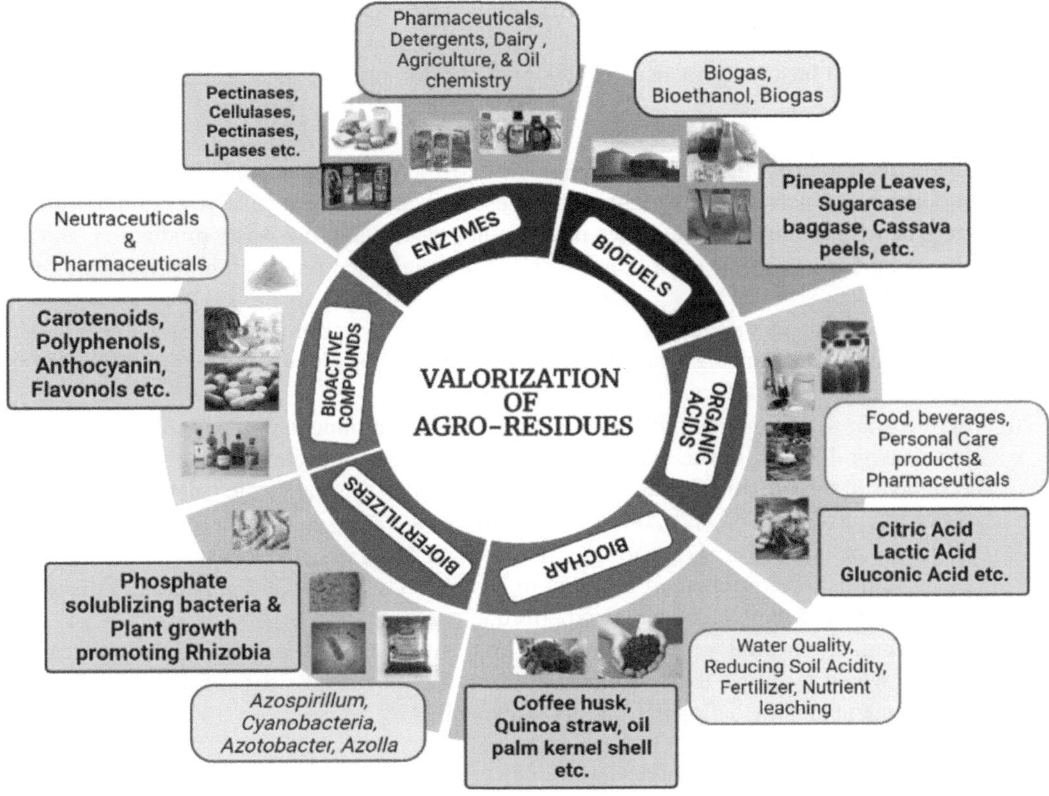

FIGURE 14.2 Conversion of agro-residues into value-added products.

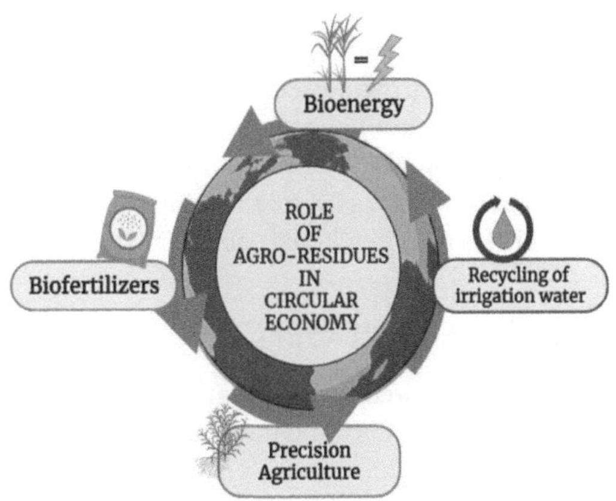

FIGURE 14.3 Role of agro-residues in the circular economy.

14.6 TECHNO-ECONOMIC ASPECTS OF AGRO-RESIDUES

The most efficient utilization of agro-residues is our generation's most crucial technological and economic problem. The agro-residues' utilization industries will need to solve this problem if they are to be effective, balanced and implemented on a large scale. The variety of characteristics that distinguish different agricultural wastes adds to the complexity of managing these leftovers in business practices. As a result, it is critical to concentrate on this process by employing certain strategies at different organizational levels (Nitsenko et al., 2019; Ostapenko et al., 2020). The incorporation of agro-residues into business practices necessitates the importance of thorough research on the following fronts:

(a) theoretical, conceptual and methodological aid for the implementation of agro-residues' production, which includes exploring the essence and concept of agro-residues' production, demonstrating the differences between non-waste and low-waste production, and categorizing agro-residues according to various criteria based on different legislations across the world.

(b) an examination of the current condition and trends in agricultural market development, involving resource-saving innovations, existing market resources, and an assessment of the agro-residues' product competitiveness;

(c) agro-residues' product output is ensured by investment, innovation, and marketing, which is an integral aspect of the technical and organizational decision-making process in agro-residues' transformation into valuable commodities. Small-scale bioenergy systems, according to Ruan et al. (2008), have various advantages over large-scale bioenergy systems, including lower operating costs, reduced feedstock and transportation costs, and better returns for consumers. Furthermore, small-scale systems are projected to make biomass application both practicable and affordable.

(d) enhancing the overall social, economic and environmental outcomes of agro-residues output, which is critical to agricultural waste to wealth business organizations' competitive development and the national economy in general.

As a result, the aim is to select the most appropriate economical, societal and ecologically resilient criteria for agro-waste output and to identify strategies to improve it. Breakthroughs open up new possibilities for agricultural production systems to be modified based on technological advancements along with the most efficient and sustainable use of material, human and financial resources. The growth of scientific cognition is marked by a persistent diversification of the range of theoretical and practical topics, as well as a stronger emphasis on the methodologies and technologies of scientific and practical activity. There is a growing recognition that obtaining substantial outcomes is directly dependent on the underlying theoretical perspective, more specifically, on a principle-based approach to challenges and defining common scientific study objectives. Science and practice should join forces to help agriculture transition to waste-free production by improving technological, economic and organizational capabilities (Andreichenko etal, 2021).

14.7 CHALLENGES AND FUTURE PERSPECTIVES

Agricultural waste is the best natural source of different essential nutrients. On the other side, its improper disposal can pose significant environmental risks and disrupt the natural cycle. A suitable alternative is to convert nutrition residues into useful consumer items, antioxidants, chemicals and fuels for sustainability, economics, social and ecological consequences. Their easy availability, abundance and low cost are all characteristics that may be used to generate valuable substances economically. The current issue of employing agro-based wastes is that other than the target product, unwanted chemicals are produced. Furthermore, the required product amount must be scaled up to an industrial scalethroughvarious parameters and selecting a single or a consortium of microorganism.

Appropriate pre-treatment procedures can also improve the yield of value-added chemicals. For biofuel generation, microbial pre-treatment of lignocellulosic biomass is more effective and ecologically friendly than chemical pre-treatment. Various research has resulted in substantial advances in utilizing agricultural waste as a cost-effective option. With the increasing necessity of enzymes, biofuels, and bioactive chemicals, improving their synthesis through low-cost residues is critical for future work.

14.8 CONCLUSION

The wastes generated during cultivation and harvesting of crops is usually burnt to clear the land for successive plantation, which not only degrades the soil and land quality but also contributes towards serious climate change issues like global warming. Therefore, sustainable measures are required for effective management of agro-wastes so that their negative effects can be brought down before releasing them in the environment. The concept of 'waste to wealth' can be successfully implemented on agro-residues taking into consideration their physical and chemical properties. The nutritional rich chemical composition of the agro-residues (after suitable processing) serves as a reservoir from which an array of value-added products such as biofuels, biofertilizers, biochar, organic acids, enzymes, secondary metabolites, etc., can be generated without exploiting any other essential natural resource. Utilization of agro-residues as raw materials also helps in achieving the principles of circular economy: minimizes reliance on conventional fossil fuels thereby protecting the environment from unnecessary carbon emissions; effectively manages the wastes by extracting its full potential as a raw material and circumvents the use of non-renewable resources.Different nations across the globe have realized their role in attaining Sustainable Development Goals and are working effortlessly to convert such wastes into productive raw materials thereby conserving and preserving the environment.

ACKNOWLEDGMENTS

The authors are thankful to Counsellor, Science and Technology, Embassy of India, Moscow (Russia) for an opportunity to be a part of Indo-Russia Webinar, 2021. We are also highly grateful to Director, CSIR-Indian Institute of Toxicology Research (CSIR-IITR), Lucknow (India) for his valuable encouragement and intra institutional financial support (MLP002). This is CSIR-IITR manuscript no. IITR/SEC/MS/2022/45.

REFERENCES

Abu Yazid, N., Barrena, R., Komilis, D. and Sánchez, A., 2017. Solid-state fermentation as a novel paradigm for organic waste valorization: A review. *Sustainability*, 9(2), p.224.

Agrawal, T., Jadhav, S.K. and Quraishi, A., 2019. Bioethanol production from an agro-waste, deoiled rice bran by Saccharomyces cerevisiae MTCC 4780 via optimization of fermentation parameters. *The International Journal by the Thai Society of Higher Education Institutes on Environment, Environment Asia*, 12(1), pp. 20–24.

Ahmad, A., Chowdhary, P., Khan, N., Chaurasia, D., Varjani, S., Pandey, A. and Chaturvedi, P., 2022. Effect of sewage sludge biochar on the soil nutrient, microbial abundance, and plant biomass: A sustainable approach towards mitigation of solid waste. *Chemosphere*, 287, p.132112.

Ahmad, A., Khan, N., Giri, B. S., Chowdhary, P. and Chaturvedi, P., 2020. Removal of methylene blue dye using rice husk, cow dung and sludge biochar: Characterization, application, and kinetic studies. *Bioresource Technology*, 306, p.123202.

Ali, E. A. M., Abd Ellatif, S., and Abdel Razik, E., 2020. Production, purification, characterization and immobilization of laccase from *Phomabetae*and its application in synthetic dyes decolorization. *EgyptianJournal ofBotany*. 60, pp. 301–312.

Alias, N. H., and Abbas, Z., 2017. Preliminary investigation on the total phenolic content and antioxidant activity of pineapple wastes via microwave-assisted extraction at fixed microwave power. *Chemical Engineering Transactions*. 56, pp. 1675–1680.

Andreichenko, A., Andreichenko, S. and Smentyna, N., 2021. Ensuring biosphere balance in the context of agricultural waste management. *Philosophy & Cosmology*, 26, p.46.

Bajracharya, A. M., 2019. Plant growth promoting rhizobacteria (PGPR): Biofertiliser and biocontrol agent-Review article. *Journal of Balkumari College*, 8, pp. 42–45.

Baruah, J., Nath, B. K., Sharma, R., Kumar, S., Deka, R. C., Baruah, D. C. and Kalita, E., 2018. Recent trends in the pretreatment of lignocellulosic biomass for value-added products. *Frontiers in Energy Research*, 6, p.141.

Bharathiraja, S., Suriya, J., Krishnan, M., Manivasagan, P. and Kim, S. K., 2017. Production of enzymes from agricultural wastes and their potential industrial applications. In Advances in food and nutrition research (Vol. 80, pp. 125–148). Academic Press.

Bhattacharjee, R. and Dey, U., 2014. Biofertilizer, a way towards organic agriculture: A review. *African Journal of Microbiology Research*, 8(24), pp. 2332–2343.

Bhuvaneshwari, S., Hettiarachchi, H. and Meegoda, J. N., 2019. Crop residue burning in India: Policy challenges and potential solutions. *International Journal of Environmental Research and Public Health*, 16(5), p.832.

Bibi, F., Ilyas, N., Arshad, M., Khalid, A., Saeed, M., Ansar, S. and Batley, J., 2022. Formulation and efficacy testing of bio-organic fertilizer produced through solid-state fermentation of agro-waste by Burkholderiacenocepacia. *Chemosphere*, 291, p.132762.

Briones-Labarca, V., Plaza-Morales, M., Giovagnoli-Vicuña, C., Jamett, F.,2015. High hydrostatic pressure and ultrasound extractions of antioxidant compounds, sulforaphane and fatty acids from Chilean papaya (*Vasconcelleapubescens*) seeds: effects of extraction conditions and methods. *LWT*, 60(1), pp. 525–534. doi: 10.1016/j.lwt.2014.07.057.

Buntić, A. V., Milić, M. D., Stajković-Srbinović, O. S., Rasulić, N. V., Delić, D. I. and Mihajlovski, K. R., 2019. Cellulase production by Sinorhizobiummeliloti strain 224 using waste tobacco as substrate. *International Journal of Environmental Science and Technology*, 16(10), pp. 5881–5890.

Cai, W., Dai, L. and Liu, R., 2018. Catalytic fast pyrolysis of rice husk for bio-oil production. *Energy*, 154, pp. 477–487.

Carvalho, E. A., Nunes, L. V., Goes, L. M. D. S., Silva, E. G. P. D., Franco, M., Gross, E., Uetanabaro, A. P. T., Costa, A. M. D., 2018. Peach-palm (Bactris gasipaesKunth.) waste as substrate for xylanase production by *Trichoderma stromaticum*AM7. *Chemical Engineering Communication*, 205, pp. 975–985.

Cassou, E., 2018. Field Burning (English). Agricultural Pollution.Washington, DC: World Bank Group.

Chaurasia, D., Singh, A., Shukla, P. and Chaturvedi, P., 2022. Biochar: A sustainable solution for the management of agri-wastes and environment. In *Biochar in Agriculture for Achieving Sustainable Development Goals* Daniel, C. W. T. and Sik Ok, Y., (eds.) (pp. 361–379). Academic Press.

Chen Y., Wen J., Deng Z., Pan X., Xie X., Peng C., 2020. Effective utilization of food wastes: Bioactivity of grape seed extraction and its application in food industry. *Journal of Functional Foods,* 73, 104113. doi: 10.1016/j.jff.2020.104113

Chen, D., Cen, K., Jing, X., Gao, J., Li, C. and Ma, Z., 2017. An approach for upgrading biomass and pyrolysis product quality using a combination of aqueous phase bio-oil washing and torrefaction pretreatment. *Bioresource Technology*, 233, pp. 150–158.

Chhaya, U., Gupte, A., 2019. Studies on a better laccase-producing mutant of *Fusarium incarnatum*LD-3 under solid substrate tray fermentation. *3 Biotech*, 9, p.100.

Chihoub W., Dias M. I., Barros L., et al., 2019. Valorisation of the green waste parts from turnip, radish and wild cardoon: Nutritional value, phenolic profile and bioactivity evaluation. *Food Research International*, 126, 108651. doi: 10.1016/j.foodres.2019.108651.

Chintagunta, A. D., Ray, S. and Banerjee, R., 2017. An integrated bioprocess for bioethanol and biomanure production from pineapple leaf waste. *Journal of Cleaner Production*, 165, pp. 1508–1516.

Cruz, G. J., Mondal, D., Rimaycuna, J., Soukup, K., Gómez, M. M., Solis, J. L. and Lang, J., 2020. Agrowaste derived biochars impregnated with ZnO for removal of arsenic and lead in water. *Journal of Environmental Chemical Engineering*, 8(3), p.103800.

de Castro, R. J. S. and Sato, H. H., 2015. Enzyme production by solid state fermentation: general aspects and an analysis of the physicochemical characteristics of substrates for agro-industrial wastes valorization. *Waste and Biomass Valorization,* 6(6), pp. 1085–1093.

Demirbas, A., 2009. Biorefineries: Current activities and future developments. *Energy Conversion and Management*, 50(11), pp. 2782–2801.

Elliott, D. C., Biller, P., Ross, A. B., Schmidt, A. J. and Jones, S. B., 2015. Hydrothermal liquefaction of biomass: Developments from batch to continuous process. *Bioresource Technology*, 178, pp. 147–156.

Europe, B., 2019. Biomass for energy: Agricultural residues & energy crops. *Eingesehen am*, 30, p.2020. https://bioenergyeurope.org/article/204-bioenergy-explained-biomass-forenergyagricultural-residues-energy-crops. html

Gao, Y., Guo, X., Liu, Y., Fang, Z., Zhang, M., Zhang, R., You, L., Li, T. and Liu, R. H., 2018. A full utilization of rice husk to evaluate phytochemical bioactivities and prepare cellulose nanocrystals. *Scientific Reports*, 8(1), pp. 1–8.

Garcia-Castello, E. M., Rodriguez-Lopez, A. D., Mayor, L., Ballesteros, R., Conidi, C. and Cassano, A., 2015. Optimization of conventional and ultrasound assisted extraction of flavonoids from grapefruit (Citrus paradisi L.) solid wastes. *LWT-Food Science and Technology*, 64(2), pp. 1114–1122.

Gayathri, R., Gopinath, K. P. and Kumar, P. S., 2021. Adsorptive separation of toxic metals from aquatic environment using agro waste biochar: Application in electroplating industrial wastewater. *Chemosphere*, 262, p.128031.

Georgieva, V. G., Gonsalvesh, L. and Tavlieva, M. P., 2020. Thermodynamics and kinetics of the removal of nickel (II) ions from aqueous solutions by biochar adsorbent made from agro-waste walnut shells. *Journal of Molecular Liquids*, 312, p.112788.

Ginni, G., Kavitha, S., Kannah, Y., Bhatia, S. K., Kumar, A., Rajkumar, M., Kumar, G., Pugazhendhi, A. and Chi, N.T.L., 2021. Valorization of agricultural residues: Different biorefinery routes. *Journal of Environmental Chemical Engineering*, 9(4), p.105435.

Gu, Y., Dai, L., Zhou, X. and Xu, Y., 2022. Multifactorial effects of gluconic acid pretreatment of waste straws on enzymatic hydrolysis performance. *Bioresource Technology*, 346, p.126617.

Hao, B., Xu, D., Jiang, G., Sabri, T. A., Jing, Z. and Guo, Y., 2021. Chemical reactions in the hydrothermal liquefaction of biomass and in the catalytic hydrogenation upgrading of biocrude. *Green Chemistry*, 23(4), pp. 1562–1583.

He, Z., Xu, D., Wang, S., Zhang, H. and Jing, Z., 2018. Catalytic upgrading of water-soluble biocrude from hydrothermal liquefaction of chlorella. *Energy & Fuels*, 32(2), pp. 1893–1899.

Hofvendahl, K. and Hahn–Hägerdal, B., 2000. Factors affecting the fermentative lactic acid production from renewable resources1. *Enzyme and Microbial Technology*, 26(2–4), pp. 87–107.

Hosseini, M., Dincer, I. and Rosen, M.A., 2012. Steam and air fed biomass gasification: Comparisons based on energy and exergy. *International Journal of Hydrogen Energy*, 37(21), pp. 16446–16452.

Huang, X., Wang, C., Liu, Q., Zhu, Z., Lynn, T.M., Shen, J., Whiteley, A.S., Kumaresan, D., Ge, T., Wu, J., 2018. Abundance of microbial CO2-fixing genes during the late rice season in a long-term management paddy field amended with straw and straw-derived biochar. *Canadian Journal of Soil Science,* 98, pp. 306–316.

Ibarra-Gonzalez, P. and Rong, B. G., 2019. A review of the current state of biofuels production from lignocellulosic biomass using thermochemical conversion routes. *Chinese Journal of Chemical Engineering*, 27(7), pp. 1523–1535.

Iqbal, M. W. and Kang, Y., 2021. Waste-to-energy supply chain management with energy feasibility condition. *Journal of Cleaner Production*, 291, p.125231

Ishchi, T., Sibi, G., 2019. Ragi husk as substrate for cellulase production under temperature mediated solid state fermentation by *Streptomyces* sp. *American Journal of BioScience*, 7, pp. 77–81.

Jang, H. M. and Kan, E., 2019. Engineered biochar from agricultural waste for removal of tetracycline in water. *Bioresource Technology*, 284, pp. 437–447.

Kakde, P. R., and Aithal, S. C., 2020. Effect of pretreatment on agro-residues for cellulase production by solid state fermentation using fungi. *International Journal of Recent Scientific Research,* 11(01), pp. 36903–36906

Kaur, S., and Kaur, S., 2017. Production of pectinolytic enzymes pectinase and pectin lyase by Bacillus subtilis SAV-21 in solid state fermentation. Annals of microbiology, 67, pp. 333–342.

Khan, N., Chowdhary, P., Ahmad, A., Giri, B. S. and Chaturvedi, P., 2020. Hydrothermal liquefaction of rice husk and cow dung in Mixed-Bed-Rotating Pyrolyzer and application of biochar for dye removal. *Bioresource Technology*, 309, p.123294.

Khan, N., Chowdhary, P., Gnansounou, E. and Chaturvedi, P., 2021. Biochar and environmental sustainability: Emerging trends and techno-economic perspectives. *Bioresource Technology*, 332, p.125102.

Khoshnevisan, B., Tsapekos, P., Zhang, Y., Valverde-Pérez, B. and Angelidaki, I., 2019. Urban biowaste valorization by coupling anaerobic digestion and single cell protein production. *Bioresource Technology*, 290, p.121743.

Kirchherr, J., Reike, D. and Hekkert, M., 2017. Conceptualizing the circular economy: An analysis of 114 definitions. *Resources, Conservation and Recycling*, 127, pp. 221–232.

Kloepper, J. W., Schroth, M. N. and Miller, T. D., 1980. Effects of rhizosphere colonization by plant growth-promoting rhizobacteria on potato plant development and yield. *Phytopathology*, 70(11), pp. 1078–1082.

Koul, B., Yakoob, M. and Shah, M. P., 2022. Agricultural waste management strategies for environmental sustainability. *Environmental Research*, 206, p.112285.

Kuhnert, S., Lehmann, L. and Winterhalter, P., 2015. Rapid characterisation of grape seed extracts by a novel HPLC method on a diol stationary phase. *Journal of Functional Foods*, 15, pp. 225–232.

Kumar, A., Naqvi, S. D. Y., Kaushik, P., Khojah, E., Amir, M., Alam, P. and Samra, B. N., 2022. Rhizophagusirregularis and nitrogen fixing azotobacter enhances greater yam (Dioscoreaalata) biochemical profile and upholds yield under reduced fertilization. *Saudi Journal of Biological Sciences*, 29(5), pp. 3694–3703.

Kuvvet, C., Uzuner, S. and Cekmecelioglu, D., 2019. Improvement of pectinase production by co-culture of Bacillus spp. using apple pomace as a carbon source. *Waste and Biomass Valorization*, 10(5), pp. 1241–1249.

Lamina, O. G., 2022. The Use of Decomposers and Detritivores in Solid Waste Management: A Review Paper. *Available at SSRN 4014235*.

Lemes, A. C., Egea, M. B., de Oliveira Filho, J. G., Gautério, G. V., Ribeiro, B. D. and Coelho, M. A. Z., 2022. Biological approaches for extraction of bioactive compounds from agro-industrial by-products: A review. *Frontiers in Bioengineering and Biotechnology*, 9, p.802543.

Leng, L., Zhang, W., Peng, H., Li, H., Jiang, S. and Huang, H., 2020. Nitrogen in bio-oil produced from hydrothermal liquefaction of biomass: A review. *Chemical Engineering Journal*, 401, p.126030.

Lohan, S. K., Jat, H. S., Yadav, A. K., Sidhu, H. S., Jat, M. L., Choudhary, M., Jyotsna, Kiran P., and Sharma, P. C.,2018. Burning issues of paddy residue management in north-west states of India. *Renewable and Sustainable Energy Reviews*, 81, 693–706. doi: 10.1016/j.rser.2017.08.057.

Lyu, D., Backer, R., Robinson, W. G. and Smith, D. L., 2019. Plant growth-promoting rhizobacteria for cannabis production: yield, cannabinoid profile and disease resistance. *Frontiers in Microbiology*, 10, p.461387.

Mahmoodi, M., Najafpour, G. D. and Mohammadi, M., 2019. Bioconversion of agroindustrial wastes to pectinases enzyme via solid state fermentation in trays and rotating drum bioreactors. *Biocatalysis and Agricultural Biotechnology*, 21, p.101280.

Maurya, N.K. ed.,2022. *Pharmacognosy & Nutrition Volume 2*. Virgin Sahityapeeth.

Mettu, S., Halder, P., Patel, S., Kundu, S., Shah, K., Yao, S., Hathi, Z., Ong, K. L., Athukoralalage, S., Choudhury, N. R. Dutta, N. K., and Lin, C. S. K 2020. Valorisation of Agricultural Waste Residues. *Waste Valorisation: Waste Streams in a Circular Economy*, pp. 51–85.

Mohammed, N. I., Kabbashi, N. and Alade, A., 2018. Significance of agricultural residues in sustainable biofuel development. In *Agricultural Waste and Residues* Aladjadjiyan, A., (ed.) (pp. 71–88). http://dx.doi.org/10. 5772/intechopen.78374

Moldes, A. B., Alonso, J. L. and Parajo, J. C., 2000. Multi-step feeding systems for lactic acid production by simultaneous saccharification and fermentation of processed wood. *Bioprocess Engineering*, 22(2), pp. 175–180.

Myo, E. M., Liu, B., Ma, J., Shi, L., Jiang, M., Zhang, K. and Ge, B., 2019. Evaluation of Bacillus velezensis NKG-2 for bio-control activities against fungal diseases and potential plant growth promotion. *Biological Control*, 134, pp. 23–31.

Neogi, S., Sharma, V., Khan, N., Chaurasia, D., Ahmad, A., Chauhan, S., Singh, A., You, S., Pandey, A. and Bhargava, P. C., 2022. Sustainable biochar: A facile strategy for soil and environmental restoration, energygeneration, mitigation of global climate change and circular bioeconomy. *Chemosphere*, 293, p.133474.

Nguyen, D. L. T., Binh, Q. A., Nguyen, X. C., Nguyen, T. T. H., Vo, Q. N., Nguyen, T. D., Tran, T. C. P., Nguyen, T. A. H., Kim, S. Y., Nguyen, T. P. and Bae, J., 2021a. Metal salt-modified biochars derived from agro-waste for effective congo red dye removal. *Environmental Research*, 200, p.111492.

Nguyen, X. C., Nguyen, T. T. H., Nguyen, T. H. C., Van Le, Q., Vo, T. Y. B., Tran, T. C. P., La, D. D., Kumar, G., Nguyen, V. K., Chang, S. W. and Chung, W. J., 2021b. Sustainable carbonaceous biochar adsorbents derived from agro-wastes and invasive plants for cation dye adsorption from water. *Chemosphere*, 282, p.131009.

Nigam, P. S., 2009. Production of bioactive secondary metabolites. In *Biotechnology for Agro-Industrial Residues Utilisation: Utilisation of Agro-Residues* Singh nee' Nigam, P., Pandey, A., (eds.) (pp. 129–145) Springer, Dordrecht. https://doi.org/10.1007/978-1-4020-9942-7_7.

Nitsenko, V. S., Mardani, A., Streimikis, J., Ishchenko, M., Chaikovsky, M., Stoyanova-Koval, S. and Arutiunian, R., 2019. *Automatic Information System of Risk Assessment for Agricultural Enterprises of Ukraine.* Montenegrin Journal of Economics, 15(2), pp. 139–152.

Okur, N., 2018. A review-bio-fertilizers-power of beneficial microorganisms in soils. *Biomedical Journal of Scientific & Technical Research*, 4(4), pp. 4028–4029.

Oladejo, D., Jegede, B., Regina, J. P., Akinbomi, J. G. and Bakare, L. O., 2022. Pretreatment and optimization strategy for biogas production from agro-based wastes. *The International Journal of Engineering and Science (IJES)*, 11(3), pp. 14–20.

Onuguh, I. C., Ikhuoria, E. U. and Obibuzo, J. U., 2022. Comparative studies on bioethanol production from some starch based agricultural waste peels. *International Journal of Research in Science & Engineering (IJRISE) ISSN: 2394-8299*, 2(01), pp. 7–12.

Ostapenko, R., Herasymenko, Y., Nitsenko, V., Koliadenko, S., Balezentis, T. and Streimikiene, D., 2020. Analysis of production and sales of organic products in Ukrainian agricultural enterprises. *Sustainability*, 12(8), p.3416.

Pais, S., Hebbar, A. and Selvaraj, S., 2022. A critical look at challenges and future scopes of bioactive compounds and their incorporations in the food, energy, and pharmaceutical sector. *Environmental Science and Pollution Research*, 29(24), pp. 35518–35541.

Pan, S. Y., Du, M. A., Huang, I. T., Liu, I. H., Chang, E. E. and Chiang, P. C., 2015. Strategies on implementation of waste-to-energy (WTE) supply chain for circular economy system: a review. *Journal of Cleaner Production*, 108, pp. 409–421.

Pattanaik, L., Pattnaik, F., Saxena, D. K. and Naik, S. N., 2019. Biofuels from agricultural wastes. In *Second and Third Generation of Feedstocks* Basile, A., Dalena, F., (eds.) (pp. 103–142). Elsevier.

Pereira, A. D. S., Fontes-Sant'Ana, G. C., Amaral, P. F., 2019. Mango agro-industrial wastes for lipase production from *Yarrowialipolytica* and the potential of the fermented solid as a biocatalyst. *Food and Bioproducts Process,* 115, pp. 68–77.

Prabakaran, S., Mohanraj, T., Arumugam, A. and Sudalai, S., 2022. A state-of-the-art review on the environmental benefits and prospects of Azolla in biofuel, bioremediation and biofertilizer applications. *Industrial Crops and Products*, 183, p.114942.

Putri, D. N., Khootama, A., Perdani, M. S., Utami, T. S., Hermansyah, H., 2020. Optimization of *Aspergillus niger*lipase production by solid state fermentation of agro-industrial waste. *Energy Reports,* 6, pp. 331–335.

Rahman, M. M., Bibi, S., Rahaman, M. S., Rahman, F., Islam, F., Khan, M. S., Hasan, M. M., Parvez, A., Hossain, M. A., Maeesa, S. K. and Islam, M. R., 2022. Natural therapeutics and nutraceuticals for lung diseases: Traditional significance, phytochemistry, and pharmacology. *Biomedicine & Pharmacotherapy*, 150, p.113041.

Ramesh, D., Kiruthika, T. and Karthikeyan, S., 2021. Sustainable biorefinery technologies for agro-residues: challenges and perspectives. In *Sustainable Bioeconomy* Venkatramanan, V., Shah, S. and Ram Prasad, R., (eds.) (pp. 101–130). Springer, Singapore.

Ramesh, D., Muniraj, I. K., Thangavelu, K. and Karthikeyan, S., 2019. Chemicals and fuels production from agro residues: A biorefinery approach. In *Sustainable Approaches for Biofuels Production Technologies* Neha, S., Manish, S., Mishra, P. K., Upadhyay, S. N., Pramod W. R. and Vijaj Kumar G., (eds.) (pp. 47–71). Springer, Cham.

Ravindran, R., Hassan, S. S., Williams, G. A. and Jaiswal, A. K., 2018. A review on bioconversion of agro-industrial wastes to industrially important enzymes. *Bioengineering*, 5(4), p.93.

Rezende, Y. R. R. S., Nogueira, J. P., and Narain, N., 2017. Comparison and optimization of conventional and ultrasound assisted extraction for bioactive compounds and antioxidant activity from agro-industrial acerola (*Malpighia emarginata* DC) residue. *LWT*, 85, pp. 158–169. doi: 10.1016/j.lwt.2017.07.020.

Ritchie, H., and Roser, M., 2021. Deforestation and forest loss. Retrieved from https://ourworldindata.org/deforestation?module=inline&pgtype=article.

Rohela, G. K. and Saini, P., 2022. Nitrogen-fixing biofertilizers and biostimulants. In *Microbial Biostimulants for Sustainable Agriculture and Environmental Bioremediation* Inamuddin, Charles Oluwaseun, A., Mohd Imran, A., Tariq, A., (eds.) (pp. 83–100). CRC Press.

Rossi, R. C., da Rosa, S. R., Weimer, P., Moura, J. G. L., de Oliveira, V. R. and de Castilhos, J., 2020. Assessment of compounds and cytotoxicity of Citrus deliciosa Tenore essential oils: from an underexploited by-product to a rich source of high-value bioactive compounds. *Food Bioscience*, 38, p.100779.

Ruales Salcedo A. V., Rojas González A. F., Cardona Alzate C. A., 2017. Obtaining phenolic compounds from isabella grape (Vitis labrusca) residues Biotechnology in the Agricultural and Agroindustrial Sector, 15(SPE2), pp. 72–79. doi: 10.18684/bsaa(15).2004.

Ruan, R. R., Chen, P., Hemmingsen, R., Morey, V. and Tiffany, D., 2008. Size matters: small distributed biomass energy production systems for economic viability. *International Journal of Agricultural and Biological Engineering*, 1(1), pp. 64–68.

Sadh, P. K., Duhan, S. and Duhan, J. S., 2018. Agro-industrial wastes and their utilization using solid state fermentation: A review. *Bioresources and Bioprocessing*, 5(1), pp. 1–15.

Saini, A., Panwar, D., Panesar, P.S. and Bera, M. B., 2019. Bioactive compounds from cereal and pulse processing byproducts and their potential health benefits. *Austin Journal of Nutrition & Metabolism*, 6(2), p.1068.

Sakuragi, K., Li, P., Otaka, M. and Makino, H., 2016. Recovery of bio-oil from industrial food waste by liquefied dimethyl ether for biodiesel production. *Energies*, 9(2), p.106.

Salazar, L. N., Dal Maso, S. S., Ogimbosvski, T. A., Daronch, N. A., Zeni, J., Valduga, E., Toniazzo Backes, G.and Cansian, R.L., 2019. Production, partial characterization and application of cellulases by newly isolated *Penicillium* sp. using agro-industrial substrate solid-state fermentation. *Industrical Biotechnology*, 15, pp. 79–88.

Salim, A. A., Grbavčić, S., Šekuljica, N., Stefanović, A., Tanasković, S. J., Luković, N. and Knežević-Jugović, Z., 2017. Production of enzymes by a newly isolated Bacillus sp. TMF-1 in solid state fermentation on agricultural by-products: The evaluation of substrate pretreatment methods. *Bioresource Technology*, 228, pp. 193–200.

Salomão, G. S. B., Agnezi, J. C., Paulino, L. B., Hencker, L. B., de Lira, T. S., Tardioli, P. W. and Pinotti, L. M., 2019. Production of cellulases by solid state fermentation using natural and pretreated sugarcane bagasse with different fungi. *Biocatalysis and Agricultural Biotechnology*, 17, pp. 1–6.

Santana-Méridas, O., González-Coloma, A. and Sánchez-Vioque, R., 2012. Agricultural residues as a source of bioactive natural products. *Phytochemistry Reviews*, 11(4), pp. 447–466.

Sasidharan, S., Chen, Y., Saravanan, D., Sundram, K. M. and Latha, L. Y., 2011. Extraction, isolation and characterization of bioactive compounds from plants' extracts. *African Journal of Traditional, Complementary and Alternative Medicines*, 8(1), pp. 1–10.

Savino, S., Bulgari, D., Monti, E. and Gobbi, E., 2021. Agro-industrial wastes: A substrate for multi-enzymes production by cryphonectriaparasitica. *Fermentation*, 7(4), p.279.

Sethi, B.K., Nanda, P.K., Sahoo, S., 2016. Enhanced production of pectinase by *Aspergillus terreus*NCFT 4269.10 using banana peels as substrate. *3 Biotech* 6, p. 36.

Shahbaz, M., Yusup, S., Inayat, A., Patrick, D. O. and Pratama, A., 2016. Application of response surface methodology to investigate the effect of different variables on conversion of palm kernel shell in steam gasification using coal bottom ash. *Applied Energy*, 184, pp. 1306–1315.

Shamina, M., 2022. A review on cyanobacterial biofertilizer for organic rice cultivation: Technology, improvement and future prospects. *International Journal on Algae*, 24(1), pp. 89–103.

Sharma, A., Vivekanand, V. and Singh, R. P., 2008. Solid-state fermentation for gluconic acid production from sugarcane molasses by Aspergillus niger ARNU-4 employing tea waste as the novel solid support. *Bioresource Technology*, 99(9), pp. 3444–3450.

Sharma, M., Gupta, D., Parhi, M. P., Gurung, S., Parasar, H. and Kumar, A., 2021 Phosphate solubilizing microbes: A sustainable tool for healthy agriculture. *Eco. Env. & Cons.*, 27, pp. (S303–S310).

Shi, T., Liu, Y., Zhang, L., Hao, L. and Gao, Z., 2014. Burning in agricultural landscapes: an emerging natural and human issue in China. *Landscape Ecology*, 29(10), pp. 1785–1798.

Shukla, P., Giri, B. S., Mishra, R. K., Pandey, A. and Chaturvedi, P., 2021. Lignocellulosic biomass-based engineered biochar composites: A facile strategy for abatement of emerging pollutants and utilization in industrial applications. *Renewable and Sustainable Energy Reviews*, 152, p.111643.

Sindhu, R., Gnansounou, E., Rebello, S., Binod, P., Varjani, S., Thakur, I. S., Nair, R. B. and Pandey, A., 2019. Conversion of food and kitchen waste to value-added products. *Journal of Environmental Management*, 241, pp. 619–630.

Singh, T.A., Sharma, M., Sharma, M., Sharma, G. D., Passari, A. K. and Bhasin, S., 2022. Valorization of agro-industrial residues for production of commercial biorefinery products. *Fuel*, 322, p.124284.

Smuda, S. S., Mohsen, S. M., Olsen, K. and Aly, M. H., 2018. Bioactive compounds and antioxidant activities of some cereal milling by-products. *Journal of Food Science and Technology*, 55(3), pp. 1134–1142.

Srinivasan, P., Selvankumar, T., Kamala-Kannan, S., Mythili, R., Sengottaiyan, A., Govarthanan, M., Senthilkumar, B., and Selvam, K., 2019. Production and purification of laccase by *Bacillus* sp. using millet husks and its pesticide degradation application. *3 Biotech*, 9, p.396.

Srivastava, N., Srivastava, M., Manikanta, A., Singh, P., Ramteke, P. W., Mishra, P. K. and Malhotra, B. D., 2017. Production and optimization of physicochemical parameters of cellulase using untreated orange waste by newly isolated Emericellavariecolor NS3. *Applied Biochemistry and Biotechnology*, 183(2), pp. 601–612.

Thakrar, F., Goswami, D., Singh, S.P., 2020. Production of an alkaline protease from *nocardiopsis alba* om-4, a haloalkaliphilic actinobacteria in solid-state fermentation using agricultural waste products (April 12, 2020). Proceedings of the National Conference on Innovations in Biological Sciences (NCIBS) 2020, Available at SSRN: https://ssrn.com/abstract=3586506 or http://dx.doi.org/10.2139/ssrn.3586506

Tripathi, N., Hills, C. D., Singh, R. S. and Atkinson, C. J., 2019. Biomass waste utilisation in low-carbon products: harnessing a major potential resource. *NPJ Climate and Atmospheric Science*, 2(1), pp. 1–10.

Tseng, M. L., Tran, T. P. T., Ha, H. M., Bui, T. D. and Lim, M. K., 2021. Sustainable industrial and operation engineering trends and challenges Toward Industry 4.0: A data driven analysis. *Journal of Industrial and Production Engineering*, 38(8), pp. 581–598.

Udeh, B. A., 2018. Bio-waste transesterification alternative for biodiesel production: A combined manipulation of lipase enzyme action and lignocellulosic fermented ethanol. *Asian Journal of Biotechnology and Bioresource Technology*, 3(3), pp. 1–9.

Ullah, K., Sharma, V. K., Dhingra, S., Braccio, G., Ahmad, M. and Sofia, S., 2015. Assessing the lignocellulosic biomass resources potential in developing countries: A critical review. *Renewable and Sustainable Energy Reviews*, 51, pp. 682–698.

Vasudevan, Y., Govindharaj, D., Udayakumar, G. P., Ganesan, A. and Sivarajasekar, N., 2020. A review on the production of biogas from biological sources. In *Sustainable Development in Energy and Environment* Sivasubramanian, V. Pugazhendhi, A., and Ganesh Moorthy, I., (eds.) (pp. 1–12) Springer Proceedings in Energy. Springer, Singapore. https://doi.org/10.1007/978-981-15-4638-9_1.

Ventorino, V., Robertiello, A., Cimini, D., Argenzio, O., Schiraldi, C., Montella, S., Faraco, V., Ambrosanio, A., Viscardi, S. and Pepe, O., 2017. Bio-based succinate production from Arundo donax hydrolysate with the new natural succinic acid-producing strain Basfiasucciniciproducens BPP7. *BioEnergy Research*, 10(2), pp. 488–498.

Vodnar D. C., Călinoiu L. F., Dulf F. V., Ştefănescu B. E., Crişan G., and Socaciu C., 2017. Identification of the bioactive compounds and antioxidant, antimutagenic and antimicrobial activities of thermally processed agro-industrial waste. *Food Chemistry*, 231, pp. 131–140. doi: 10.1016/j.foodchem.2017.03.131.

Wang, W., Liu, Y. and Zhang, L., 2013. The spatial distribution of cereal bioenergy potential in China. *Gcb Bioenergy*, 5(5), pp. 525–535.

Wehaidy, H. R., Wahab, W. A. A., Kholif, A. M., Elaaser, M., Bahgaat, W. K., Abdel-Naby, M. A., 2020. Statistical optimization of *B. subtilis* MK775302 milk clotting enzyme.

Yadav, J. P. and Singh, B. R., 2011. Study on comparison of boiler efficiency using husk and coal as fuel in rice mill. *SAMRIDDHI: A Journal of Physical Sciences, Engineering and Technology*, 2(02), pp. 1–15.

Zhang, Y. and Chen, W. T., 2018. Hydrothermal liquefaction of protein-containing feedstocks. In *Direct Thermochemical Liquefaction for Energy Applications* Rosendahl, L., (ed.) (pp. 127–168). Woodhead Publishing.

Zhao, G., Zhang, R., Dong, L., Huang, F., Liu, L., Deng, Y., Ma, Y., Zhang, Y., Wei, Z., Xiao, J. and Zhang, M., 2018. A comparison of the chemical composition, in vitro bioaccessibility and antioxidant activity of phenolic compounds from rice bran and its dietary fibres. *Molecules*, 23(1), p.202.

Zhao, H. M., Guo, X. N. and Zhu, K. X., 2017. Impact of solid state fermentation on nutritional, physical and flavor properties of wheat bran. *Food Chemistry*, 217, pp. 28–36.

Zhou, Y., Li, Z., Ji, L., Wang, Z., Cai, L., Guo, J., Song, W., Wang, Y. and Piotrowski, A. M., 2022. Facile preparation of alveolate biochar derived from seaweed biomass with potential removal performance for cationic dye. *Journal of Molecular Liquids*, 353, p.118623.

15 Rational Synthetic Approaches to the Promising Molecules and Materials

Grigory V. Zyryanov

15.1 UTILIZATION OF CO/CO$_2$ AND NO$_x$ GASES

One of the above-mentioned approaches involves the utilization of such common greenhouse gases, as hazardous and highly toxic NOx gases or CO$_2$ into regio- or chemoselective reagents [5] or media/reagents for organic synthesis [6, 7].

Greenhouse gases, including carbon dioxide (CO$_2$) have a significant effect on climate change and global warming, which has become a widespread concern in recent years. Carbon capture, utilization and sequestration strategy appears to be effective in decreasing the carbon dioxide level in the atmosphere [8]. Utilization of ionic liquids (ILs) for CO$_2$ capture has appreciably attracted researchers' attention. Properties of ILs such as negligible vapour pressure and their affinity to capture CO$_2$ molecules make them a feasible alternative for currently available solvents [9]. On another hand, inorganic absorbents such as zeolites have significant potential for CO$_2$ capture owing to their high porosity, ultra-small pores, structural diversity, high stability, and excellent recyclability and chemical reactivity [10]. Zeolites could be obtained from fly ash [11], common coal combustion by-products, which are estimated in around 115 million tons per year USA and EU [12]. And the combination of both ILs and zeolites could be beneficial for the creation of advanced materials for CO$_2$ capturing and/or utilization.

For the above-mentioned purposes we have prepared different IL polymers namely poly(1-vinylimidazole), poly(1-vinyl-3-ethylimidazolium bromide), poly(1-vinyl-3-ethylimidazolium bis(trifluoro methylsulfonyl))imide materials through the free radical polymerization [12] (Figure 15.1). Keeping in mind the large microporous structure, wide abundance, hydrophobic character, and high CO$_2$ sorption affinity natural zeolite (zeolite 13X) we also prepared a zeolite incorporated poly(1-vinyl-3-ethylimidazolium bis(trifluoro methylsulfonyl)imide. Among the synthesized IL polymers the zeolite incorporated poly(1-vinyl-3-ethylimidazolium bis(trifluoro methylsulfonyl)imide was found to be the most efficient for the CO$_2$ separation.

The CO$_2$ sorption capacity of IL polymer materials followed the trend as zeolite incorporated poly(1-vinyl-3-ethylimidazolium bis(trifluoro methylsulfonyl)imide > poly(1-vinyl-3-ethylimidazolium bis(trifluoro methylsulfonyl)imide > poly(1-vinyl-3-ethylimidazolium bromide) > poly(1-vinylimidazole). After 24 hrs of adsorption experiment, the CO$_2$ sorption was found to be 1.58, 1.12, 0.82 and 0.52 mol/mol in accordance with the above sequence.

One of the important applications of CO$_2$ is associated with its use in a form of sub- [13] or supercritical (*sc*)CO$_2$ [14] as a non-convenient solvent media for organic reactions or extraction processes. In our recent study, we have developed a convenient method for the synthesis of 5-methyl-1,2,4-triazolo[1,5-a]pyrimidine-7(4H)-one (a key intermediate for the preparation of antiviral drug Triazid®) by carrying out the cyclocondensation reaction between 5-amino-3-H-1,2,4-triazole with

DOI: 10.1201/9781003327646-15

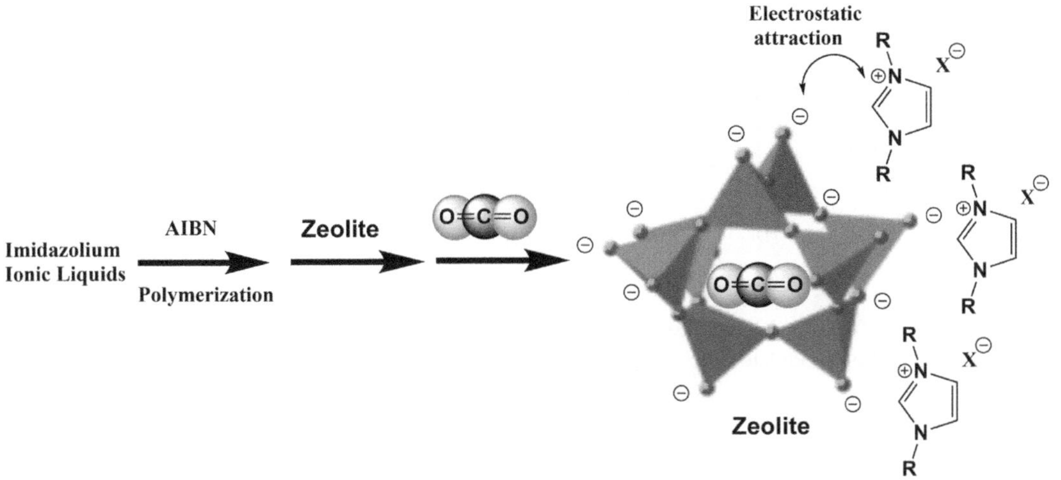

FIGURE 15.1 Proposed electrostatic interaction between zeolite-imidazolium ILs and subsequent CO_2 adsorption.

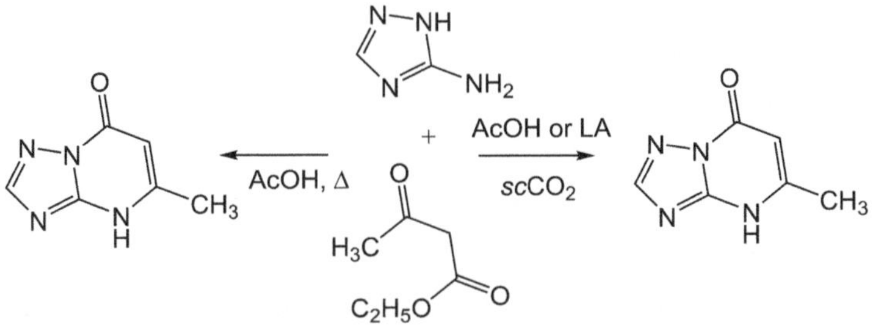

FIGURE 15.2 Synthesis of 5-methyl-1,2,4-triazolo[1,5-a]pyrimidine-7(4H)-one in $scCO_2$.

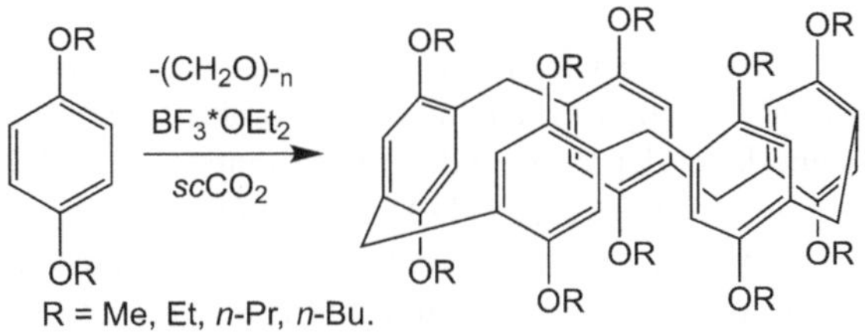

FIGURE 15.3 Synthesis of pillar[5]arene in $scCO_2$.

acetoacetic ester in $scCO_2$ (Figure 15.2). The reaction was carried out in the presence of various so-called CO_2-philic components, such as acetic acid, Lewis acids ($ZnCl_2$, TiO_2 and SiO_2 were studied). According to the product yields, the best conditions for the reaction were $scCO_2$ in the presence of acetic acid (89%) or $ZnCl_2$ (88%) at a temperature of 150–190°C.

For another application, we have developed a method for the preparation of pillar[5]arenes by carrying out the cyclocondensation reaction between paraformaldehyde and 1,4-dialcoxybenzene in $scCO_2$ in the presence of $BF_3{*}OEt_2$ as Lewis acid (Figure 15.3) [15].

As the C1 component, CO and CO_2 may participate in carbonylation [16, 17] or CO-insertion reactions [18, 19] with various synthons including arynes. Thus, by using CO_2 gas and simplest

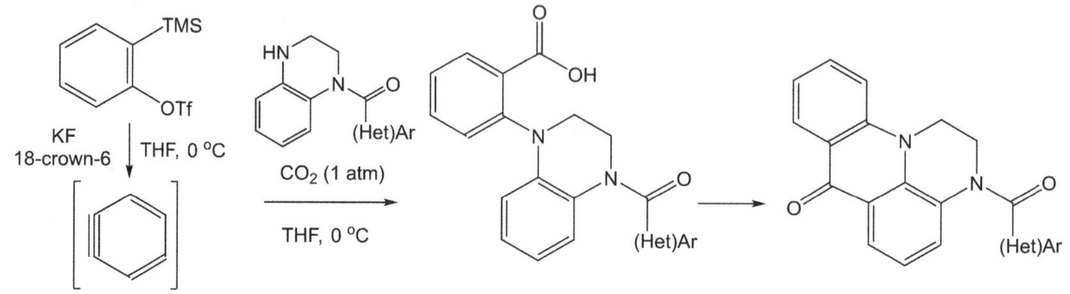

FIGURE 15.4 CO_2 and 1,2-dehydrobenzene approach to 2,3-dihydropyrazino [3,2,1-de]acridin-7(1H)-ones – anticancer drug candidates.

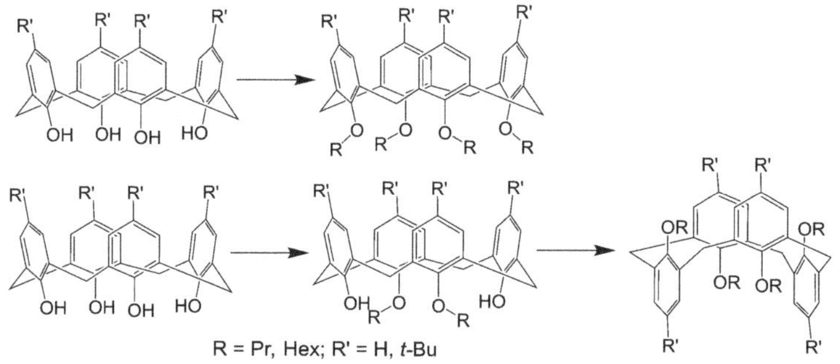

FIGURE 15.5 Synthesis of calix[4]arenes as NOx encapsulating agents.

aryne, 1,2-dehydrobenzene, generate *in situ*, we have developed a convenient method for the preparation of derivatives of anthranilic acid, including 2-(4-benzoyl-3,4-dihydroquinoxalin-1(2H)-yl)benzoic acid [20]. The last one is considered as a key precursor for the synthesis of 2,3-dihydropyrazino [3,2,1-*de*]acridin-7(1H)-ones – anticancer drug candidates, which were recently reported by us [21] (Figure 15.4).

NOx is a common name for the mixture of nitrogen oxides, containing N_2O_4, NO_2, NO, N_2O, etc. The emission of these gases are mainly formed during the combustion of fossil fuels. These gases are heavily involved in the ozone-layer depletion and in the formation of acid rains as well as contamination of water sources. Among the NOx gases N_2O_4, NO_2, NO are known to form stable complexes with calix[4]arenes. As a very first example, Kochi et al. [22] reported the formation of stable nitrosonium complexes between calix[4]arenes and NO in the presence of oxidants or with NO^+ salts. We have prepared stable nitrosonium complexes upon interaction of *cone-* and *1,3-alternate* calix[4]arenes (Figure 15.5) with gaseous NO_2/N_2O_4 in the presence of $SnCl_4$. In our experiments, we have observed the formation of stable complexes in 1:1 stoichiometry upon bubbling of $NO_2/$ $N2O_4$ through the solutions of calixarenes in organic solvents (Figure 15.6). These complexes could be isolated as solid compounds (Figure 15.7), and they are quite stable and could be stored for weeks without decomposition [23–25].

On the other hand, the reaction of these complexes with water, alcohols, thiols and secondary amides (which can be considered as simple models of peptides) results in the formation of nitrosated species with the complete release of un-changed calixarenes (Figures 15.8–15.9). For the secondary amides the N-nitrosation provides good chemoselectivity, and only N-methyl amides were N-nitrosated. Moreover, we were able to carry out the N-nitrozation of racemic mixtures of amides by

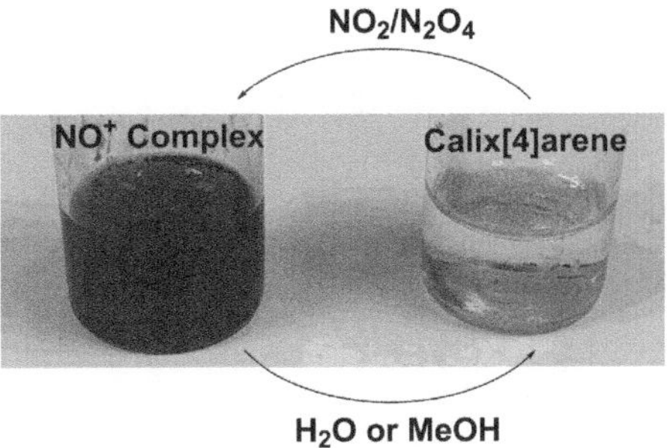

FIGURE 15.6 Formation of NO⁺ complexes in solutions.

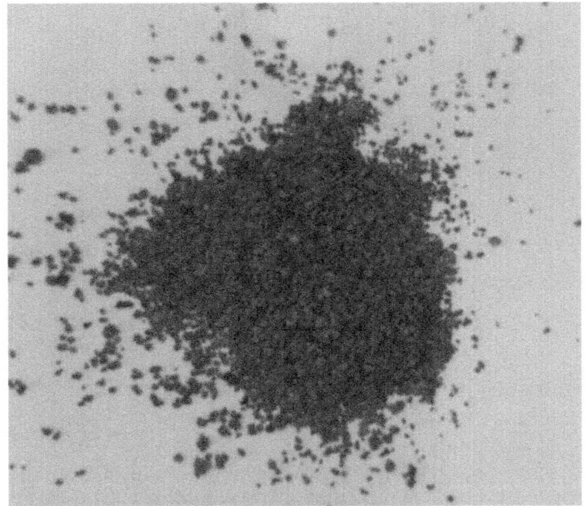

FIGURE 15.7 Solid calix*NO⁺ complex.

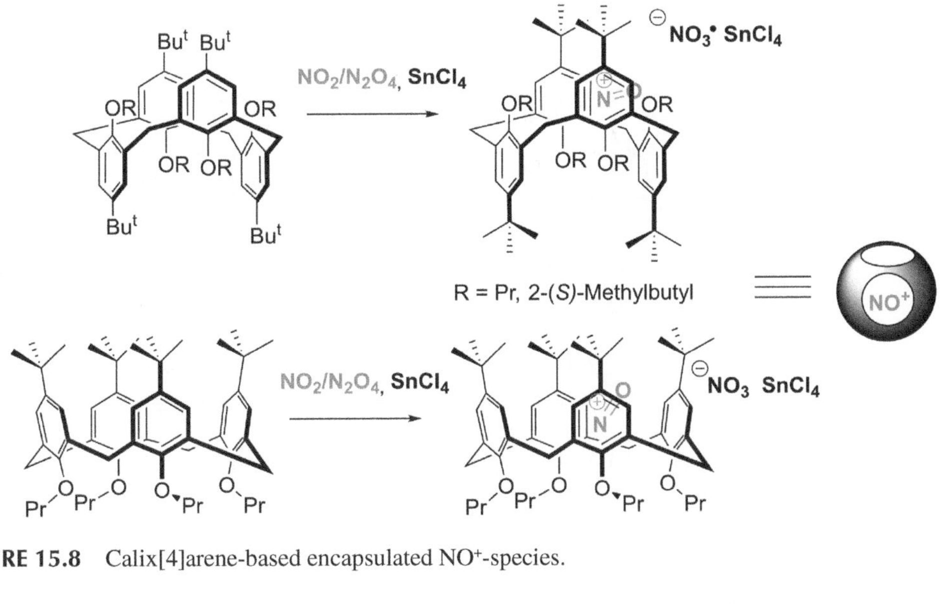

FIGURE 15.8 Calix[4]arene-based encapsulated NO⁺-species.

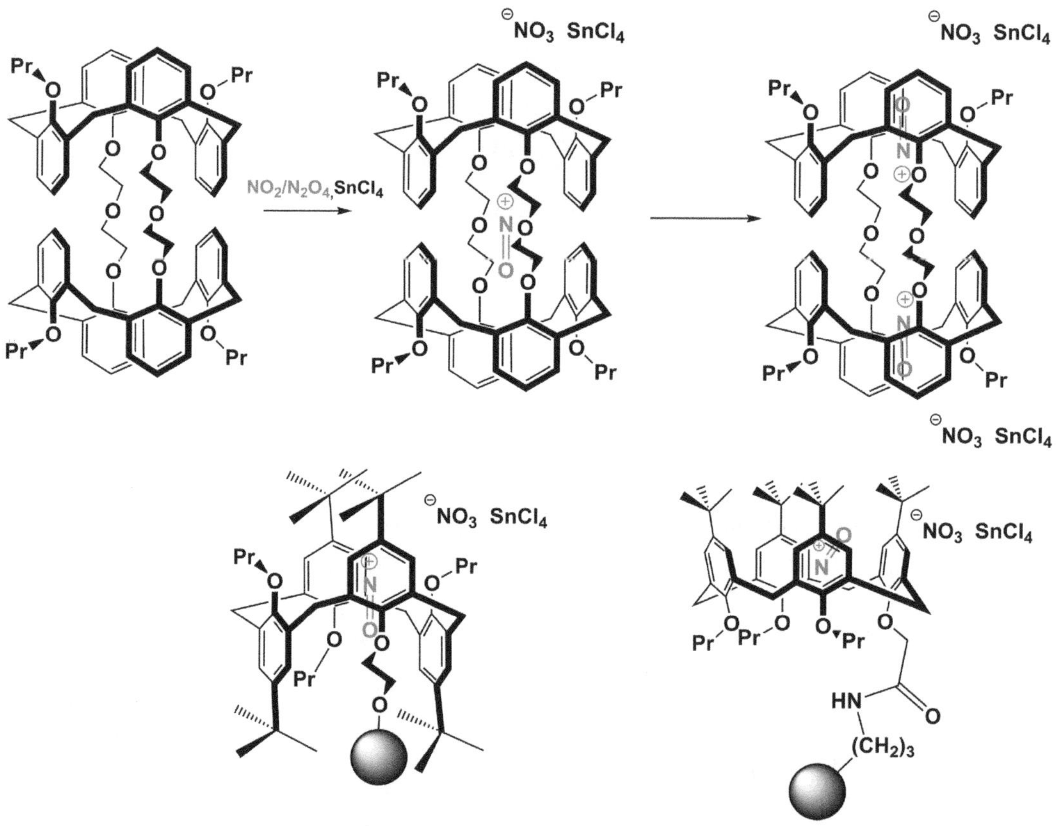

FIGURE 15.9 Nitrosonium complexes and their use for the N-nitrosation of secondary amides.

FIGURE 15.10 Other calix[4]arene based materials for the utilization and storage of NOx-gases.

using chiral nitrosation agents, and after one cycle the enrichment of the racemic mixture resulted in up to 15% enantiomeric excess [26].

By using calix[4]arenes, connected with ethylenglycol units some synthetic nanotubes can be prepared. These nanotubes can be used for the encapsulation and gradual release of NO^+ species, for instance, by reacting with secondary amides to afford N-nitrosoamides. Additionally, some solid-supported materials were prepared for the storage of NO^+ species and for its release in reactions of N-nitrosation of amides [26] (Figure 15.10).

15.2 UTILIZATION OF SOME COMPONENTS OF TRANSFORMER OILS *VIA* ARYNES

Polychlorinated biphenyls (PCBs), a group of 209 organic chlorine compounds with 1 to 10 Cl atoms with the formula $C_{12}H_{10}-xClx$, are highly toxic components of transformer oils [27], and some PCBs possess structural similarity and toxic mode of action with dioxins [28] or could form dioxins *via* chemical transformations. One of the possible ways for the chemical utilizations of PCBs is their dehydrochlorination in the presence of strong bases and the following interaction of isomeric bis-arynes with nucleophiles (Figure 15.11).

One of the applications of such a reaction we have used in our research aimed to obtain isomeric dicarbazolyl-substituted diphenyls, such as 2,2'- and 3,3'-di(9*H*-carbazol-9-yl) biphenyls. These compounds are promising host materials for the construction of OLEDs due to the higher energy of the triplet state [29] due to the lower degree of the conjugation of (hetero)aromatic rings. We have developed a method for the preparation of isomeric symmetric 3,3'-, 4,4'-di(9*H*-carbazol-9-yl)biphenyls as well as 3,4'-di(9*H*-carbazol-9-yl)biphenyl *via* the interaction of the corresponding biphenyl-arynes, generated *in situ* from 4,4'-dihalogenbiphenyls, with carbazole. Depending on the nature of the halogen atoms the reaction afforded 4,4'-di(9*H*-carbazol-9-yl)biphenyl in up to 90% [30] (Figure 15.12).

Additionally, we have developed an 'aryne approach' to the chemosensors for nitro-analytes [31–33]. The dehydrohalogenetaion reaction between the corresponding mono- or bishalogen(het)arenes under the action of *t*-BuOK and the following Diels-Alder cycloaddition between thus forming arynes and anthracene afforded 1,4-di(thiophen-2-yl)-9,10-dihydro-9,10-[1,2]benzenoanthracene and 6,13-R-5,7,12,14-tetrahydro-5,14:7,12-bis([1,2]benzeno)pentacenes substituted with various (het)arene moieties *via* the (Figure 15.13). These iptycenes have exhibited an intensive '*turn-off*' fluorescence response towards nitro-explosive components both in solution and in a vapour phase.

15.3 UTILIZATION OF COMPONENTS OF AGRICULTURAL WASTES

Agricultural wastes produced by food and agricultural industries are rich in bioactive compounds and can be used as an alternative source for the production of raw materials for both lab researches and industrial processes. In our studies we have developed several approaches towards utilization of some components of agricultural wastes.

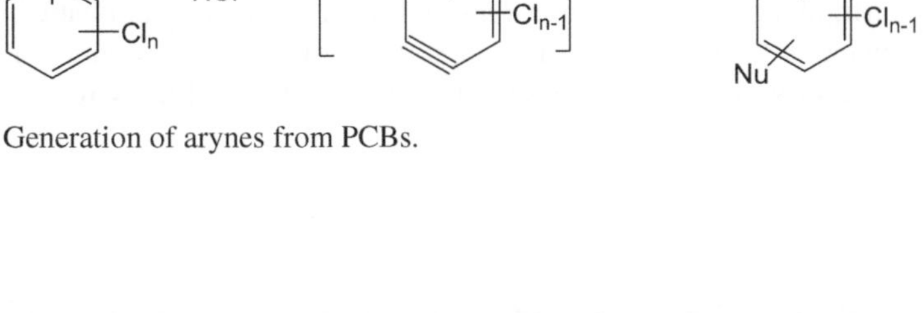

FIGURE 15.11 Generation of arynes from PCBs.

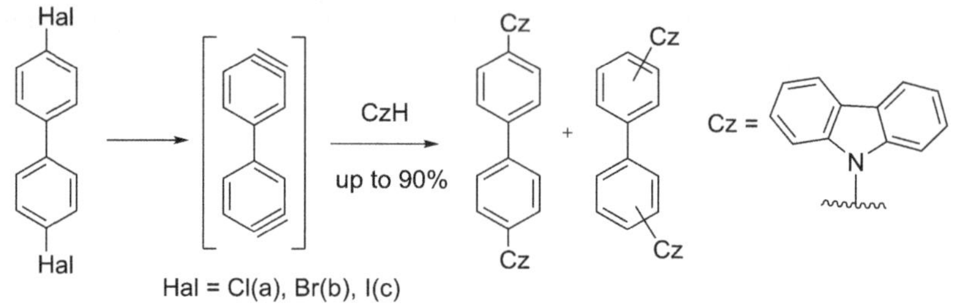

FIGURE 15.12   Preparation of isomeric di(9*H*-carbazol-9-yl)biphenyls.

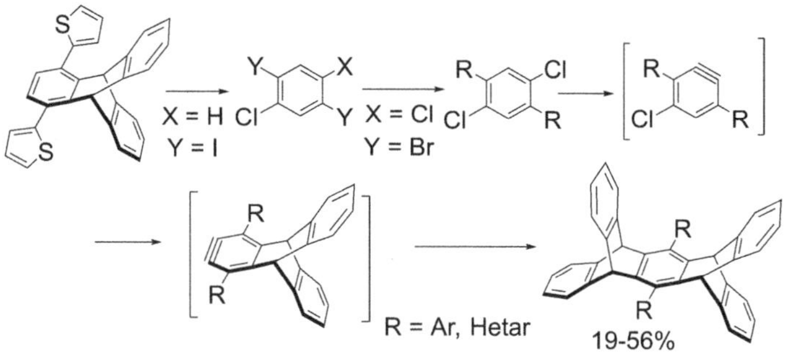

FIGURE 15.13   'Aryne approach' towards iptycene-based chemosensors for nitro-analytes.

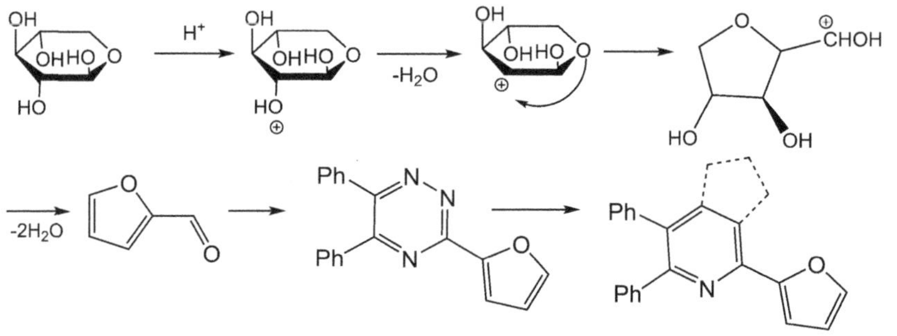

FIGURE 15.14   Preparation of furfural and its use for the synthesis of 1,2,4-triazines.

Furfural is a main source for the production of furfuryl alcohol, an industrial product used world-wide [34]. And sunflower husk is one of the main sources for the production of furfural, as well as other non-petroleum-derived chemicals and solvents, including furan and tetrahydrofuran. Among the pentosans containing fibrous materials, which can theoretically be used as raw materials for the production of furfural, only sunflower husk contains up to 25% pentosans. In our research, we simulated furfural production from the pre-ball-milled sunflower husk with a particle sizes of 6–10 μm. As a result, the furfural was obtained in up to 60% yield and it can be further used to prepare 5,6-diphenyl-3-(furan-2-yl)-1,2,4-triazines and 1-(furan-2-yl)-3,4-diphenyl-6,7-dihydro-5*H*-cyclopenta[*c*]pyridine [35] (Figure 15.14).

In addition to furfural, agricultural sources are rich in aromatic and aliphatic alcohols. The first ones include polyphenols [36] and the common sources of these compounds are tea and coffee extracts. There are reports about using these extracts as good stabilizers for the green synthesis

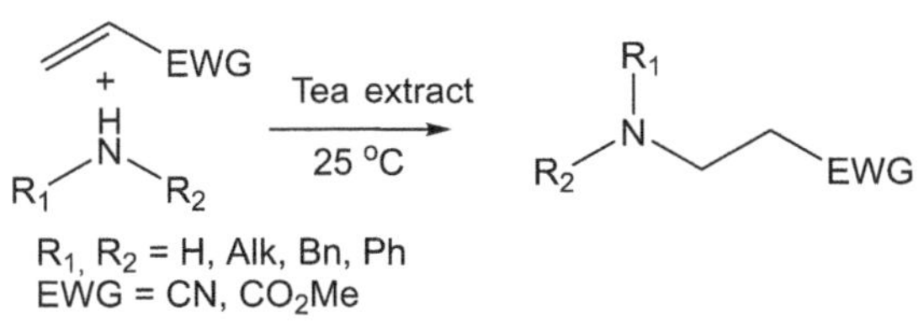

FIGURE 15.15 Tea extract-catalyzed conjugate addition of amines to vinyl derivatives.

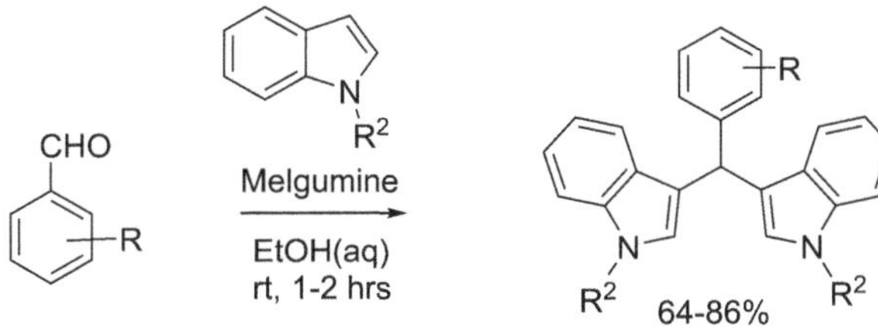

FIGURE 15.16 Melgumine – catalyzed synthesis of bis-indoles.

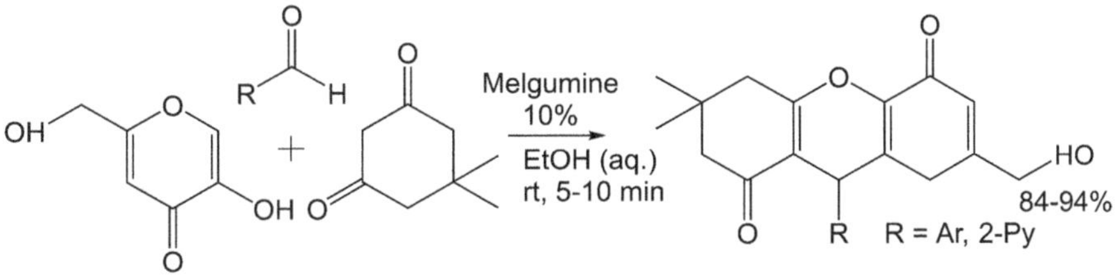

FIGURE 15.17 Meglumine-catalyzed synthesis of dihydropyrano[3, 2-*b*]chromenediones.

of silver and palladium nanoparticles [37]. Also, it was reported that the extracts of normal tea are acidic in nature [38]. We have demonstrated the catalytic role of tea extract for the important organic reactions, for instance, for the conjugate addition of a variety of amines to different Michael acceptors, such as EWG-substituted vinyl derivatives, affording a great variety of α,β-ethylenic derivatives in 75–96% yields [39] (Figure 15.15).

Meglumine is a sugar alcohol derived from glucose. So far, there are only a very few examples of using meglumine derivatives as catalyzts in organic transformations. For instance, we developed a method for the preparation of bis-indoles by using meglumine sulfate as the catalyzt [40] (Figure 15.16). Some of the obtained compounds exhibited potent antioxidant activity *in vitro* [41].

In addition, we have developed a convenient approach to dihydropyrano[3, 2-*b*]chromenediones *via* three-component reaction catalyzed by meglumine [42] (Figure 15.17).

As a way to utilize some other naturally occurring chiral alcohols (such as borneols, cholesterol, *L*-menthol, etc.) their reactions as *O*-nucleophiles with 1,2,4-triazin-5(4H)-ones were used to prepare several 1,2,4-triazines as chiral ligands and potential drug candidates [43] (Figure 15.18). It worthwhile to mention that most alcohols afforded a racemic mixture of 6-alkoxy-1,2,4-triazines along with 5-hydroxy-substituted ones, while *L*-menthol afforded mainly *S*-isomer [44, 45].

Additionally, an efficient way for the preparation of α-*O*-alkyl-2,2'-bipyridine push-pull fluorophores was developed *via ipso*-substitution of cyanogroup in 5-cyano-1,2,4-triazines under the action of alkyl alcohols and the following *aza*-Diels-Alder reaction [46] (Figure 15.19):

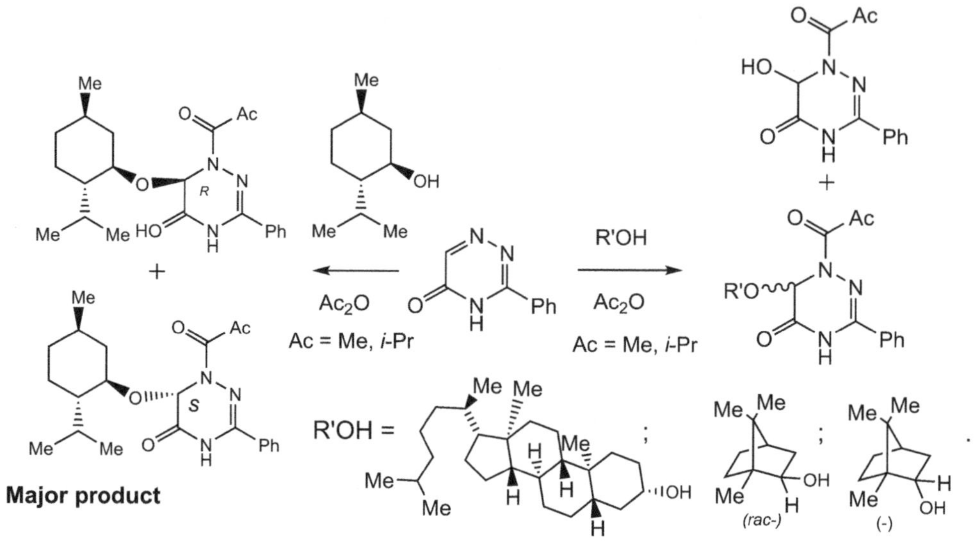

FIGURE 15.18 Chiral alcohols as synthons for the preparation of azine-ligands/drug candidates.

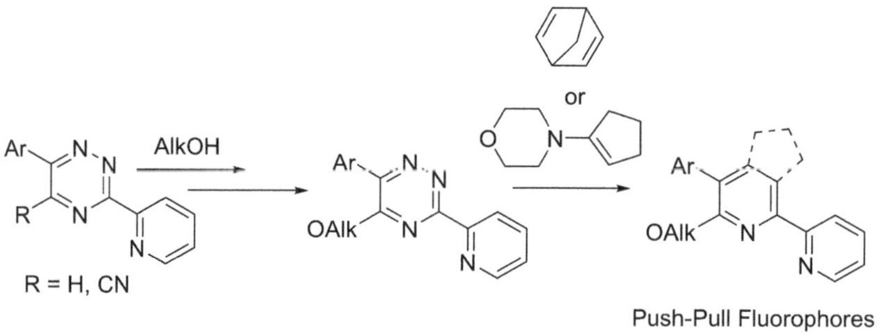

FIGURE 15.19 Preparation of α-O-alkyl-2,2'-bipyridine push-pull fluorophores.

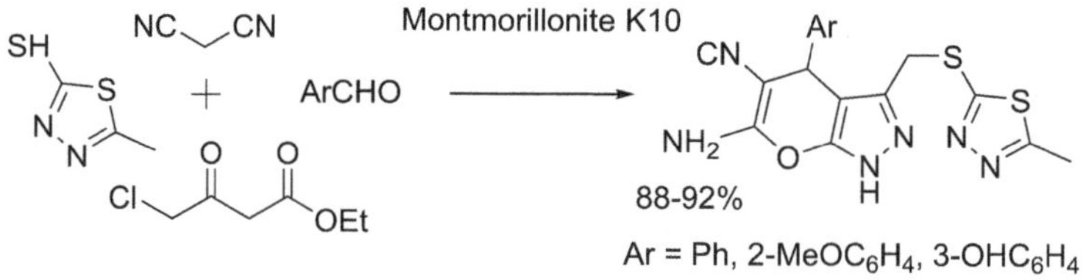

FIGURE 15.20 Montmorillonite K10 catalyzed synthesis of multi-substituted pyranopyrazoles.

15.4 UTILIZATION OF CONSTRUCTION WASTES

Bentonite is an absorbent swelling clay consisting of montmorillonite. One of the possible applications of bentonite clay is a partial substitute of cement in concrete in construction materials for various purposes [47]. On the other hand, montmorillonites could be used as catalyzts for multicomponent reactions. Thus, we have developed a montmorillonite K10 catalyzed four-component synthesis of multi-substituted pyranopyrazoles, which showed promising antimicrobial activity [48] (Figure 15.20).

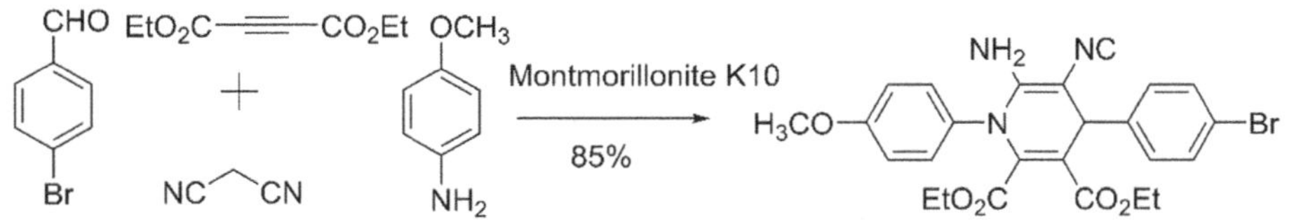

FIGURE 15.21 One-pot multicomponent synthesis of 1,4-dihydropyridines.

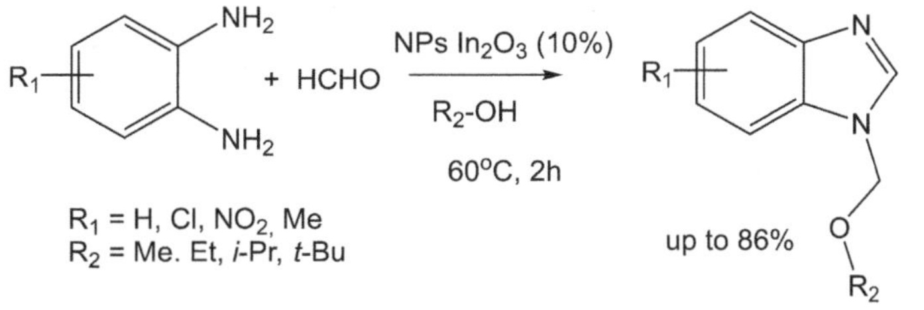

FIGURE 15.22 In_2O_3 NPs-catalyzed synthesis of 1-(alkoxymethyl)-1H-benzo[d]imidazoles.

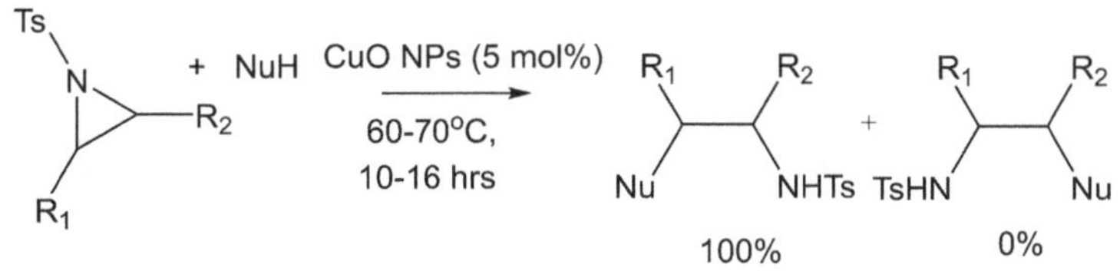

FIGURE 15.23 CuO NPs-catalyzed nucleophilic ring opening of N-Ts aziridines.

We also demonstrated an environmentally benign one-pot multi-component synthesis of 1,4-dihydropyridine derivatives using montmorillonite K10 [49] (Figure 15.21).

Metal nanoparticles have already received wide recognition as a useful catalyzt for organic synthesis [50]. We have demonstrated in our studies that In(III) and Cu(II) oxide nanoparticles can be used as catalyzts for the synthesis of 1-(alkoxymethyl)-1H-benzo[d]imidazoles *via* cyclocondenacation reaction between *o*-phenylene diamine and formaldehyde in alcohol solutions [51] (Figure 15.22) or for the highly regioselective nucleophilic ring opening reaction in N-Ts aziridines [52] (Figure 15.23).

It worthwhile to mention that, in addition to naturel sources [53], construction, industrial, electric/electronic, and plastic wastes can be used as a source for the synthesis of nanoparticles [54].

15.5 IONIC LIQUIDS AND MOLTEN SALTS AS ALTERNATIVE SOLVENTS

Organic solvents, especially chlorinates ones, are among major environment pollutants [55]. Use of a non-conventional reaction media could be beneficial in terms of lowering environmental impact, caused by organic solvents. In our researches we are utilizing two approaches: use of ionic liquids/molten salts as media for the organic transformations or using solvent-less or solvent-free reaction conditions [56]. The main benefits of the ionic liquids/molten salts are their reusability, low toxicity, inflammability, water solubility and, finally, their suitability for dissolving wide ranges of substrates at room temperature [57]. In our studies we have developed several synthetic approaches towards important organic synthons or important drug candidates by using ionic liquids/molten salts. Thus, we have developed a method for the highly regioselective nucleophilic ring opening in

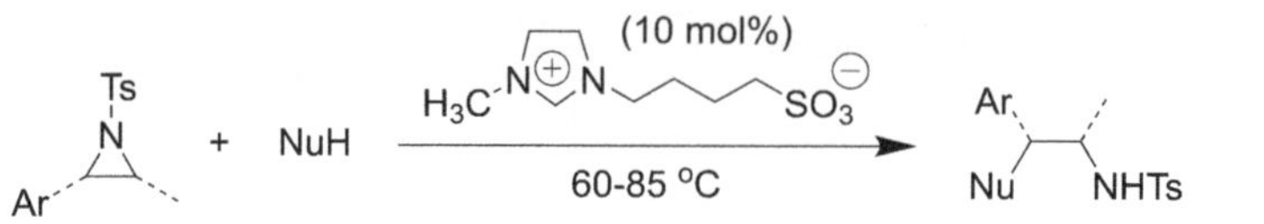

NuH = indole, pyrrole, methanol, ethanol, acetic acid, *N,N*-diisopropylamine

FIGURE 15.24   MBS-catalyzed regioselective nucleophilic ring opening in N-Ts aziridines.

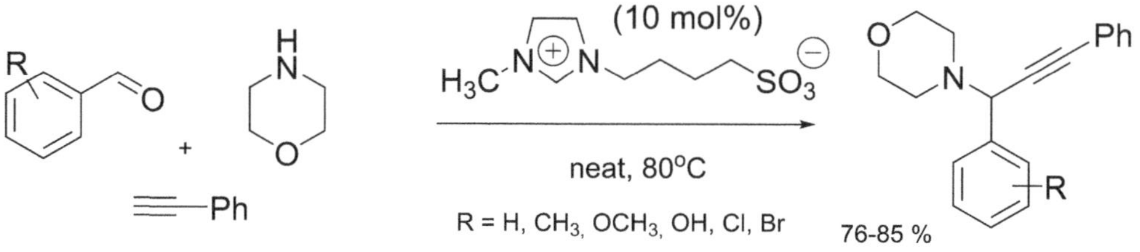

R = H, CH$_3$, OCH$_3$, OH, Cl, Br   76-85 %

FIGURE 15.25   MBS-catalyzed three-component synthesis of 4-(1,3-diarylprop-2-ynyl)morpholines.

R = H, CH$_3$, OCH$_3$, NO$_2$, Cl, Br   80-86 %

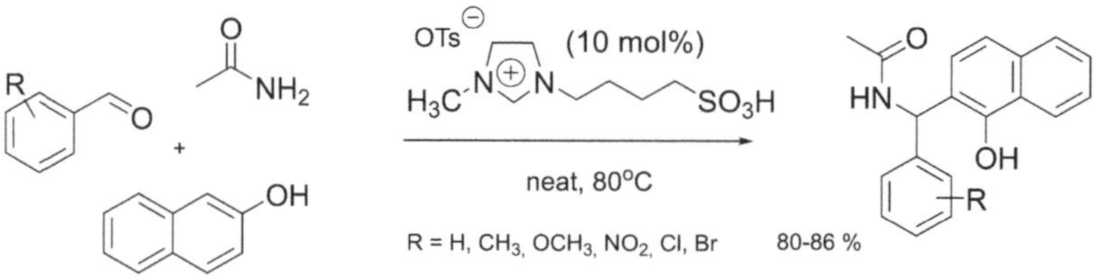

FIGURE 15.26   BAILs-catalyzed synthesis of N-((1-hydroxynaphthalen-2-yl)(aryl)methyl)acetamides.

N-Ts aziridines in the presence of 4-(3-methylimidazolium)butane sulfonate (MBS) in neat [58] (Figure 15.24).

Using the same molten salt a three-component synthesis of 4-(1,3-diarylprop-2-ynyl)morpholines has been developed to afford target products in 76–85% yields [59] (Figure 15.25).

In a similar manner as using Brønsted acid ionic liquids (BAILs) in neat, we have developed efficient approaches towards new derivatives of N-((1-hydroxynaphthalen-2-yl)(aryl)methyl) acetamides [60] (Figure 15.26), quinazolin-4(3*H*)-ones [61, 62] (Figure 15.27) and 2-arylbenzo[*d*] thiazoles [63] (Figure 15.28).

15.6   SOLVENT-FREE REACTIONS FOR THE PREPARATION OF SOME IMPORTANT MACRO- AND HETEROCYCLES

In order to estimate how 'green' the chemical reaction is one needs to calculate its E-factor, namely the ratio of total weight of byproducts, reactants, solvents, catalyzts and catalyzt supports, and any-thing else counted as wastes to the total weight of the target product(s) [64]. And the removing solvents, as the main contributors (in the E-factor), from the reaction media would reduce the E-factor dramatically, especially for the chemistry of macrocycles. Thus, we have developed solvent-free and solvent-less methods for the synthesis of calix[8]arenes [65] (Figure 15.29), thiacalix[4] arenes [65] (Figure 15.30) [65] as well as pillar[6]arenes [65, 66] (Figure 15.31).

Along with macrocycles, some important heterocycles can be prepared. By using solvent-free solid-state co-grinding between N-tosylaziridines and nitriles in the presence of HClO$_4$ as a catalyzt the high-yield method for the synthesis of 2-imidazolines has been developed [67] (Figure 15.32).

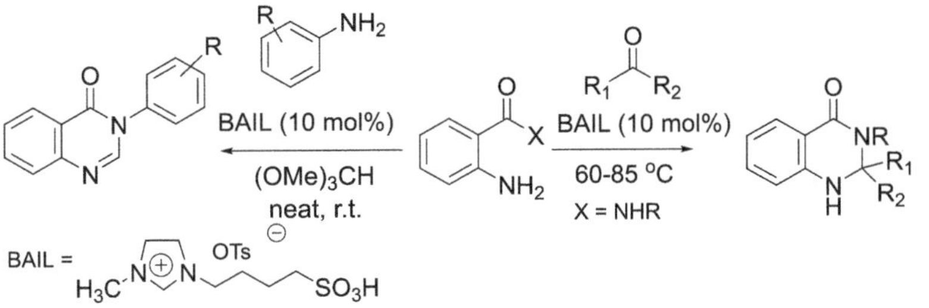

FIGURE 15.27   BAILs-catalyzed synthesis of quinazolin-4(3*H*)-ones.

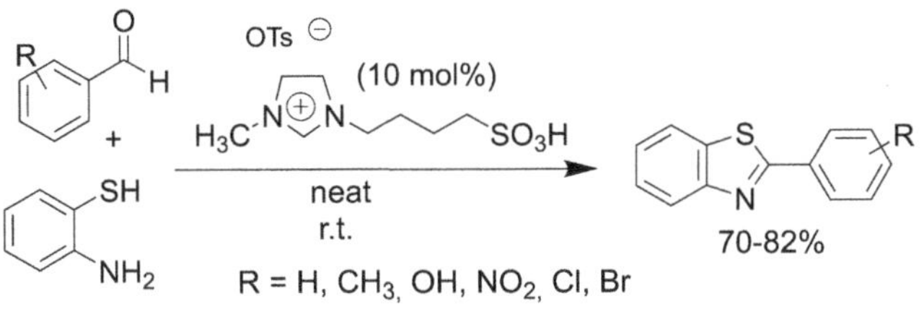

FIGURE 15.28   BAILs-catalyzed synthesis of 2-arylbenzo[*d*]thiazoles.

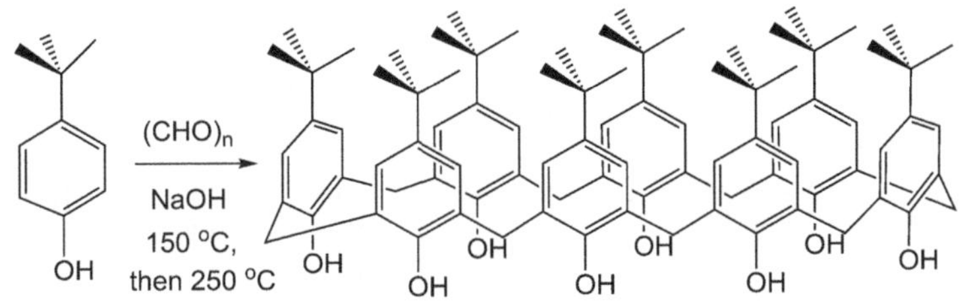

FIGURE 15.29   Solvent-less synthesis of calix[8]arenes.

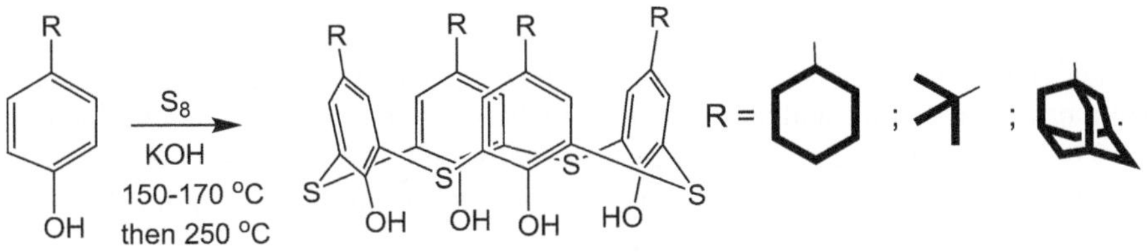

FIGURE 15.30   Solvent-free synthesis of thiacalix[4]arenes.

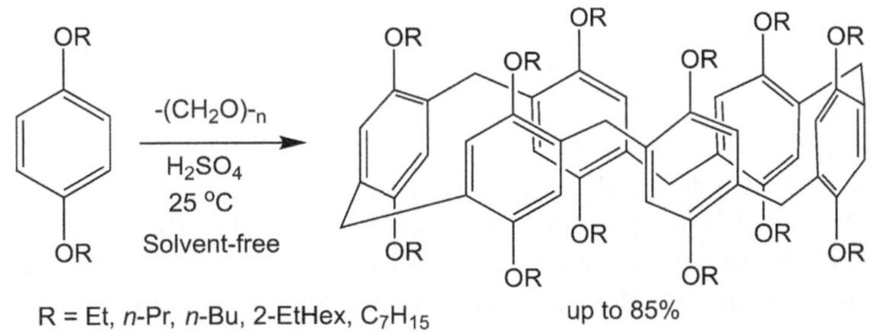

FIGURE 15.31   Solvent-free synthesis of pillar[6]arenes.

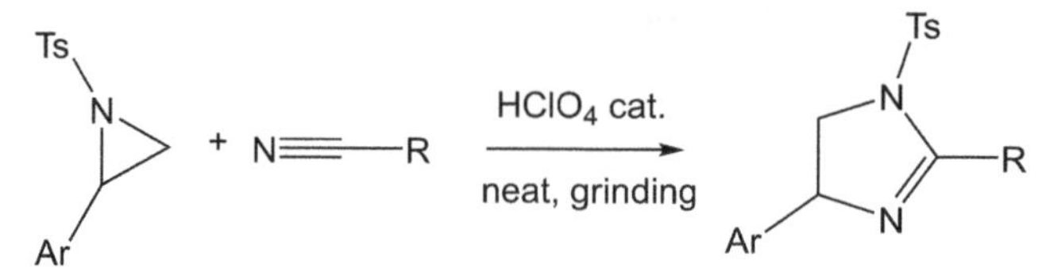

FIGURE 15.32   Solvent-free synthesis of 2-imidazolines.

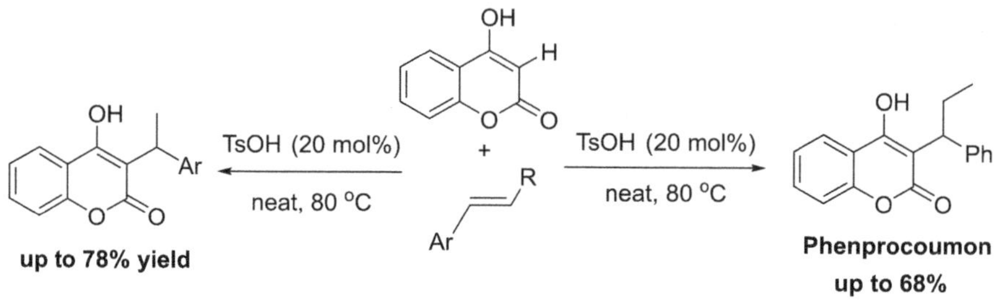

FIGURE 15.33   Solvent-free synthesis of 4-hydroxy-3-(1-arylalkyl)-2*H*-chromen-2-ones.

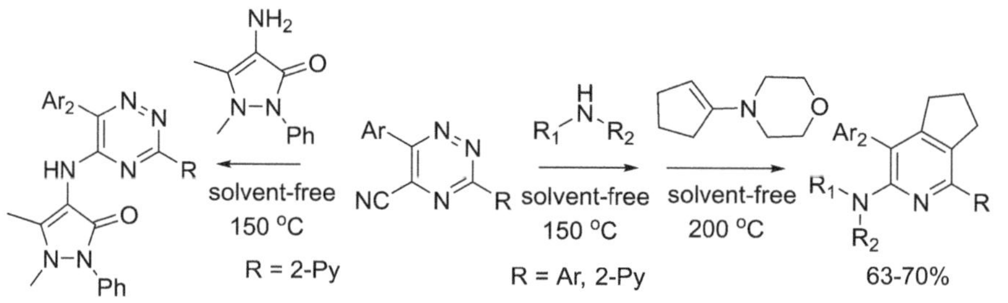

FIGURE 15.34   Solvent-free synthesis of 5-amino-1,2,4-triazines and α-aminopyridines.

As another approach, we have developed an efficient way to prepare ligands/push-pull fluorophores or drug candidates by using solvent-free heating of coumarins with styrenes in the presence of *p*-toluene sulphonic acid as the catalyzt to afford 4-hydroxy-3-(1-arylalkyl)-2*H*-chromen-2-ones [68] (Figure 15.33) or by means of catalyzt-free reaction between 5-cyano-1,2,4-triazines and arylamines [69] or hetarylamines, such as known drug 4-aminoantipyrine [70] (Figure 15.34), to afford 5-amino-1,2,4-triazines as precursors of α-aminopyridines.

15.7   LIGANDS/EXTRACTANTS FOR METAL CATIONS

Metal ions are widely involved in many industrial fields and, as a result, they are widely dispersed into the environment as pollutants [71, 72]. Therefore, efficient and affordable materials for the metal cations recognition and extraction as well as efficient methods for their preparation are on high demand. In our studies, we are developing efficient methods for the preparation of 2,2'-bipyridine-based ligands for the recognition transition metal cations, such as Zn^{2+}, Cd^{2+}, etc. (Figure 15.35), or DTTA-bipyridine chelates for lanthanide cations, such as Eu^{3+} and Tb^{3+} (Figure 15.36) [7]. Some of the obtained DTTA-appended ligands have demonstrated efficient sensibilization of Eu^{3+} and Tb^{3+} cations emission with photoluminescence quantum yields (PLQYs) up to 27.1%.

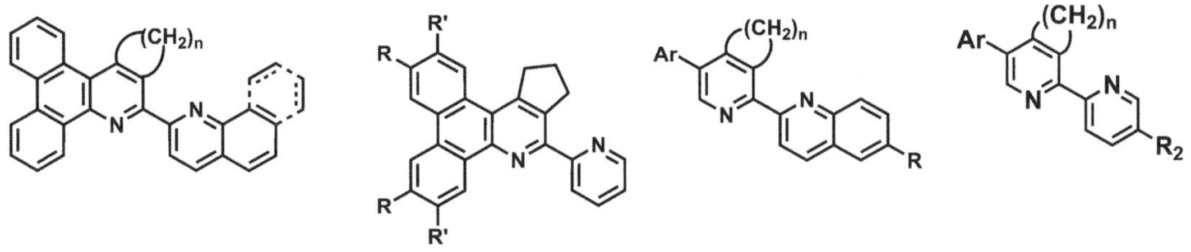

FIGURE 15.35 2,2'-Bipyridine-based ligands for the transition metal cations.

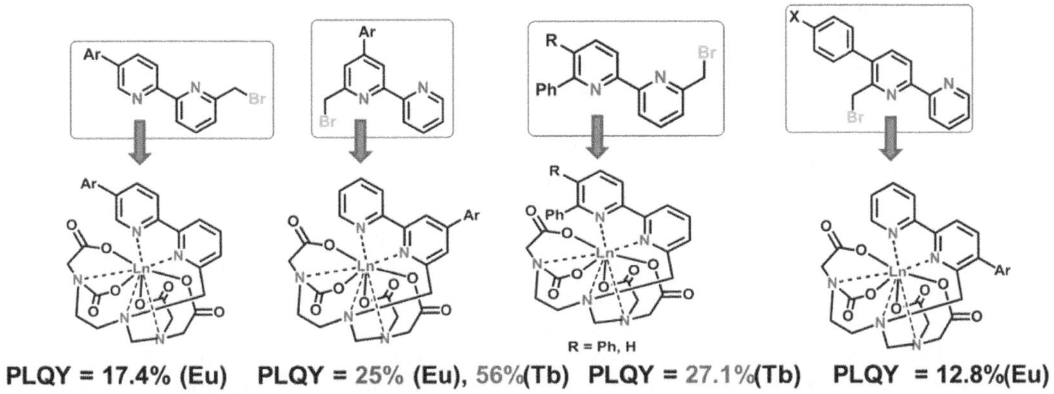

PLQY = 17.4% (Eu)   PLQY = 25% (Eu), 56%(Tb)   PLQY = 27.1%(Tb)   PLQY = 12.8%(Eu)

FIGURE 15.36 DTTA-appended ligands for lanthanide cations.

15.8 CONCLUSION

In conclusion, we have reported herein our approaches and methods for the preparation of some promising molecules and materials. These approaches might be helpful in solving some problems of green and sustainable chemistry, such as development of resource-saving and environmentally friendly synthetic methodologies while involving some common components of industrial, agricultural and construction wastes. As a result, a wide range of small molecules and functional materials were prepared for possible use as drug candidates, selective reagents for organic synthesis, ligands for metal cations, fluorophores, sorbents for greenhouse gases and other applications.

ACKNOWLEDGEMENTS

This work was supported by Grants Council of the President of the RF (Ref. # NSh-1223.2022.1.3) & Russian Foundation for Basic Research (Ref. # 19-53-55002).

REFERENCES

1. F. Savini "The economy that runs on waste: Accumulation in the circular city," *Journal of Environmental Policy & Planning* 21(6) (2019): 675–691.
2. Z. Kish *Perspectives on Waste-to-Energy Technologies* (Quasar Science Tech, March 2016), 5. https://img1.wsimg.com/blobby/go/757a9198-b0ad-4152-be42-2f9c86003574/downloads/Perspectives%20on%20Waste-to-Energy%20Technologies.pdf?ver=1556046403667
3. O. Awogbemi, D. V. Von Kallon "Valorization of agricultural wastes for biofuel applications," *Heliyon* 8(10) (2022): e11117.
4. J. Cullen "Circular economy: Theoretical benchmark or perpetual motion machine?" *Journal of Industrial Ecology* 21 (3) (2017): 483–486.

5. C. Zhuo, Y. A. Levendis "Upcycling waste plastics into carbon nanomaterials: A review," *Journal of Applied Polymer Science* 131 (2014): 39931.

6. D. M. Rudkevich, Y. Kang, A. V. Leontiev, V. G. Organo, G. V. Zyryanov "Molecular containers for NOx-gases," *Supramolecular Chemistry* 17(1–2) (2005): 93–99.

7. G. V. Zyryanov, D. S. Kopchuk, I. S. Kovalev, S. Santra, M. Rahman, A. F. Khasanov, A. P. Krinochkin, O. S. Taniya, O. N. Chupakhin, V. N. Charushin "Rational synthetic methods in creating promising (hetero)aromatic molecules and materials," *Mendeleev Communications* 30(5) (2020): 537–554.

8. K. C. O'Brien, Y. Lu, V. Patel, S. Greenberg, R. Locke, M. Larson, K. R. Krishnamurthy, M. Byron, J. Naumovitz, D. S. Guth, S. J. Bennett "Creating markets for captured carbon: Retrofit of Abbott power plant and future utilization of captured CO_2," *Energy Procedia* 114 (2017): 1263–1272.

9. M. Aghaie, N. Rezaei, S. Zendehboudi "A systematic review on CO_2 capture with ionic liquids: Current status and future prospects," *Renewable and Sustainable Energy Reviews* 96 (2018): 502–525.

10. S. Kumar, R. Srivastav, J. Koh "Utilization of zeolites as CO_2 capturing agents: Advances and future perspectives," *Journal of CO_2 Utilization* 41 (2020): 101251.

11. G. Steenbruggen, G. G. Hollman "The synthesis of zeolites from fly ash and the properties of the zeolite products," *Journal of Geochemical Exploration* 62(1–3) (1998): 305–309.

12. X. Querol, N. Moreno, J. C. Umaña, A. Alastuey, E. Hernández, A. López-Soler, F. Plana "Synthesis of zeolites from coal fly ash: An overview," *International Journal of Coal Geology* 50(1–4) (2002): 413–423.

13. C. X. Tan, G. H. Chong, H. Hamzah, H. M. Ghazalia "Comparison of subcritical CO_2 and ultrasound-assisted aqueous methods with the conventional solvent method in the extraction of avocado oil," *The Journal of Supercritical Fluids* 135 (2018): 45–51.

14. M. Marchionni, G. Bianchi, S. A. Tassou "Review of supercritical carbon dioxide (sCO_2) technologies for high-grade waste heat to power conversion," *SN Application Science* 2 (2020): 611.

15. A. V. Baklykov, G. L. Rusinov, G. V. Zyryanov, D. S. Kopchuk, V. N. Charushin, G. A. Artem'ev, V. L. Rusinov, A. F. Khasanov "Comparison of methods of synthesis of 5-methyl-1,2,4-triazolo[1,5-a]pyrimidin-7(4H)-one in supercritical carbon dioxide," *AIP Conference Proceedings* 2280 (2020): 040005.

16. S. Santra, M. Rahman, I. S. Kovalev, D. S. Kopchuk, G. V. Zyryanov, V. N. Charushin, O. N. Chupakhin "Synthesis of pillar[5]arenes in supercritical carbon dioxide," *Book of Abstracts XX Mendeleev Congress on General and Applied Chemistry*, 26–30 September 2016, Yekaterinburg, Russia.

17. D. Bhattacherjee, M. Rahman, S. Ghosh, A. K. Bagdi, G. V. Zyryanov, O. N. Chupakhin, P. Das, A. Hajra "Advances in transition-metal catalyzed carbonylative Suzuki-Miyaura coupling reaction: An update," *Advanced Synthesis and Catalysis* 363(6) (2021): 1597–1624.

18. D. Das, B. M. Bhanage "Double carbonylation reactions: Overview and recent advances," *Advanced Synthesis and Catalysis* 362(15) (2020): 3022–3058.

19. H. Yoshida, H. Fukushima, J. Ohshita, A. Kunai "CO_2 incorporation reaction using arynes: Straightforward access to benzoxazinone," *Journal of the American Chemical Society* 128(34), (2006): 11040–11041.

20. W.-J. Yoo, T. V. Q. Nguyen, S. Kobayashi "Synthesis of isocoumarins through three-component couplings of arynes, terminal alkynes, and carbon dioxide catalyzed by an NHC-copper complex," *Angewandte Chemie International Edition* 53 (2014): 10213–10217.

21. R. Veligeti, R. B. Madhu, J. Anireddy, V. R. Pasupuleti, V. K. R. Avula, K. S. Ethiraj, S. Uppalanchi, S. Kasturi, Y. Perumal, H. S. Anantaraju, N. Polkam, M. R. Guda, S. Vallela, G. V. Zyryanov "Synthesis of novel cytotoxic tetracyclic acridone derivatives and study of their molecular docking, ADMET, QSAR, bioactivity and protein binding properties," *Scientific Reports* 10(1) (2020): 20720.

22. R. Rathore, S. V. Lindeman, K. S. S. P. Rao, D. Sun, J. K. Kochi "Guest penetration deep within the cavity of Calix[4]Arene hosts: The tight binding of nitric oxide to distal (Cofacial) aromatic groups," *Angewandte Chemie International Edition* 39 (2000): 2123–2127.

23. D. M. Rudkevich, G. V. Zyryanov "Solid-state materials and molecular cavities and containers for the supramolecular recognition and storage of NO_x-species: A review," *Comments on Inorganic Chemistry* 35(3) (2015): 128–178.

24. G. V. Zyryanov, Y. Kang, D. M. Rudkevich "Sensing and fixation of NO2/N2O4 by Calix[4]Arenes," *Journal of the American Chemical Society* 125 (2003): 2997–3007.

25. G. V. Zyryanov, Y. Kang, S. P. Stampp, D. M. Rudkevich "Supramolecular fixation of NO_2 with calix[4]arenes," *Chemical Communication* 61 (2002): 2792–2793.

26. Y. Kang, G. V. Zyryanov, D. M. Rudkevich "Towards supramolecular fixation of NOx gases: Encapsulated reagents for nitrosation," *Chemistry –– A European Journal* 11(6), (2005): 1924–1932.

27. Robertson, Larry W.; Hansen, Larry G., eds. (2004). *PCBs: Recent Advances in Environmental Toxicology and Health Effects* (Lexington, KY, USA: University Press of Kentucky, ISBN 978-0813122267), 11.

28. P. J. Sauer, M. Huisman, C. Koopman-Esseboom, D. C. Morse, A. E. Smits-van Prooije, K. J. van de Berg, L. G. Tuinstra, C. G. van der Paauw, E. R. Boersma, N. Weisglas-Kuperus, J. H. C. M. Lammers, B. M. Kulig, A. Brouwer "Effects of Polychlorinated Biphenyls (PCBs) and dioxins on growth and development," *Human & Experimental Toxicology* 13(12) (1994): 900–906.

29. S. Gong, X. He, Y. Chen, Z. Jiang, C. Zhong, D. Ma, J. Qin, C. Yang "Simple CBP isomers with high triplet energies for highly efficient blue electrophosphorescence," *Journal of Materials Chemistry* 22 (2012): 2894–2899.

30. I. S. Kovalev, D. E. Pavlyuk, V. A. Zaripov, G. V. Zyryanov, D. S. Kopchuk, V. L. Rusinov, O. N. Chupakhin "Synthesis of substituted 4,4´-dihalobiphenyls and their use for the preparation of isomeric bis(carbazolyl)biphenyls," *Russian Chemical Bulletin* 64 (2015): 1978–1981.

31. G. V. Zyryanov, M. A. Palacios, P. Anzenbacher "Simple molecule-based fluorescent sensors for vapor detection of TNT," *Organic Letters* 10 (2008): 3681–3684.

32. L. Mosca, P. Koutník, V. M. Lynch, G. V. Zyryanov, N. A. Esipenko, P. Anzenbacher, "Host–guest complexes of pentiptycene receptors display edge-to-face interaction," *Crystals Growth and Design* 12 (2012): 6104–6109.

33. P. Anzenbacher, L. Mosca, M. A. Palacios, G. V. Zyryanov, P. Koutnik "Iptycene-based fluorescent sensors for nitroaromatics and TNT," *Chemistry –– A European Journal* 18 (2012): 12712–12718.

34. A. I. Casoni, P. M. Hoch, M. A. Volpe, V. S. Gutierrez "Catalytic conversion of furfural from pyrolysis of sunflower seed hulls for producing bio-based furfuryl alcohol," *Journal of Cleaner Production* 178 (2018): 237–246.

35. O. S. Taniya, L. K. Sadieva, I. S. Kovalev, G. V. Zyryanov, D. S. Kopchuk, V. L. Rusinov, O. N. Chupakhin "Synthesis of furfural from pre-ball-milled sunflower husks," *AIP Conference Proceedings* 2280 (2020): 0018848.

36. S. Vijayalaxmi, S. K. Jayalakshmi, and K. Sreeramulu "Polyphenols from different agricultural residues: Extraction, identification and their antioxidant properties," *Journal of Food Science and Technology* 52(5) (2015): 2761–2769.

37. N. N. Mallikarjuna, R. S. Varma "Green synthesis of silver and palladium nanoparticles at room temperature using coffee and tea extract," *Green Chemistry* 10(8) (2008): 859–862.

38. Q. V. Vuong, J. B. Golding, C. E. Stathopoulos, P. D. Roach "Effects of aqueous brewing solution *pH* on the extraction of the major green tea constituents," *Food Research International* 53(2) (2013): 713–719.

39. A. Mukherjee, R. Chatterjee, A. De, S. Samanta, S. Mahato, N. C. Ghosal, G. V. Zyryanov, A. Majee "Conjugated addition of amines to electron deficient alkenes: A green approach," *Chimica Techno Acta* 4(2) (2017): 140–147.

40. G. Sravya, G. V. Zyryanov, M. R. M. Reddy, V. N. Ratnakaram, N. B. Reddy "Meglu-mine sulphate: An efficient catalyst for the synthesis of bis(indolyl)methanes," *AIP Conference Proceedings* 2280 (2020): 040048.

41. B. R. Nemallapudi, G. V. Zyryanov, B. Avula, M. R. Guda, S. Gundala "Meglumine as a green, efficient and reusable catalyst for synthesis and molecular docking studies of bis(indolyl)methanes as antioxidant agents," *Bioorganic Chemistry* 87 (2019): 465–473.

42. I. N. Egorov, G. V. Zyryanov, P. A. Slepukhin, T. A. Tseitler, V. L. Rusinov, O. N. Chupakhin "Reaction of 3-phenyl-1,2,4-triazin-5(4H)-one under acylating conditions with natural alcohols containing an asymmetric carbon atom," *Chemistry of Heterocyclic Compounds* 48 (2012): 625–633.

43. G. Sravya, G. Suresh, Grigory V. Zyryanov, A. Balakrishna, K. M. K. Reddy, C. S. Reddy, C. Venkataramaiah, W. Rajendra, N. B. Reddy "A meglumine catalyst–based synthesis, molecular docking, and antioxidant studies of dihydropyrano[3, 2-b]chromenedione derivatives," *Journal of Heterocyclic Chemistry* 57(1) (2020): 355–369.

44. O. N. Chupakhin, G. V. Zyryanov, V. L. Rusinov, V. P. Krasnov, G, L. Levit, M. A. Korolyova, M. I. Kodess "Direct diastereoselective addition of l-menthol to activated 1,2,4-triazin-5(4H)-one," *Tetrahedron Letters* 42 (2001): 2393–2395.

45. G. V. Zyryanov, V. L. Rusinov, O. N. Chupakhin, V. P. Krasnov, G. L. Levit, M. I. Kodess, T. S. Shtukina "Direct diastereoselective introduction of l-menthol residue into 1,2,4-triazin-5(4H)-one," *Russian Chemical Bulletin* 53 (2004): 1290–1294.

46. M. I. Savchuk, A. F. Khasanov, D. S. Kopchuk, A. P. Krinochkin, I. L. Nikonov, E. S. Starnovskaya, Y. K. Staitz, I. S. Kovalev, G. V. Zyryanov, O. N. Chupakhin "New push-pull fluorophores on the basis of 6-alkoxy-2,2'-bipyridines: Rational synthetic approach and photophysical properties," *Chemistry of Heterocyclic Compounds* 55 (2019): 554–559.

47. A. O. Adeboje, W. K. Kupolati, E. R. Sadiku, J. M. Ndambuki, C. Kambole "Experimental investigation of modified bentonite clay-crumb rubber concrete," *Construction and Building Materials* 233 (2020): 117187.

48. G. M. Reddy, J. R. Garcia, G. V. Zyryanov, G. Sravya, N. B. Reddy "Pyranopyrazoles as efficient antimicrobial agents: Green, one pot and multicomponent approach," *Bioorganic Chemistry* 82 (2019): 324–331.

49. V. H. Reddy, A. K. Kumari, G. M. Reddy, Y. V. R. Reddy, J. R. Garcia, G. V. Zyryanov, N. B. Reddy, A. Rammohan "Environmentally benign one-pot multicomponent synthesis of 1,4-dihydropyridine derivatives applying montmorillonite K10 as reusable catalyst," *Chemistry of Heterocyclic Compounds* 55(1) (2019): 60–65.

50. N. K. Ojha, G. V. Zyryanov, A. Majee, V. N. Charushin, O. N. Chupakhin, S. Santra, "Copper nanoparticles as inexpensive and efficient catalyst: A valuable contribution in organic synthesis," *Coordination Chemistry Reviews* 353 (2017): 1–57.

51. S. Samanta, S. Mahato, R. Chatterjee, S. Santra, G. V.Zyryanov, A. Majee "Nano indium oxide-catalyzed domino reaction for the synthesis of N-alkoxylated benzimidazoles," *Tetrahedron Letters* 61(32) (2020): 152177.

52. R. Chatterjee, S. Santra, N. C. Ghosal, K. Giri, G. V. Zyryanov, A. Majee "CuO nanoparticles as a simple and efficient green catalyst for the aziridine ring-opening: Examination of a broad range of nucleophiles," *ChemistrySelect* 5(15) (2020): 4525–4529.

53. J. Singh, T. Dutta, K.-H. Kim, M. Rawat, P. Samddar, P. Kumar "Green" synthesis of metals and their oxide nanoparticles: Applications for environmental remediation," *Journal of Nanobiotechnology* 16 (2018): 84.

54. S. M. Abdelbasir, K. M. McCourt, Cindy M. Lee, Diana C. Vanegas "Waste-derived nanoparticles: Synthesis approaches, environmental applications, and sustainability considerations," *Frontiers in Chemistry* 8 (2020): 782. https://doi.org/10.3389/fchem.2020.00782

55. M. Tobiszewski, J. Namieśnik, F. Pena-Pereira "Environmental risk-based ranking of solvents using the combination of a multimedia model and multi-criteria decision analysis," *Green Chemistry* 19 (2017): 1034–1042.

56. A. Loupy "Solvent-free Reactions" In: Knochel, P. (eds) *Modern Solvents in Organic Synthesis. Topics in Current Chemistry* vol 206 (Berlin-Heidelberg: Springer, 1999) https://doi.org/10.1007/3-540-48664-X_7.

57. J. Flieger, M. Flieger "Ionic liquids toxicity—Benefits and threats," *International Journal of Molecular Sciences* 21(17) (2020): 6267.

58. N. Chakraborty Ghosal, S. Santra, S. Das, A. Hajra, G. V. Zyryanov, A. Majee "Organocatalysis by an aprotic imidazolium zwitterion: Regioselective ring-opening of aziridines and applicable to gram scale synthesis," *Green Chemistry* 18 (2016): 565–574.

59. S. Sarkar, R. Chatterjee, A. Mukherjee, S. Santra, G. V. Zyryanov, A. Majee "Zwitterionic molten salt: An efficient organocatalyst for the one-pot synthesis of propargylamines," *AIP Conference Proceedings* 2280 (2020): 050049.

60. R. Chatterjee, A. Mukherjee, Satyajit Pal, S. Santra, G. V. Zyryanov, A. Majee "Brønsted acidic ionic liquid promoted synthesis of amidoalkyl naphthols under solvent-free conditions," *AIP Conference Proceedings* 2280 (2020): 040009.

61. S. Das, S. Santra, S. Jana, G. V. Zyryanov, A. Majee, A. Hajra "The remarkable cooperative effect of a Brønsted-acidic ionic liquid in the cyclization of 2-aminobenzamides with ketones," *European Journal of Organic Chemistry* 33 (2017): 4955–4962.

62. R. Chatterjee, S. Sarkar, S. Santra, G. V. Zyryanov, A. Majee "Brønsted acidic ionic liquid-catalyzed one-pot synthesis of 4(3H)-quinazolinones under solvent-free conditions," *AIP Conference Proceedings* 2280 (2020): 050013.

63. S. Pal, R. Chatterjee, S. Santra, G. V. Zyryanov, A. Majee "Ionic liquid catalyzed simple and green synthesis of benzothiazole under neat condition," *AIP Conference Proceedings* 2280 (2020): 050037.

64. R. A. Sheldon "The E factor 25 years on: The rise of green chemistry and sustainability," *Green Chemistry* 19 (2017): 18–43.

65. M. Rahman, S. Santra, I. S. Kovalev, D. S. Kopchuk, G. V. Zyryanov, A. Majee, O. N. Chupakhin "Green synthetic approaches for practically relevant (hetero)macrocycles: An overview," *AIP Conference Proceedings* 2280 (2020): 0018078.

66. S. Santra, D. S. Kopchuk, I. S. Kovalev, G. V. Zyryanov, A. Majee, V. N. Charushin, O. N. Chupakhin "Solvent-free synthesis of pillar[6]arenes," *Green Chemistry* 18 (2016): 423–426.

67. A. De, S. Santra, I. S. Kovalev, D. S. Kopchuk, G. V. Zyryanov, O. N. Chupakhin, V. N. Charushin, A. Majee," Synthesis of 2-imidazolines by co-grinding of N-tosylaziridines and nitriles," *Mendeleev Communications* 30(2) (2020): 188–189.

68. R. Chatterjee, A. Mukherjee, S. Santra, G. V. Zyryanov, O. N. Chupakhin, A. Majee "An expedient solvent-free C-benzylation of 4-hydroxycoumarin with styrenes," *Mendeleev Communications* 31(1) (2021): 123–124.

69. D. S. Kopchuk, N. V. Chepchugov, I. S. Kovalev, S. Santra, M. Rahman, K. Giri, G. V. Zyryanov, A. Majee, V. N. Charushin, O. N. Chupakhin "Solvent-free synthesis of 5-(aryl/alkyl)amino-1,2,4-triazines and α-arylamino-2,2′-bipyridines with greener prospects," *RSC Advances* 7 (2017): 9610–9619.

70. A. Rammohan, G. M. Reddy, A. P. Krinochkin, D. S. Kopchuk, M. I. Savchuk, Y. K. Shtaitz, G. V. Zyryanov, V. L. Rusinov, O. N. Chupakhin "A facile synthesis of triazine integrated antipyrine derivatives through ecofriendly approach," *Synthetic Communications* 51(2) (2021): 256–262. DOI: 10.1080/00397911.2020.1823993.

71. Y. Wu, H. Panget, Y. Liu, X. Wang, S. Yu, D. Fu, J. Chen, X. Wang "Environmental remediation of heavy metal ions by novel-nanomaterials: A review," *Environmental Pollution* 246 (2019): 608–620. DOI: 10.1016/j.envpol.2018.12.076.

72. S. Babel, T. A. Kurniawan "Low-cost adsorbents for heavy metals uptake from contaminated water: A review," *Journal of Hazardous Materials* 97 (2003): 219–243.

16 Hybrid Advanced Oxidation Processes for Treatment of Wastewater from the Pharmaceutical Industry

Pubali Mandal and Brajesh Kumar Dubey

16.1 INTRODUCTION

Pharmaceuticals are considered as one of the significant classes of emerging contaminants owing to its adverse impacts on the environment and human health. Pharmaceutical compounds can be classified into the following categories depending upon its usage; antibiotics, anti-inflammatory drugs, lipid regulators, analgesics, diuretics, steroids, beta blockers, stimulant drugs, anti-microbial, antiseptics and hormones [1, 2]. Improper usage and disposal of these pharmaceutical compounds may lead to serious consequences. The increased concentration of these pharmaceutical compounds in surface water bodies contributes to antimicrobial resistance to the hosts consuming the water. As a result, the hosts can no longer be treated by antibiotics. The sources of these compounds entering water bodies are effluents from municipality, hospital, pharmaceutical industries and shops discarding expired drugs in an inappropriate way. The presence of pharmaceuticals in groundwater, surface water, soil, sludge and sediments are reported. The compounds may undergo conversion to conjugates and metabolites. The wastewater treatment plants fail in degrading the parent compounds, conjugates or metabolites of pharmaceuticals due to the inherent properties such as recalcitrance and stability. Although most of the non-steroidal anti-inflammatory drugs show higher removal efficiency of 90%, removal efficiency of several antibiotics, such as sulphonamides, fluoroquinolones and tetracyclines were in the range of 40 to 75% [3]. Therefore, implementation of alternative treatment technologies is inevitable to bring down the risk of transferring the emerging contaminants to the aquatic environment.

Advanced oxidation processes are mainly preferred by researchers too degrade the pharmaceutical compounds to non-toxic products. To date, AOPs are one of the successful technologies for removing these compounds. AOP is mainly a process which involves highly reactive and non-selective hydroxyl radicals [4]. Hydroxyl radicals can degrade almost all types of organic contaminants including recalcitrant compounds. Hybrid AOPs are mainly processes involving integration or simultaneous application of two separate AOPs to obtain faster reaction rate and higher degree of pollutant mineralization [5]. Among the different AOPs used, Fenton and photo-Fenton, ultraviolet and H_2O_2 combination, O_3, sonolysis and photocatalysis are significant based on their sources of hydroxyl radical production [6].

Among different pharmaceutical compounds, the majority of the degradation study involves analgesics, anti-inflammatory drugs, antibiotics and antiepileptic drugs. Around 90% of the used antibiotics consumed by humans are excreted as the parent form [7]. Therefore, antibiotics are generally found in its parent compound form in the water environment. Nonsteroidal anti-inflammatory drugs, used for the treatment of inflammation and pain, are also abundantly found in water bodies.

DOI: 10.1201/9781003327646-16

Among the anti-inflammatory drugs, paracetamol, ibuprofen, diclofenac, naproxen, etc., are few common names, which are detected in wastewater treatment plants [8]. An antiepileptic drug, carbamazepine is widely detected in all types of aquatic matrices. Other than that, hormones and several other pharmaceuticals are also reported to be present [1].

In this chapter, the pathways by which pharmaceutical compounds enter the aquatic environment and the major sources are highlighted. The risk associated with the discharge of pharmaceuticals towards environment and human health is addressed. Furthermore, hybrid AOP technologies used for the treatment of pharmaceutically active compounds (both synthetic and real wastewater) are documented and discussed. A thorough analysis of the advantages and drawback of hybrid AOPs are also discussed.

16.2 PATHWAYS OF WATER BODIES' POLLUTION FROM PHARMACEUTICALS AND OCCURRENCES

Pharmaceuticals are inevitable in today's world due to the advancement of medical fields and growing health concern. However, improper management may pollute surface water, groundwater and other sources of water used for human consumption. Among the wastewater generated by human activities such as pharmaceutical industries, farms, municipal and hospital effluents are the significant sources of pharmaceuticals [9]. The pharmaceutical compounds may find their way into the aquatic ecosystem covering rivers, lakes and even groundwater once passed unaltered through the conventional wastewater treatment plants. Human excreta and secretions are the main sources of pharmaceuticals found in sewage effluent due to the consumption of drugs by human and household animals. Antibiotics, hormones and analgesics may enter the water bodies through human urine and faeces [10]. Hospital activities and pharmaceutical industries also contribute to several pharmaceutical compounds in water bodies due to their seldom usage. Hospital wastewater is a significant source of β-blockers, antibiotics and analgesics.

The occurrences of pharmaceuticals of steroid and non-steroid categories were reported in surface water bodies in countries like India, UK, USA, Australia, Canada, China, Nigeria, Germany, Switzerland, Spain, Romania, Turkey, South Korea, etc. [11]. The pharmaceuticals which are most commonly found in the environment can be categorized into several categories such as antibiotics, anti-inflammatory drugs, hormones, steroids, lipid regulators, beta blockers, antidepressants, diuretics, cancer therapeutics, antiepileptic and tranquillisers [8, 10]. The pharmaceuticals originate from hospitals, pharmaceutical manufacturing companies, household consumption and veterinary clinics, and find their way to the municipal wastewater treatment plants eventually. For example, a significant presence of antibiotics, hormones, antiepileptic, stimulant, antiseptic, antihypersensitive, antilipidemic, analgesic were reported in five wastewater treatment plants located in Ulsan city, Korea [11]. Acetaminophen, lincomycin and atenolol were the most dominant pharmaceuticals and observed concentrations were 7.46±3.2, 8.17±6.5, and 7.8±2.9 µg/L, respectively [12]. Several pharmaceuticals categorized under anti-inflammatory, antibiotics, analgesic, anti-hypertensive, gastrointestinal, antidepressants, antiepileptics, and anti-cancer groups were also found in sewage treatment works [13]. Mohapatra et al. [14] listed a total of 72 compounds generally found in municipal wastewater and also in surface water bodies. Pharmaceuticals reported in the study can be categorized as anti-inflammatory, analgesic, anti-epileptic, antibiotics, anti-histamine, lipid regulator, psychoactive, beta blockers, etc. Yearly average concentrations of pharmaceuticals in two wastewater treatment plants located in India were 537 ± 5 µg/L and 353 ± 9 µg/L. In terms of seasonal variations, the highest concentration was observed in summer followed by winter and monsoon seasons [14]. Tran et al. [15] reported the presence of 25 pharmaceuticals in water reclamation plant, which falls under the category of beta blockers, lipid regulator, neuroactive, nonsteroidal anti-inflammatory drugs and hormones.

16.3   THE COMPETENCE OF CONVENTIONAL TREATMENT PROCESSES

Several studies have been conducted on biological treatment process performance evaluation on pharmaceuticals' removal, which involves a conventional as well as a relatively advanced system. A comparative evaluation between conventional activated sludge and membrane bioreactor revealed higher removal performance of the later one [15]. Some pharmaceuticals such as atenolol, acetaminophen, fenoprofen, ibuprofen and indomethacin shows >90% removal efficiency, whereas several come out as persistent (carbamazepine, iopamido and diclofenac). The criteria for obtaining high removal efficiencies of pharmaceuticals by biological processes are existence of molecules as cation species, low (<3.5) octanol-water distribution coefficient, and electron donating groups greater than withdrawing groups in terms of both number and strength [15]. The combination of cyclic activated sludge and chlorination was able to remove 85% of total pharmaceuticals. However, screening, grit chamber and facultative aerated lagoon were found to be less effective, only 59% of total pharmaceuticals (especially for antibiotics) removal efficiency was obtained [14]. The primary treatment has minor contribution in removing pharmaceuticals from wastewater, the majority of the removal happens during the secondary treatment process. The antibiotics (sulfamethazine, metoprolol, and carbamazepine) exhibited low removal efficiencies (<30%) [12]. Therefore, the primary and secondary treatment system appears to be inefficient for significant pharmaceuticals' removal. Membrane treatment processes applied as tertiary treatment technology were able to remove 60% of diclofenac and naproxen whereas removal of carbamazepine was even less [16]. Another study involves granular activated carbon adsorption as a tertiary treatment process where sewage was pre-treated by activated sludge process [13]. Steroidal estrogens' reductions were varied from 43 to 64%. The reductions of some of the widely used pharmaceuticals (carbamazepine and propranolol, etc.) were much less, i.e., 17 to 23%. Apart from the lower removal efficiency obtained for several pharmaceuticals, significant reductions of 84 to 99% were observed for pharmaceuticals such as mebeverine.

Therefore, primary treatment processes are incapable in reducing pharmaceutical compounds significantly [10]. The hydrophilic nature of the compounds adds to the problem. The secondary treatment processes including the activated sludge process has been reported to be ineffective in the significant removal of pharmaceutical compounds. Membrane filtration, adsorption by granular activated carbon, and advanced oxidation processes are reported to exhibit higher removal percentages for pharmaceutical compounds' removal [10]. Out of these three treatment technologies, membrane filtration and adsorption are merely separation techniques and the parent compounds remain the same. Degradation is considered better than transfer of pollutants from one media to another, especially for persistent and toxic contaminants. Advanced oxidation is a degradation process able to transform the compounds throughout the treatment duration.

16.4   IMPACTS ON THE ENVIRONMENT AND HUMAN HEALTH

During drug usage, it is not completely metabolized by the organisms and therefore these pharmaceutically active compounds may reach the environment and cause adverse impacts. The compounds are also known as active pharmaceutical ingredients (APIs). The APIs may change their structural properties and undergo a number of reactions such as oxidation, biotransformation, photolysis, etc., in the environment and thus deviate from the original compounds [17, 18]. Hospital wastewater may contain microorganisms resistant to antibiotics. The phenomenon is of great concern and associated with serious global health consequences [19].

Since conventional wastewater treatment processes are not designed to remove these compounds, many APIs find their way into surface, ground water and affect different aquatic species. β-blockers are a commonly available group of pharmaceuticals in the environment. β-blockers are responsible for immobilization, mortality, growth inhibition and embryonic development problems in several aquatic species, such as fishes, green algae, etc. [20–23]. Psychoactive pharmaceuticals may cause anaesthesia, analgesia, excitement, anxiety, inability to concentrate, and mania in humans

[24]. Non-steroidal anti-inflammatory drug, diclofenac, may cause adverse impacts on population growth, liver, gill, prostate gland, and kidney damage of certain species such as *Gyps* vultures and *Salmo trutta f. fario* [25,26]. Ibuprofen, another anti-inflammatory drug may change the breeding pattern of *Oryzias latipes* [27]. Usage of anti-inflammatory and analgesics drugs affect invertebrates and plant growth [28, 29]. Several antibiotics find their way into water bodies and act as photosynthesis inhibitors thus affecting algae and aquatic plants [30, 31]. Another problem associated with the presence of antibiotics in water bodies is that several microorganisms are developing resistance against antibacterial substances through prolonged exposure [32]. Low to high acute toxicity of antibiotics were reported on invertebrates and fish species [33]. Another group of pharmaceuticals, which are abundantly present in wastewater and seriously affect human health, is antiretrovirals. These compounds may cause serious ecotoxicological problems once they pass through wastewater treatment works. Prolonged exposure of the concerned drug may create resistant strains of HIV in human body [34, 35]. The non-selective property of anticancer drugs is another growing concern. Since these drugs are designed to remove fast-growing cells, it also attacks healthy cells. The drugs may impart genotoxic, cytotoxic, teratogenic and mutagenic impacts [36, 37]. Hormones may also cause adverse health impacts as hormones and metabolites cause cancer by damaging DNA [38]. Certain pharmaceuticals show equal or greater toxicity than original compounds upon incomplete degradation. Formation of by-products which shows similar or higher toxicity than the parent compound is of serious concern for ozonation of pharmaceutically active compounds containing water [39].

16.5 TREATMENT BY HYBRID ADVANCED OXIDATION PROCESSES

16.5.1 ANALGESICS AND NONSTEROIDAL ANTI-INFLAMMATORY DRUG

Adityosulindro et al. [40] investigated degradation of Ibuprofen using the hybrid advanced oxidation process (AOP) of sonolysis and Fenton. For an initial Ibuprofen concentration of 20 mg/L, 97% degradation efficiency was achieved within 2 h of treatment time. Detailed experimental conditions are presented in Table 16.1. Although the results were impressive, the efficiency of the sonolysis process get reduced for real effluent due to alkaline pH. Durán et al. [41] studied the removal of antipyrine using H_2O_2/UV/Fe/ultrasound. The technology was successful in achieving 92% of TOC removal efficiency. The degradation mechanism of antipyrine was predicted as the formation of aromatic acids followed by organic acids, which may be decomposed further to CO_2. Sonocatalytic degradation was applied to remove phenazopyridine from aquatic solution by Eskandarloo et al. [42]. Sm-doped ZnO and periodate combination was selected as the best in degrading phenazopyridine (90% removal) after evaluating several other oxidants in lieu of periodate, such as H_2O_2, HSO_5^-, and $S_2O_8^{2-}$. Another sono-hybrid AOP, sonoelectrochemical oxidation (combination of an ultrasound transducer and boron-doped diamond electrodes) was implemented for the reduction of diclofenac [43]. The drug decomposition followed pseudo first-order kinetics with a rate constant of 0.505/min. Güyer and Ince [44] also studied the degradation of anti-inflammatory drug diclofenac using a sonocatalysis process. High-frequency ultrasound with the addition of Fe-species were attempted to improve the degradation efficiency. Three different forms of Fe-species, i.e., iron superoxide nanoparticles (divalent and zero-valent iron) were used and the obtained efficiencies can be presented as 43% for divalent, 30% for superoxide, and 22% for zero-valent iron species. The study showed applicability of iron nanoparticles in removing pharmaceuticals.

Four hybrid AOPs (photoelectrocatalysis using UVA and sunlight, combination of photoelectrocatalysis and photoelectro-Fenton using UVA and solar irradiation) were compared in a study for the degradation of paracetamol solutions [45]. Although 100% removal efficiency was obtained within 30 min using photoelectrocatalysis and photoelectro-Fenton with solar irradiation, mineralization efficiency was quite lower (24% in 180 min), indicating intermediates' formation during degradation. Im et al. [46] evaluated the performance of sonocatalysis using single-walled

TABLE 16.1

Treatment of Pharmaceutical Compounds Containing Wastewater by Hybrid AOPs

Pharmaceuticals	Hybrid AOPs Used	Treatment Conditions	Removal Efficiency	References
Ibuprofen	Sonolysis+Fenton	Initial concentration (C_0): 20 mg/L, sonication frequency: 20 kHz, pH: 2.6, time: 120 min, H_2O_2: 6.4 mM, Fe^{2+}: 0.134 Mm	97%	[40]
Antipyrine	H_2O_2/UV/Fe/Ultrasound	C_0: 50 mg/L, time: 50 min, H_2O_2: 1500 mg/L, pH: 2.7	TOC: 92%	[41]
Phenazopyridine	Sonocatalysis (Sm-doped ZnO and periodate)	C_0: 10 mg/L, frequency: 20 kHz, power: 700 W, catalyst dose: 1 g/L, pH 6.1, IO_4^-: 0.68 g/L	90%	[42]
Diclofenac	Sonoelectrochemical degradation (Boron-doped diamond)	C_0: 0.157 μmol, time: 5 min, frequency: 850 kHz, Na_2SO_4: 30 mmol/L	93%	[43]
Diclofenac	Sonocatalysis	C_0: 30 μM, time: 90 min, frequency: 861 kHz, power: 100 ± 3 W, pH: 3.0, iron superoxide: 0.001 mM, zero-valent Fe = 8.9 mM, divalent Fe = 0.01 mM	30 (divalent), 22 (zero-valent), 43% (superoxide)	[44]
Paracetamol	Photo electrocatalysis using UVA and sunlight, combination of photo electrocatalysis and photoelectro-Fenton using UVA and solar irradiation	Au-TiO_2 photoanode/carbon cloth, air-diffusion electrode, C_0: 78.6 mg/l, pH: 3.0, Na_2SO_4: 0.05 M, anodic potential: 0.82 V vs Ag/AgCl without irradiation, $FeSO_4$: 0.5 mM	Degradation:100% at 25 min, TOC: 24% after 180 min	[45]
Acetaminophen and naproxen	Sonocatalysis (single-walled carbon nanotube)	C_0: 5 μM, time: 10 min, frequency: 28 kHz, pH 6, catalyst dose: 45 mg/L	44, 90%	[46]
Acetaminophen and naproxen	Sonocatalysis (powdered activated carbon and biochar)	C_0: 5 μM, time: 10 min, frequency: 580 kHz, pH 6, catalyst dose: 10 mg/L	100, 100%	[47]
Acetaminophen and naproxen	US/Fenton/TiO_2 NT process	C_0: 3 μM, time: 30 min, pH: 3, Fe^{2+}:H_2O_2: 20:4, frequency: 1000 kHz	85.3, 96%	[48]
Ibuprofen	UV/H_2O_2, UV/$S_2O_8^{2-}$	C_0: 1 μM (reverse osmosis treated water), H_2O_2: 1 mM, UV fluence: ~220 mJ/cm²	~90% by UV/H_2O_2	[49]
Ibuprofen	Sonophotocatalysis (US + UV +TiO_2)	C_0: 0.09 mM, time: 3 h, TiO_2: 1 g/L	TOC: 92%	[50]
Diclofenac	Sonophotocatalysis (US + UV +TiO_2)	C_0: 0.07 mM, time: 3 h, TiO_2: 1 g/L	Reaction rate: 64.4×10^{-7} M/min, TOC: 95%	[51]
Ibuprofen	Sonophoto-Fenton, sonophotocatalysis, and TiO_2/Fe^{2+}/sonolysis	C_0: 0.019 mM, time: 240 min, ultrasonic power: 80 W, frequency: 300 kHz, TiO_2: 10 mg/L, Fe(II): 100 mg/L	Mineralization: 92% (TiO_2/Fe^{2+}/sonolysis)	[52]
Diclofenac and ibuprofen	Sonophotocatalysis	C_0: 10 mg/L, pH: 6.5, frequency: 20 kHz, TiO_2: 100 mg/L, power density: 8.4 W/cm²	Degradation rate: 49×10^{-3}/min, 40×10^{-3}/min	[53]

Pollutant	Process	Conditions	Removal	Ref.
Diclofenac (spiked wastewater sample)	$US+O_3$	C_0: 40 mg/L, frequency: 20 kHz, time: 40 min, power density: 400 W/L, O_3: 31 g/h	Mineralization~40%	[54]
Paracetamol	Anodic photoelectrochemical oxidation and cathodic electro-Fenton	Anode: bismuth vanadate – bismuth oxyiodide/fluorine doped tin oxide glass, cathode: carbon felt, C_0: 0.1 mM, $FeSO_4.7H_2O$: 0.2 mM, pH: 3, time: 4 h, current density: 10 mA/cm^2	TOC: 92%	[55]
Paracetamol	Sonophotolysis	Catalyst: Pd/Au-TiO_2, C_0: 35 µM, pH: 6.5, time: 1 h, TiO_2: 5 mg/L, light fluence: 107.4 W/cm^2, power intensity: 1.14 W/cm^2 (sono)	60%	[56]
Ibuprofen	Sonoelectrochemical oxidation	Anode: platinum, cathode: stainless-steel, C_0: 2 mg/L, voltage: 30 V, time: 1 h, ultrasound frequency: 1000 kHz, power density: 100 W/L	88.70%	[57]
Enrofloxacin and ciprofloxacin	O_3/UV/H_2O_2/Fe(II)	C_0: 0.15 mM, time: 30 min, H_2O_2: 10 mM, Fe(II): 0.5 mM, O_3: 8.5 mg/L, pH: 3	TOC: 93%, 99%	[58]
Aamoxicillin, ampicillin, and cloxacillin	UV/TiO_2/H_2O_2	C_0: 103-105 mg/L, time: 30 min, TiO_2: 1 g/L, pH: ~5, H_2O_2: 100 g/L	100%	[59,60]
Tetracycline	Sonocatalysis	C_0: 75 mg/L, time: 75 min, frequency: 35 kHz, power: 80 W, TiO_2: 0.1-0.5 g/L, H_2O_2: 20-100 mg/L	TOC: 100%	[61]
Tetracycline	Sonocatalysis (Fe_3O_4/$Na_2S_2O_8$/US)	C_0: 100 mg/L, time: 90 min, pH: 3.7, frequency: 20 kHz, power: 80 W, $Na_2S_2O_8$: 200 mM, magnetite: 1 g/L, US: 80 W	89%	[62]
Sulfamethoxazole	US/O_3	Ultrasonic power density: 600 W/L, C_0: 100 mg/L, O_3: 3 g/h, pH: 9	Reaction rate constant: 0.5/min	[63,64]
Amoxicillin	Ultrasound+ozonation	C_0: 25 mg/L, time: 90 min, frequency: 575 kHz, dissolved ozone concentration: 0.13 mg/L, pH 10	>99%, mineralization: 10%	[65]
Sulfadiazine	Sono-Fenton	C_0: 25 mg/L, time: 15 min, ultrasonic power: 22 W, frequency: 580 kHz, pH: 3, Fe(II): 2 mg/L, H_2O_2: 122 mg/L	89%	[66]
Ofloxacin	Sonoelectrochemical	Anode: Ti/RuO_2, cathode: stainless steel, C_0: 20 mg/L, pH: 6.3, ultrasonic power: 54 W, current density: 214 A/m^2, and Na_2SO_4: 2.0 g/L	91.2%, COD: 70.12%	[67]
Sulfadiazine	Hydrodynamic cavitation-based heterogeneous Fenton/ persulfate system	C_0: 20 mg/L, pH: 4, inlet pressure: 10 atm, H_2O_2: 0.95 mL/L, $Na_2S_2O_8$: 348.5 mg/L, α-Fe_2O_3 catalyst: 181.8 mg/L	81%	[68]
Tetracycline	US/goethite/O_3	C_0: 100 mg/L, time: 20 min, frequency: 20 kHz, power density: 85.7 W/L, O_3: 13.8 mg/L, goethite: 0.5 g/L, gas flow rate: 30 L/h	99.2%, TOC: 4%	[69]
Tetracycline	$US+O_3$	C_0: 400 mg/L, time: 90 min, frequency: 20 kHz, power density: 142.8 W/L, O_3: 45.6 mg/L, gas flow rate: 35 L/h	COD: 91%	[70]
Azithromycin	Graphene oxide @Fe_3O_4/ZnO/SnO_2 nanocomposite	C_0: 30 mg/L, time: 120 min, pH: 3, catalyst: 1 g/L	90.06%	[71]
Azithromycin	UV/H_2O_2	C_0: 1 mg/L, irradiance: 500 W/m^2, time: 120 min, H_2O_2: 482 mg/L	~100%	[72]

(continued)

TABLE 16.1 (Continued)
Treatment of Pharmaceutical Compounds Containing Wastewater by Hybrid AOPs

Pharmaceuticals	Hybrid AOPs Used	Treatment Conditions	Removal Efficiency	References
Cephalexin	WO_3 photocatalysis and ozone	C_0: 0.1 mM, time:120 min, O_3: 45 mg/L, gas flow rate: 100 mL/min, WO_3: 0.1 g/L	TOC: 53%	[73]
Sulfamethazine	Sonophotolytic goethite/oxalate Fenton-like system	C_0: 25 mg/L, time: 1 h, ultraviolet intensity: 7.7 ± 0.1 mW/cm^2, frequency: 20 kHz, power: 330 W, oxalic acid: 0.8 mM, goethite: 0.5 g/L	~70%	[74]
Sulfamethazine	Sonochemical Fe^0-catalyzed persulfate	C_0: 25 mg/L, pH: 7.00, Fe^0: 0.94 mM, persulfate: 1.90 mM, ultrasound: 20 W	90%	[75]
Sulfamethazine	Sono-Fe^0/persulfate Fenton-like system	C_0: 20 mg/L, time: 1 h, pH: 7.00, Fe^0: 0.92 mM, persulfate: 1.84 mM, input power: 40 W	95.70%	[76]
Carbamazepine	Sonophotocatalysis (TiO_2)	C_0: 10 mg/L, time: 120 min, frequency: 20 kHz, pH 6.5, power density: 640 W/L, ultrasound intensity: 8.4 W/cm^2, catalyst dose: 100 mg/L	82%, DOC: 47%	[77]
Carbamazepine	Sonoelectrochemical oxidation	Anode: Ti/PbO_2, cathode: Ti, C_0: 10 mg/L, current density: 4.86 A, time: 177 min, ultrasound power: 38.29 W	90%	[78]
Real effluent	Acoustic cavitation (AC) AC/H_2O_2, AC/Fe(II)/H_2O_2, AC/potassium persulphate and AC/O_3	C_0: 500 mg/L, time: 60 min, ultrasonic power: 125 W, frequency: 22 kHz, pH: 3	85.53% for AC/Fe(II)/H_2O_2	[79]
Pharmaceutical wastewater	Electrochemical AOP+Fenton+Flotation	COD: 1198 mg/L, Ti/Pt-Ir anode, carbon cloth cathode, electrolyte: 4000 mg/L, ferrous sulfate: 0.2 mM, pH: 2.5	60% in 3 h, 90% in 6 h	[80]
Pharmaceutical wastewater	Electrochemical AOP+Fenton+Flotation	COD: 1198 mg/L, Ti/Pt-Ir anode, carbon cloth cathode, electrolyte: 4000 mg/L, ferrous sulfate: 0.2 mM, pH: 2.5	60% in 3 h, 90% in 6 h	[81]
Real pharmaceutical wastewater	Ultrasound combined with a dual oxidant system (H_2O_2 and activated persulphate)	COD: 10667 mg/L, time: 150 min, ultrasound: 30 kHz, pH: 3, H_2O_2: 5 g/L, ammonium persulphate: 5 g/L, iron swarf: 3 g/L	~100%	[82]
Diclofenac, sulfamethoxazole and carbamazepine	US+O_3	C_0: 40 mg/L, frequency: 20 kHz, time: 40 min, power density: 370 W/L, O_3: 3.3 g/h	90%, ~54%, 44%	[83]
Municipal wastewater	H_2O_2+O_3	Pharmaceuticals (32 numbers): 3-2100 ng/L, time: 30 min, H_2O_2: 0.15 mL, 30% weight/volume in 5-L solution every 5 min, gas flow rate: 0.36 N m^3/h, gas ozone: 45.9 g/N m^3	Pharmaceuticals (32): <0.1-153 ng/L	[84]
Pharmaceutical effluent	Solar photo-Fenton and ozonation	C_0: time: 1 to 3 h, pH: 3.2-9, O_3: 0.1 g/min, Fe: 10-100 mg/L	TOC: 10-24%	[85]
Carbamazepine	Ozone and UV	C_0: 5 mg/L, O_3: 14.4 mg/h, ultraviolet: 4.4 mW/cm^2, time: 3 h	TOC: 91%	[86]
Carbamazepine, 17-α-ethinylestradiol	UV/O_3/H_2O_2	C_0: 200 μg/L, time: 30 min, pH: 7, O_3: 0.5 mg/L, H_2O_2: 0.5 mM	90%, 70%	[87]

carbon nanotube. The degradation efficiencies at optimized conditions were obtained as 44% for acetaminophen and 90% for naproxen. Using ultrasound, the single-walled carbon nanotube acted as nuclei and enhanced the formation of H_2O_2. The same research group also applied two types of adsorbent materials (powdered activated carbon and biochar), which acted as catalyzts during sonolysis [47] and 100% degradation efficiencies were achieved for both the pharmaceuticals (acetaminophen and naproxen). TiO_2 nanotubes were also investigated for enhancing H_2O_2 production with sonolysis to degrade acetaminophen and naproxen [48]. The TiO_2 nanotubes under ultrasound irradiation and in combination with Fenton reaction was proved to be successful as the hybrid AOP process evident from the higher degradation efficiencies obtained for acetaminophen (85.3%) and naproxen (96%). A comparative evaluation between UV/H_2O_2 and $UV/S_2O_8^{2-}$ processes was conducted for ibuprofen removal [49] in three different kinds of wastewater (wastewater effluent, microfiltration treated water and reverse osmosis treated water). For both processes, removal of ibuprofen from reverse osmosis treated water were higher than the other two sources. The removal efficiency observed in the UV/H_2O_2 process was higher and approximately 90%. The study also reported that ibuprofen removal may be suppressed by the presence of HCO_3^-, Cl^-, SO_4^{2-}, NO_3^-, humic acid, fulvic acid and organic compounds. Ibuprofen and diclofenac removal from aqueous solution was investigated by Madhavan et al. [50, 51] using the sonophotocatalysis process, which involves TiO_2. In both cases, >90% of removal efficiencies were obtained using ultrasound (US), ultraviolet (UV) and TiO_2 together. Three hybrid systems, sonophoto-Fenton, sonophotocatalysis and TiO_2/Fe^{2+} /sonolysis were employed for ibuprofen removal [52]. Ibuprofen degradation efficiency of 100% and mineralization efficiency of 92% were obtained within 240 min of treatment duration using TiO_2/Fe_{2+}/sonolysis process. Sonophotocatalysis using TiO_2 as the catalyzt was applied to remove diclofenac and ibuprofen [53]. The degradation rates were observed as 49×10^{-3}/min for diclofenac and 40×10^{-3}/min for ibuprofen. Studies with a spiked wastewater sample such as diclofenac has also been conducted [54]. Combination of ultrasound and ozonation was used to treat urban wastewater sample with diclofenac concentration of 40 mg/L and 40% mineralization efficiency was obtained within 40 min of treatment duration. Photoelectrochemical oxidation and electro-Fenton process was coupled in a single reactor for the mineralization of paracetamol [55]. Total mineralization efficiency of 71% for paracetamol was achieved within 4 h of treatment duration. The hybrid system exhibited higher removal performance than that of standalone photoelectrochemical oxidation and electro-Fenton process. Sonophotocatalytic treatment showed excellent decay of paracetamol [56]. Two types of catalyzts, i.e., $Pd-TiO_2$ and $Pd/Au-TiO_2$ were used. Although $Pd/Au-TiO_2$ showed higher removal of paracetamol (60%), $Pd-TiO_2$ was best in terms of reuse and recovery efficiencies. Sonoelectrochemical oxidation using platinum anode exhibited quite high removal efficiency of 88.7% for ibuprofen within 1 h of treatment duration [57]. Three types of electrolytes were used during degradation and among them, NaOH showed higher but comparable efficiency with deionized water.

16.5.2 Antibiotics

Comparative assessments among several hybrid AOPs were also performed for enrofloxacin and ciprofloxacin. Among the applications of UV/H_2O_2, $UV/H_2O_2/Fe(II)$, O_3/UV, $O_3/UV/H_2O_2$ and $O_3/UV/H_2O_2/Fe(II)$, maximum total organic carbon (TOC) removal efficiencies were achieved for enrofloxacin (93%) and ciprofloxacin (99%) when photo-assisted Fenton and ozonation were applied [58]. Degradation of three antibiotics (amoxicillin, ampicillin and cloxacillin) in the concentration range of 103–105 mg/L were attempted using the $UV/TiO_2/H_2O_2$ process [59,60]. Although 100% removal of the antibiotics were obtained, lower dissolved organic carbon (DOC) removal efficiency (14%) indicates formation of the degradation products. Not only in terms of organic carbon degradation, the hybrid AOP costs around $260/kg DOC, which is economically not preferable. Sonocatalysis was also used by Hoseini et al. [61] for the degradation of tetracycline. The process was proved to be quite effective by removing 100% of TOC within 75 min when initial tetracycline concentration was 75 mg/L. Another hybrid AOP method, Fe_3O_4 activated $Na_2S_2O_8$ in the presence of sonolysis was also used for tetracycline removal [62] and 89% of degradation efficiency was achieved. Guo et al. [63, 64] investigated the performance of sono-ozonation in the removal of sulfamethoxazole antibiotic. After studying through several important process parameters, such as initial contaminant concentration, pH, O_3 dose, and power density, maximum reaction rate constant of 0.5/min was achieved under best operating conditions. Kidak and Doğan [65] applied a combination of ultrasound and ozone for the removal of amoxicillin antibiotic. The removal profile followed pseudo first-order kinetics (0.04/min) and >99% of removal was observed at 575 kHz ultrasonic frequency, pH of 10, and dissolved ozone concentration of 0.13 mg/L after 90 min. However, the mineralization efficiency was as low as 10%. The inhibition study revealed that the presence of humic acid and alkalinity species may reduce the process efficiency. Kidak and Doğan [65] applied a combination of ultrasound and ozone for the removal of amoxicillin antibiotic. The removal profile followed pseudo first-order kinetics (0.04/min) and >99% of removal was observed at 575 kHz ultrasonic frequency, pH of 10, and dissolved ozone concentration of 0.13 mg/L after 90 min. However, the mineralization efficiency was as low as 10%. The inhibition study revealed that the presence of humic acid and alkalinity species may reduce the process efficiency. Large-scale ultrasound reactor also exhibited acoustic cavitation-based $Fe(II)/H_2O_2$ as most efficient. A hybrid AOP, sono-Fenton process was used to remove sulfadiazine [66]. The integrated process achieved 89% of sulfadiazine removal. However, the formation of by-products during the degradation process was not analyzed. A sonoelectrochemical reactor was used to treat ofloxacin by Patidar and Srivastava [67]. The hybrid process showed degradation efficiency of ~95%, which is higher than the standalone sonolysis and electrochemical process. The synergistic effect between electrochemical and sonolysis process was reported as 34%. Hydrodynamic cavitation-based heterogeneous Fenton/persulfate system was implemented to degrade sulfadiazine [68]. For the purpose, two catalysts were prepared in the nanoparticles form (α-Fe_2O_3 and Fe_3O_4) out of which α-Fe_2O_3 showed higher removal efficiency of 81%. Tetracycline antibiotics' removal ozonation enhanced by ultrasound was used by Wang et al [69]. For this purpose, an air-lift reactor was designed for ultrasound enhanced ozonation. Although 99.2% removal of tetracycline was observed, only 4% of TOC removal was obtained. A further increase in treatment duration from 20 to 90 min increased TOC removal efficiency to 30%. Tetracycline degradation was also attempted using ultrasound and ozonation combination [70]. The process exhibited COD removal efficiency of 91% after 90 min of treatment duration. Although various toxic intermediates were formed during the first few minutes, the products were transformed into less toxic products at the end of treatment. Degradation of azithromycin antibiotic was investigated by Sayadi et al. [71] using graphene oxide @Fe_3O_4/ZnO/ SnO_2 nanocomposite. The efficient nanocomposite achieved degradation efficiency of 90.06% within 2 h. Azithromycin removal using UV/H_2O_2 has been attempted by Cano et al. [72] and almost complete degradation was achieved within 2 h. Yang et al. [73] investigated degradation of cephalexin using photocatalysis combined with ozonation process. WO_3 was used as catalyzts under visible light irradiation. The hybrid AOP showed significant increase in cephalexin mineralization than standalone treatments. TOC removal efficiency of 53% was obtained after 120 min of treatment duration. Sonophotolytic goethite/oxalate Fenton-like system was used by Zhou et al. [74] to degrade sulfamethazine. The system showed 70% removal of the antibiotic within 1 h of treatment time [74]. Sonochemical Fe^0-catalyzed persulfate was also used for the treatment and approximately 90% removal efficiency was observed at optimized conditions. The study also investigated effects of inorganic anions and chelating agents on removal performances. The results exhibited that inorganic ions may create inhibitory effects of removal efficiencies whereas cheating agents may act as pollutants [75]. Sulfamethazine decay using sono-Fe^0/persulfate Fenton-like system was attempted by Zou et al. [76] and 95.7% removal efficiency was obtained within 1 h of treatment duration.

16.5.3 ANTIEPILEPTIC

Jelic et al. [77] attempted removal of antiepileptic drug, carbamazepine using sonophotocatalysis. Ultrasound and UVA irradiation were applied together for the sonophotocatalysis process. The obtained DOC removal efficiency or mineralization was 47%. High mineralization efficiency of 91% was observed in the case of spiked distilled water with carbamazepine for the combination of ozone and UVC irradiation (4.4 mW/cm²). However, for spiked wastewater the degradation rate was lower because of the presence of organics. Sonoelectrochemical oxidation has also been used for the degradation of carbamazepine using Ti/PbO$_2$ as the anode [78]. Under the constant current intensity of 4.86 A, 90% removal was obtained. For real wastewater containing 10 µg/L of carbamazepine, degradation efficiency of 93% and mineralization efficiency of 60% was obtained.

16.5.4 MIXTURE OF PHARMACEUTICALS

Lakshmi et al. [79] investigated several acoustic and hydrodynamic cavitation-based hybrid AOPs for treating pharmaceutical compounds containing wastewater. Among the acoustic cavitation-based AOPs (in combination with H$_2$O$_2$, Fe(II)/H$_2$O$_2$, potassium persulphate and O$_3$), higher chemical oxygen demand (COD) removal was achieved as 83.53% when cavitation-based Fe(II)/H$_2$O$_2$ was used. COD removal efficiency of 87.4% was obtained for the same system when hydrodynamic cavitation was applied. Mukimin and Vistanty [80] attempted a hybrid AOP setup, which provides the benefits of electrochemical oxidation process, Fenton and flotation process during the treatment of pharmaceutical wastewater. 90% COD reduction was obtained in 6 h using Ti/Pt-Ir anode, carbon cloth cathode, ferrous sulphate, and electrolyte. The research group also investigated removal of rifampicin from pharmaceutical wastewater [81]. The hybrid AOP process consisting of three main mechanisms (electrochemical oxidation, Fenton and flotation) of pollutant removal showed 100% of COD reduction indicating complete conversion of pollutant to CO$_2$. Nachiappan and Muthukumar [82] tried to treat a real pharmaceutical wastewater having a very high COD content of 10667 mg/L. The hybrid AOP system, ultrasound combined with a dual oxidant system (H$_2$O$_2$ and activated persulphate) proved efficient in 100% degradation within 150 min. When a mixture of diclofenac, sulfamethoxazole and carbamazepine was treated by the same hybrid technology, removal efficiencies achieved were in the ascending order of diclofenac (90%)> sulfamethoxazole> carbamazepine (44%) [83]. Biologically pre-treated municipal wastewater containing pharmaceuticals was degraded with the combination of two oxidants, i.e., O$_3$ and H$_2$O$_2$ [84]. The combined process exhibited mineralization efficiency of 90% in comparison with only 15% during ozonation process. The study also listed around 30 pharmaceutical compounds originally present in municipal wastewater. O$_3$ in combination with H$_2$O$_2$ was applied for pharmaceutical wastewater treatment [85]. The removal efficiencies of various pharmaceutical compounds varied from 10 to 24% for TOC. The treatment duration was varied from 1 to 3 h. Spiked synthetic water and secondary treated municipal wastewater containing pharmaceuticals were degraded by photo-assisted ozonation process [86]. Liu et al. [87] analysed the efficiency of UV/O$_3$/H$_2$O$_2$ hybrid AOP process in removing carbamazepine and 17-α-ethinylestradiol. Although satisfactory removal of 90% was observed for carbamazepine, only 70% of removal was achieved for 17-α-ethinylestradiol by UV/O$_3$/H$_2$O$_2$ process.

16.6 CONCLUSIONS AND PERSPECTIVE

Pharmaceutical compounds are hydrophilic and persistent in nature, which resist them from getting removed by conventional treatment processes easily. Several pharmaceutical compounds exhibit toxicity and adverse impacts on human health. The common wastewater treatment processes are ineffective for several types of pharmaceuticals removal, which increase the need of AOPs. Although the hybrid AOPs showed excellent removal efficiencies, the mineralization efficiencies were quite lower for several classes of pharmaceutical compounds. Therefore, it can be said that complete mineralization to CO_2 was not achieved and intermediate compounds can be formed. The future studies involving AOPs and hybrid AOPs should focus on mineralization efficiency rather than removal efficiency. The toxicity of the water matrices needs to be measured before and after degradation process to detect formation of any intermediates more toxic than parent compounds. Most of the studies were conducted using synthetic water. However, several inorganic and organic compounds could be present in the case of real effluent. Only a few studies were conducted where the inhibitory effects of several inorganic ions and organic compounds were analyzed. More studies on real pharmaceutical wastewater or on spiked wastewater needs to be investigated. The cost of AOPs is generally higher than conventional treatment processes. Therefore, the future studies need to incorporate cost analysis. The performance evaluation of AOPs in combination with other low-cost technologies may solve the problem. Besides mineralization efficiency, by-products' formation and cost analysis should also be incorporated.

REFERENCES

1. Taoufik, N., Boumya, W., Achak, M., Sillanpää, M., and Barka, N. (2021). Comparative overview of advanced oxidation processes and biological approaches for the removal pharmaceuticals. *Journal of Environmental Management,* 288, 112404.
2. de Jesus Gaffney, V., Almeida, C. M., Rodrigues, A., Ferreira, E., Benoliel, M. J., and Cardoso, V. V. (2015). Occurrence of pharmaceuticals in a water supply system and related human health risk assessment. *Water Research,* 72, 199–208.
3. Meijide, J., Lama, G., Pazos, M., Sanromán, M. A., and Dunlop, P. S. M. (2022). Ultraviolet-based heterogeneous advanced oxidation processes as technologies to remove pharmaceuticals from wastewater: An overview. *Journal of Environmental Chemical Engineering,* 10, 107630.
4. Wang, J. L., and Xu, L. J. (2012). Advanced oxidation processes for wastewater treatment: formation of hydroxyl radical and application. *Critical Reviews in Environmental Science and Technology,* 42(3), 251–325.
5. Chauhan, R., Dinesh, G. K., Alawa, B., and Chakma, S. (2021). A critical analysis of sono-hybrid advanced oxidation process of ferrioxalate system for degradation of recalcitrant pollutants. *Chemosphere,* 277, 130324.
6. Pandis, P. K., Kalogirou, C., Kanellou, E., Vaitsis, C., Savvidou, M. G., Sourkouni, G., Zorpas, A. A., and Argirusis, C. (2022). Key points of advanced oxidation processes (AOPs) for wastewater, organic pollutants and pharmaceutical waste treatment: A mini review. *ChemEngineering,* 6(1), 8.
7. Rayaroth, M. P., Aravind, U. K., and Aravindakumar, C. T. (2016). Degradation of pharmaceuticals by ultrasound-based advanced oxidation process. *Environmental Chemistry Letters,* 14(3), 259–290.
8. Nikolaou, A., Meric, S., and Fatta, D. (2007). Occurrence patterns of pharmaceuticals in water and wastewater environments. *Analytical and Bioanalytical Chemistry,* 387(4), 1225–1234.
9. Ghazal, H., Koumaki, E., Hoslett, J., Malamis, S., Katsou, E., Barcelo, D., and Jouhara, H. (2022). Insights into current physical, chemical and hybrid technologies used for the treatment of wastewater contaminated with pharmaceuticals. *Journal of Cleaner Production,* 361, 132079.
10. Yang, Y., Ok, Y. S., Kim, K. H., Kwon, E. E., and Tsang, Y. F. (2017). Occurrences and removal of pharmaceuticals and personal care products (PPCPs) in drinking water and water/sewage treatment plants: A review. *Science of the Total Environment,* 596, 303–320.
11. Ebele, J. A., Abdallah, M. A., and Harrad, S. (2017). Pharmaceuticals and personal care products (PPCPs) in the freshwater aquatic environment. *Emerging Contaminants,* 3(1), 1–16.
12. Behera, K. S., Woo, H., Oh, J., and Park, H. (2011). Occurrence and removal of antibiotics, hormones and several other pharmaceuticals in wastewater treatment plants of the largest industrial city of Korea. *Science of the Total Environment,* 409(20), 4351–4360.
13. Grover, D. P., Zhou, J. L., Frickers, P. E., and Readman, J. W. (2011). Improved removal of estrogenic and pharmaceutical compounds in sewage effluent by full scale granular activated carbon: Impact on receiving river water. *Journal of Hazardous Materials,* 185(2–3), 1005–1011.
14. Mohapatra, S., Huang, C. H., Mukherji, S., and Padhye, L. P. (2016). Occurrence and fate of pharmaceuticals in WWTPs in India and comparison with a similar study in the United States. *Chemosphere,* 159, 526–535.

15. Tran, N. H., and Gin, K. Y. H. (2017). Occurrence and removal of pharmaceuticals, hormones, personal care products, and endocrine disrupters in a full-scale water reclamation plant. *Science of the Total Environment,* 599, 1503–1516.

16. Röhricht, M., Krisam, J., Weise, U., Kraus, U. R., and Düring, R. A. (2009). Elimination of carbamazepine, diclofenac and naproxen from treated wastewater by nanofiltration. *CLEAN–Soil, Air, Water,* 37(8), 638–641.

17. Ortúzar, M., Esterhuizen, M., Olicón-Hernández, D. R., González-López, J., and Aranda, E. (2022). Pharmaceutical pollution in aquatic environments: A concise review of environmental impacts and bioremediation systems. *Frontiers in Microbiology,* 13, 869332.

18. Wu, S., Zhang, L., and Chen, J. (2012). Paracetamol in the environment and its degradation by microorganisms. *Applied Microbiology and Biotechnology,* 96(4), 875–884.

19. Buelow, E., Bayjanov, J. R., Majoor, E., Willems, R. J., Bonten, M. J., Schmitt, H., and van Schaik, W. (2018). Limited influence of hospital wastewater on the microbiome and resistome of wastewater in a community sewerage system. *FEMS Microbiology Ecology,* 94(7), fiy087.

20. Godoy, A. A., de Oliveira, Á. C., Silva, J. G. M., de Jesus Azevedo, C. C., Domingues, I., Nogueira, A. J. A., and Kummrow, F. (2019). Single and mixture toxicity of four pharmaceuticals of environmental concern to aquatic organisms, including a behavioral assessment. *Chemosphere,* 235, 373–382.

21. Fonseca, V. F., Duarte, I. A., Duarte, B., Freitas, A., Pouca, A. S. V., Barbosa, J., Gillanders, B. M., and Reis-Santos, P. (2021). Environmental risk assessment and bioaccumulation of pharmaceuticals in a large urbanized estuary. *Science of the Total Environment,* 783, 147021.

22. Ferrari, B., Mons, R., Vollat, B., Fraysse, B., Paxēaus, N., Giudice, R. L., Pollio, A., and Garric, J. (2004). Environmental risk assessment of six human pharmaceuticals: are the current environmental risk assessment procedures sufficient for the protection of the aquatic environment? *Environmental Toxicology and Chemistry: An International Journal,* 23(5), 1344–1354.

23. Bittner, L., Teixido, E., Seiwert, B., Escher, B. I., and Klüver, N. (2018). Influence of pH on the uptake and toxicity of β -blockers in embryos of zebra fish, *Danio rerio. Aquatic Toxicology,* 201, 129–137.

24. Jin, H., Yang, D., Wu, P., and Zhao, M. (2022). Environmental occurrence and ecological risks of psychoactive substances. *Environment International,* 158, 106970.

25. Cuthbert, R., Parry-Jones, J., Green, R. E., and Pain, D. J. (2007). NSAIDs and scavenging birds: potential impacts beyond Asia's critically endangered vultures. *Biology Letters,* 3(1), 91–94.

26. Hoeger, B., Köllner, B., Dietrich, D. R., and Hitzfeld, B. (2005). Water-borne diclofenac affects kidney and gill integrity and selected immune parameters in brown trout (*Salmo trutta f. fario*). *Aquatic Toxicology,* 75(1), 53–64.

27. Flippin, J. L., Huggett, D., and Foran, C. M. (2007). Changes in the timing of reproduction following chronic exposure to ibuprofen in Japanese medaka, *Oryzias latipes. Aquatic Toxicology,* 81(1), 73–78.

28. Parolini, M. (2020). Toxicity of the Non-Steroidal Anti-Inflammatory Drugs (NSAIDs) acetylsalicylic acid, paracetamol, diclofenac, ibuprofen and naproxen towards freshwater invertebrates: A review. *Science of the Total Environment,* 740, 140043.

29. Wijaya, L., Alyemeni, M., Ahmad, P., Alfarhan, A., Barcelo, D., El-Sheikh, M. A., and Pico, Y. (2020). Ecotoxicological effects of ibuprofen on plant growth of *Vigna unguiculata L. Plants,* 9(11), 1473.

30. Brausch, J. M., Connors, K. A., Brooks, B. W., and Rand, G. M. (2012). Human pharmaceuticals in the aquatic environment: A review of recent toxicological studies and considerations for toxicity testing. *Reviews of Environmental Contamination and Toxicology,* 218, 1–99.

31. Nie, X. P., Liu, B. Y., Yu, H. J., Liu, W. Q., and Yang, Y. F. (2013). Toxic effects of erythromycin, ciprofloxacin and sulfamethoxazole exposure to the antioxidant system in *Pseudokirchneriella subcapitata. Environmental Pollution,* 172, 23–32.

32. Wang, S., Ma, X., Liu, Y., Yi, X., Du, G., and Li, J. (2020). Fate of antibiotics, antibiotic-resistant bacteria, and cell-free antibiotic-resistant genes in full-scale membrane bioreactor wastewater treatment plants. *Bioresource Technology,* 302, 122825.

33. Minguez, L., Pedelucq, J., Farcy, E., Ballandonne, C., Budzinski, H., and Halm-Lemeille, M. P. (2016). Toxicities of 48 pharmaceuticals and their freshwater and marine environmental assessment in north-western France. *Environmental Science and Pollution Research,* 23(6), 4992–5001.

34. Ncube, S., Madikizela, L. M., Chimuka, L., and Nindi, M. M. (2018). Environmental fate and ecotoxicological effects of antiretrovirals: A current global status and future perspectives. *Water Research,* 145, 231–247.

35. Mlunguza, N. Y., Ncube, S., Mahlambi, P. N., Chimuka, L., and Madikizela, L. M. (2020). Determination of selected antiretroviral drugs in wastewater, surface water and aquatic plants using hollow fibre liquid phase microextraction and liquid chromatography-tandem mass spectrometry. *Journal of Hazardous Materials,* 382, 121067.

36. Chari, R. V. (2008). Targeted cancer therapy: Conferring specificity to cytotoxic drugs. *Accounts of Chemical Research,* 41(1), 98–107.

37. Johnson, A. C., Jürgens, M. D., Williams, R. J., Kümmerer, K., Kortenkamp, A., and Sumpter, J. P. (2008). Do cytotoxic chemotherapy drugs discharged into rivers pose a risk to the environment and human health? An overview and UK case study. *Journal of Hydrology,* 348(1–2), 167–175.

38. Yasuda, M. T., Sakakibara, H., and Shimoi, K. (2017). Estrogen- and stress-induced DNA damage in breast cancer and chemoprevention with dietary flavonoid. *Genes and Environment,* 39(1), 1–9.

39. Westerhoff, P., Yoon, Y., Snyder, S., and Wert, E. (2005). Fate of endocrine-disruptor, pharmaceutical, and personal care product chemicals during simulated drinking water treatment processes. *Environmental Science & Technology,* 39(17), 6649–6663.

40. Adityosulindro, S., Barthe, L., González-labrada, K., Haza, J. J., Delmas, H., and Julcour, C. (2017). Sonolysis and sono-Fenton oxidation for removal of ibuprofen in (waste)water. *Ultrasonics–Sonochemistry,* 39, 889–896.

41. Durán, A., Monteagudo, J. M., and Sanmartín, I. (2013). Sonophotocatalytic mineralization of antipyrine in aqueous solution. *Applied Catalysis B: Environmental,* 138–139, 318–325.

42. Eskandarloo, H., Badiei, A., Behnajady, M. A., and Ziarani, G. M. (2016). Ultrasonic-assisted degradation of phenazopyridine with a combination of Sm-doped ZnO nanoparticles and inorganic oxidants. *Ultrasonics Sonochemistry,* 28, 169–177.

43. Finkbeiner, P., Franke, M., Anschuetz, F., Ignaszak, A., Stelter, M., and Braeutigam, P. (2015). Sonoelectrochemical degradation of the anti-inflammatory drug diclofenac in water. *Chemical Engineering Journal,* 273, 214–222.

44. Güyer, G. T., and Ince, N. H. (2011). Degradation of diclofenac in water by homogeneous and heterogeneous sonolysis. *Ultrasonics Sonochemistry,* 18(1), 114–119.

45. Hernández, R., Olvera-Rodríguez, I., Guzmán, C., Medel, A., Escobar-Alarcón, L., Brillas, E., Sirés, I., and Esquivel, K. (2018). Microwave-assisted sol-gel synthesis of an Au-TiO$_2$ photoanode for the advanced oxidation of paracetamol as model pharmaceutical pollutant. *Electrochemistry Communications,* 96, 42–46.

46. Im, J. K., Heo, J., Boateng, L. K., Her, N., Flora, J. R., Yoon, J., Zoh, K. D., and Yoon, Y. (2013). Ultrasonic degradation of acetaminophen and naproxen in the presence of single-walled carbon nanotubes. *Journal of Hazardous Materials,* 254, 284–292.

47. Im, J. K., Boateng, L. K., Flora, J. R., Her, N., Zoh, K. D., Son, A., and Yoon, Y. (2014). Enhanced ultrasonic degradation of acetaminophen and naproxen in the presence of powdered activated carbon and biochar adsorbents. *Separation and Purification Technology,* 123, 96–105.

48. Im, J. K., Yoon, J., Her, N., Han, J., Zoh, K. D., and Yoon, Y. (2015). Sonocatalytic-TiO$_2$ nanotube, Fenton, and CCl$_4$ reactions for enhanced oxidation, and their applications to acetaminophen and naproxen degradation. *Separation and Purification Technology,* 141, 1–9.

49. Kwon, M., Kim, S., Yoon, Y., Jung, Y., Hwang, T. M., Lee, J., and Kang, J. W. (2015). Comparative evaluation of ibuprofen removal by UV/H$_2$O$_2$ and UV/S$_2$O$_8^{2-}$ processes for wastewater treatment. *Chemical Engineering Journal,* 269, 379–390.

50. Madhavan, J., Grieser, F., and Ashokkumar, M. (2010). Combined advanced oxidation processes for the synergistic degradation of ibuprofen in aqueous environments. *Journal of Hazardous Materials,* 178(1–3), 202–208.

51. Madhavan, J., Kumar, P. S. S., Anandan, S., Zhou, M., Grieser, F., and Ashokkumar, M. (2010). Ultrasound assisted photocatalytic degradation of diclofenac in an aqueous environment. *Chemosphere,* 80(7), 747–752.

52. Mendez-Arriaga, F., Torres-Palma, R. A., Pétrier, C., Esplugas, S., Gimenez, J., and Pulgarin, C. (2009). Mineralization enhancement of a recalcitrant pharmaceutical pollutant in water by advanced oxidation hybrid processes. *Water Research,* 43(16), 3984–3991.

53. Michael, I., Achilleos, A., Lambropoulou, D., Torrens, V. O., Pérez, S., Petrović, M., Barceló, D., and Fatta-Kassinos, D. (2014). Proposed transformation pathway and evolution profile of diclofenac and

ibuprofen transformation products during (sono) photocatalysis. *Applied Catalysis B: Environmental,* 147, 1015–1027.

54. Naddeo, V., Ricco, D., Scannapieco, D., and Belgiorno, V. (2012). Degradation of antibiotics in wastewater during sonolysis, ozonation, and their simultaneous application: operating conditions effects and processes evaluation. *International Journal of Photoenergy,* 2012, 624270.

55. Orimolade, B. O., Zwane, B. N., Koiki, B. A., Rivallin, M., Bechelany, M., Mabuba, N., Lesage, G., Cretin, M., and Arotiba, O. A. (2020). Coupling cathodic electro-Fenton with anodic photo-electrochemical oxidation: A feasibility study on the mineralization of paracetamol. *Journal of Environmental Chemical Engineering,* 8(5), 104394.

56. Ziylan-Yavas, A., Mizukoshi, Y., Maeda, Y., and Ince, N. H. (2015). Supporting of pristine TiO_2 with noble metals to enhance the oxidation and mineralization of paracetamol by sonolysis and sonophotolysis. *Applied Catalysis B: Environmental,* 172, 7–17.

57. Thokchom, B., Kim, K., Park, J., and Khim, J. (2015). Ultrasonically enhanced electrochemical oxidation of ibuprofen. *Ultrasonics Sonochemistry,* 22, 429–436.

58. Bobu, M., Yediler, A., Siminiceanu, I., Zhang, F., and Schulte-Hostede, S. (2013). Comparison of different advanced oxidation processes for the degradation of two fluoroquinolone antibiotics in aqueous solutions Comparison of different advanced oxidation processes for the degradation of two fluoroquinolone antibiotics in aqueous solutions. *Journal of Environmental Science and Health, Part A,* 48(3), 251–262.

59. Elmolla, E. S., and Chaudhuri, M. (2010). Comparison of different advanced oxidation processes for treatment of antibiotic aqueous solution. *Desalination,* 256(1–3), 43–47.

60. Elmolla, E. S., and Chaudhuri, M. (2010). Photocatalytic degradation of amoxicillin, ampicillin and cloxacillin antibiotics in aqueous solution using UV/TiO_2 and $UV/H_2O_2/TiO_2$ photocatalysis. *Desalination,* 252(1–3), 46–52.

61. Hoseini, M., Safari, G. H., Kamani, H., Jaafari, J., Ghanbarain, M., and Mahvi, A. H. (2013). Sonocatalytic degradation of tetracycline antibiotic in aqueous solution by sonocatalysis. *Toxicological & Environmental Chemistry,* 95(10), 1680–1689.

62. Hou, L., Zhang, H., and Xue, X. (2012). Ultrasound enhanced heterogeneous activation of peroxydisulfate by magnetite catalyst for the degradation of tetracycline in water. *Separation and Purification Technology,* 84, 147–152.

63. Guo, W. Q., Yin, R. L., Zhou, X. J., Du, J. S., Cao, H. O., Yang, S. S., and Ren, N. Q. (2015). Sulfamethoxazole degradation by ultrasound/ozone oxidation process in water: Kinetics, mechanisms, and pathways. *Ultrasonics Sonochemistry,* 22, 182–187.

64. Guo, W. Q., Yin, R. L., Zhou, X. J., Cao, H. O., Chang, J. S., and Ren, N. Q. (2016). Ultrasonic-assisted ozone oxidation process for sulfamethoxazole removal: Impact factors and degradation process. *Desalination and Water Treatment,* 57(44), 21015–21022.

65. Kıdak, R., and Doğan, Ş. (2018). Medium-high frequency ultrasound and ozone based advanced oxidation for amoxicillin removal in water. *Ultrasonics Sonochemistry,* 40, 131–139.

66. Lastre-Acosta, A. M., Cruz-González, G., Nuevas-Paz, L., Jáuregui-Haza, U. J., and Teixeira, A. C. S. C. (2015). Ultrasonic degradation of sulfadiazine in aqueous solutions. *Environmental Science and Pollution Research,* 22(2), 918–925.

67. Patidar, R., and Srivastava, V. C. (2020). Mechanistic insight into ultrasound-induced enhancement of electrochemical oxidation of ofloxacin: Multi-response optimization and cost analysis. *Chemosphere,* 257, 127121.

68. Roy, K., and Moholkar, V. S. (2020). Sulfadiazine degradation using hybrid AOP of heterogeneous Fenton/persulfate system coupled with hydrodynamic cavitation. *Chemical Engineering Journal,* 386, 121294.

69. Wang, Y., Zhang, H., and Chen, L. (2011). Ultrasound enhanced catalytic ozonation of tetracycline in a rectangular air-lift reactor. *Catalysis Today,* 175(1), 283–292.

70. Wang, Y., Zhang, H., Chen, L., Wang, S., and Zhang, D. (2012). Ozonation combined with ultrasound for the degradation of tetracycline in a rectangular air-lift reactor. *Separation and Purification Technology,* 84, 138–146.

71. Sayadi, M. H., Sobhani, S., and Shekari, H. (2019). Photocatalytic degradation of azithromycin using GO@ $Fe_3O_4/ZnO/SnO_2$ nanocomposites. *Journal of Cleaner Production,* 232, 127–136.

72. Cano, P. A., Jaramillo-Baquero, M., Zúñiga-Benítez, H., Londoño, Y. A., and Peñuela, G. A. (2020). Use of simulated sunlight radiation and hydrogen peroxide in azithromycin removal from aqueous solutions: optimization & mineralization analysis. *Emerging Contaminants, 6*, 53–61.

73. Yang, J., Xiao, J., Cao, H., Guo, Z., Rabeah, J., Brückner, A., and Xie, Y. (2018). The role of ozone and influence of band structure in WO_3 photocatalysis and ozone integrated process for pharmaceutical wastewater treatment. *Journal of Hazardous Materials, 360*, 481–489.

74. Zhou, T., Wu, X., Zhang, Y., Li, J., and Lim, T. (2013). Synergistic catalytic degradation of antibiotic sulfamethazine in a heterogeneous sonophotolytic goethite/oxalate Fenton-like system. *Applied Catalysis B: Environmental, 136–137*, 294–301.

75. Zhou, T., Zou, X., Mao, J., and Wu, X. (2016). Decomposition of sulfadiazine in a sonochemical Fe^0-catalyzed persulfate system: Parameters optimizing and interferences of wastewater matrix. *Applied Catalysis B: Environmental, 185*, 31–41.

76. Zou, X., Zhou, T., Mao, J., and Wu, X. (2014). Synergistic degradation of antibiotic sulfadiazine in a heterogeneous ultrasound-enhanced Fe^0/persulfate Fenton-like system. *Chemical Engineering Journal, 257*, 36–44.

77. Jelic, A., Michael, I., Achilleos, A., Hapeshi, E., Lambropoulou, D., Perez, S., Petrovic, M., Fatta-Kassinos, D., and Barcelo, D. (2013). Transformation products and reaction pathways of carbamazepine during photocatalytic and sonophotocatalytic treatment. *Journal of Hazardous Materials, 263*, 177–186.

78. Tran, N., Drogui, P., and Brar, S. K. (2015). Sonoelectrochemical oxidation of carbamazepine in waters: optimization using response surface methodology. *Journal of Chemical Technology & Biotechnology, 90*(5), 921–929.

79. Lakshmi, N. J., Agarkoti, C., Gogate, P. R., and Pandit, A. B. (2022). Acoustic and hydrodynamic cavitation-based combined treatment techniques for the treatment of industrial real effluent containing mainly pharmaceutical compounds. *Journal of Environmental Chemical Engineering, 10*(5), 108349.

80. Mukimin, A., and Vistanty, H. (2019). Hybrid advanced oxidation process (HAOP) as an effective pharmaceutical wastewater treatment. In *E3S Web of Conferences* (Vol. 125, p. 03007). EDP Sciences.

81. Mukimin, A., Vistanty, H., and Zen, N. (2020). Hybrid advanced oxidation process (HAOP) as highly efficient and powerful treatment for complete demineralization of antibiotics. *Separation and Purification Technology, 241*, 116728.

82. Nachiappan, S., and Muthukumar, K. (2013). Treatment of pharmaceutical effluent by ultrasound coupled with dual oxidant system. *Environmental Technology, 34*(2), 209–217.

83. Naddeo, V., Uyguner-Demirel, C. S., Prado, M., Cesaro, A., Belgiorno, V., and Ballesteros, F. (2015). Enhanced ozonation of selected pharmaceutical compounds by sonolysis. *Environmental Technology, 36*(15), 1876–1883.

84. Rosal, R., Rodríguez, A., Perdigón-Melón, J. A., Mezcua, M., Hernando, M. D., Letón, P., García-Calvo, E., Agüera, A., and Fernández-Alba, A. R. (2008). Removal of pharmaceuticals and kinetics of mineralization by O_3/H_2O_2 in a biotreated municipal wastewater. *Water Research, 42*(14), 3719–3728.

85. Sanchis, S., Meschede-Anglada, L., Serra, A., Simon, F. X., Sixto, G., Casas, N., and Garcia-Montano, J. (2018). Solar photo-Fenton with simultaneous addition of ozone for the treatment of real industrial wastewaters. *Water Science and Technology, 77*(10), 2497–2508.

86. Somathilake, P., Dominic, J. A., Achari, G., Langford, C. H., and Tay, J. H. (2018). Degradation of carbamazepine by photo-assisted ozonation: influence of wavelength and intensity of radiation. *Ozone: Science & Engineering, 40*(2), 113–121.

87. Liu, Z., Hosseinzadeh, S., Wardenier, N., Verheust, Y., Chys, M., and Hulle, S. V. (2019). Combining ozone with UV and H_2O_2 for the degradation of micropollutants from different origins: Lab-scale analysis and optimization. *Environmental Technology, 40*(28), 3773–3782.

17 Social and Economic Impacts of Plastics in the Environment

Satya Brot Sen and Vinay Yadav

17.1 INTRODUCTION

Plastics, which were once termed wonder materials, are now a great cause of concern for all kinds of life and nature. We are dependent and addicted to the use of plastics in our daily lives. If we thoughtfully observe, we will find that plastics are everywhere, and it is tough to replace them. Production and consumption of plastics have incredibly increased after World War II, replacing many items which are comparatively more expensive, such as glass, metals, animal matter and paper [1]. Some of the reasons why plastics are so extensively used in our daily lives are their property of strength, durability, light-weight, versatility and all these properties with low cost of production [2]. Thus, the production of plastics has increased extensively, but they have a low recycling rate as to the amount of waste generated, and they depend majorly on primary sources [3]. The modern-day plastics are of polymers sourced mostly (more than 99%) from petroleum and other fossil fuels, all of which are non-renewable, not environmentally friendly, unsustainable resources, and at the same time are a major driver of ecological problems, such as climate change or biological loss [4]. Plastics are difficult to dispose of or degrade because of the same chemicals that make these polymers durable and versatile. As a result they stay in the environment for centuries. Plastics in the environment can be found as macro-plastics,[1] microplastics[2] and nano-plastics[3] [5]. Plastics degrade very slowly; some types of plastics take over hundreds to thousands of years [6]. This degradation process of plastics is a huge concern. When plastics in the environment are exposed to solar radiation and other natural weather conditions like winds and currents, they break down into microscopic particles micro-plastics, which pollutes our air, ocean and terrestrial ecosystem. Micro-plastics do not always originate from the degradation of plastics, but some micro-plastics also originate from the manufacture of such particles for diverse purposes. Micro-plastics are present everywhere, easily transported by the wind across large distances. It has even entered the human body through food consumption. The presence of micro-plastics has many adverse impacts on all kinds of life, the natural environment and the ecosystem. Production of plastics, plastic wastes and their losses to the environment has seen continuous growth over the years [7]. Today the production of plastic wastes has reached about 300 million tonnes every year, and shockingly this is almost equivalent to the weight of the entire human population [3]. According to a report published by Dalberg advisors for World Wide Fund For Nature (WWF), if plastic straws are laid lengthwise, they can wrap around the world 2.8 million times approximately [8]. Plastics that are a colossal pollution concern is man-made and the waste originates mostly in land, which also escapes to the marine environment through various mediums like river, atmospheric transport, aquaculture, shipping, and fishing-related activities. Among the different types of plastics available, the single-use plastics, which are the significant chunk of plastics produced each year (approximately 50% of plastics) [9], are the most

DOI: 10.1201/9781003327646-17

251

concerning to which we mostly have become addicted. These single-use plastics are discarded after being used only once, mostly within a year of manufacturing. It includes plastic bags, sachets, straws, drinking bottles, wrappers, disposable cutlery and coffee-cup lids. It has been estimated that globally 1 million plastic drinking bottles are purchased every minute, and 5 trillion plastic bags are used annually [3]. Researchers have estimated that from the 1950s until 2015, more than 8 billion tonnes of plastics have been produced, out of which almost 79% has ended up either in landfill or in the natural environment [7].

The harmful impacts of plastics in the environment are manifolds ranging from but not limited to social impacts, economic impacts, physical impacts, ecological impacts, environmental impacts and impacts on the life of any forms in the land, water or air. From among all these huge and complex impacts of plastics, we have focused in this chapter on the social and economic impacts of plastics in the environment. Figure 17.1 summarizes the major social and economic impacts of plastics in the environment with details of global production, the share of its use and fate from 1950 to 2015, and the share of mismanaged plastics, 2019. The average estimated decomposition rate of typical marine debris items is also included in the figure to help us understand the magnitude of the problem.

The social and economic impact of plastics in the environment are very alarming, and thus it needs proper comprehensive understanding to undertake mitigation efforts from multiple stakeholders, including the consumers. This chapter presents a holistic approach to understanding these complex and drastic impacts of plastics in the environment in the social and economic aspects.

17.2　SOURCES OF PLASTICS IN THE ENVIRONMENT

According to a report by United Nations Environment Programme (UNEP), from 1950–2017, an estimate of 9200 million tonnes of plastics was produced, out of which an approximate 7000 million tonnes have turned to waste [12]. Out of these huge amounts of waste, 75% have been either discarded and placed in landfills or have ended in open areas, water bodies like rivers, seas and oceans [7]. Research has claimed that the material flow of plastics from landfills is linear and not cyclic [13], which means that none of the material resources used in plastics that are collected from landfills is recovered. A vast majority of plastics ever made are still present in the environment in some form [14]. Contamination of plastics in the land is estimated to be annually 4 to 23 fold larger from that of in the ocean [15]. Plastics are found abundantly in both terrestrial and marine environments. An approximate 3 million tonnes of micro-plastics and 5.3 million tonnes of macro-plastics are estimated to be annually lost to the environment [16]. Per year an estimate of 0.013 to 25 million metric tons of micro- and nano-plastics are transported within the marine atmosphere and deposited in the oceans [17].

Macro-plastics end up in the environment through both land-based pathways and ocean-based pathways. Approximately 6.2 million tonnes of macro-plastics were found to be lost to the environment in 2015 [18]. Mismanaged municipal solid waste was identified as the major source of macro-plastics losses to the environment [18]. The primary source of macro-plastics losses to the marine environment is also mismanaged municipal solid waste, including open dumping estimated to be 3.9 million tonnes [16]. Other major pathways include littering, estimated to be 0.8 million tonnes [16], losses from industry (construction, demolition, industrial facilities), improper waste treatment/disposal, unprotected landfills, especially when it is nearer to coasts or watercourses, shoreline infrastructure, untreated municipal sewage water, storm-water, losses during transport, waste disposal to open sea, losses from fishing industry (boats, fishing equipment, aquaculture) estimated to be 0.6 million tonnes [16], lost cargo, losses from shipping vessels, research vessels, offshore oil and gas platforms. A large proportion of macro-plastics in the environment consist of disposable plastic products such as food wrappers, bottles and lids, bags, sanitary items, cotton bud sticks, cups [19]. These macro-plastics can be further transported into aquatic, terrestrial or atmospheric

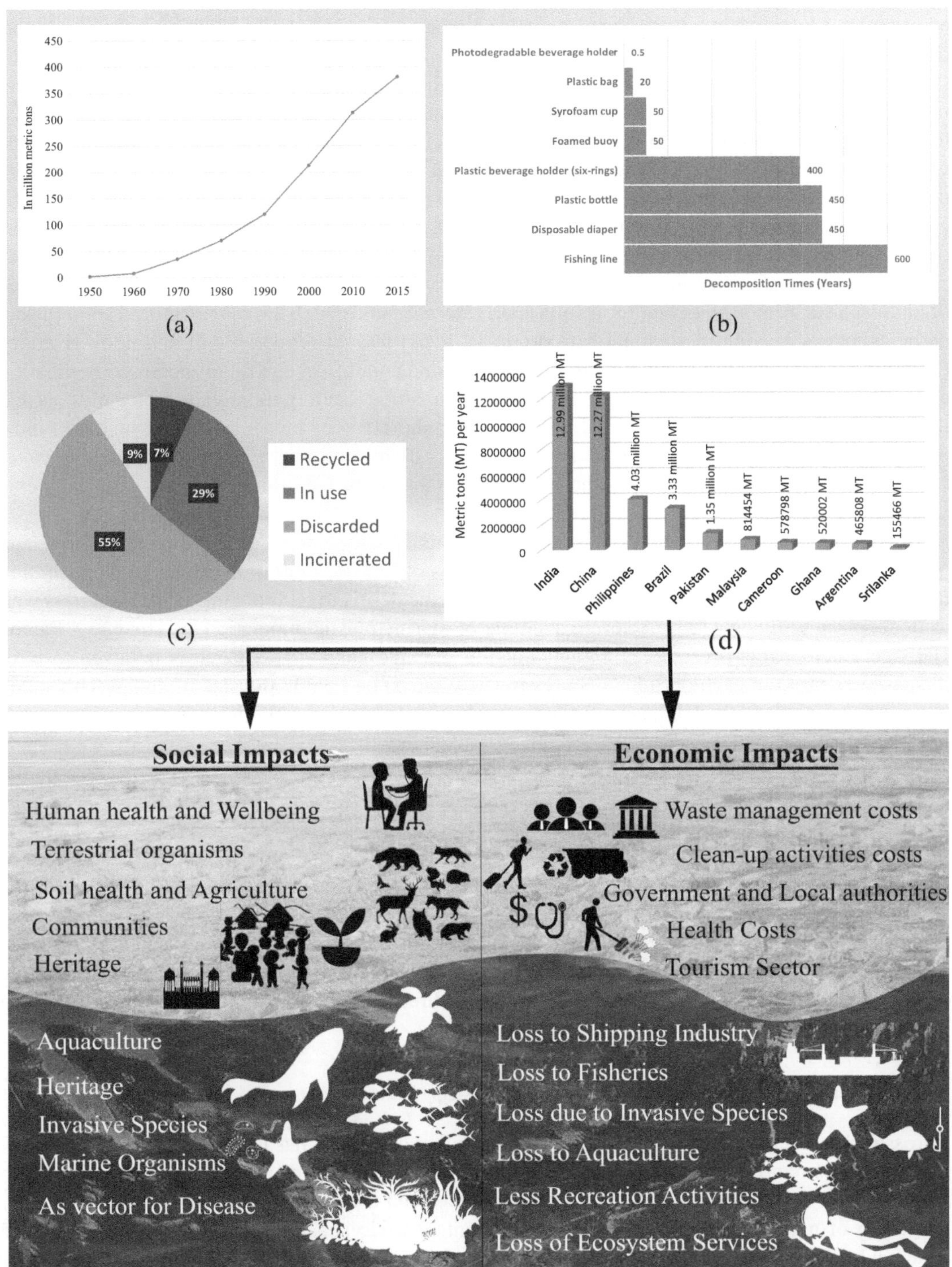

FIGURE 17.1 This figure depicts the major social and economic impacts of plastics present in the environment. (a) Global plastics production (polymer resin and fibers) in million metric tons [7]. (b) Average estimated decomposition rate of typical marine plastic debris items [10]. (c) Share of global production, use, and the fate of polymer resins, synthetic fibers, and additives (1950 to 2015) [7]. (d) Share of mismanaged plastic waste for some countries, 2019 [11].

environments by wind and water currents, anthropogenic activities such as plowing in farmed lands, or flood events.

For micro-plastics, approximately 3.0 million tonnes were found to be lost to the environment in 2015 [18]. Micro-plastics end up in the environment majorly through abrasion of tires estimated to be 1.4 million tonnes and in the form of general city dust estimated to be 0.65 million tonnes, which include plastics losses from exterior paints, abrasion of road markings (0.6 million tonnes), domestic consumer products [16]. Some of the micro-plastics ending up in the soil (including agricultural soils) are lost to the environment, and the various pathways include anthropogenic activities, organic fertilization, use of mulch films, plastic covers, water hoses, flooding of areas, application of sewage sludge for fertilization, atmospheric deposition. Organic fertilizers are a significant source of micro-plastics entering agricultural soils [20]. Micro-beads, which are manufactured solid plastic particles primarily used in personal care products, if not treated properly, will end up in rivers or oceans. Space for landfills is also getting scarce in some countries because of increased population and urbanization. This increased population and urbanization result in more demand for plastics and thus more plastics in the environment. Due to these shortages or unavailability of landfills, most plastics either end up in the open land as litter, are locally burned, or are dumped into nearby water bodies or drainage systems. Many of these plastics are lost to the environment, causing huge, diverse and harmful impacts.

According to a report published by the International Union for Conservation of Nature and Natural Resources (IUCN), at least 14 million tonnes of plastics end up in oceans every year, and thus plastic debris makes up to around 80% of all marine debris to be found in surface water as well as in deep-sea sediments [21]. A severe example of marine plastic population is 'The Great Pacific Garbage Patch', which resembles a giant floating garbage island. This patch is believed to have increased '10 fold in each decade' since 1945 [22], and to be double the size of Texas. This patch has grown to such a scale that it is now claimed to be infeasible for any country to clean up the patch [23]. Another such patch is the North Atlantic Garbage Patch.

17.3 SOCIAL IMPACTS OF PLASTICS IN THE ENVIRONMENT

Societal impact is the effect of plastics beyond their market cost. It can be stated as the effect or changes that society experience from the presence of plastics in the environment. Plastics in the environment has evidence of posing many direct or indirect, adverse and deleterious impacts and consequences on biodiversity, human and the ecosystem. Plastics' life-cycle does not end once it becomes waste but is far beyond extending for thousands of years. Plastics across their life-cycle impose a burden on societies and government, an indirect cost impacting the environment, human health and well-being.

17.3.1 HUMAN HEALTH AND HUMAN WELL-BEING

Across its life-cycle from production to waste and beyond, plastics has far reaching and detrimental impacts on human health and well-being. Even before plastics turn into waste, it has impacts on human health. Plastics majorly come from fossil fuels, and the extraction of oils and gas, hydraulic fracturing of natural gas release significant amount of toxic substances into the atmosphere and water, which has vast impacts on human health [24]. According to a report by the Center for International Environmental Law (CIEL), over 170 chemicals that are used in the production of feedstocks for plastics cause problems to human health, some of which include cancer, neurotoxicity [25]. Research has also found that higher fracking wells are resulting in increased hospitalization with cardiac or problems related to neurological issues [26]. Plastics, when it turns into waste, degrade into micro-plastics, the majority of which get lost in the environment, and human exposure to micro-plastics is substantial [27]. Micro-plastics are everywhere, in the air we breathe,

the food we take, the water we drink, and other consumable products. It is a fact that humans are ingesting micro-plastic [27]. Human beings are also subjected to the potential risk of consuming fish or shellfish, which are chemically contaminated from plastics, or there can be a possible transfer of pathogens. Although the scientific study of the health implications that these micro-plastics has on human health is limited, the presence and ingestion of micro-plastics by humans (infants, young child, and older people) are alarming as these micro-plastics, which generally have chemical additives, could release harmful chemicals or disrupt the normal functioning of the body [28]. The presence of macro-plastics on beaches where people visit for leisure can cause them injury, and such a degraded environment can further cause entanglement of individuals with plastics resulting in well-being loss, and mental health impacts [21].

Not only these, open burning of plastics in the environment or in uncontrolled burning releases harmful substances, which can adversely impact human health, including cancers, aggravates respiratory problems, increase the risk of heart disease, nausea and damage to nervous systems [29]. Heavy metals and toxic chemicals like dioxins, furans, mercury and polychlorinated biphenyls can be released into the atmosphere, posing a threat to human health, animal health and vegetation [29]. These substances can also settle in crops or waterways, thereby entering our food chain. In many rural or underdeveloped areas across the globe, the plastics present in the environment are commonly and regularly disposed of by open burning by their local people [30]. This can lead to the release of many hazardous gas and chemicals into the air polluting it for breathing. It can also generate acrid odours and fumes that causes an immediate effect of a burning sensation in the eyes leading to momentary loss of vision. The toxins released on open burning can stay in the environment and can even travel to long distances [31]. These impacts mentioned here are only some of the many implications of plastic pollution on human health and well-being.

17.3.2 Soil Health and Agriculture

Plastics, when dumped in open space, can contaminate soil, and some can also release chemicals and additives, which affect the soil health and soil fertility, making it unsuitable for vegetation or habitation [32]. Food security and human health have specific critical linkages to soil health, function and soil fertility (especially agricultural soils), as it is the crucial element in the production of food. Micro-plastics affect the soil structure, soil water balance, chemistry and life of the soil, and also affect the root and tissue of plants, thereby bearing the possibility of entering into the human food chain [33]. Thus, contamination of soil with plastics (micro-plastics or nano-plastics) has consequences on the soil health, plant performance and, at large, has impacts on agro-ecosystem and terrestrial biodiversity [34].

17.3.3 Microbial Ecotoxicity

Micro-plastics and nano-plastics found in soil and air can be taken up by microbial communities causing impacts on their growth and functioning [35, 36]. Some marine microorganisms play a crucial role in the marine food web, in photosynthesis (primary producer), carbon cycling in oceans, and are also involved in other biogeochemical cycles such as nitrogen, sulphur, iron, calcium, silicate and phosphorus cycles [35]. The presence of plastics at sea may impact these critical functions of microorganisms [35]. Additives of plastics can leach out and impact the growth of a variety of microorganisms.

17.3.4 Groundwater Contamination

Plastics which might also include chlorinated plastics,[4] when dumped in soil or landfills, can release certain chemicals causing geo-chemical changes in soil and can also seep deep into the groundwater or might leach out to the surrounding water-bodies [32]. This can result in many harmful consequences on any species consuming this water. Also, certain additives from plastics such as phthalates and Bisphenol A (BPA) can leach out, which are known to cause hormonal system disruption of certain species [32].

17.3.5 DISEASE SPREAD AND INVASIVE SPECIES

Plastics lost in the marine environment has the ability to hold water, providing a conducive breeding ground for insects, such as mosquitoes, leading to a greater chance of transmission of diseases such as dengue, malaria or Zika [37]. This is more severe in slums and tropical areas. Micro-plastics in the soil environment are found to be a persistent reservoir and vector for fungal pathogens [38]. Floating plastics in the marine environment also provide a conducive environment for the spread of pathogens [39]. Various microorganisms, including the pathogenic ones, are found to be present on plastics' surfaces, thereby acting as carriers of these pathogenic bacteria and viruses [40]. Studies have confirmed plastic surfaces to be colonized by microorganisms due to their hydrophobic surface and longer half-life compared to other floating marine substrates [41].[5] Of these diverse microorganisms, some are virulence genes having a high potential risk of infectious disease in both humans and animals [41]. Micro-plastics lost in the environment stay in the environment for thousands of years, and thus they act as a vector for such pathogens. A coral draped in plastics has 20 times increased the likelihood of being sick [42]. A range of different algae living on floating plastic litter can also cause certain diseases in corals [43]. The small plastic fragments under the influence of wind or rain may also act as carriers of these harmful pathogen species (such as bacteria, viruses or protists), making its effect widespread especially in coastal communities residing near these water bodies.

Plastics, with their physical properties along with their buoyancy, provide a unique habitat for the diverse microbial hitchhikers (such as Anthropoda, Annelida, Mollusca, and cnidaria phyla), algae, invertebrates, and fish to travel through long distances to non-native regions being attached to its surface [44]. Some of these organisms act as invasive alien species, which has the ability to cause great damage.

17.3.6 TERRESTRIAL AND AQUATIC ORGANISMS

Plastics lost in the environment can be of varied sizes and have detrimental impacts on terrestrial and aquatic organisms. These can be ingested by terrestrial animals, including milch animals such as cows, goats or buffaloes, causing severe effects to their internal organs or can get accumulated inside their gut leading to various health issues such as ruminal impaction and sometimes even death [45]. These milch animals then are unable to give milk, and even if they do, it might contain harmful substances and chemicals, which can be consumed by humans when drinking milk [45]. Several other wild animals in search of food ingest these plastics leading to many health implications, including death. Ingestion of plastics in the environment can cause pseudo satiation, blockage in the digestive tract, or some abrasion or irritation of mucosa [46]. Even if these organisms largely excrete plastics after ingestion, there is evidence that such micro-plastics remain in the gut or even enter other body tissues [5]. Bees as pollinators are vital to interconnected ecosystems and the environment. There are reports of bees incorporating micro-plastics into their nest and even ingestion of such micro- and nano-plastics from the environment, causing effects such as decreased diversity of honey bee gut microbiota, gene expression changes, detoxification and immunity [48].

Plastics in the marine environment can be taken up by the aquatic organisms via ingestion, absorption or ventilation [5]. These can cause the transfer of harmful chemicals from plastic additives to fish, corals, oysters and phytoplankton [49, 42, 50, 51]. Derelict fishing gears, plastic bags, six-pack rings and other marine debris can cause entanglement of marine animals such as turtles, whales, sea lions and sea birds, causing them to drown, starve, infections from debris cutting their flesh, compromising their ability to escape from predators or to reproduce [52]. Ingestion of plastics and micro-plastics by these animals can cause internal abrasion or blockages [52]. According to a study, when high-density polyethylene (HDPE) particles were taken in by mussels by the gills and were transported to their stomach and digestive gland, they got accumulated in lysosomal and evoked inflammatory responses [53]. Micro-plastics may escape to various tissues and organs via the haemolymph of some organisms, thereby accumulating and causing harmful impacts [5]. Predation of such marine organisms contaminated by micro-plastics may lead to the transfer of plastics and their harmful consequences along the food chain.

17.3.7 Greenhouse Effect

Greenhouse gases of earth trap heat in the atmosphere, warming the planet and causing the greenhouse effect, which is the primary driver of climate change [54]. Anthropogenic activities inducing global warming are found to be presently increasing per decade at a rate of 0.2°C [55]. Plastics are found to be emitting a significant amount of greenhouse gases at every stage throughout their life-cycle at the extraction and manufacturing stage of plastics, waste-management stage, and from mismanaged plastics [56]. Research has confirmed that production processes account for 91% of greenhouse gas emissions from plastics [8]. This implies that plastics at the stage of production, which is even before turning to waste, impose a high societal cost.

17.3.8 Heritage and Tourism

Different important and charismatic organisms such as seabirds, turtles, lions, whales, dolphins, flamingos, earthworms has significant value in the nature and to the individuals [57, 58, 59]. Human experience well-being from such organisms present in the nature and globally there are many organizations working and putting efforts to conserve these species in land and water. Earthworms are extremely important to the health of soil as they increase aeration of soil, in plant growth, in decomposition of organic matter, and in nutrient cycling [60]. When these earthworms encounter the presence of micro-plastics in soil, it affects their overall fitness, their life cycle and also studies have confirmed that they make their burrows differently in the presence of micro-plastics in soil. The presence of micro-plastics, which has additives, might also have a toxicological threat to wildlife [62]. It can cause damage to several functions of the body of animals including their reproductive process [63]. When people dump a huge amount of plastics in unprotected wild land, it highly increases the chances of wild animals encountering such plastics. Animals in forest or in the wild spend most of their time searching for food and when such hungry animals encounter plastic containers or bags with the smell of food in it, they can ingest it which can lead to intestinal blockages and its outcome can be fatal [64]. Wild animals and birds in forest or in open areas, out of curiosity or in search for food often gets entangled in plastics, which sometimes may cause them to suffocate and to overheat, leading to their dehydration, starvation and eventually death [65]. In some cases when some animals are entangled with plastics, they become less agile increasing their vulnerability to predators [66]. Plastics in the environment thus has significant impacts to our wildlife and forests. Also marine species are impacted through entanglement and ingestion of plastics causing varied range of impacts including reduced re-productivity and mortality [67]. Such wildlife, forest, marine ecosystem and its organisms holds significant value to human beings and plastics in the environment causing such widespread negative effects brings losses to human well-being and ecosystem.

Mismanaged plastics has serious and major impacts also on the tourism sector, as presence or build up of plastic debris or litter damages the aesthetics of tourist destinations [21]. Litter on vacation places or destinations or on beaches is disliked by tourists and they end up spending less time or avoiding such environments [68]. These tourists when they have such a negative experience on vacation, might spread negative feedback on social media or in some travel forums, thereby defaming those destinations, resulting in a further decrease of tourist visits. The presence of plastics as litter on a tourist destination can also have direct consequences on their mental and physical well-being [21]. Tourists can get infected by some disease, if these plastics contain some infectious agents or they can get themselves injured by cutting themselves on sharp items or the presence of litter can also have a detrimental effect on their mood and mental well-being [69, 21]. Livelihood of many coastal communities are dependent on tourists and when the number of tourists visiting the beaches falls down because of plastic debris, it can cause a huge impact on the survival of these coastal communities. Societal cost of plastics in the environment thus can be observed to be huge and concerning. Also, all these can have a negative impact on the country's overall global influence and image, when people from different country collectively experience and share such experiences with others or write articles and share photographs on the presence of plastic debris on their vacation spot.

17.3.9 ADDITIONAL RISKS

Apart from all the risks mentioned above, causing some direct or indirect social consequences, there are other diverse impacts of plastics in the environment. Plastics present in the land can be transported by wind or rainwater or dumped directly in some rivers, which can cause clogging of such rivers, disrupting their natural flow and can also cause floods in the nearby areas [70]. Wind can carry plastic debris and deposit it in many places, increasing the presence of litter. Such debris can get stuck on traffic lights, trees or towers resulting in serious concerns.

Production of plastics and the presence of plastics, which is now abundant in the environment, with other pollutants can aggravate conditions such as climate change and environmental degradation [21]. Plastics with their toxic additives, can have multiple adverse and direct as well as indirect impacts on human health and well-being [24]. Certain chemicals used in plastics are carcinogenic and also have impacts on the childhood development process [71]. Exposure to Bisphenol A (BPA), which is mainly used to produce polycarbonate plastics such as storage containers and baby bottles, has multiple health effects like brain and behaviour problems in infants and young children, disruption of hormonal levels, heart problems and even may harm human fertility [72]. Plastics along with other environmental stressors such as overexploitation of resources, ocean acidification, and changing temperature, may result in far more significant damage than suggested. Plastic production is also a major consumer of fossil fuels, with over 99% of plastics sourced from fossil fuels [4]. Thus, the list of direct or indirect societal impacts of plastics in the environment is very diverse, and there are possibilities of other such adverse impacts apart from those mentioned so far in this chapter.

17.4 ECONOMIC IMPACTS OF PLASTICS IN THE ENVIRONMENT

The economic impact can be understood as the various direct and indirect financial effects of plastics in the environment. Plastics, from their production to turning into waste, getting released into the environment and far beyond, stay in the environment for thousands of years [6]. Across this life-cycle, its negative impacts imposes costs much more than the market cost of these plastics. According to a report by Dalberg advisors commissioned by World Wide Fund for Nature (WWF), for the plastics produced in the year 2019 alone, the lifetime cost of it has been revealed at US$3.7 trillion (lower bound: US$2.7 trillion and upper bound: US$4.8 trillion), which is more than the GDP of India [8]. According to a study conducted by "The Ocean Clean-up" in collaboration with Deloitte, the economic cost of the presence of marine plastics alone is estimated to be US$6 to US$19 billion[6] yearly [73].

The negative economic impacts of plastics have some direct and some indirect costs, which places a huge burden on societies and the government. The plastic degradation process is very lengthy, and as it degrades, it breaks down into further smaller particles, which makes the cost associated with these plastics in the environment much more complex and diverse [74]. Some of these costs are quantifiable by nature, while others are not [75]. The major costs that occur across the plastic life-cycle include the following: greenhouse gas emission related costs, health costs, direct waste management costs to government, indirect waste management costs to corporate and citizens, costs related to informal waste management, and costs due to unmanaged wastes. The cost of unmanaged wastes itself has many verticals; some of them worth discussing are the cost of lost ecosystem services due to plastic pollution on marine and terrestrial ecosystems, revenue reductions in fisheries, tourism, and related clean-up activity costs.

17.4.1 Greenhouse Gas Emissions (GHG)

Greenhouse gases are released into the environment during the production process of plastics, from the waste management process of the plastics, and also from uncontrolled plastics in the environment [76]. Plastics in the environment thus account for a significant portion of global greenhouse gas emissions [77]. These, along with the other sources of GHG emissions, are contributing hugely to climate change [77]. For sustainable development, climate change is the greatest challenge. Effects of climate change are very alarming, and this can be evidenced by 'SDG goal: 13 Climate Action', which requires countries to take urgent action to combat climate change and its impacts. The costs associated with GHG emissions are carbon prices, related costs for carbon commitments, and costs related to undertaking activities to fulfill commitment under the Paris Agreement to reduce GHG emissions. Government bodies, corporate entities, and industries are continuously undertaking various initiatives by investing huge money, resources, and time to reduce their carbon footprints or are buying carbon credits to reduce their net carbon emissions.

17.4.2 Health Cost

Plastics in the environment have distinct health implications for humans due to exposure to microplastics, toxic chemical additives, and sometimes hazardous and sharp plastic litter [24]. The direct health impacts of plastics can significantly increase in the case of co-morbidity conditions. Plastic fragments are also studied to act as the carrier of disease-causing organisms, impacting human health [41]. There is limited information, but such health issues may cause economic loss through mortality, treatment cost or productivity.

17.4.3 Management Cost of Plastics

Plastic products are cheaper to produce than to manage after it turns into waste and when they are lost in the environment [8]. Formal and informal sectors are continuously making efforts to manage the plastics in the environment to some extent. The cost involved in the management of plastics is huge. A report published in 2021 by Dalberg advisors has mentioned that the cost of managing plastic waste alone is US$32 billion [8]. It included collection cost, sorting cost, recycling cost or cost to dispose of the plastics by both formal and informal sectors [8]. Formal sector management of plastics is the responsibility of the country's solid-waste management authorities, whereas individuals or the independent private sector conducts informal sector management of plastics. A massive amount of funds are needed for these management approaches. For formal sector management, in some countries, producers of plastics are required to invest resources and money, and also in many countries, some portion of public funds are utilized for the management of plastics in the environment [8]. This fund used for the management of plastics in the environment could be otherwise used for other important sectors such as education, health and food security if such huge quantities of plastics would not be there in the environment. Moreover, the fund utilized for waste management is an indirect cost to corporate bodies or citizens in terms of taxes or extended producer responsibility.

17.4.4 Cost of Ecosystem Services Lost

Ecosystem service is an important aspect, which refers to the benefits that are obtained by humans from the ecosystems and these services are the result of interactions within the ecosystem[7] [78]. Different ecosystems like forests, mangroves and urban areas offer different services to society. These services provided by the ecosystems directly affect humans, and these include provisioning (e.g., food, water), regulating (e.g., climate regulation, waste treatment, flood regulation), supporting (e.g., nutrient cycle, oxygen production), and cultural services (e.g., recreational, spiritual benefits) [79]. Research has suggested substantial negative impacts of plastics in the environment on all ecosystem services [80]. Lost of ecosystem services on both terrestrial and marine ecosystems on account of plastics has a huge cost associated with it [75]. In terrestrial ecosystems, plastics reduce the carbon sequestration ability, which is one of the primary ecosystem services of the soil ecosystem, to maintain the carbon cycle [81, 82]. Ocean affects every human life and is one of the most important resources. For the year 2011, the worth of annual ecosystem services provided by the marine ecosystem alone is estimated to be US$61.3 trillion [8]. As a result of the presence of marine plastics in the ocean in 2011, research has postulated a reduction of 1–5% of marine ecosystem service delivery [75]. This decline equates to a loss of $500 to $2500 billion worth of benefits derived from marine ecosystem services annually [75]. Plastics, throughout their life-cycle, will incur a cost as it further degrades into smaller particles and each tonne of plastics entering into ocean incurs over its lifetime a cost of minimum US$204,270 to US$408,541 [8]. Suppose we observe the same effect of reducing ecosystem services on terrestrial ecosystems due to plastics. In that case, these figures of loss will further rise, but due to the unavailability of such information, we cannot incorporate them in this chapter. This means each tonne of plastics in the environment (marine and terrestrial) is likely to have a much greater cost.

17.4.5 Tourism Sector

Tourism has huge importance for any country across the world. It contributes significantly to the economy of the country in terms of revenues being earned from the tourists [83]. Companies operating around tourist destinations can bring revenue to the government in the form of taxes and fees to be paid. Tourism also facilitates and boosts the export of local goods, including gifts, clothing and souvenirs, which is a good source of income for the local people [84]. Millions of jobs are generated because of tourism, employing a high share of women and young adults. Globally around 80% of tourism enterprises consist of small- and medium-sized enterprises (SMEs), and these SMEs mostly employ women and young adults [86]. Coastal tourism earns a massive revenue in the global tourism industry [87]. International tourism receipts of the entire world as of 2019 is US$1481 billion [88]. In 2019, the contribution of the travel and tourism sector to the global GDP was 10.4% [83]. Thus, it can be seen how significant is the contribution of the tourism sector to the economy of a country. The tourism sector also encourages infrastructural development in the country, thereby creating more jobs [89]. Not only this, it promotes cultural exchange between the global community [90]. But the presence of plastics in the tourist spots, along the coastlines, or in the marine environment can impact the inflow of visitors to such places who might end up spending less time or might avoid certain sites if they anticipate them to be littered [91, 92, 93]. The sharp debris can cause injury or other health issues to the tourists. For the tourism industry, this is a huge loss because of the reduction in revenues as the number of tourists would fall, especially during the peak season. The majority of the revenue in the tourism sector comes from the coastal regions, almost 80% [94].

Revenues are not only earned because of tourists visit to a particular place but also hugely depend on the length of their stay [95]. Tourists may reduce their length of stay at a particular destination if it has environmental pollution, and thus they spend less [95]. In 2011, the marine debris pollution event on Geojedo Island (South Korea), where large quantities of debris were washed up to the beaches, affected hugely the island's tourism with a estimated revenue loss of US\$29 to US\$37 million [96]. The impact of marine debris on the Asian Pacific Economic Cooperation (APEC) region, according to the APEC Ocean and Fisheries Working Group, is estimated to be a loss of US\$1.2 billion [97].

17.4.6 FISHERIES AND AQUACULTURE

Oceans, seas and other such marine environments are the source of various aquatic foods, which are renewable resources and, if managed sustainably, can feed people indefinitely [98]. Aquatic foods, including invertebrates, fish, and plants captured or cultured in water, have high nutritional values and low carbon footprints [99]. Also, the livelihood of millions of people, estimated to be over 3 billion (most of them being women comprising over 70% of the workforce) globally, is dependent on this marine and coastal biodiversity [100]. According to Organisation for Economic Co-operation and Development (OECD), the annual contribution of oceans in value-added is US\$1.5 trillion to the overall economy [101]. The presence of marine plastic pollution poses a significant risk to such fishing and aquaculture activities. Marine plastics, or the micro-plastics, which are found in abundance, affect and contaminate aquaculture and reduce the quality of farmed fishes [102]. The quality of water also gets affected because of the presence of such debris or micro-plastics, and this, in turn, affects the survivability of fish larvae [103]. Thus, fish production gets impacted, and subsequently, fish catch for a particular period declines, impacting the revenues for fisheries and aquaculture [104, 105]. Plastics in the marine environment can act as a vector for many disease-causing organisms and non-native parasites, which can cause an outbreak and bring losses to fisheries and aquaculture [44]. Marine plastic pollution thus has direct and indirect impacts on fisheries. Economic losses also include damages to fishing gears and nets because of plastic debris and its related cost, lost earnings due to spending more time clearing litter from nets, fewer quantities of fish per catch, damages to fishing vessels, or entanglement of plastic debris with the engine. According to a report by United Nations Environment Programme (UNEP) published in the year 2014, the yearly revenue loss due to the presence of plastics for fisheries and aquaculture activities is estimated to be US\$794 and 7 million [106], which has now definitely increased to much higher levels considering the current amount of plastics in the marine environment.

17.4.7 CLEAN-UP ACTIVITIES

The cost of clean-up activities is a major economic impact of plastics or litter present in the environment requiring additional resources and expenditure. The use of plastics and subsequent presence of plastics in both terrestrial and marine environments have been increasing [7, 3], while the already present plastics in the environment are taking thousands of years to degrade [6], causing severe and multiple impacts. It is thus important and necessary to undertake regular clean-up activities of such plastics, especially in those places where its impact is immediate, widespread, and severe. Most of the clean-up activities are undertaken on sites such as coastlines which are inhabited, beaches, rivers, ports, marinas, parks, monuments, tourist destinations and forests [8]. As per a report published by Deloitte (2019), a significant cost as high as US\$15 billion annually is incurred for clean-up activities by governments, NGOs and concerned citizens [8, 107]. Local municipalities and governments undertake the setting up of management infrastructure, and also they are required to invest in the value chain of collection and treatment. In most countries, these variable costs related to cleanup activities are reserved as some percentage of the fiscal budget [107]. Apart from the government-sanctioned clean-up activities, various volunteers led by NGOs, concerned citizens, industries and

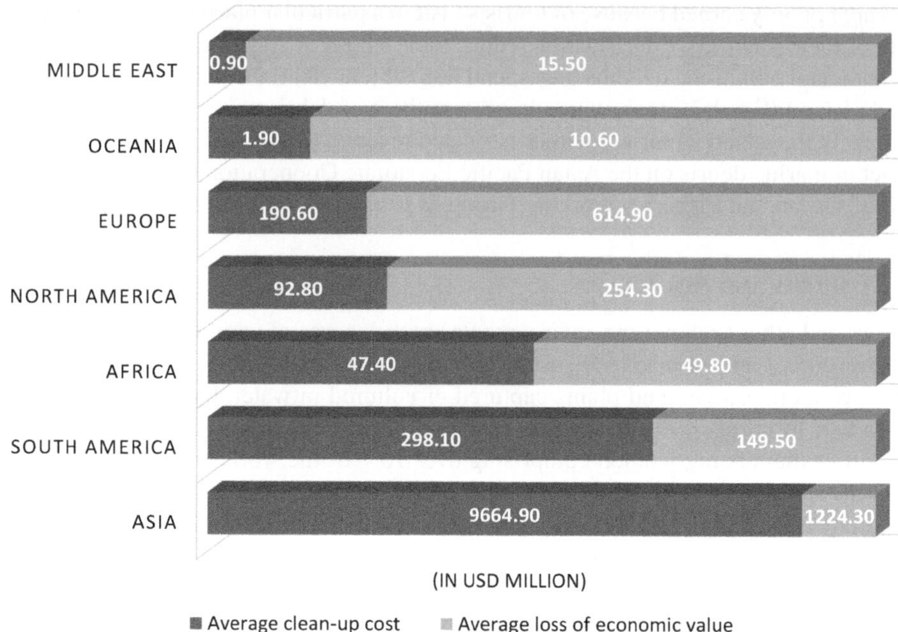

FIGURE 17.2 The graph shows data for 2018, region-wise costs of only marine tourism and fisheries and agriculture. This cost is segregated into clean-up cost and loss of economic value. The clean-up costs are limited to clean-up of stranded or floating plastic debris along coastlines and in waterways. Data used in this graph is as per a report by Deloitte [107].

corporate entities are also involved in various clean-up activities of such plastics in the environment [107], which has some economic impacts associated with it. Direct cost associated with clean-up activities includes government and NGO funding. In contrast, some indirect clean-up costs include unpaid volunteers' time and potential risk to health from sometimes hazardous and sharp debris.

We have taken the data published in a report by Deloitte [107] to show an overview of region-wise costs of only marine tourism, and fisheries and aquaculture in Figure 17.2. These costs are segregated into two components average clean-up costs and average loss of economic value and do not include any other costs. Further in this chapter, the tourism cost, governmental clean-up cost, and fisheries and aquaculture cost of marine plastic pollution for some countries in land-based water sources are also shown.

17.4.8 COMMERCIAL SHIPPING

Maritime transportation has historical significance and is one of the most cost-effective modes of transportation, requiring very low operating cost [108]. It refers to the movement of people and goods across oceans, seas, canals, rivers, or any other medium of water transportation. It is a cost-effective mode of transport for large, perishable, and heavy goods across long distances [109]. Maritime transportation is considered to be the backbone of international trade and the global economy [110]. But this commercial shipping or vessels used in the transportation of people and goods is the source of marine debris and also is affected by it [111]. Plastic dumping in commercial shipping is observed to be both accidental and deliberate dumping [111]. According to a report of the World Shipping Council (WSC), between 2008 to 2019, the estimated average number of containers lost at sea every year is 1382 [112], which results in the discharge of enormous volumes of plastics into the sea. If we see the history, throwing plastics overboard was a common practice

[113]. With the increase in shipping traffic globally and the rise of plastic wastes generated, the issue of such litter accumulating in the marine environment has increased significantly. According to a report of the United Nations Conference on Trade and Development (UNCTAD), global trade carried by sea is around 80 percent by volume, and by the value, it is over 70 percent [110]. There are many laws and conventions internationally, nationally and regionally providing legislation to reduce marine litter from shipping vessels into the sea. One of the main international conventions is the International Convention for the Prevention of Pollution from Ships (MARPOL), which aims to prevent operational and accidental marine pollution by ships. But there are many studies arguing that despite all these legislations, millions of plastics from ships still enter the oceans every year. Cruise ships, which are used for vacationing, carry a large number of passengers. Thus they can generate a significant amount of waste estimated to be typically 70 times as much as a cargo vessel [114]. Cruise ships, although they form a minimal percentage of the global merchant fleet, the marine litter discharge that they can generate is huge and significant [114]. These marine litters and their associated impacts and risks are also experienced by nearby coastal areas especially those of the busiest cruise destinations like Miami and Alaska.

Marine litter generated from shipping vessels and the already present plastics brings high costs to the commercial shipping sector [113]. Due to the presence of plastic debris in the water, collisions may happen with the marine litter, which can be sharp and heavy, resulting in damage to the vessels [115]. Marine litter, mainly the small plastics, can cause fouling of the ship propulsion systems or the cooling systems [115]. This can have a high cost to the shipping industry, and in some cases, containers might get lost, or it can cause harm to life or injury. The lost containers are of typically huge sizes and can stay afloat for several weeks [116]. The lost containers may contain many valuable and vulnerable items like laptops, mobile phones, cars and perishable items, resulting in further financial loss. The supply chain might also get disrupted when ships collide with marine debris, and accidents or blockages happen, obstructing routes for other ships to travel, especially in narrow canals like Suez Canal. Such delays or disruptions can result in loss of productivity and revenue, which is also an economic impact. The service deliverability, brand image or economic output of industries depend on supply chains, and thus the resultant costs resulting from its disruptions will further multiply. Other economic costs related to the impacts of marine litter on the shipping industry are the cost of repairing the vessels being damaged by sharp plastic debris, cost related to rescue efforts of people and goods on vessels, indirect costs related to operational costs, and disruption of services, and public image. The marine litters increasing over the years need to be cleaned up. Many port authorities are actively cleaning the marine debris for safety and aesthetics. The cost of clean-up activities is also indirectly paid by the shipping sectors. According to a report by APEC, the economies of its members lost more than US$1 billion yearly because of the impacts (clean-up and damage to vessels) of marine litter [11].

17.4.9 Other Economic Impacts

Other economic costs of plastics can include the cost of loss to the coral reef ecosystem, contamination of plastics in the deep sea, loss of food sources and livelihoods of coastal communities, and cost of damaged or lost fishing gear. Lost fishing gear has some direct economic impact, including the loss of commercial species by entrapment, the replacement cost of lost gear, and the cost of its disposal. Thus, the economic impacts mentioned so far in this section could be just the tip of the iceberg, and there might exist some other direct or indirect economic impacts of the presence of plastics in the environment.

Figure 17.3 shows the cost for some countries, only the impact of marine plastic pollution on tourism, fisheries and aquaculture, and governmental clean-up activities cost in land-based water sources, as per the data published by 'The Ocean Clean-up' [73]. These costs are limited to only these three costs of marine plastic pollution in land-based water sources. They do not include other

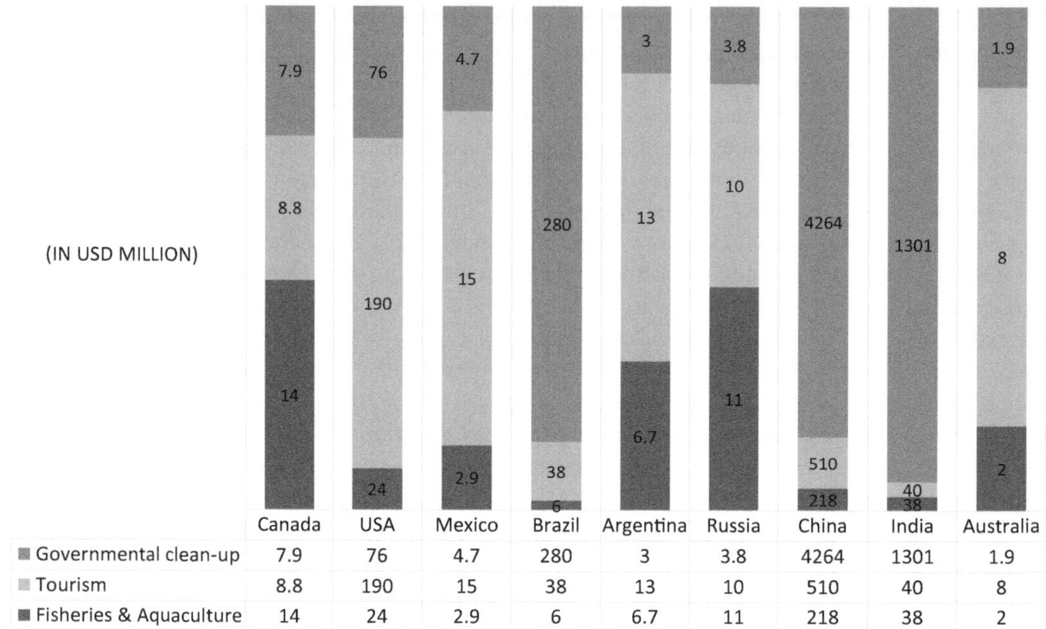

	Canada	USA	Mexico	Brazil	Argentina	Russia	China	India	Australia
■ Governmental clean-up	7.9	76	4.7	280	3	3.8	4264	1301	1.9
▨ Tourism	8.8	190	15	38	13	10	510	40	8
■ Fisheries & Aquaculture	14	24	2.9	6	6.7	11	218	38	2

FIGURE 17.3 Economic cost emerging from marine plastic pollution in land-based water sources for some countries. The economic impact of tourism revenue loss, fisheries and aquaculture loss, and cost of governmental clean-up activities only are shown in USD million. Data taken in this graph is as per a report published by 'The Ocean clean-up' [73].

components such as the cost of human health, marine ecosystem service loss, or any additional cost of terrestrial plastic pollution.

17.5 DISCUSSION

In this chapter, while summarizing all the social and economic impacts of plastics in the environment, we have undertaken an extensive literature review. We have observed the most recent contributions of plastics in the environment towards the societal and economic costs. The relationship between the proliferation of plastics in the environment and the negative societal and economic impacts of these plastics in the environment, globally or locally, is direct. Among the different social and economic costs, a clear and significant impact is observed on human health, well-being, fisheries, aquaculture, tourism, and clean-up cost of these plastics. The cost involved has many other components, which need further understanding, and thus this implies that the actual social and economic impacts would be much greater than those mentioned here. Fewer data and research are available on the impacts of terrestrial plastics in the environment. In contrast, terrestrial plastics are the major chunk of plastics in the environment and a major source of plastics in the marine ecosystem. There are also less data available on the direct impacts of micro-plastics on human health and the human food chain contamination.

Considering the huge societal and economic impacts of plastics in the environment, it is recommended that all stakeholders, including the government, consider the per tonne social and economic cost of plastics and make collective concrete efforts to reduce and reuse these plastics in the environment. Whether global or local, policies and regulations are observed to be the most effective approach to address this alarming issue of rising plastics in the environment. The public should be aware of the substantial negative impacts of plastics on society, economy, ecology and environment.

Usage of plastics is increasing over the years, and so is the waste generated from it, the plastics lost into the environment, and the resultant multiple cost. Although plastics are a cheap alternative, their life-cycle cost is enormous and diverse. Reduction in plastics in the environment can provide multiple benefits to humans, the environment, the economy and society.

NOTES

1 Macro-plastics are those particles, fibres or fragments of any type of plastics, which are greater than 5 millimeters.
2 Micro-plastics are those particles, fibres, or fragments of any type of plastics which are smaller than 5 millimeters.
3 Nano-plastics results from the degradation of plastics or from the manufacturing of plastic objects and are less than 0.1μm.
4 Chlorinated plastic is a thermoplastic produced by chlorination of the resins such as Chlorinated polyvinyl chloride (CPVC), which has more flexible usage and the ability to withstand higher temperatures than its standard version. But it has many adverse environmental consequences, including releasing of carcinogenic dioxins.
5 Substrate refers to the base on which an organism lives.
6 This cost included only the impacts on tourism, fisheries and aquaculture, and governmental clean-ups.
7 Ecosystem means a dynamic complex consisting of plants, animals, microorganisms and the features of the physical environment that interact with each other.

BIBLIOGRAPHY

[1] Laura Parker. The world's plastic pollution crisis explained. *Environment Explainer*, 2019. URL www.nationalgeographic.com/environment/article/plastic-pollution.
[2] Rabin Tuladhar and Shi Yin. 21 sustainability of using recycled plastic fiber in concrete. In Fernando Pacheco-Torgal, Jamal Khatib, Francesco Colangelo, and Rabin Tuladhar, editors, *Use of Recycled Plastics in Ecoefficient Concrete*, Woodhead Publishing Series in Civil and Structural Engineering, pp. 441–460. Woodhead Publishing, 2019. ISBN 978-0-08-102676-2.
[3] United Nations Environment Programme. Beat Plastic Pollution, 2021. URL www.unep.org/interactive/beat-plastic-pollution/.
[4] CIEL. Fossil fuels and plastics, 2021. URL www.ciel.org/issue/fossil-fuels-plastic/.
[5] Karen Duis and Anja Coors. Microplastics in the aquatic and terrestrial environment: Sources (with a specific focus on personal care products), fate and effects. *Environmental Sciences Europe*, 28: 2, 2016. doi: 10.1186/s12302015-0069-y.
[6] Ali Chamas, Hyunjin Moon, Jiajia Zheng, Yang Qiu, Tarnuma Tabassum, Jun Hee Jang, Mahdi Abu-Omar, Susannah L. Scott, and Sangwon Suh. Degradation rates of plastics in the environment. *ACS Sustainable Chemistry & Engineering*, 8(9): 3494–3511, 2020.
[7] Roland Geyer, Jenna R. Jambeck, and Kara Lavender Law. Production, use, and fate of all plastics ever made. *Science Advances*, 3(7): e1700782, 2017.
[8] Wijnand Dewit, Erin Towers Burns, Jean-Charles Guinchard, and Nour Ahmed. Plastics: The costs to society, the environment and the economy, 2021. URL https://media.wwf.no/assets/attachments/Plastics-the-costto-society-the-environment-and-the-economy-WWF-report.pdf.
[9] Plastic Oceans International. The facts, 2021. URL https://plasticoceans.org/the-facts/.
[10] U.S. National Park Service; Mote Marine Lab, Sarasota, FL and "Garbage In, Garbage Out," Audubon magazine, Sept/Oct 1998.
[11] Lourens J. J. Meijer, Tim van Emmerik, Ruud van der Ent, Christian Schmidt, and Laurent Lebreton. More than 1000 rivers account for 80% of global riverine plastic emissions into the ocean. *Science Advances*, 7(18): eaaz5803, 2021.

[12] United Nations Environment Programme. From pollution to solution: A global assessment of marine litter and plastic pollution, 2021. URL https://wedocs.unep.org/bitstream/handle/20.500.11822/36963/POLSOL.pdf.

[13] Jefferson Hopewell, Robert Dvorak, and Edward Kosior. Plastics recycling: Challenges and opportunities. *Philosophical Transactions of the Royal Society of London. Series B, Biological Sciences*, 364(1526): 2115–2126, 2009.

[14] Laura Parker. Here's how much plastic trash is littering the earth. *Planet or Plastic?* 2018. URL www.nationalgeographic.com/science/article/plastic-produced-recycling-waste-ocean-trash-debris-environment.

[15] Alice A. Horton, Alexander Walton, David J. Spurgeon, Elma Lahive, and Claus Svendsen. Microplastics in freshwater and terrestrial environments: Evaluating the current understanding to identify the knowledge gaps and future research priorities. *Science of the Total Environment*, 586: 127–141, 2017.

[16] Morten W. Ryberg, Alexis Laurent, and Michael Hauschild. *Mapping of Global Plastics Value Chain and Plastics Losses to the Environment (with a Particular Focus on Marine Environment)*. United Nations Environment Programme, 2018. URL https://wedocs.unep.org/20.500.11822/26745.

[17] Deonie Allen, Steve Allen, Sajjad Abbasi, Alex Baker, Melanie Bergmann, Janice Brahney, Tim Butler, Robert A. Duce, Sabine Eckhardt, Nikolaos Evangeliou, Tim Jickells, Maria Kanakidou, Peter Kershaw, Paolo Laj, Joseph Levermore, Daoji Li, Peter Liss, Kai Liu, Natalie Mahowald, Pere Masque, Du˘san Materi´c, Andrew G. Mayes, Paul McGinnity, Iolanda Osvath, Kimberly A. Prather, Joseph M. Prospero, Laura E. Revell, Sylvia G. Sander, Won Joon Shim, Jonathan Slade, Ariel Stein, Oksana Tarasova, and Stephanie Wright. Microplastics and nanoplastics in the marineatmosphere environment. *Nature Reviews Earth & Environment*, 3: 393–405, 2022.

[18] Morten W. Ryberg, Michael Z. Hauschild, Feng Wang, Sandra AverousMonnery, and Alexis Laurent. Global environmental losses of plastics across their value chains. *Resources, Conservation and Recycling*, 151: 104459, 2019. ISSN 0921-3449.

[19] Vinay Yadav, M.A. Sherly, Pallav Ranjan, Rafael O. Tinoco, Alessio Boldrin, Anders Damgaard, Alexis Laurent. Framework for quantifying environmental losses of plastics from landfills. Resources, Conservation and Recycling, 161, 104914, 2020.

[20] Jiajia Zhang, Xuexia Wang, Wentao Xue, Li Xu, Wencheng Ding, Meng Zhao, Shanjiang Liu, Guoyuan Zou, and Yanhua Chen. Microplastics pollution in soil increases dramatically with long-term application of organic composts in a wheat–maize rotation. *Journal of Cleaner Production*, 356, 131889, 2022. ISSN 0959-6526.

[21] International Union for Conservation of Nature and Natural Resources. Marine Plastic Pollution, 2021. URL www.iucn.org/resources/issues-briefs/marine-plastic-pollution.

[22] Chris Maser. *Interactions of land, ocean and humans: A global perspective*, volume 1. CRC Press, 2014. doi: https://doi.org/10.1201/b17529.

[23] Office of Response and NOAA Restoration. Garbage patches, 2022. URL https://marinedebris.noaa.gov/info/patch.html.

[24] CIEL. Plastic and human health: A lifecycle approach to plastic pollution, 2022. URL www.ciel.org/project-update/plastic-and-human- health-a-lifecycle-approach-to-plastic-pollution/.

[25] CEIL. Plastic & climate: The hidden costs of a plastic planet, 2019. URL www.ciel.org/reports/plastic-health-the-hidden-costs- of-a-plastic-planet-may-2019/.

[26] Thomas Jemielita, George L. Gerton, Matthew Neidell, Steven Chillrud, Beizhan Yan, Martin Stute, Marilyn Howarth, Poun´e Saberi, Nicholas Fausti, Trevor M. Penning, Jason Roy, Kathleen J. Propert, and Reynold A. Panettieri Jr. Unconventional gas and oil drilling is associated with increased hospital utilization rates. *PLOS One*, 10(8): e0137371, 2015.

[27] Kurunthachalam Kannan and Krishnamoorthi Vimalkumar. A review of human exposure to microplastics and insights into microplastics as obesogens. *Frontiers in Endocrinology*, 12, 724989, 2021. doi: 10.3389/fendo.2021.724989.

[28] Qiqing Chen, Annika Allgeier, Daqiang Yin, and Henner Hollert. Leaching of endocrine disrupting chemicals from marine microplastics and mesoplastics under common life stress conditions. *Environment International*, 130: 104938, 2019. ISSN 0160-4120.

[29] UNEP. Plastic bag bans can help reduce toxic fumes, 2019. URL www.unep.org/news-and-stories/story/plastic-bag-bans-can-help-reduce-toxic-fumes.

[30] Navarro Ferronato and Vincenzo Torretta. Waste mismanagement in developing countries: A review of global issues. *International Journal of Environmental Research and Public Health*, 16: 1060, 2019. doi: 10.3390/ijerph16061060.

[31] Alexander Cogut. Open burning of waste: A global health disaster, October 2016. URL https://region s20.org/wp-content/uploads/2016/08/OPENBURNING-OF-WASTE-A-GLOBAL-HEALTH-DISAS TER R20-Research-Paper Final 29.05.2017.pdf.

[32] United Nations Environment Programme. Plastic planet: How tiny plastic particles are polluting our soil, 2021. URL www.unep.org/news-and-stories/story/plastic-planet-how-tiny-plastic-particles-are-polluting-our-soil.

[33] Anderson Abel de Souza Machado, Chung W. Lau, Jennifer Till, Werner Kloas, Anika Lehmann, Roland Becker, and Matthias C. Rillig. Impacts of microplastics on the soil biophysical environment. *Environmental Science and Technology*, 52: 9656–9665, 2018.

[34] Anderson Abel de Souza Machado, Chung W. Lau, Werner Kloas, Joana Bergmann, Julien B. Bachelier, Erik Faltin, Roland Becker, Anna S. G¨orlich, and Matthias C. Rillig. Microplastics can change soil properties and affect plant performance. *Environmental Science and Technology*, 53: 6044–6052, 2019.

[35] Justine Jacquin, Jingguang Cheng, Charl`ene Odobel, Caroline Pandin, Pascal Conan, Mireille Pujo-Pay, Val´erie Barbe, Anne-Leila Meistertzheim, and Jean-Fran¸cois Ghiglione. Microbial ecotoxicology of marine plastic debris: A review on colonization and biodegradation by the "plastisphere". *Frontiers in Microbiology*, 10, 865 2019. ISSN 1664-302X.

[36] Meredith E. Seeley, Bongkeun Song, Renia Passie, and Robert C. Hale. Microplastics affect sedimentary microbial communities and nitrogen cycling. *Nature Communications*, 11: 2372, 2020. doi: 10.1038/s41467-020-16235-3.

[37] Benjamin MacCormack-Gelles, Antonio S. Lima Neto, Geziel S. Sousa, Osmar J. Nascimento, Marcia M. T. Machado, Mary E. Wilson, and Marcia C. Castro. Epidemiological characteristics and determinants of dengue transmission during epidemic and non-epidemic years in fortaleza, brazil: 2011-2015. *PLOS Neglected Tropical Diseases*, 12(12): 1–30, 2018. doi: 10.1371/journal.pntd.0006990.

[38] Gerasimos Gkoutselis, Stephan Rohrbach, Janno Harjes, Martin Obst, Andreas Brachmann, Marcus A. Horn, and Gerhard Rambold. Microplastics accumulate fungal pathogens in terrestrial ecosystems. *Scientific Reports*, 13214, 2021. doi: 10.1038/s41598-021-92405-7.

[39] Robyn J. Wright, Gabriel Erni-Cassola, Vinko Zadjelovic, Mira Latva, and Joseph A. Christie-Oleza. Marine plastic debris: A new surface for microbial colonization. *Environmental Science & Technology*, 54(19): 11657–11672, 2020. doi: 10.1021/acs.est.0c02305.

[40] Jian Meng, Qun Zhang, Yifan Zheng, Guimei He, and Huahong Shi. Plastic waste as the potential carriers of pathogens. *Current Opinion in Food Science*, 41: 5, 2021. doi: 10.1016/j.cofs.2021.04.016.

[41] Erik R. Zettler, Tracy J. Mincer, and Linda A. Amaral-Zettler. Life in the "plastisphere": Microbial communities on plastic marine debris. *Environmental Science & Technology*, 47(13): 7137–7146, 2013. doi: 10.1021/es401288x.

[42] Joleah B. Lamb, Bette L. Willis, Evan A. Fiorenza, Courtney S. Couch, Robert Howard, Douglas N. Rader, James D. True, Lisa A. Kelly, Awaludinnoer Ahmad, Jamaluddin Jompa, and C. Drew Harvell. Plastic waste associated with disease on coral reefs. *Science*, 359(6374): 460–462, 2018.

[43] Bednarz V, Leal M, B´eraud E, Ferreira Marques J, and Ferrier-Pages C. The invisible threat: How microplastics endanger corals. *Front. Young Minds*, 9: 574637, 2021. doi: 10.3389/frym.2021.574637.

[44] Jos´e Carlos Garc´ıa-G´omez, Marta Garrig´os, and Javier Garrig´os. Plastic as a vector of dispersion for marine species with invasive potential. A review. *Frontiers in Ecology and Evolution*, 9: 629756, 2021.

[45] M. Priyanka and S. Dey. Ruminal impaction due to plastic materials an increasing threat to ruminants and its impact on human health in developing countries. *Veterinary World*, 11: 1307–1315, 2018. doi: 10.14202/vetworld.2018.1307-1315.

[46] Kay Critchell and Mia O. Hoogenboom. Effects of microplastic exposure on the body condition and behaviour of planktivorous reef fish (Acanthochromis polyacanthus). *PLoS ONE*, 13: 3, 2018. doi: https://doi.org/10.1371/journal.pone.0193308.

[47] Keng-Lou James Hung, Jennifer M. Kingston, Matthias Albrecht, David A. Holway, and Joshua R. Kohn. The worldwide importance of honey bees as pollinators in natural habitats. *Proceedings of the Royal Society B*, 285: 20172140, 2018.

[48] Yahya Al Naggar, Markus Brinkmann, Christie M. Sayes, Saad N. ALKahtani, Showket A. Dar, Hesham R. El-Seedi, Bernd Gru¨newald, and John P. Giesy. Are honey bees at risk from microplastics? *Toxics*, 9: 109, 2021. doi: 10.3390/toxics9050109.

[49] Chelsea M. Rochman, Akbar Tahir, Susan L. Williams, Dolores V. Baxa, Rosalyn Lam, Jeffrey T. Miller, Foo-Ching Teh, Shinta Werorilangi, and Swee J. Teh. Anthropogenic debris in seafood: Plastic debris and fibers from textiles in fish and bivalves sold for human consumption. *Scientific Reports*, 5: 14340, 2015.

[50] Sasha G. Tetu, Indrani Sarker, Verena Schrameyer, Russell Pickford, Liam D. H. Elbourne, Lisa R. Moore, and Ian T. Paulsen. Plastic leachates impair growth and oxygen production in prochlorococcus, the ocean's most abundant photosynthetic bacteria. *Communications Biology*, 2: 184, 2019.

[51] Rossana Sussarellu, Marc Suquet, Yoann Thomas, Christophe Lambert, Caroline Fabioux, Marie Eve Julie Pernet, Nelly Le Go¨ıc, Virgile Quillien, Christian Mingant, Yanouk Epelboin, Charlotte Corporeau, Julien Guyomarch, Johan Robbens, Ika Paul-Pont, Philippe Soudant, and Arnaud Huvet. Oyster reproduction is affected by exposure to polystyrene microplastics. *Proceedings of the National Academy of Sciences*, 113(9): 2430–2435, 2016. doi: 10.1073/pnas.1519019113.

[52] Murray R. Gregory. Environmental implications of plastic debris in marine settings—entanglement, ingestion, smothering, hangers-on, hitch-hiking and alien invasions. *Philosophical Transactions of the Royal Society of London. Series B, Biological Sciences*, 364(1526): 2013–2025, 2009.

[53] Nadia von Moos, Patricia Burkhardt-Holm, and Angela K¨ohler. Uptake and effects of microplastics on cells and tissue of the blue mussel mytilus edulis l. After an experimental exposure. *Environmental Science & Technology*, 46(20): 11327–11335, 2012.

[54] UCAR Center of Science Education. The greenhouse effect, 2024. URL https://scied.ucar.edu/learning-zone/how-climate-works/greenhouseeffect.

[55] Athanasios Valavanidis. Global warming and climate change. Fossil fuels and anthropogenic activities have warmed the earth's atmosphere, oceans, and land. *Scientific Reviews*, 2: 1–42 2022.

[56] Maocai Shen, Wei Huang, Ming Chen, Biao Song, Guangming Zeng, and Yaxin Zhang. (Micro)plastic crisis: Un-ignorable contribution to global greenhouse gas emissions and climate change. *Journal of Cleaner Production*, 254: 120138, 2020. ISSN 0959 6526.

[57] Victoria Gonz´alez Carman, Pablo Denuncio, Martina Vassallo, Mar´ıa Paula Ber´on, Karina C. A´lvarez, and Sergio Rodriguez-Heredia. Charismatic species as indicators of plastic pollution in the r´ıo de la plata estuarine area, sw atlantic. *Frontiers in Marine Science*, 8: 699100, 2021. doi: 10.3389/fmars.2021.699100.

[58] Leopoldo Torres-Cristiani, Salima Machkour-M'Rabet, Sophie Calm´e, Holger Weissenberger, and Griselda Escalona-Segura. Assessment of the american flamingo distribution, trends, and important breeding areas. *PLOS ONE*, 15: e0244117, 2020. doi: 10.1371/journal.pone.0244117.

[59] Dara E. Seidl and Peter Klepeis. Human dimensions of earthworm invasion in the adirondack state park. *Human Ecology*, 39(5): 641–655, 2011. ISSN 03007839, 15729915.

[60] A. Tyler Labenz. Earthworm activity increases soil health, 2021. URL www.nrcs.usda.gov/wps/portal/nrcs/detail/ks/newsroom/features /?cid=stelprdb1242736.

[61] Dongdong Cao, Xiao Wang, Xianxiang Luo, Guocheng Liu, and Hao Zheng. Effects of polystyrene microplastics on the fitness of earthworms in an agricultural soil. *IOP Conference Series: Earth and Environmental Science*, 61: 012148, 2017.

[62] Anna K¨arrman, Christine Sch¨onlau, and Magnus Engwall. Exposure and effects of microplastics on wildlife a review of existing data. *DIVA Portal*, 3: 921211, 2016. URL www.divaportal.org/smash/get/diva2:921211/FULLTEXT01.pdf.

[63] Chris E. Talsness, Anderson J. M. Andrade, Sergio N. Kuriyama, Julia A. Taylor, and Frederick S. vom Saal. Components of plastic: Experimental studies in animals and relevance for human health. *Philosophical Transactions of the Royal Society of London. Series B, Biological Sciences*, 364 (1526): 2079–2096, 2009.

[64] Office of Response Marine Debris Program and NOAA Restoration. Why is marine debris a problem? Ingestion, 2022. URL https://marinedebris.noaa.gov/why-marine-debris-problem/ingestion.

[65] Deborah Sullivan Brennan. Report tallies wildlife entangled or choked by plastic waste. *The San Diego Union-Tribune*, November 2020. URL www.sandiegouniontribune.com/news/environment/story/2020- 11-19/report-tallies-animals-choked-entangled-in-ocean-plastics.

[66] Plastc Soup Foundation. Animals stuck in plastic, 2021. URL www.plasticsoupfoundation.org/en/plastic-problem/plastic-affect-animals/animal-stuck-plastic/.

[67] Chelsea M. Rochman, Tomofumi Kurobe, Ida Flores, and Swee J. Teh. Early warning signs of endocrine disruption in adult fish from the ingestion of polyethylene with and without sorbed chemical pollutants from the marine environment. *Science of the Total Environment*, 493: 656–661, 2014. ISSN 0048-9697.

[68] Allan Paul Krelling, Allan Thomas Williams, and Alexander Turra. Differences in perception and reaction of tourist groups to beach marine debris that can influence a loss of tourism revenue in coastal areas. *Marine Policy*, 85: 87–99, 2017.

[69] Amy Krystosik, Gathenji Njoroge, Lorriane Odhiambo, Jenna E. Forsyth, Francis Mutuku, and A. Desiree LaBeaud. Solid wastes provide breeding sites, burrows, and food for biological disease vectors, and urban zoonotic reservoirs: A call to action for solutions-based research. *Frontiers in Public Health*, 7: 405, 2020. ISSN 2296-2565. doi: 10.3389/fpubh.2019.00405.

[70] Vinay Yadav, M.A. Sherly, Pallav Ranjan, Vindhyawasini Prasad, Rafael O Tinoco, Alexis Laurent. Risk of plastics losses to the environment from Indian landfills. *Resources, Conservation and Recycling*, 187: 106610, 2022.

[71] Claudia Campanale, Carmine Massarelli, Ilaria Savino, Vito Locaputo, and Vito Felice Uricchio. A detailed review study on potential effects of microplastics and additives of concern on human health. *International Journal of Environmental Research and Public Health*, 17: 1212, 2020.

[72] Yvette Brazier. How does bisphenol a affect health? 2021. URL www.medicalnewstoday.com/articles/221205.

[73] The Ocean Clean-up. The price tag of plastic pollution, 2021. URL https://theoceancleanup.com/the-price-tag-of-plastic-pollution/.

[74] David Azoulay (CIEL), Priscilla Villa (Earthworks), Yvette Arellano (TEJAS), Miriam Gordon (UPSTREAM), Doun Moon (GAIA), Kathryn Miller, and Kristen Thompson (Exeter University). Plastic and health: The hidden costs of a plastic planet, 2019. URL www.ciel.org/wp- content/uploads/2019/02/Plastic-and-Health-The-Hidden-Costs-of-aPlastic-Planet-February-2019.pdf.

[75] Nicola J. Beaumont, Margrethe Aanesen, Melanie C. Austen, Tobias B¨orger, James R. Clark, Matthew Cole, Tara Hooper, Penelope K. Lindeque, Christine Pascoe, and Kayleigh J. Wyles. Global ecological, social and economic impacts of marine plastic. *Marine Pollution Bulletin*, 142: 189–195, 2019. doi: 10.1016/j.marpolbul.2019.03.022.

[76] Ive Vanderreydt (VITO), Tom Rommens (VITO), Anna Tenhunen (VTT), Lars Fogh Mortensen, and Ida Tange. Greenhouse gas emissions and natural capital implications of plastics (including biobased plastics). *European Topic Centre Waste and Materials in a Green Economy*, 2021. URL www.eionet.europa.eu/etcs/etc-wmge/products/etc-wmge-reports/greenhouse-gas-emissions-and-natural-capital-implications-of-plastics-including-biobasedplastics/@@download/file/ETC2.1.2.1. GHGEmissionsOfPlastics FinalReport v7.0 ED.pdf.

[77] Helen V. Ford, Nia H. Jones, Andrew J. Davies, Brendan J. Godley, Jenna R. Jambeck, Imogen E. Napper, Coleen C. Suckling, Gareth J. Williams, Lucy C. Woodall, and Heather J. Koldewey. The fundamental links between climate change and marine plastic pollution. *Science of the Total Environment*, 806: 150392, 2022. ISSN 0048-9697.

[78] Patricia Balvanera, Sandra Quijas, Daniel S. Karp, Neville Ash, Elena M. Bennett, Roel Boumans, Claire Brown, Kai M.A. Chan, Rebecca ChaplinKramer, Benjamin S. Halpern, Jordi Honey-Ros´es, Choong-Ki Kim, Wolfgang Cramer, Maria Jos´e Mart´ınez-Harms, Harold Mooney, Tuyeni Mwampamba, Jeanne Nel, Stephen Polasky, Belinda Reyers, Joe Roman, Woody Turner, Robert J. Scholes, Heather Tallis, Kirsten Thonicke, Ferdinando Villa, Matt Walpole, and Ariane Walz. *Ecosystem Services*, pp. 39–78. Springer International Publishing, 2017. ISBN 978-3-319-27288-7. doi: 10.1007/978-3-319-27288-7 3.

[79] Millennium Ecosystem Assessment. *Ecosystems and Human Well-being: Opportunities and Challenges for Business and Industry*. World Resources Institute, 2005. URL www.millenniumassessment.org/documents/document.353.aspx.pdf.

[80] Rakesh Kumar, Anurag Verma, Arkajyoti Shome, Rama Sinha, Srishti Sinha, Prakash Kumar Jha, Ritesh Kumar, Pawan Kumar, Shubham, Shreyas Das, Prabhakar Sharma, and P. V. Vara Prasad. Impacts of plastic pollution on ecosystem services, sustainable development goals, and need to focus on circular economy and policy interventions. *Sustainability*, 13(17), 2021. ISSN 2071-1050. doi: 10.3390/su13179963.

[81] Ruili Li, Lingyun Yu, Minwei Chai, Hailun Wu, and Xiaoshan Zhu. The distribution, characteristics and ecological risks of microplastics in the mangroves of southern china. *Science of the Total Environment*, 708: 135025, 2020. ISSN 0048-9697.

[82] Jing-Jie Guo, Xian-Pei Huang, Lei Xiang, Yi-Ze Wang, Yan-Wen Li, Hui Li, Quan-Ying Cai, Ce-Hui Mo, and Ming-Hung Wong. Source, migration and toxicology of microplastics in soil. *Environment International*, 137: 105263, 2020. ISSN 0160-4120.

[83] World Travel & Tourism Council. Economic impact reports, 2021. URL https://wttc.org/Research/Economic-Impact.

[84] R. Torres. Linkages between tourism and agriculture in mexico. *Annals of Tourism Research*, 30(3): 546–556, 2003. doi: 10.1016/S01607383(02)00103-2.

[85] International Labour Organization. Sustainable tourism: A driving force of job creation, economic growth and development, 2016. URL www.ilo.org/global/about-the-ilo/newsroom/news/WCMS_480824/lang--en/index.htm.

[86] M.P. Bezbaruah. View: Travel, tour after stepping out again, 2020. URL https://economictimes.indiatimes.com/news/economy/policy/viewtravel-tour-after-stepping-out-again/articleshow/75497286.cms?from=mdr.

[87] Yehuda L. Klein, Jeffrey P. Osleeb, and Mariano R. Viola. Tourism generated earnings in the coastal zone: A regional analysis. *Journal of Coastal Research*, 20(4): 1080–1088, 2004. ISSN 07490208, 15515036.

[88] UNWTO. 2020 Edition. *International Tourism Highlights*, 2020. URL www.e-unwto.org/doi/pdf/10.18111/9789284422456.

[89] Wolfgang Weinz and Lucie Servoz. *Poverty Reduction through Tourism*. International Labour Office. 2013. URL www.ilo.org/media/456231/download

[90] Gavin Jack and Alison Phipps. *Tourism and Intercultural Exchange: Why Tourism Matters*. Channel View Publications, 12, 2005. ISBN 9781845410193. doi: 10.21832/9781845410193.

[91] Dorothy H. Anderson and Perry J. Brown. The displacement process in recreation. *Journal of Leisure Research*, 16(1): 61–73, 1984.

[92] D. T. Tudor and A. T. Williams. A rationale for beach selection by the public on the coast of wales, UK. *Area*, 38(2): 153–164, 2006. doi: 10.1111/j.1475-4762.2006.00684.x.

[93] World Health Organization. *Guidelines for Safe Recreational Water Environments*. Volume 1, Coastal and Fresh Waters, 2003.

[94] United Nations Environment Programme. Paradise lost? travel and tourism industry takes aim at plastic pollution but more action needed, 2019. URL www.unep.org/news-and-stories/story/paradise-lost-travel-and-tourism-industry-takes-aim-plastic-pollution-more.

[95] Mengmeng Qiang, Manhong Shen, and Huiming Xie. Loss of tourism revenue induced by coastal environmental pollution: A length-of-stay perspective. *Journal of Sustainable Tourism*, 28(4): 550–567, 2020.

[96] Yong Chang Jang, Sunwook Hong, Jongmyoung Lee, Mi Jeong Lee, and Won Joon Shim. Estimation of lost tourism revenue in geoje island from the 2011 marine debris pollution event in south korea. *Marine Pollution Bulletin*, 81(1): 49–54, 2014. doi: 10.1016/j.marpolbul.2014.02.021.

[97] APEC Oceans & Fisheries Working Group. Cleaning up apec rivers and oceans of plastic pollution. *Asia-Pacific Economic Cooperation*, 2019. URL www.apec.org/press/news-releases/2019/0822marine.

[98] Ben Belton. Fishing and aquaculture: Underestimated as a source of income and food. *World Food Journal*, August 2021. URL www.welthungerhilfe.org/news/latest-articles/2021/fishing-and-aquaculture-as-a-source-of-income-and-food/.

[99] Jim Leape, Fiorenza Micheli, Michelle Tigchelaar, Edward H. Allison, Xavier Basurto, Abigail Bennett, Simon R. Bush, Ling Cao, Beatrice Crona, Fabrice DeClerck, Jessica Fanzo, Jessica A. Gephart, Stefan Gelcich, Christopher D. Golden, Christina C. Hicks, Avinash Kishore, J. Zachary Koehn, David C. Little, Rosamond L. Naylor, Elizabeth R. Selig, Rebecca E. Short, U. Rashid Sumaila, Shakuntala H. Thilsted, Max Troell, and Colette C.C. Wabnitz. The vital roles of blue foods in the global food system. United Nations Food Systems Summit 2021, 2021. URL https://sc-fss2021.org/wpcontent/uploads/2021/04/FSS Brief Blue Economy MT.pdf.

[100] Convention on Biological Diversity. People depend on marine and coastal biodiversity for their livelihoods, 2018. URL www.cbd.int/article/food-2018-11-21-09-29-49.

[101] OECD. *The Ocean Economy in 2030*. OECD Publishing, April 2016. URL http://dx.doi.org/10.1787/9789264251724-en.

[102] Amy L. Lusher and Natalie A. C. Welden. *Microplastic Impacts in Fisheries and Aquaculture*, pp. 1–28. Springer International Publishing, 2020. doi: 10.1007/978-3-030-10618-8 30-1.

[103] Merlin N. Issac and Balasubramanian Kandasubramanian. Effect of microplastics in water and aquatic systems. *Environmental Science and Pollution Research*, 28: 19544–19562, 2021.

[104] Courtney Arthur, Ariana E. Sutton-Grier, Peter Murphy, and Holly Bamford. Out of sight but not out of mind: Harmful effects of derelict traps in selected U.S. coastal waters. *Marine Pollution Bulletin*, 86(1): 19–28, 2014. ISSN 0025-326X.

[105] Donna Marie Bilkovic, Kirk Havens, David Stanhope, and Kory Angstadt. Derelict fishing gear in chesapeake bay, virginia: Spatial patterns and implications for marine fauna. *Marine Pollution Bulletin*, 80(1): 114–123, 2014. ISSN 0025-326X.

[106] Julie Raynaud (Trucost). Valuing plastic, 2014. URL https://issuu.com/unpublications/docs/9789210575669.

[107] Deloitte. The price tag of plastic pollution: An economic assessment of river plastic, 2019. URL www2.deloitte.com/content/dam/Deloitte/nl/Documents/strategy-analytics-and-ma/deloitte-nl-strategy-analytics-and-ma-the-price-tag-of-plastic-pollution.pdf.

[108] Riley E.J. Schnurr and Tony R. Walker. Marine transportation and energy use. In *Reference Module in Earth Systems and Environmental Sciences*. Elsevier, 2019. ISBN 978-0-12-409548-9. doi: https://doi.org/10.1016/B978-0-12-409548-9.09270-8.

[109] James J. Corbett. Marine transportation and energy use. In Cutler J. Cleveland, editor, *Encyclopedia of Energy*, pp. 745–758. Elsevier, New York, 2004. ISBN 978-0-12-176480-7. doi: https://doi.org/10.1016/B0-12176480-X/00193-5.

[110] UNCTAD. Review of maritime transport 2018, 2018. URL https://unctad.org/webflyer/review-maritime-transport-2018/.

[111] UNEP, 2009. *Marine Litter: A Global Challenge*. Nairobi: UNEP. p. 232.

[112] World Shipping Council. Containers lost at sea – 2020 update, 2020. URL https://static1.squarespace.com/static/5ff6c5336c885a268148bdcc/t/60ccbd4acca7252f8cd2c2c5/1624030539346/Containers_Lost_at_Sea_-_2020_Update_FINAL_.pdf/

[113] Kathryn J. O'Hara, Suzanne Iudicello, and Rose Bierce. *A citizen's guide to plastics in the ocean: More than a litter problem*. 2nd ed., p. 15, 17. Center for Marine Conservation, 1988, ISBN : 0-9615294-2- URL https://files.eric.ed.gov/fulltext/ED312152.pdf.

[114] European Commission. Sustainable cruise prototypes and approaches for raising the waste hierarchy on board and certifying it, 2014. URL https://webgate.ec.europa.eu/life/publicWebsite/index.cfm?fuseaction=search.dspPage&n proj id=3933.

[115] John Mouat, Rebeca Lopez Lozano, and Hannah Bateson. Economic impacts of marine litter, 2010. URL www.kimointernational.org/wp/wp-content/uploads/2017/09/KIMO_Economic-Impacts-of-Marine-Litter.pdf

[116] Are Solum. What happens to containers lost overboard? how long do they float? February 2021. URL www.gard.no/web/updates/content/31171690/what-happens-to-containers-lost-overboard-how-long-do-they-float.

[117] Alistair McIlgorm, Harry F. Campbell, and Michael J. Rule. The economic cost and control of marine debris damage in the asia-pacific region. *Ocean & Coastal Management*, 54(9): 643–651, 2011. doi: 10.1016/j.ocecoaman.2011.05.007.

18 Modelling of Industrial Wastewater Treatment Using Membrane Distillation with Crystallization

Denis Kalmykov, Sergey Makaev, Georgy Golubev, Ilia Eremeev, Jianfeng Song, Tao He, Andrea Bernardes, Lueta-Ann de Kock, Swachchha Majumdar, Sourja Ghosh and Alexey Volkov

18.1 INTRODUCTION

In the metallurgical industry, water contaminated with heavy metal ions is one of the most serious environmental problems [1, 2]. Heavy metal ions in wastewater are harmful to soil, groundwater and surface water, with devastating effects on the environment, animal health and human health if not treated properly [3, 4]. On the other hand, the purification and separation of heavy metal ions for reuse is an additional source of resources to ensure the sustainability of the industry [5]. In general, zinc, copper, chromium, nickel and iron ions are meant as heavy metal ions in industry. At the same time, standard wastewater treatment technologies for the removal of heavy metals, such as chemical precipitation, ion exchange, solution adsorption and extraction [6–10], suffer from low recovery efficiency, large area, high energy consumption, and production of secondary liquid or solid waste [5]. The development of an environment friendly, highly energy-efficient separation process is of great importance.

Electroplating or metal processing require significant water consumption, which leads to the formation of a large amount of wastewater at the output of this production. Thus, the water consumption in some cases can reach 400 litres per 1 m^2 of the treated metal surface, while when optimizing electroplating processes (using modern equipment and the optimal amount of high-quality reagents), water consumption can be reduced to 10 litres/m^2 [11]. Several toxic compounds are used in the industry, such as nickel (Ni) compounds, which are used to coat various materials; nickel improves corrosion resistance and also provides decorative characteristics [12]. The Watts galvanic bath, consisting of nickel chloride, nickel sulfate and boric acid, is most widely used for this process, while organic compounds, such as trichloroethylene, Synthanol DS-10, TMS-31 and others are often added to improve the surface treatment characteristics. As a result of this process, wastewater with nickel salts and organic additives is formed, requiring further purification and processing.

Currently, industrial enterprises treat heavy metal wastewater by lime deposition [13], which significantly shifts the problem to large volumes of sludge containing heavy metals, and does not provide sufficient removal efficiency (metal content less than 1 mg/l), and also requires secondary treatment [14, 15]. Membrane separation processes are the main methods of traditional treatment, and they include reverse osmosis, micro-, ultra- and nanofiltration to produce recirculating water

DOI: 10.1201/9781003327646-18

[16–19]. These methods are used not only to solve the problem of technogenic pollution of the environment with heavy metals, but also should be aimed at allocating valuable metals for secondary use in the electroplating industry, which is becoming increasingly expensive due to the declining quality of metal ores [15, 20]. Therefore, an economical method is needed not only to remove heavy metals from wastewater, but also to extract these metals.

The electrochemical extraction of nickel from the rinse water is of interest because the pure metal can be recovered for recycling. Processes such as electrodialysis (ED), electrolysis, and electrodeionization are alternative methods of wastewater treatment [21]. ED is one of the most recent technologies that is used to extract compounds used in the electroplating industry from rinse water [21–23]. This process is based on the selective transport of water ions through ion-exchange membranes as a result of the action of an electromotive force [24]. The advantages of ED are that a low concentration of heavy metals can be concentrated, and the other part of the wastewater can be diluted for reuse [14]. This method not only concentrates the metals from the washing water, but also helps to maintain the quality of the electroplating bath [20].

Generally, organic compounds such as citric, lactic and succinic acids are widely used in the nickel plating process, and nickel exists as organometallic complexes rather than in a hydrated form. The presence of a wide range of co-existing metal cations and organic compounds in the electroplating solution causes the problem of nickel processing [25]. There are studies that apply ED and deal with the formation of Ni-organic complexes in order to extract one preferred metal from a multicomponent solution [22], using the difference in solubility constants and the size of the complexes. Since most organic substances present in natural waters and wastewater have a negative charge, the ED process almost always results in anion-selective membrane fouling [26]. Ni-organic complexes that are negatively charged can be separated as anions, through an anionic membrane, or they can block this membrane. There is little information about the use of efficient and economical methods for extracting nickel from Ni^{2+}-organic acid complexes [25].

Relatively recently, a new approach has been proposed, namely, membrane crystallization, an integrated process of separating water by the membrane method and subsequent dissolved compounds' separation in solid form as a result of their crystallization from saturated solutions [27]. The advantage of this approach is its compactness, reduced metal consumption and flexibility in process control [27]. This approach was used for the crystallization of inorganic salts (for example, sodium or lithium salts [28, 29]) and organic compounds (primarily biologically active components [30]) using baromembrane concentration processes (reverse osmosis, nanofiltration), evaporation through the membrane processes (pervaporation, membrane distillation) and direct osmosis (this approach is mainly used to obtain crystals of organic compounds).

At the same time, there are no works on the membrane crystallization of complex solutions containing heavy metals in the world literature. Taking into account the tasks set, the most optimal method for concentrating dissolved salts is membrane distillation (MD). Membrane distillation is one of the most promising methods for the concentration of unsaturated solutions from the electroplating baths' wastewater and other industrial waters [27]. This method allows one to obtain a highly concentrated salt solution without significant resource consumption. At the same time, the solubility of vitriol compounds, so often used in electroplating baths, significantly depends on the temperature of the solution. In this case, cooling of a saturated solution will lead to its supersaturation with further precipitation from the solution, which can easily be used when combining the membrane distillation part with the crystallizer. In continuation of the development of MD with an air gap, a new method has recently been proposed, namely, MD with a porous condenser [31], where the porous membrane acts as a surface for the condensation of water vapour and as the interface between the air gap and the coolant circuit. The water vapour passed through the membrane enters the air gap, condenses and is transferred through the pores of the porous membrane to the refrigerant/permeate circuit, which allows one to reduce the width of the air gap to 0.1 mm [31]. This version of the process simplifies the design, reduces the size of the membrane module, allows the use of the membrane module regardless of orientation in space and scaling up of the process for industrial applications [31].

Before implementing the integrated technologies, the entire process should be modelled to test the feasibility of the proposed processes. Process modelling has become a recognized and widely used tool for calculating process performance, designing and optimizing of the process parameters. Thus, the purpose of this work was to simulate the operation of the membrane crystallization process (combining the MD process with a membrane condenser with simultaneous crystallization) when concentrating a complex solution of galvanic wastewater with further precipitation of copper and nickel salts. The model was calculated in the Simulink MATLAB software package, which allows one combining various installation modules and obtaining dynamically changing results.

18.2 MATERIALS AND METHODS

A laboratory porous condenser membrane distillation unit used for the concentration of an aqueous solution of sodium chloride is described in [31, 32]. The initial salt solution was circulated in the membrane distillation module at 60 or 80°C with linear velocity of 0.012 m/s. The temperature of the cooling liquid (distilled water) in the condenser with a porous membrane was maintained at 20°C (flow rate 0.3 l/min). The MD module with an active surface area of 146 cm² was equipped with a commercial microfiltration membrane MFFK-1 (STC 'Vladipor', Vladimir, Russia); a porous condensator with a thickness of 200 μm and a porosity of 30% was made of sintered stainless steel (LLC 'VMZ-Techno', Moscow, Russia). MFFK-1 (pore size 0.15 μm, total porosity 85%) consisted of a porous top layer based on F42L fluoropolymer (a copolymer of tetrafluoroethylene and vinylidene fluoride) and a non-woven support (polypropylene/lavsan). The air gap between the MFFK-1 and the porous capacitor was set to 3 mm. Principal scheme of the installation is shown in Figure 18.1.

The integrated operations of membrane distillation and salt crystallization were modelled using several subsystems in Simulink/MATLAB. The salt solution was fed into a 'solution tank with a heating system' to be heated to 60°C, and this temperature was maintained by the system within ± 1°C. The hot salt solution was fed to the 'membrane distillation module' unit, where part of the

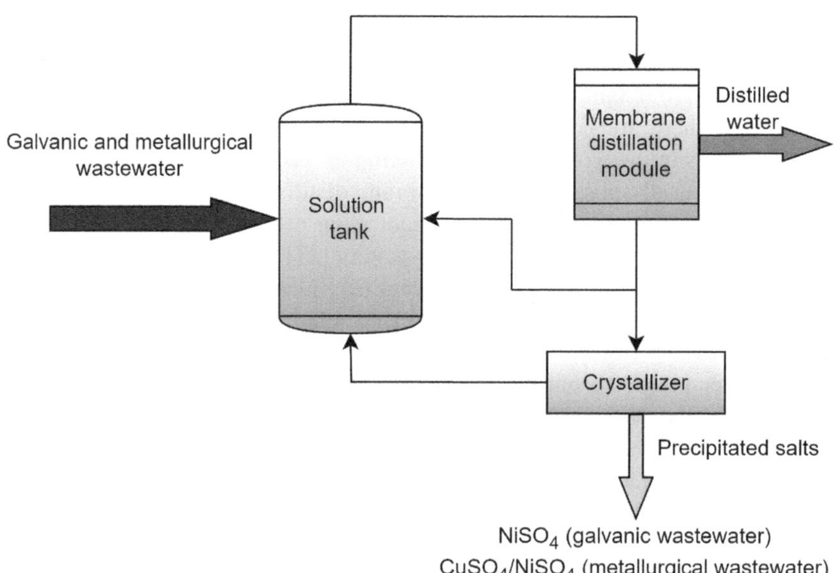

FIGURE 18.1 Schematic illustration of membrane crystallization process.

water was evaporated, and then part of the solution (~ 6 kg/h) was fed to the 'crystallizer' unit, where, after cooling to 20°C, the corresponding amount of salt (NaCl) was deposited from the solution. The remaining hot solution was recirculated back from the 'membrane distillation module' to the 'solution tank with heating system'. A 'crystallizer' containing 90 kg of solution was considered a black box, assuming that an excess of salts exceeding the saturation concentration at 20°C was precipitated. The crystallization time was 10 min [29, 30]. The salt solution was recirculated back to the 'solution tank with heating system' after the crystallizer. The main purpose of the 'membrane distillation module' was to concentrate the salt content to the saturated solution concentration at 20°C for the subsequent comfortable extraction of salts using a crystallizer.

As an example, Figure 18.2 shows the Simulink code for the 'membrane distillation module' subsystem, which calculates the amount of solvent (water) evaporated from the salt solution based on a number of parameters. The amount of water evaporated from the salt solution over the specific time [kg/min] can be described as follows (Eq. 18.1):

$$m = J\left(T_f\right) \cdot S \cdot K \cdot Z\left(t\right) \cdot 1/60, \tag{18.1}$$

where $J(T_f)$ is a polynomial equation that defines the distilled water flow as a function of the feed material temperature; T_f is the feed temperature [°C]; S is the active surface area of one membrane module [m²]; K is the number of membrane modules, $Z(t)$ is a parameter describing the deterioration of the membrane characteristics over time due to scale/contamination. During the simulation, the solution temperature changed to about 60°C, but the coolant/permeate temperature was kept constant and equal to 20°C. The polynomial equation $J(T_f)$ was determined from experimental data and was assumed for modelling as (Eq. 18.2):

$$J\left(T_f\right) = 0.1071 \cdot T_f - 2.0757 \tag{18.2}$$

The parameter $Z(t)$ was determined from the experimental data as $Z(t) = 5.2908e^{-0.001 \cdot t}$ and was used to describe the flow drop when the membrane module was used once for 4.5 h before the membrane was washed (0.5 h). The recovery rate of the membrane performance after each wash was set to 99%. As soon as the performance of the membrane module after the next washing was restored

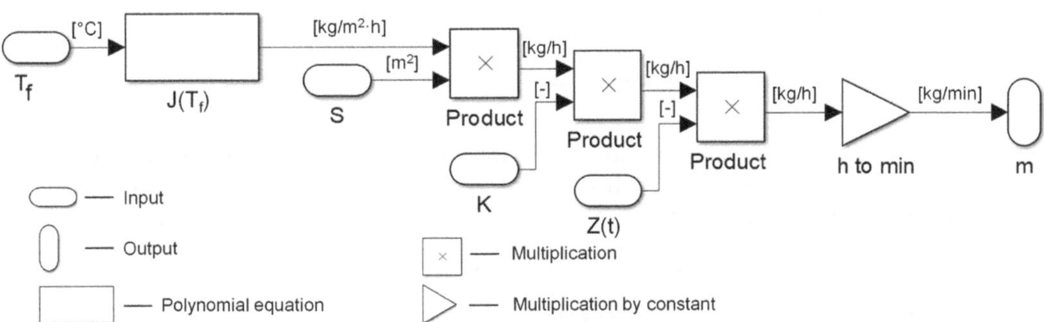

FIGURE 18.2 Representation of the membrane module in Simulink: m is the amount of water evaporated in MD module per minute [kg/min], $J(T_f)$ is the polynomial equation that defines the flux of distilled water as a function of the feed temperature [kg/m²·h]; T_f is the feed temperature [°C]; S is the active surface area of one membrane module [m²], K is the number of membrane modules, $Z(t)$ is the non-linear function describing the deterioration of membrane performance due to scaling/fouling over time.

to 50% of the original one, the membrane module was replaced with a new one. The internal operation of the system was modelled using the well-known principle of proportional-integral-derivative (PID) control. Simulink performs calculations based on the input signals at each stage, which allows to avoid the need to pre-define a mathematical description for the long-term operation of the entire system.

18.3   RESULTS AND DISCUSSION

In this study, the MD-MC process was modelled for the sources presented in the papers [33, 34]. The literature data on the solubility limits are presented in Figure 18.3, and these values were used for the calculations. The solubilities of both $NiSO_4$ and $CuSO_4$ are higher than that of NaCl, and it may affect the performance of the real-life unit, but according to the literature data and our experiments, the distillation rate mainly depends on the temperature and not on the composition of the solution.

In the first case, the source of the salts was the wastewater from the treatment of electroplating baths containing nickel sulphate, and in the second case, the source of the salts was industrial wastewater containing copper sulphate. The total amount of solution in the system was 50 kg, and the feed flow of fresh salt solution with the composition shown in Table 18.1 to the 'solution tank with heating system' was set at about 0.2 kg/min and varied to maintain the mass balance of the main components relative to the evaporated water and crystallized salts.

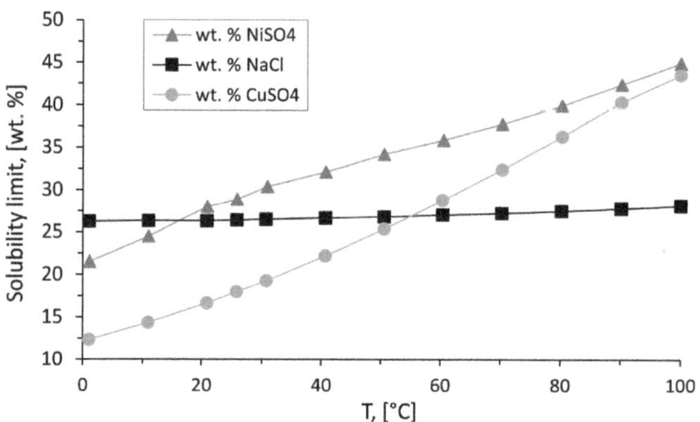

FIGURE 18.3   Dependence of the solubility limits of several salts on the temperature of the solution [35].

TABLE 18.1
Results of Modelling of the Concentration of Salt Solutions. Simulation Mode: 2 Months of Operation in a Stable Mode after the Start of Crystallization. Working Time Between Washes 4.5 h, Washing Time 30 Min

		Build-up		2 Months of Work			
Salt	Initial Concentration, [g/l]	Time, [h]	Number of Membrane Washes	Number of Membrane Replacements	Evaporated Water, [kg]	Precipitated Salt, [kg]	Source
$NiSO_4$	2.75	682	136	6	4373	25	[33]
$CuSO_4$	20.7	41	8	4	265	197	[34]
$NiSO_4$	6.1	341	68	5	2190	58	[34]

The main problem of membrane distillation of diluted solutions is relatively long build-up times. For example, it would require 682 h to concentrate $NiSO_4$ effluent from the electroplating plant and start the precipitation of the crystallized heptahydrate. In two months of operation it would only produce 25 kg of salts. Of course, the increase of the membrane distillation surface will increase the fluxes of water and crystallized salt, but will also increase the price. Another option would be the treatment of the diluted nickel effluent from the industrial electroplating plant using electrodyalisis, which was successfully performed in [34]. This resulted in the production of an effluent with Ni composition of about 40 mg/l, which is still higher than the safe limits. The total removal of Ni from the effluent can be performed with the existing absorption technologies [36]. And the subsequent treatment of the concentrated solution can be performed using membrane distillation unit, which will drastically reduce the build-up duration and the cost of the treatment. The total cost of combining two methods will depend on the optimization of the conditions, minimizing the energy and material consumption, and it may be even more profitable to pretreat the initial solution using microfiltration membranes, as it was done in [37].

Using the industrial wastewaters with the composition given in [34], it will take from 41 to 341 hours before the solution will become saturated, and the salts will begin to precipitate. A large amount of membrane replacements is another drawback, which can be mitigated using more advanced membranes. But the real-life wastewaters very rarely contain only one component, and in that case the membrane distillation lacks certain appeal because membrane distillation itself cannot separate the precipitated salts. One of the ways to deal with this is membrane extraction. For example, in [34] it was shown that using a special membrane extraction unit it is possible to extract copper from such industrial wastewater with basically 100% selectivity. This resulted in a solution of copper sulphate, which can be used for further membrane distillation.

Combination of this two methods (selective membrane extraction and membrane distillation) was also modelled using Simulink/MATLAB. According to [34], the membrane flux in the membrane extraction unit decreases over time, as is shown on Figure 18.4. That means that in the combined model the composition of the feed flow is changing over time (blue arrow on Figure 18.1), unlike our previous works on the modelling of the membrane distillation units. This dependence was interpolated, and the equation was used for modelling the composition of the feed flow.

Scatter plot graph with a trend line showing the dependence of the copper (Cu^{2+}) flux on time within a membrane extraction unit. The x-axis is labeled 't_1[min]' and represents time in minutes, ranging from 0 to 500 minutes. The y-axis is labelled 'Cu^{2+} flux (10^8 mol/(cm²·s))' and measures the flux of

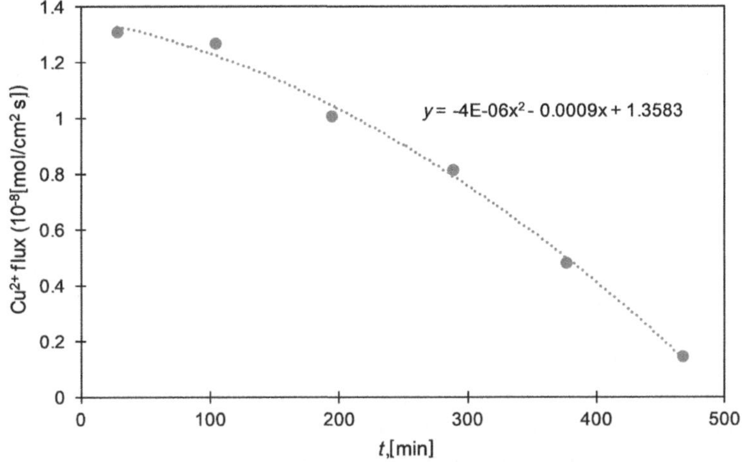

FIGURE 18.4 Dependence of the Cu flux in the membrane extraction unit [34].

TABLE 18.2

Results of Modelling of the Concentration of $CuSO_4$ solutions. Simulation Mode: 2 Months of Operation in a Stable Mode After the Start of Crystallization. Working Time of the Membrane Distillation Unit Between Washes 4.5 h, Washing Time 30 min. Working Time of the Membrane Extraction Unit Between Washes 8 h, Washing Time 1 h

Concentration in the Feed	Initial Concentration, [g/l]	Build-up		2 Months of Work		
		Time, [h]	Number of Membrane Washes	Number of Membrane Replacements	Evaporated Water, [kg]	Precipitated Salt, [kg]
Stable	7.2	341	68	5	2388	58
Variable	7.2	371	74	5	2191	66

copper ions in 10^8 moles per square centimeter per second. The plot shows individual data points, marked as blue dots, that decrease in a nonlinear fashion as time progresses. The trend line, dotted and coloured in blue, represents a quadratic fit to the data, with the equation 'y = -4E-06x² - 0.0009x + 1.3583' provided on the graph, indicating a downward parabolic trend. The overall trend suggests that the flux of copper ions through the membrane decreases over time. After 8 hours of the extraction, the Cu flux drops tenfold, which requirs membrane washing for 1 hours. This essentially leads to two different membrane units with different parameters for washing. In order to incorporate this changes in our model it was decided to increase the solution tank to 100 kg (instead of 50 kg). This helped to mitigate the problem of the tank overflow when membrane distillation unit was being washed and the solution depletion when feed stopped for membrane extraction unit washing.

The concentration in the membrane extraction unit increases almost linearly until about 2.5–3 g/l of Cu, and then starts to slow down; therefore the membrane distillation unit in our model starts when the concentration of the solution reaches 2.8 g/l Cu (or 7.2 g/l $CuSO_4$). After that, the concentration in the feed fluctuates because of the membrane performance drop. Table 18.2 contains the results of the modelling.

The modelling compares two modes of the feed compostion, the first being the stable concentration when the feed solution has constant composition (7.2 h/l $CuSO_4$). In the second mode the concentration varies accroding to Cu flux (Figure 18.4). The stable mode is obviously better because it produces about 13% more salt during the same duration and has about 8% less build-up time, which can be attributed to higher average concentration of the feed solution. It stands to reason that the increase of the feed concentration is crucial for the membrane distillation process, and it can be implemented using, for example, electrodialysis process. It can be seen from the obtained data that membrane distillation of such wastewaters can be a perspective way to treat industrial wastewater in order to produce clean water and concentrate the solution.

18.4 CONCLUSION

In this work, the modelling of treatment of industrial wastewaters from two different sources was modelled. The industrial wastewaters from electroplating plants were used for modelling of the membrane distillation process and the modelling demonstrated that the membrane distillation process alone can be perspective for purification of rinse water. Another way to treat it is electrodialysis, which has an advantage in the energy consumption; combination of these two methods can significantly improve the process of wastewater treatment.

The treatment of the metallurgical wastewater containing nickel and copper was also modelled. Membrane distillation of such wastewater leads to production of a mix of salts, but combining the membrane distillation process with membrane extraction process results in the production of pure $CuSO_4 \cdot 5H_2O$. The additional stages for better performance may include the pretreatment of the solution using microfiltration membranes and adsorption of the trace amounts of salts to produce pure water.

ACKNOWLEDGEMENTS

BRICS consortium was funded by Scientific and Technological Development, grant number CNPq/BRICS-STI-2-442229/2017-8; by Russian Foundation for Basic Research (grant number 18-58-80031) and by the Ministry of Science and Higher Education of the Russian Federation (Project 13.2251.21.0166, Agreement MNTS BRICS 075-15-2022-1218); by Department of Science and Technology, grant number DST/IMRCD/BRICS/PC2/From waste to resources/2018 (G)); by National Natural Science Foundation of China, grant number 51861145313, 52011530031, 21978315, and CAS International Collaboration, grant number GJHZ2080; by National Research Foundation, grant number 116020.

REFERENCES

[1] J. Cui, L. Zhang, Metallurgical recovery of metals from electronic waste: A review, *J. Hazard. Mater.* 158 (2008) 228–256. https://doi.org/10.1016/j.jhazmat.2008.02.001.

[2] M.A. Barakat, New trends in removing heavy metals from industrial wastewater, *Arab. J. Chem.* 4 (2011) 361–377. https://doi.org/10.1016/j.arabjc.2010.07.019.

[3] N. Jamali-Zghal, B. Lacarrière, O. Le Corre, Metallurgical recycling processes: Sustainability ratios and environmental performance assessment, *Resour. Conserv. Recycl.* 97 (2015) 66–75. https://doi.org/10.1016/j.resconrec.2015.02.010.

[4] Z. Liu, H. Li, Metallurgical process for valuable elements recovery from red mud—A review, *Hydrometallurgy.* 155 (2015) 29–43. https://doi.org/10.1016/j.hydromet.2015.03.018.

[5] F. Fu, Q. Wang, Removal of heavy metal ions from wastewaters: A review, *J. Environ. Manage.* 92 (2011) 407–418. https://doi.org/10.1016/j.jenvman.2010.11.011.

[6] L. Monser, N. Adhoum, Modified activated carbon for the removal of copper, zinc, chromium and cyanide from wastewater, *Sep. Purif. Technol.* 26 (2002) 137–146. https://doi.org/10.1016/S1383-5866(01)00155-1.

[7] D.S. Kim, The removal by crab shell of mixed heavy metal ions in aqueous solution, *Bioresour. Technol.* 87 (2003) 355–357. https://doi.org/10.1016/S0960-8524(02)00259-6.

[8] M.J. Santos Yabe, E. de Oliveira, Heavy metals removal in industrial effluents by sequential adsorbent treatment, *Adv. Environ. Res.* 7 (2003) 263–272. https://doi.org/10.1016/S1093-0191(01)00128-9.

[9] G. Golubev, I. Eremeev, S. Makaev, M. Shalygin, V. Vasilevsky, T. He, E. Drioli, A. Volkov, Thin-film distillation coupled with membrane condenser for brine solutions concentration, *Desalination.* 503 (2021) 114956. https://doi.org/10.1016/j.desal.2021.114956.

[10] F.J. Alguacil, A. Cobo, Solvent extraction with LIX 973N for the selective separation of copper and nickel, *J. Chem. Technol. Biotechnol.* 74 (1999) 467–471. https://doi.org/10.1002/(SICI)1097-4660(199905)74:5<467::AID-JCTB60>3.0.CO;2-7.

[11] A. Telukdarie, Brouckaert, Y. Haung, A case study on artificial intelligence based cleaner production evaluation system for surface treatment facilities, *J. Clean. Prod.* 14 (2006) 1622–1634. https://doi.org/10.1016/j.jclepro.2006.01.007.

[12] M. Schario, Troubleshooting decorative nickel plating solutions (Part I of III installments): Any experimentation involving nickel concentration must take into account several variables, namely the temperature, agitation, and the nickel-chloride mix, *Met. Finish.* 105 (2007) 34–36. https://doi.org/10.1016/S0026-0576(07)80584-8.

[13] T.A. Kurniawan, G.Y.S. Chan, W.-H. Lo, S. Babel, Physico–chemical treatment techniques for wastewater laden with heavy metals, *Chem. Eng. J.* 118 (2006) 83–98. https://doi.org/10.1016/j.cej.2006.01.015.

[14] C. Peng, Y. Liu, J. Bi, H. Xu, A.-S. Ahmed, Recovery of copper and water from copper-electroplating wastewater by the combination process of electrolysis and electrodialysis, *J. Hazard. Mater.* 189 (2011) 814–820. https://doi.org/10.1016/j.jhazmat.2011.03.034.

[15] J. Shao, S. Qin, J. Davidson, W. Li, Y. He, H.S. Zhou, Recovery of nickel from aqueous solutions by complexation-ultrafiltration process with sodium polyacrylate and polyethylenimine, *J. Hazard. Mater.* 244–245 (2013) 472–477. https://doi.org/10.1016/j.jhazmat.2012.10.070.

[16] C. Aydiner, M. Bayramoglu, B. Keskinler, O. Ince, Nickel removal from waters using a surfactant-enhanced hybrid powdered activated carbon/microfiltration process. II. The influence of process variables, *Ind. Eng. Chem. Res.* 48 (2009) 903–913. https://doi.org/10.1021/ie8004308.

[17] Z. Wang, G. Liu, Z. Fan, X. Yang, J. Wang, S. Wang, Experimental study on treatment of electroplating wastewater by nanofiltration, *J. Membr. Sci.* 305 (2007) 185–195. https://doi.org/10.1016/j.memsci.2007.08.011.

[18] P. Baticle, C. Kiefer, N. Lakhchaf, O. Leclerc, M. Persin, J. Sarrazin, Treatment of nickel containing industrial effluents with a hybrid process comprising of polymer complexation–ultrafiltration–electrolysis, *Sep. Purif. Technol.* 18 (2000) 195–207. https://doi.org/10.1016/S1383-5866(99)00063-5.

[19] V. Coman, B. Robotin, P. Ilea, Nickel recovery/removal from industrial wastes: A review, *Resour. Conserv. Recycl.* 73 (2013) 229–238. https://doi.org/10.1016/j.resconrec.2013.01.019.

[20] A. Mahmoud, A.F.A. Hoadley, An evaluation of a hybrid ion exchange electrodialysis process in the recovery of heavy metals from simulated dilute industrial wastewater, *Water Res.* 46 (2012) 3364–3376. https://doi.org/10.1016/j.watres.2012.03.039.

[21] K.N. Njau, M. vd Woude, G.J. Visser, L.J.J. Janssen, Electrochemical removal of nickel ions from industrial wastewater, *Chem. Eng. J.* 79 (2000) 187–195. https://doi.org/10.1016/S1385-8947(00)00210-2.

[22] N. Tzanetakis, W.M. Taama, K. Scott, R.J.J. Jachuck, R.S. Slade, J. Varcoe, Comparative performance of ion exchange membranes for electrodialysis of nickel and cobalt, *Sep. Purif. Technol.* 30 (2003) 113–127. https://doi.org/10.1016/S1383-5866(02)00139-9.

[23] T.A. Green, S. Roy, K. Scott, Recovery of metal ions from spent solutions used to electrodeposit magnetic materials, *Sep. Purif. Technol.* 22–23 (2001) 583–590. https://doi.org/10.1016/S1383-5866(00)00141-6.

[24] M.A.S. Rodrigues, F.D.R. Amado, J.L.N. Xavier, K.F. Streit, A.M. Bernardes, J.Z. Ferreira, Application of photoelectrochemical–electrodialysis treatment for the recovery and reuse of water from tannery effluents, *J. Clean. Prod.* 16 (2008) 605–611. https://doi.org/10.1016/j.jclepro.2007.02.002.

[25] L. Song, X. Zhao, J. Fu, X. Wang, Y. Sheng, X. Liu, DFT investigation of Ni(II) adsorption onto MA-DTPA/PVDF chelating membrane in the presence of coexistent cations and organic acids, *J. Hazard. Mater.* 199–200 (2012) 433–439. https://doi.org/10.1016/j.jhazmat.2011.11.046.

[26] V. Lindstrand, G. Sundström, A.-S. Jönsson, Fouling of electrodialysis membranes by organic substances, *Desalination.* 128 (2000) 91–102. https://doi.org/10.1016/S0011-9164(00)00026-6.

[27] E. Drioli, A. Ali, F. Macedonio, Membrane distillation: Recent developments and perspectives, *Desalination.* 356 (2015) 56–84. https://doi.org/10.1016/j.desal.2014.10.028.

[28] E. Drioli, G. Di Profio, E. Curcio, Progress in membrane crystallization, *Curr. Opin. Chem. Eng.* 1 (2012) 178–182. https://doi.org/10.1016/j.coche.2012.03.005.

[29] W. Ye, J. Lin, J. Shen, P. Luis, B. Van der Bruggen, Membrane crystallization of sodium carbonate for carbon dioxide recovery: Effect of impurities on the crystal morphology, *Cryst. Growth Des.* 13 (2013) 2362–2372. https://doi.org/10.1021/cg400072n.

[30] G.D. Profio, E. Curcio, A. Cassetta, D. Lamba, E. Drioli, Membrane crystallization of lysozyme: kinetic aspects, *J. Cryst. Growth.* 257 (2003) 359–369. https://doi.org/10.1016/S0022-0248(03)01462-3.

[31] A.V. Volkov, E.G. Novitsky, I.L. Borisov, V.P. Vasilevsky, V.V. Volkov, Porous condenser for thermally driven membrane processes: Gravity-independent operation, *Sep. Purif. Technol.* 171 (2016) 191–196. https://doi.org/10.1016/j.seppur.2016.07.038.

[32] D. Kalmykov, S. Makaev, G. Golubev, I. Eremeev, V. Vasilevsky, J. Song, T. He, A. Volkov. Operation of three-stage process of lithium recovery from geothermal brine: simulation. *Membranes.* 11 (2021) 175–194 https://doi.org/10.3390/membranes11030175.

[33] T. Benvenuti, R.S. Krapf, M.A.S. Rodrigues, A.M. Bernardes, J. Zoppas-Ferreira, Recovery of nickel and water from nickel electroplating wastewater by electrodialysis, *Sep. Purif. Technol.* 129 (2014) 106–112. https://doi.org/10.1016/j.seppur.2014.04.002.

[34] J. Song, X. Niu, X.-M. Li, T. He, Selective separation of copper and nickel by membrane extraction using hydrophilic nanoporous ion-exchange barrier membranes, *Process Saf. Environ. Prot.* 113 (2018) 1–9. https://doi.org/10.1016/j.psep.2017.09.008.

[35] D.R. Lide, *CRC Handbook of Chemistry and Physics*, 85th Edition. (2004). UK: Taylor & Francis.

[36] C.L. Dlamini, L.-A. De Kock, K. K. Kefeni, B. B. Mamba, T. A. M. Msagati, Novel hybrid metal loaded chelating resins for removal of toxic metals from acid mine drainage, *Water Sci. Technol.* 81, 12 (2020) 2568–2584. https://doi.org/10.2166/wst.2020.285.

[37] P. Bhattacharya, S. Ghosh, A. Mukhopadhyay, Efficiency of combined ceramic microfiltration and biosorbent based treatment of high organic loading composite wastewater: An approach for agricultural reuse, *J. Environ. Chem. Eng.*, 1–2 (2013) 38–49. https://doi.org/10.1016/j.jece.2013.03.002.

19 Recovery of Precious Metals and Rare Earth Elements from e-Waste

Hema Jha, Upakrishnam S.V. Sanskrit and Brajesh Kumar Dubey

19.1 INTRODUCTION

The industrial revolution followed by digitalization of the world has changed the lifestyle of human beings drastically. Devices are available for about everything that serves the purpose of the day-to-day activity of a person including professional engagements, household chores and entertainment. Typically, any electrical and electronic equipment is composed of metals, polymers and ceramics. Heterogeneity in the product shape, size and composition is a result of sophisticated design of devices. All these devices require electricity to function and after the utility is over, the devices are called a waste of electrical and electronic equipment (WEEE or e-waste) and require special end-of-life (EoL) management. This consistently growing stream of discarded material opens an opportunity to dig for the material, becoming the secondary source of mining.

At present, the value of metals and rare earth elements (REEs), which is present in the e-waste, has already been realized. REEs include 17 elements including 15 elements from the lanthanide group, scandium, and yttrium (Navarro & Zhao, 2014). They are also known as technology critical elements (TCEs) due to their fundamental role in specific applications, no alternative in the short term, and monopoly of reserve (Favot & Massarutto, 2019). Figure 19.1 represents the lifecycle of REEs in manufacturing of devices and their recovery from e-waste.

Recovery of precious metals and REEs are essential and serves manifold objectives of human development. For example, geological reserves of REEs are scarce, though they are significant for shifting to renewable sources of energy like wind and solar, as 171 kg of REE is required per MW of electricity produced by a wind turbine (Grandell & Höök, 2015; Navarro & Zhao, 2014). During the virgin mining of metals and REEs, a significant quantity of solid and liquid waste is produced, which imposes an environmental burden (Edahbi et al., 2019), also, the leaching of heavy metals from e-waste piled up in dumpsites causes additional risk to human health and environment (Murthy & Ramakrishna, 2022). Likewise, pollution arising from improper and informal recycling of e-waste can potentially cause genetic mutation and induce cytogenic damage (Robinson, 2009). Informal recycling implies the treatment of e-waste where the workers try to make their living by just extracting a little value from this waste without using any protective equipment and thereafter getting exposed to the hazard that is generated during the processing of waste material. Establishing a circular economy demands the materials from products EoL to be back in the supply chain, and REEs are critical for a high technology, low-carbon economy (Dang et al., 2021). Currently, dumping of e-waste in developing countries from developed countries is a known practice, and a perspective of urban mining to recover resources from this waste is widespread. Notably, extracting the metal and REEs from e-waste also requires sophisticated technologies, alongside an appropriate policy framework in the country.

DOI: 10.1201/9781003327646-19

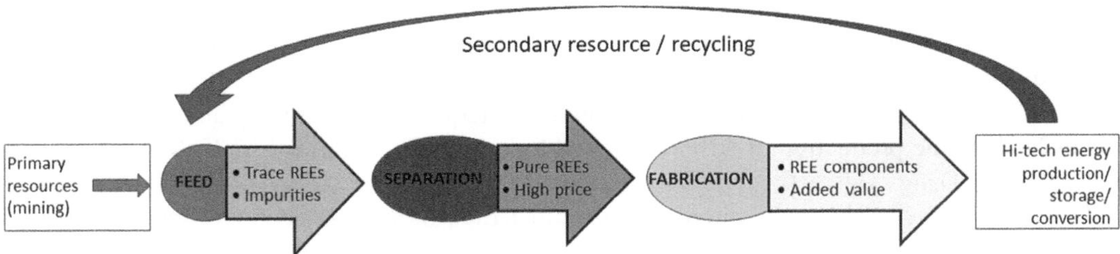

FIGURE 19.1 Lifecycle of REEs including mining, conversion and energy application, and recycling (Patil et al., 2022).

Owing to the multitudinous ambitions involved with the recovery and recycling of metals and REEs, the domain of e-waste management has drawn attention from economists, environmentalists, policymakers, and other experts in geopolitical matters. A holistic approach is required to deal with e-waste as it is a crucial field from a global perspective, and can play a pivotal role in socioeconomic and environmental paradigms. In this chapter, insight is provided on the potential resource that resides in e-waste and includes an overview and basics of the technologies for extracting metals and REEs followed by a discussion on the environmental and economic assessment of these technologies. Lastly, the policies and status of e-waste management around the globe are discussed with concluding remarks to take as future perspectives on the research in this domain.

19.2 E-WASTE AS E-RESOURCE

The generation of e-waste is swiftly increasing with a growth rate of about 3–5% annually (Ramprasad et al., 2022). In 2019, the global generation of e-waste was 53.6 Mt, a 21% increase from the previous five years, and expected to grow up to 74 Mt by 2030 (Murthy & Ramakrishna, 2022). The generation of e-waste is significantly different in developed and developing countries. As reported in 2014, the per capita generation was 22.1 kg/year in the US alone, while 5.4 kg/year worldwide (Tansel, 2017). However, as globalization and digitalization are at a rapid pace, and owing to the large population share of developing countries, the overall generation of e-waste is high enough to not only regulate the EoL management, though considering these quantities for urban mining. It is important to note here, that the data related to e-waste generation are estimates based on sales and the life span of product, though the tendency of storing the e-waste at homes makes it difficult to design the overall collection and management scheme (Pariatamby & Victor, 2013). Models, which employ geographic information systems (GIS) and material flow analysis (MFA), could help the local estimation become more reliable (Liu et al., 2022).

Japan has set an exceptional example to prove the potential of e-waste to be an enormous resource through the Tokyo medal 2020 project, during which 78,985 tons of e-waste was collected within the country and 32 kg of gold, 3500 kg of silver and 2200 kg of bronze was recovered (Nithya et al., 2021). The approximate composition of e-waste is 50% Fe, 20% plastic materials, and 13% non-ferrous metals, which includes precious metals and REEs. For the most common fraction of e-waste, i.e., printed circuit boards (PCBs), liquid crystal displays (LCDs), etc. Tunali et al. (2021) proposed an order for the heavy metal composition inside e-waste as Cu > Fe > Zn > Ni > Pb > Al > Mn > Co > Cr > Mo > Cd. They propose precious metals and REEs to follow a descending order as follows: Ag > Au > Pd > Pt for precious metals and Nd > Pr > Dy > La for REEs.

Recently, the RREs have become essential for electric and electronic equipment design and performance. Owing to their unique characteristics, rare earth metals are finding applications in multiple domains, including new energy technologies, laser materials, secondary batteries, high-temperature superconductors, metallurgy, optical/magnetic/chemical engineering, and luminescence (Navarro

& Zhao, 2014). The uneven distribution and scarcity of reserves of these metals, globally, leads to a high value associated. A total of 96% of the global REEs reserves (120 million metric tons) are concentrated in six countries, including China (38%), Vietnam (19%), Brazil (18%), Russia (10%), India (6%), and Australia (5%). Although, the high-tech requirement in mining and recovering these metals leads to the production of REEs being led by only four countries China, the USA, Burma, and Australia (Dang et al., 2021).

Despite the noteworthy economic importance of REEs, in a linear economy scheme, precious resources are getting lost (Ambaye et al., 2020). It is reported that less than 1% of the total REEs are recycled back to the supply chain from e-waste (Favot & Massarutto, 2019). The proposed reason for this low efficacy of recycling is the short lifespan of the product and its design complexity, which makes the recovery process complicated (Cardoso et al., 2019). This leads to the less industrial deployment of metal recovery technologies from its research phase.

An emphasis to be given here that the virgin mining of REEs has a huge environmental and health impact associated (Dang et al., 2021) and as the concentration of REEs in e-waste is higher than its natural ore (Ramprasad et al., 2022), the recycling and recovery of these valuables are essential to leap forward towards sustainability.

19.3 TECHNOLOGIES FOR THE RECOVERY OF PRECIOUS METALS AND RRES

Metal recovery is the most significant step in e-waste management owing to the high value of metals and REEs. Extraction of metals is the subsequent stage to dismantling and separating all the metal components from the plastic and ceramics in the electronic device. Metals and plastic components from electronic waste are separated by mechanical separation. The separated plastic components are either recycled or used as a source of energy through incineration, pyrolysis and gasification. The other separated part, i.e., valuable metals like gold, iron, aluminum, silver, REEs, etc., are recovered through many physical, chemical or biological processes. The main challenge, which makes metal recovery a complex process, is the heterogenous composition of various metals (Cui et al., 2017). In this section, the methods for metal recovery are discussed in the physical, chemical and biological process categories.

19.3.1 Physical Processes

The difference in physical properties of various metals is exploited to separate metals from each other or metals from non-metallic parts of e-waste. Table 19.1 highlights the physical separation process with the property that serves as the principle for separation. Physical properties like specific gravity, magnetism, electrical conductivity, etc., are serving as a base for the principles of the separation process.

These physical processes are basic in nature and require further treatment. Low capital and operating expenses are advantageous for physical separation procedures; however, these processes suffer from 10–35% of metal loss due to strong metal-plastic associations in e-waste (Tuncuk et al., 2012).

TABLE 19.1
Physical Process for Separating Metals from Other Parts in E-Waste

Methods	Separation Criteria	Separated Material	References
Gravity separation	Specific gravity	Metals and plastics	Galbraith & Devereux, 2002
Magnetic separation	Magnetic susceptibility	Ferrous and non-ferrous material	Cardoso et al., 2019(Freitas et al., 2020)
Electrostatic separation	Electrical conductivity	Metals from nonmetals	Jia Li et al., 2007
Eddy-current	Electric conductivity	Non-ferrous materials from non-metals	Jianzhi Li et al., 2004

TABLE 19.2
Recovery Methods in Hydrometallurgy

Recovery Methods	About the Process	Recovery %	Reference
Precipitation	Specific metal recovery using a chemical agent	96.17% Cr 99.39% Fe	Yue et al., 2019
Solvent Extraction	Solvent selection is for high solubility of target element	98.6% Ca 99% Mg	Guimarães & Mansur, 2017
Electrodeposition	Uses electricity to deposit metal from a solution onto the cathode	99.95 % Cu	Fogarasi et al., 2014

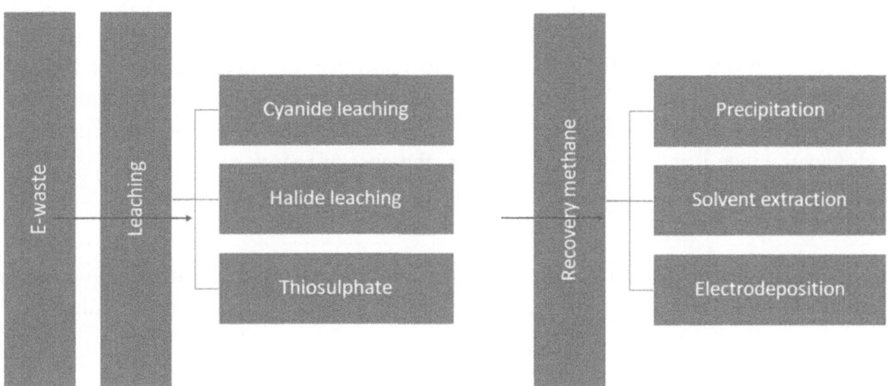

FIGURE 19.2 Hydrometallurgical route for extraction of metals from e-waste.

19.3.2 CHEMICAL PROCESSES

19.3.2.1 Hydrometallurgy

The first step in hydrometallurgy involves extracting valuable metals into solutions through leaching in an acidic or alkaline medium. The next step is to separate the targeted metal from the leached solution using methods such as precipitation, absorption, ion exchange, electrowinning, or solvent extraction. Table 19.2 discusses the different separation methods available for separating metals from solution. The complete route for the following hydrometallurgy is shown in Figure 19.2.

Hydrometallurgical processes are easier to control when used to recover REEs from e-waste and are preferable due to selective recovery without energy input and toxic gases being released (Sahu et al., 2022). Leaching of REEs from light-emitting diodes (LED) was investigated through different routes and it was mentioned that overall effectiveness depends upon the separation techniques of metals from leached solution, which is the same scenario encountered when virgin mining is done (de Oliveira et al., 2021).

19.3.2.2 Pyrometallurgy

Pyrometallurgy refers to the thermal treatment of e-waste to recover metals. It is considered a traditional method for the recovery of non-ferrous and precious metals (Tuncuk et al., 2012). Smelting is one of the most common types of pyrometallurgy processes. For smelting, dismantled and shredded material is fed to a smelter and smelted to get a copper bullion, which is further refined electrolytically to get the pure copper. The slimes from the refining are further treated to get metals like Ag, Au, Pd, Pt, etc. Smelting requires very pure feed (rich in copper and precious metals) and high energy and cost are also other necessities of the process (Tuncuk et al., 2012). Another disadvantage is that it only recovers copper and precious metals, although the recovery rate is good, i.e., around 95% (Li et al., 2019).

Vacuum pyrometallurgy can re-cover REEs from electronic wastes with the difference in pressure using sublimation or distillation (Ambaye et al., 2020). A range of REEs has been recovered from e-waste such as NdFeB magnets, waste printed circuit boards, and lamp fluorescent powder by pyrometallurgical process (Brewer et al., 2019).

19.3.2.3 Electrometallurgy

This method utilizes electrical energy to separate metals through an electrolysis mechanism. In this process, an electro-generated oxidant is used for leaching of the metal, and recovery is done by the ion exchange process. Kim et al. (2011) used chlorine as an oxidant to leach gold from mobile phone PCBs. The leaching reactor is shown in Figure 19.3. Due to the high reactivity of REEs, they cannot be directly recovered on the cathode. A range of electrochemical techniques including electrocoagulation, electrodeposition, electrodialysis, and electro sorption are investigated for REEs (Ambaye et al., 2020; Brewer et al., 2019). The process has energy efficiency, minimum chemical use, and low environmental effect, electrochemical technology is a viable way to recover precious metals from e-waste. However, since it is a multi-step process, the cost of recovery increases.

19.3.3 Biological Process

A variety of microorganisms, such as chemolithotrophic, heterotrophic bacteria, and fungi, participate in the natural process of bioleaching, from e-waste. This property of certain microbes becomes

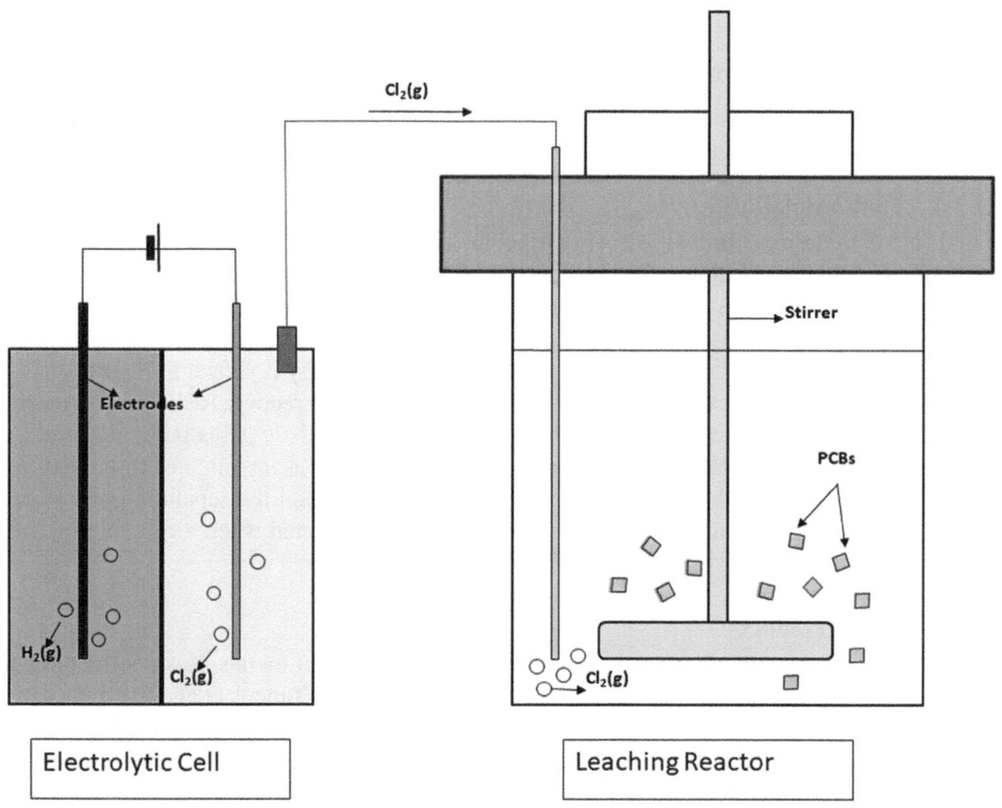

FIGURE 19.3 Leaching apparatus used to recover metals from e-waste (Kim et al., 2011).

a potential method for metal extraction from e-waste. Examples are *Acidithiobacillus ferrooxidans, Acidithiobacillus thiooxidans, Leptospirillum ferrooxidans*, and *Sulfolobus sp.* Metals are frequently extracted by microorganisms from the sulfide- and/or iron-containing ores and mineral concentrate. The insoluble metal sulfides of copper, nickel and zinc are converted by microbial oxidation of iron and sulphur to ferric ions and sulphuric acid, which then make the soluble metal sulfates that are easily recoverable from the solution (Pant et al., 2012). Acidolysis, complexolysis, redoxolysis and bioaccumulation are the common mechanisms involved in bioleaching (Bosshard et al., 1996). The efficiency of bioleaching largely depends on the microorganism itself however there are several other factors influencing the leaching rate nutrients available, O_2 and CO_2 concentrations in the atmosphere, pH, temperature, etc. The REEs' recovery through bioleaching and biosorption has been investigated by multiple bacterial strains, algae, activated carbon, charcoal, leaf powder, Sargassum biomass, *S. fluitans, Pseudomonas sp.* and *Agrobacterium sp.* (Ambaye et al., 2020; Brewer et al., 2019; Shahbaz, 2022).

19.4 ECONOMIC AND ENVIRONMENTAL ASSESSMENT OF RECOVERY TECHNOLOGIES

Evaluation of the economic aspects and environmental impact of current e-waste disposal and recovery methods is essential. The ultimate goal is to identify ways to optimize e-waste management to maximize economic value while minimizing negative environmental effects. The economy of RRE recovery is influenced by the price of elements and hence EoL management of e-waste should be centered around the more valuable REEs. The recycling and recovery methods for REEs are in their early stage (Gaustad et al., 2021). Evidently, only 2% of REEs are recycled as compared to 90% of Fe and steel (Patil et al., 2022).

Hydrometallurgy and pyrometallurgy rank among the most efficient treatment processes for metal recovery, however, the consumption of chemicals and energy, and the generation of solid, liquid and gaseous wastes encourage the development of greener methods for e-waste recycling. Electrochemical processes are possible alternatives that lower chemical and energy consumption, enhanced control and an overall lower environmental footprint when compared to hydrometallurgical and pyrometallurgical processes by LCA investigations (Li et al., 2019). Biological routes for recovery are getting increased attention, and bio-metallurgy is said to be a promising and cost-effective strategy for REE recovery (Ambaye et al., 2020; Dodson et al., 2015), however slow kinetics of the process is a major bottleneck in its successful installations (Baniasadi et al., 2019). It is also highlighted as bioleaching of sulphide ores had been conventional methods, it cannot directly be implemented on e-waste due to underlying chemistry and requires other novel process development to minimize the additional energy requirements of microbes (Işıldar et al., 2019). Other novel methods which could serve as part of treatment process and cater to the limitations of the above-mentioned methods have been investigated, for example, the use of carbon-based nanocomposites for sorption (Baniasadi et al., 2019; Cardoso et al., 2019), porous materials with metal-organic frameworks (Iftekhar et al., 2022), etc. These methods are environmentally benign, though techno-economic evaluation is needed to ensure the economic viability of the processes. A comprehensive review of conventional and novel methods for REEs' recovery from e-waste was done by Ramprasad et al. (2022) and the advantages and limitations of all the methods were listed.

In the wake of dynamic shifting of technologies in today's world, which are designed based on the characteristics of these precious metals and REEs, the research in recovery technologies also seeks the pace shift. A portfolio of solutions can be evaluated for understanding techno-economic and environmental impacts (Gaustad et al., 2021). Diligent use of LCA and life cycle costing (LCC) tools should be done to perform sustainability assessment (Işıldar et al., 2019). An iterative approach is to be followed when testing various techniques for cost, process optimization using tools like response surface methodology (RSM), and designing the process parameters for pilot testing (Ramprasad et al., 2022). Systematic, though incessant research in this domain is required to establish a substantial approach moving forward.

19.5 POLICIES AND STATUS OF GLOBAL E-WASTE MANAGEMENT

As e-waste contains hazardous substances, indirect regulations on e-waste were imposed by the Basel Convention (1992), which is a multilateral treaty among countries controlling the transboundary movement of hazardous substances. However, directives provided by the European Union (EU) for e-waste management had a direct influence on the EU and simultaneously evolving policy frameworks across the globe, specifically for e-waste management in the last two decades. A comprehensive and detailed review is done by Shittu et al. (2021), which elaborates on the evolution of global treaties and national policies with a timeline for the movement and management of e-waste.

Despite two decades of efforts from the EU, only three member countries are currently meeting the 2019 WEEE collection targets and others are far from reaching the 65% collection target in the near future (Habib et al., 2022). An investigation suggests, in Australia, residents are aware of the negative implications of improper disposal of e-waste, however, about 50% of the residents are not aware of collection centers and collection drives (Islam et al., 2016). Australia and Brazil introduced the guidelines about a decade ago, however, Brazil is lagging more behind in implementation due to a lack of follow-ups and have a dominance of informal recycling even with a registered number of 134 recyclers countrywide (Dias et al., 2022). Similar investigations in Saudi Arabia also highlight a lack of awareness among 70.1% of participants, who were included in the investigation stating they had not been educated on how e-waste poses a serious environmental problem, and 88.35% were willing to take part in managing waste after understanding the implications (Almulhim, 2022).

China and India are developing countries, still among the top three e-waste generator countries in the world. To manage the enormous quantities of e-waste, China introduced a fund-based policy linked with extended producer responsibility (EPR) in 2012. However, the policy is unsustainable due to non-closed resource use, informal recycling, and fund imbalance and requires a closed-loop resource cycle management (Tian et al., 2022). India also introduced the e-waste rules with EPR one decade ago, however, only 20% of e-waste has been recycled with the 178 registered e-waste recyclers, and the rest of the waste finding its way to informal recycling (Nithya et al., 2021).

In the case of other developing countries, the ongoing practices are less appropriate considering health and environmental hazard released by informal recycling. A few African and Asian countries had been a hub for e-waste discarded appliances from developed countries. Nigeria, Ghana and Tanzania, with Kenya, Senegal and Egypt are the main destinations for the huge amount of e-waste transported in the name of repair and reuse, and driving factors for this transport are cited to be poverty, unemployment and socioeconomic needs of Africa (Maes & Preston-Whyte, 2022). In Ghana, the hazardous and e-waste control and management Act 917 of 2016 was implemented in concern for these harmful practices and about 10% of the respondents made changes in their e-waste disposal practices (Owusu-Twum et al., 2022). Such policy frameworks and their rigorous implementation is a must to regulate improper e-waste management by creating awareness among people and providing incentives for their behaviour, and formalizing the informal sector.

In the era of globalization, regulating the transboundary movement of electronics and imposing responsibilities for its disposal to the origin country is counterintuitive. Effective e-waste management requires collaborative efforts, a shift in government attitudes, proper legislation, control of dumping, appreciation of EPR, and technology transfer between countries. Efforts to end informal dismantling and recycling sectors should be a priority for governments (Kahhat & Williams, 2012;

Nnorom & Osibanjo, 2008). Considering the public's awareness of the risks associated with e-waste recycling activities can benefit organizations and reduce potential negative impacts on society. Incentives may be necessary to encourage proper e-waste disposal at collection points rather than discarding it with household waste (Chakrabarty & Nandi, 2021). Lists of collection centres or companies involved in the formal collection and recycling of e-waste must be provided by the manufacturer, which includes the contact details such as an address, telephone numbers and e-mail. (Rajesh et al., 2022). Implementation of circular economy strategies is essential for achieving appliance sustainability due to the high environmental load of landfilling (Nakamura & Kondo, 2006). The efficient design of devices and appliances for material compatibility having homogeneity in different parts, for example, mono plastic is utilized and those combinations, which are not suitable for recycling should be avoided (Zeng et al., 2017). Proper documentation of e-waste flows, both formal and informal, is important for all the stakeholders within the e-waste management system, including policymakers, researchers, producer compliance schemes, and recyclers (Ramprasad et al., 2022).

19.6 CONCLUSION AND RECOMMENDATIONS

The inevitable growth in e-waste quantities demands technological disruptions and supportive policy frameworks. The recovery of precious metals and REEs is crucial considering the scarce availability of REEs, which are key elements to electronic and energy appliances. Also, detrimental effects of metal leaching to aquatic ecosystems have been observed, if REEs are not managed and recovered (Freitas et al., 2020). The available technologies and their assessment for efficient and eco-friendly recovery of REEs have been summarized in this chapter. It can be concluded that pyrometallurgical and hydrometallurgical routes for recovery can provide efficient yield, though they have been associated with negative environmental impacts. Electrochemical and biological processes are being given more attention as alternative methods and can be sustainable pathways in lieu of conventional methods. Other novel and green alternatives are being investigated, though the technological development requires efforts in establishing a substantial base for future investigations.

Moreover, developed countries with systematic recycling such as the separation, collection, and storage of e-waste at source are ideal for the recovery of REEs and other materials, while in developing countries e-waste is often mingled with other solid wastes such as biodegradable material, scrap metals and construction debris (Ramprasad et al., 2022). However, the literature indicates that both developed and developing nations are facing challenges while implementing the guidelines for e-waste management. The practice of dumping discarded e-waste from developed nations to developing nations explains the global exigency of effective technologies and management strategies. Effective e-waste management requires collaborative efforts, a shift in government attitudes, proper legislation, control of dumping, appreciation of EPR, and technology transfer between countries. Informal recycling is a major challenge to address to protect the livelihood of low-income groups and their exposure to health hazards when taking further stringent actions towards formalizing the recycling sector. Managerial elements of e-waste EoL disposal, for example, developing robust collection infrastructure and public awareness have to go hand in hand with advancements in materials' recovery technology and sustainable product design.

REFERENCES

Almulhim, A. I. (2022). Household's awareness and participation in sustainable electronic waste management practices in Saudi Arabia. *Ain Shams Engineering Journal, 13*(4), 101729. https://doi.org/10.1016/j.asej.2022.101729

Ambaye, T. G., Vaccari, M., Castro, F. D., Prasad, S., & Rtimi, S. (2020). Emerging technologies for the recovery of rare earth elements (REEs) from the end-of-life electronic wastes: A review on progress, challenges, and perspectives. *Environmental Science and Pollution Research*, *27*(29), 36052–36074. https://doi.org/10.1007/s11356-020-09630-2

Baniasadi, M., Vakilchap, F., Bahaloo-Horeh, N., Mousavi, S. M., & Farnaud, S. (2019). Advances in bioleaching as a sustainable method for metal recovery from e-waste: A review. *Journal of Industrial and Engineering Chemistry*, *76*, 75–90. https://doi.org/10.1016/j.jiec.2019.03.047

Bosshard, P. P., Bachofen, R., & Brandl, H. (1996). Metal leaching of fly ash from municipal waste incineration by Aspergillus niger. *Environmental Science and Technology*, *30*(10), 3066–3070. https://doi.org/10.1021/es960151v

Brewer, A., Dohnalkova, A., Shutthanandan, V., Kovarik, L., Chang, E., Sawvel, A. M., Mason, H. E., Reed, D., Ye, C., Hynes, W. F., Lammers, L. N., Park, D. M., & Jiao, Y. (2019). Microbe encapsulation for selective rare-earth recovery from electronic waste leachates. *Environmental Science and Technology*, *53*(23), 13888–13897. https://doi.org/10.1021/ACS.EST.9B04608/SUPPL_FILE/ES9B04608_SI_001.PDF

Cardoso, C. E. D., Almeida, J. C., Lopes, C. B., Trindade, T., Vale, C., & Pereira, E. (2019). Recovery of rare earth elements by carbon-based nanomaterials—A review. *Nanomaterials*, *9*(6), 1–35. https://doi.org/10.3390/nano9060814

Chakrabarty, A., & Nandi, S. (2021). Electronic waste vulnerability: Circular economy as a strategic solution. *Clean Technologies and Environmental Policy*, *23*(2), 429–443. https://doi.org/10.1007/s10098-020-01976-y

Cui, M., Nie, H., Zhang, T., Lovley, D., & Russell, T. P. (2017). Three-dimensional hierarchical metal oxide-carbon electrode materials for highly efficient microbial electrosynthesis. *Sustainable Energy and Fuels*, *1*(5), 1171–1176. https://doi.org/10.1039/c7se00073a

Dang, D. H., Thompson, K. A., Ma, L., Nguyen, H. Q., Luu, S. T., Duong, M. T. N., & Kernaghan, A. (2021). Toward the circular economy of rare earth elements: A review of abundance, extraction, applications, and environmental impacts. *Archives of Environmental Contamination and Toxicology*, *81*(4), 521–530. https://doi.org/10.1007/s00244-021-00867-7

de Oliveira, R. P., Benvenuti, J., & Espinosa, D. C. R. (2021). A review of the current progress in recycling technologies for gallium and rare earth elements from light-emitting diodes. *Renewable and Sustainable Energy Reviews*, *145*(January), 111090. https://doi.org/10.1016/j.rser.2021.111090

Dias, P. R., Cenci, M. P., Bernardes, A. M., & Huda, N. (2022). What drives WEEE recycling? A comparative study concerning legislation, collection and recycling. *Waste Management and Research*, *40*(10), 1527–1538. https://doi.org/10.1177/0734242X221081660

Dodson, J. R., Parker, H. L., García, A. M., Hicken, A., Asemave, K., Farmer, T. J., He, H., Clark, J. H., & Hunt, A. J. (2015). Bio-derived materials as a green route for precious & critical metal recovery and re-use. *Green Chemistry*, *17*(4), 1951–1965. https://doi.org/10.1039/c4gc02483d

Edahbi, M., Plante, B., & Benzaazoua, M. (2019). Environmental challenges and identification of the knowledge gaps associated with REE mine wastes management. *Journal of Cleaner Production*, *212*, 1232–1241. https://doi.org/10.1016/j.jclepro.2018.11.228

Favot, M., & Massarutto, A. (2019). Rare-earth elements in the circular economy: The case of yttrium. *Journal of Environmental Management*, *240*(March), 504–510. https://doi.org/10.1016/j.jenvman.2019.04.002

Fogarasi, S., Imre-Lucaci, F., Imre-Lucaci, Á., & Ilea, P. (2014). Copper recovery and gold enrichment from waste printed circuit boards by mediated electrochemical oxidation. *Journal of Hazardous Materials*, *273*, 215–221. https://doi.org/10.1016/j.jhazmat.2014.03.043

Freitas, R., Cardoso, C. E. D., Costa, S., Morais, T., Moleiro, P., Lima, A. F. D., Soares, M., Figueiredo, S., Águeda, T. L., Rocha, P., Amador, G., Soares, A. M. V. M., & Pereira, E. (2020). New insights on the impacts of e-waste towards marine bivalves: The case of the rare earth element Dysprosium. *Environmental Pollution*, 260, 113859. https://doi.org/10.1016/j.envpol.2019.113859

Galbraith, P., & Devereux, J. L. (2002, May). Beneficiation of printed wiring boards with gravity concentration. *IEEE International Symposium on Electronics and the Environment (Cat. No. 02CH37273)*, 242–248. IEEE. https://doi.org/10.1109/isee.2002.1003273

Gaustad, G., Williams, E., & Leader, A. (2021). Rare earth metals from secondary sources: Review of potential supply from waste and byproducts. *Resources, Conservation and Recycling*, *167*(October 2020), 105213. https://doi.org/10.1016/j.resconrec.2020.105213

Grandell, L., & Höök, M. (2015). Assessing rare metal availability challenges for solar energy technologies. *Sustainability (Switzerland)*, *7*(9), 11818–11837. https://doi.org/10.3390/su70911818

Guimarães, A. S., & Mansur, M. B. (2017). Solvent extraction of calcium and magnesium from concentrate nickel sulfate solutions using D2HEPA and Cyanex 272 extractants. *Hydrometallurgy, 173*(August), 91–97. https://doi.org/10.1016/j.hydromet.2017.08.005

Habib, H., Wagner, M., Baldé, C. P., Martínez, L. H., Huisman, J., & Dewulf, J. (2022). What gets measured gets managed – does it? Uncovering the waste electrical and electronic equipment flows in the European Union. *Resources, Conservation and Recycling, 181,* 106222. https://doi.org/10.1016/j.rescon rec.2022.106222

Iftekhar, S., Heidari, G., Amanat, N., Zare, E. N., Asif, M. B., Hassanpour, M., Lehto, V. P., & Sillanpaa, M. (2022). Porous materials for the recovery of rare earth elements, platinum group metals, and other valuable metals: A review. *Environmental Chemistry Letters, 20*(6), 3697–3746. Springer International Publishing. https://doi.org/10.1007/s10311-022-01486-x

Işıldar, A., van Hullebusch, E. D., Lenz, M., Du Laing, G., Marra, A., Cesaro, A., Panda, S., Akcil, A., Kucuker, M. A., & Kuchta, K. (2019). Biotechnological strategies for the recovery of valuable and critical raw materials from waste electrical and electronic equipment (WEEE) – A review. *Journal of Hazardous Materials, 362*(January 2018), 467–481. https://doi.org/10.1016/j.jhazmat.2018.08.050

Islam, M. T., Abdullah, A. B., Shahir, S. A., Kalam, M. A., Masjuki, H. H., Shumon, R., & Rashid, M. H. (2016). A public survey on knowledge, awareness, attitude and willingness to pay for WEEE management: Case study in Bangladesh. *Journal of Cleaner Production, 137*, 728–740. https://doi.org/10.1016/j.jclepro.2016.07.111

Kahhat, R., & Williams, E. (2012). Materials flow analysis of e-waste: Domestic flows and exports of used computers from the United States. *Resources, Conservation and Recycling, 67*, 67–74. https://doi.org/10.1016/j.resconrec.2012.07.008

Kim, E. Y., Kim, M. S., Lee, J. C., & Pandey, B. D. (2011). Selective recovery of gold from waste mobile phone PCBs by hydrometallurgical process. *Journal of Hazardous Materials, 198*, 206–215. https://doi.org/10.1016/j.jhazmat.2011.10.034

Li, Jia, Lu, H., Guo, J., Xu, Z., & Zhou, Y. (2007). Recycle technology for recovering resources and products from waste printed circuit boards. *Environmental Science and Technology, 41*(6), 1995–2000. https://doi.org/10.1021/es0618245

Li, Jianzhi, Shrivastava, P., Gao, Z., & Zhang, H. C. (2004). Printed circuit board recycling: A state-of-the-art survey. *IEEE Transactions on Electronics Packaging Manufacturing, 27*(1), 33–42. https://doi.org/10.1109/TEPM.2004.830501

Li, Z., Diaz, L. A., Yang, Z., Jin, H., Lister, T. E., Vahidi, E., & Zhao, F. (2019). Comparative life cycle analysis for value recovery of precious metals and rare earth elements from electronic waste. *Resources, Conservation and Recycling, 149*(March), 20–30. https://doi.org/10.1016/j.resconrec.2019.05.025

Liu, Y., Song, L., Wang, W., Jian, X., & Chen, W. Q. (2022). Developing a GIS-based model to quantify spatio-temporal pattern of home appliances and e-waste generation—A case study in Xiamen, China. *Waste Management, 137*(1799), 150–157. https://doi.org/10.1016/j.wasman.2021.10.039

Maes, T., & Preston-Whyte, F. (2022). E-waste it wisely: Lessons from Africa. *SN Applied Sciences, 4*(3), 72. https://doi.org/10.1007/s42452-022-04962-9

Murthy, V., & Ramakrishna, S. (2022). A review on global E-waste management: Urban mining towards a sustainable future and circular economy. *Sustainability (Switzerland), 14*(2), 647. https://doi.org/10.3390/su14020647

Nakamura, S., & Kondo, Y. (2006). A waste input-output life-cycle cost analysis of the recycling of end-of-life electrical home appliances. *Ecological Economics, 57*(3), 494–506. https://doi.org/10.1016/j.ecole con.2005.05.002

Navarro, J., & Zhao, F. (2014). Life-cycle assessment of the production of rare-earth elements for energy applications: A review. *Frontiers in Energy Research, 2*(November), 1–17. https://doi.org/10.3389/fenrg.2014.00045

Nithya, R., Sivasankari, C., & Thirunavukkarasu, A. (2021). Electronic waste generation, regulation and metal recovery: A review. *Environmental Chemistry Letters, 19*(2), 1347–1368. https://doi.org/10.1007/s10 311-020-01111-9

Nnorom, I. C., & Osibanjo, O. (2008). Overview of electronic waste (e-waste) management practices and legislations, and their poor applications in the developing countries. *Resources, Conservation and Recycling, 52*(6), 843–858. https://doi.org/10.1016/j.resconrec.2008.01.004

Owusu-Twum, M. Y., Kumi-Amoah, G., Heve, W. K., Lente, I., Owusu, S. A., Larbi, L., & Amfo-Otu, R. (2022). Electronic waste control and management in Ghana: A critical assessment of the law, perceptions and practices. *Waste Management and Research*, *40*(12), 1794–1802. https://doi.org/10.1177/073424 2X221103939

Pant, D., Joshi, D., Upreti, M. K., & Kotnala, R. K. (2012). Chemical and biological extraction of metals present in E waste: A hybrid technology. *Waste Management*, *32*(5), 979–990. https://doi.org/10.1016/j.was man.2011.12.002

Pariatamby, A., & Victor, D. (2013). Policy trends of e-waste management in Asia. *Journal of Material Cycles and Waste Management*, *15*(4), 411–419. https://doi.org/10.1007/s10163-013-0136-7

Patil, A. B., Paetzel, V., Struis, R. P. W. J., & Ludwig, C. (2022). Separation and recycling potential of rare earth elements from energy systems: Feed and economic viability review. *Separations*, *9*(3), 1–15. https://doi.org/10.3390/separations9030056

Rajesh, R., Kanakadhurga, D., & Prabaharan, N. (2022). Electronic waste: A critical assessment on the unimaginable growing pollutant, legislations and environmental impacts. *Environmental Challenges*, *7*(August 2021), 100507. https://doi.org/10.1016/j.envc.2022.100507

Ramprasad, C., Gwenzi, W., Chaukura, N., Izyan Wan Azelee, N., Upamali Rajapaksha, A., Naushad, M., & Rangabhashiyam, S. (2022). Strategies and options for the sustainable recovery of rare earth elements from electrical and electronic waste. *Chemical Engineering Journal*, *442*(P1), 135992. https://doi.org/10.1016/j.cej.2022.135992

Robinson, B. H. (2009). E-waste: An assessment of global production and environmental impacts. *Science of the Total Environment*, *408*(2), 183–191. https://doi.org/10.1016/j.scitotenv.2009.09.044

Sahu, S., Mohapatra, M., & Devi, N. (2022). Effective hydrometallurgical route for recovery of energy critical elements from E-wastes and future aspects. *Materials Today: Proceedings*, *67*, 1016–1023.. https://doi.org/10.1016/j.matpr.2022.05.491

Shahbaz, A. (2022). A systematic review on leaching of rare earth metals from primary and secondary sources. *Minerals Engineering*, *184*(February), 107632. https://doi.org/10.1016/j.mineng.2022.107632

Shittu, O. S., Williams, I. D., & Shaw, P. J. (2021). Global E-waste management: Can WEEE make a difference? A review of e-waste trends, legislation, contemporary issues and future challenges. *Waste Management*, *120*, 549–563. https://doi.org/10.1016/j.wasman.2020.10.016

Tansel, B. (2017). From electronic consumer products to e-wastes: Global outlook, waste quantities, recycling challenges. *Environment International*, *98*, 35–45. https://doi.org/10.1016/j.envint.2016.10.002

Tian, T., Liu, G., Yasemi, H., & Liu, Y. (2022). Managing e-waste from a closed-loop lifecycle perspective: China's challenges and fund policy redesign. *Environmental Science and Pollution Research*, *29*(31), 47713–47724. https://doi.org/10.1007/s11356-022-19227-6

Tunali, M., Tunali, M. M., & Yenigun, O. (2021). Characterization of different types of electronic waste: Heavy metal, precious metal and rare earth element content by comparing different digestion methods. *Journal of Material Cycles and Waste Management*, *23*(1), 149–157. https://doi.org/10.1007/s10163-020-01108-0

Tuncuk, A., Stazi, V., Akcil, A., Yazici, E. Y., & Deveci, H. (2012). Aqueous metal recovery techniques from e-scrap: Hydrometallurgy in recycling. *Minerals Engineering*, *25*(1), 28–37. https://doi.org/10.1016/j.mineng.2011.09.019

Yue, T., Niu, Z., Hu, Y., Han, H., Lyu, D., & Sun, W. (2019). Cr(III) and Fe(II) recovery from the polymetallic leach solution of electroplating sludge by Cr(III)-Fe(III) coprecipitation on maghemite. *Hydrometallurgy*, *184*(August 2018), 132–139. https://doi.org/10.1016/j.hydromet.2018.11.013

Zeng, X., Yang, C., Chiang, J. F., & Li, J. (2017). Innovating e-waste management: From macroscopic to microscopic scales. *Science of the Total Environment*, *575*, 1–5. https://doi.org/10.1016/j.scitotenv.2016.09.078

20 Environmentally Friendly Nanoparticles for Use in Agriculture

Abhisek Mondal and Brajesh Kumar Dubey

20.1 INTRODUCTION

Increase in global population along with the socio-economic development leads to an ever-increasing food demand and is expected to grow by 62% by 2050 (van Dijk, Morley, Rau, & Saghai, 2021). The agricultural industry plays the key role in achieving global food security. However, with climate change, pesticide resistance and dwindling soil fertility, achieving food security has become an uphill task for the agro-industries (Kagan, 2016). Therefore, managing these issues with scientific innovation and in an ecologically-economically viable way would pave the way toward sustainable agriculture.

Chemical fertilizers and pesticides have been aggressively applied over the past decades in fulfilling the global food demand due to its low cost, superior crop yield, and easy to use nature. However, prolonged use of these chemical heavy pesticides or fertilizers with low nutrient utilizing efficiency (NUE) lead to various environmental concerns like surface and groundwater pollution, soil microbial toxicity, eutrophication, greenhouse gas emission, and mammalian toxicity (Li, Tao, Ling, & Chu, 2017; Mekonnen, Mussone, & Bressler, 2016). Nanoparticles having high surface-to-volume ratio can be beneficially used as fertilizer or pesticide (including herbicide and insecticide) because of higher plant uptake owing to smaller size, increased NUE through controlled release, better plant protection through targeted delivery, higher mobility of nutrients through soil (Mahapatra, Satapathy, & Panda, 2022). However, with their increased efficacy these inorganic nano-formulations are also capable of exhibiting plant cytotoxicity and altering the soil ecosystem posing threats to soil biota (Ma, White, Dhankher, & Xing, 2015; Mustafa & Komatsu, 2016). Besides, consumption of foods with residual nanoparticles was found to cause acute genotoxicity in human and animals (Cox, Venkatachalam, Sahi, & Sharma, 2016; Ma, Geiser-Lee, Deng, & Kolmakov, 2010). Incorporating biodegradability or biocompatibility in these nanoparticles may provide a greener alternative in agricultural application. For this, biogenic synthesis of inorganic nanoparticles or encapsulating/supporting pesticides (including herbicides and insecticides) could be used as eco-friendly smart materials for sustainable agriculture. Therefore, this chapter briefly summarizes the biogenic or biocompatible nanoparticles used in agricultural application for plant growth promoter or plant protection. A consolidated idea about the possible routes of biogenic synthesis of these nanoparticles has also been discussed.

DOI: 10.1201/9781003327646-20

293

20.2 ENVIRONMENTALLY FRIENDLY NANOPARTICLES AND THEIR SYNTHESIS PATHWAY

Out of various phosphorus-rich nano-fertilizers, hydroxyapatite nanoparticles (nHAP) have gained considerable attention in recent years because of their biocompatible nature and enhanced phosphorus utilization efficiency. Synthesis of nHAP includes both biogenic and chemical methods such as chemical precipitation (Han, Liu, Wang, & Li, 2008; Poinern, Brundavanam, Mondinos, & Jiang, 2009), hydrothermal synthesis (Montazeri, Javadpour, Shokrgozar, Bonakdar, & Javadian, 2010), sol-gel method (Liu, Troczynski, & Tseng, 2001) and biomass pyrolysis (Carella et al., 2021). However, the morphology and crystalline properties of the synthesized nHAP largely depends on the parameters including but not limited to temperature, Ca: P ratio, and type of solvent used. Whereas chitosan and chitosan-based nano-composite materials have also been successfully utilized in agricultural application due to its non-toxicity, biocompatibility, biodegradability and antipathogenic activity. Researchers have outlined various synthesis routes of nano-sized chitosan particles (e.g., emulsion-droplet coalescence, ionic gelation, spray drying, precipitation) with variable physicochemical properties including, size, shape, surface charge and stability in different matrices (Kumaraswamy et al., 2018). Amongst them ionic gelation has been profusely adopted for producing chitosan-based nanomaterials, due to their feasibility and ease of controlling the particle morphology, essential for agricultural applications (Chandra et al., 2015; Kheiri, Jorf, Malihipour, Saremi, & Nikkhah, 2016). The ionic gelation typically involves crosslinking chitosan molecules using tripolyphosphate (TPP) or alginate (ALG) through ionic interaction between the positively charge $-NH_4^+$ of chitosan and negative centre of TPP/ALG, thereby largely dependent on factors like pH, chitosan: TPP or ALG mass ratio, molecular weight of chitosan (Antoniou et al., 2015; Maruyama et al., 2016; Sipoli, Santana, Shimojo, Azzoni, & de la Torre, 2015). Moreover, tuning the surface charge properties of these nanoparticles by controlling these factors would help in keeping them in a non aggregated form, thereby achieving the predicted functionality. Researchers have also highlighted different metal oxide nanoparticles (e.g., Fe_2O_3, ZnO) as greener alternatives of commercial fertilizer or antimicrobial agent, due to their synthesis route utilizing non-hazardous chemicals or lower environmental toxicity (Poh Yan et al., 2022). Green synthesis of iron oxide nanoparticles typically includes reduction of precursor salt with plant extracts (e.g., mango peel, tangerine peel, *Syzygium cumini* seed, *Hordeum vulgare* seed). Biomolecules (e.g., polyphenol, oxalic acid, citric acid) present in these seed/peel extract typically act as a reducing agent in converting the iron precursor to nanoparticles in aqueous medium (Makarov et al., 2014; Venkateswarlu, Rao, Balaji, Prathima, & Jyothi, 2013). Besides, the physico-chemical and magnetic properties, in turns the stability of these biogenic iron oxide nanoparticles can be tuned by regulating parameters like synthesis pH, extract concentration, type of plant selected (Ehrampoush, Miria, Salmani, & Mahvi, 2015). Researchers have also explored hydrothermal pathway for synthesis of crystalline iron oxide (i.e., Fe_2O_3, Fe_3O_4) nanoparticles at relatively lower energy cost using plant extract (Herlekar, Barve, & Kumar, 2014). Reaction temperature and residence time plays the major role in controlling the morphology of these particles (Phumying et al., 2012). Morphology controlled hydrothermally synthesized ZnO nanoparticles has also been reported as potential green antimicrobial agent due to lower residual ZnO when applied in acidic soil environment (Chang et al., 2020). Chang et al. (2020) demonstrated time-dependent morphology (i.e., 6 h – 2D plate, 12 h – multibranched flower-like structure, 24 h – 1D nanorods) of ZnO nanoparticles in hydrothermal pathway. Another promising approach is nanoscale-based formulation of biofertilizers typically by coating the biofertilizer in nanoscale polymers (Golbashy, Sabahi, Allahdadi, Nazokdast, & Hosseini, 2016).

Overall Figure 20.1 shows the overview of sustainable agriculture including different synthesis pathways adopted in the literature for the biogenic or biocompatible nanoparticles. Moreover, factors like pH, temperature, precursor concentration, encapsulating polymer concentration, and types of biomolecules present in the reducing agent play the decisive role in regulating material morphology and physico-chemical characteristics.

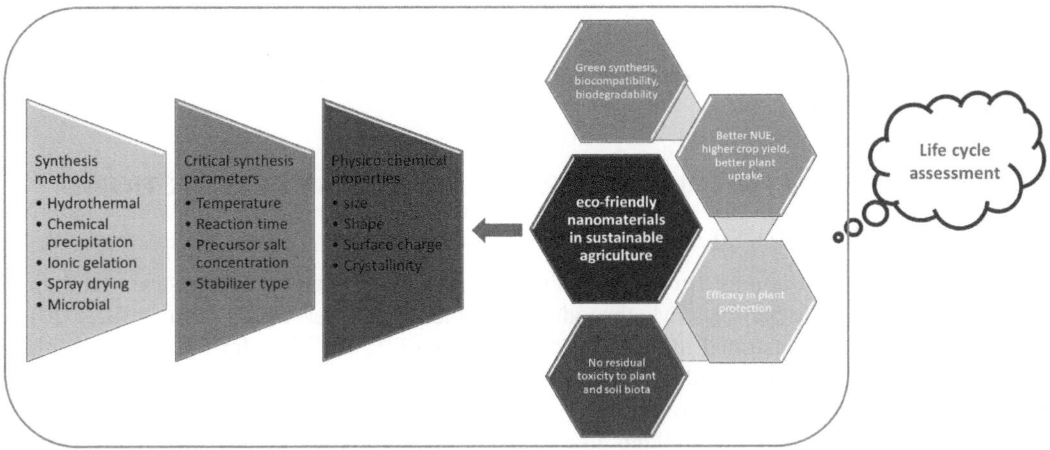

FIGURE 20.1 Overview of environmentally friendly nanoparticles towards sustainable agriculture – synthesis to application.

20.3 NANOPARTICLES AS PLANT GROWTH PROMOTER

Application of Fe_2O_3 nanoparticles have shown improved growth parameters in tomato (Shankramma, Yallappa, Shivanna, & Manjanna, 2015). Moreover, iron oxide nanoparticles not only promoted the growth of peanut by regulating phytochrome content and antioxidant enzyme activity, also averted the iron leaching from soil by binding the nanoparticles to soil grains, thereby minimizing its toxic effect (Rui et al., 2016). Another example of precision agriculture is using nHAP for controlled delivery of P, N, and macro- or micro-nutrients (Fellet, Pilotto, Marchiol, & Braidot, 2021; Kottegoda et al., 2017; Marchiol et al., 2019; Phan et al., 2019). Researchers have also highlighted the biocompatible and biodegradable nature of HAP for their expedited use in agricultural application. Yoon et al. (2020) reported improved crop yield and abiotic stress resistance for *Zea mays* in the presence of humic substance modified HAP, associated with the tunable release of phosphate ion. Similar growth improvement with better quality parameters were observed for *Camellia sinensis* (tea leaf) when exposed to urea-HAP nanohybrids due to increased nitrogen utilization efficiency (Raguraj et al., 2020). Other than chemically synthesized HAP, biogenic/waste derived HAP has also purported to be a greener option for fertilizer industry (Carella et al., 2021; Um e, Hussain, Nazir, Shafiq, & Firdaus e, 2021). Thermally synthesized nHAP from waste fish bones at a temperature range 300–900°C displayed fertilizing (for *Zea mays*) and bio-stimulating (for *Lepidium sativum*) activity on plant growth (Carella et al., 2021). Besides the combustion temperature is positively correlated with the stimulating effect, while particle physico-chemical characteristics (i.e., size, surface charge, crystallinity) underpin the scientific basis behind such effect. Chitosan encapsulated nano-formulations has also shown promising results in optimized utilization of bioactive molecules due to their sustained release behaviour. Moreover, the higher affinity of chitosan towards metals make it a suitable candidate for encapsulating micronutrients (Kumaraswamy et al., 2018). Once uptaken, the slow release of Cu from Cu-chitosan nanoparticles leads to higher growth of maize seedlings compared to the bulk chitosan or $CuSO_4$ treatment (Choudhary et al., 2017; Saharan et al., 2016). Another study with chickpea revealed positive impact of biologically synthesized chitosan nanoparticles on plant morphology (Sathiyabama & Parthasarathy, 2016). The growth-promoting nature of chitosan particles was primarily hypothesized due to the activation of hydrolytic enzymes (e.g., α-amylase and protease) resulting in degradation and mobilization of reserve food (i.e., starch and protein) (Saharan et al., 2016). Whereas several other studies have concluded chitosan-induced antioxidant enzymatic activity for better seed germination by mitigating the effect of reactive oxygen species (ROS) (Anusuya & Banu, 2016; Saharan et al., 2015). Nanoencapsulation of biofertilizer enhances their shelf life and allows controlled release of bioorganic compounds beneficial for plant growth through nitrogen fixation and organic matter solubilization (Ahemad & Kibret, 2014; Kumari & Singh, 2019).

20.4 USE OF ECO-FRIENDLY NANOPARTICLES IN PLANT PROTECTION

20.4.1 Protection Against Abiotic Stress

Protection against plant abiotic stress (i.e., salinity, heavy metal) can be achieved via land application of hydroxyapatite particles (Maghsoodi, Ghodszad, & Lajayer, 2020; Yoon et al., 2020). Li and Huang (2014) demonstrated increased plant biomass (7.97–20.23%) for Pakchoi with nHAP (5–30 g.kg^{-1}) application in a soil contaminated with cadmium (10 mg.kg^{-1}). Besides, the reduction of Cd level in plant shoot increased from 27.12 to 62.36% with increasing nHAP dose (5–30 g.kg^{-1}). Several other studies with plants like spinach, rice and ryegrass have also purported akin inhibitory role of nHAP towards metal toxicity (e.g., Pb, Cd, Zn), likely by immobilizing metal ions via precipitation on soil media (Chen et al., 2020; Gu et al., 2018; Liang, Xi, Ding, Chen, & Liu, 2019). However, the nebulous information about the nHAP uptake by plant parts keeps their defense mechanism in ambiguity. On the other hand, chitosan nanoparticles protect plants against abiotic stress (e.g., high temperature, draught, salinity) via activating defense-related metabolites or genes. Behboudi et al. (2018) reported increased proline content, and catalase (CAT) and superoxide dismutase (SOD) activity under foliar or soil application of chitosan mitigating the draught stress in barley. Similar activation of an antioxidant system under the application of chitosan nanoparticles inhibiting the effect of reactive oxygen species in wheat plant was summarized by Behboudi et al. (2019). Chitosan can also stimulate the abcisic acid (ABA) dependent anti-transpirant activity by controlling the stomatal opening in plants in stressed condition (Zhang, Wollenweber, Jiang, Liu, & Zhao, 2008). Whereas the other route of a defense mechanism through gene-induced jasmonic acid (JA) and ethylene (ET) signalling pathway can also alleviate both biotic and abiotic stress (Yin, Li, Zhao, Du, & Ma, 2006).

20.4.2 Green Nanoparticles as Fungicide and Herbicide

Application of Cu-chitosan at 0.12% (w/v) concentration inhibited both mycelia growth and spore germination in *Alternaria solani* (70.5 and 61.5%, respectively) and *Fusarium oxysporum* (73.5 and 80%, respectively) due to its antifungal property (Saharan et al., 2015). Foliar application of these nanoparticles at similar concentration was also found to be effective in controlling early blight and Fusarium wilt in tomato plants. The synergistic role of chitosan, Ag, and antracol has been found to have significantly higher antifungal activity against *Phytophthora capsici* compared to the individual component, highlighting their imperative role towards sustainable agriculture (Le et al., 2019). Sathiyabama and Parthasarathy (2016) reported antifungal activity of biologically synthesized chitosan nanoparticles in inhibiting mycelial growth in *Pyricularia grisea* (92%), *Fusarium oxysporum* f.sp. *ciceri* (87%) and *Alternaria solani* (72%), attributable to the permeable nature of such fine particles towards biological membrane. A plausible understanding behind the antifungal activity of the chitosan nanoparticles is the electrostatic interaction between positively charged particle surface and negatively charged polyanionic structures of fungi cell wall leading to cell lysis (Kheiri et al., 2016; Kong, Chen, Xing, & Park, 2010). Apart from antifungal activity, chitosan has also been used as a carrier for agricultural delivery of herbicides. Chitosan/alginate and chitosan/TPP encapsulated herbicides (i.e., imazapic and imazapyr) exhibited slow-release kinetics compared to the bulk component averting the soil microbial toxicity belonging to non-targeted use of herbicides (Maruyama et al., 2016). Not only that, the type of carrying agent (i.e., chitosan or alginate) played the deciding role in controlling the release kinetics. Similar slow release of atrazine from chitosan coated nano-capsule was attributed to the non-Fickian diffusion and relaxation of polymeric chains (Grillo, Rosa, & Fraceto, 2013). Antifungal activity of metal or metal oxide (i.e., Ag, ZnO) nanoparticles was attributed to the enhanced ROS production while flower-like morphology may lead to additional mechanical damage by puncturing cell wall (Chang et al., 2020).

20.5 PERSPECTIVES AND CONCLUSIONS

Agricultural application of nanoparticles, particularly as nano-fertilizers or antimicrobials is a booming area due to their suitability in precision farming and targeted delivery. However, the high surface area to volume ratio of these inorganic nanomaterials may lead to toxic effects on plants as well as non-targeted soil microbes, thereby posing a threat to soil biota. Biogenic synthesis or encapsulating nanoparticles with a biodegradable or biocompatible matrix may provide a promising solution to this problem. Although particles like nHAP, chitosan has been considered green due to their biocompatible/biodegradable nature, reported literature highlights their cytotoxic nature (Jiang et al., 2014). Moreover, the smaller the size of these nanoparticles the more severe the impact is, highlighting the role of particle morphology. This also highlights the fact that the uptake route of plants is largely controlled by particle morphology. Despite the improved growth parameters (i.e., better crop yield, better NUE) and better defense mechanism of the eco-friendly nanoparticles against biotic and abiotic stress, the limited toxicity assessment of them have brought forward questions as being eco-friendly. Hence, judicious assessment of nanoparticles including toxicity characteristics should be carried out before tagging them as green or eco-friendly. Furthermore, detailed life cycle assessment of these nanomaterials may provide further insight, which could lead the path towards developing smart materials for sustainable agriculture.

REFERENCES

Ahemad, M., & Kibret, M. (2014). Mechanisms and applications of plant growth promoting rhizobacteria: Current perspective. *Journal of King saud University-Science, 26*(1), 1–20.

Antoniou, J., Liu, F., Majeed, H., Qi, J., Yokoyama, W., & Zhong, F. (2015). Physicochemical and morphological properties of size-controlled chitosan–tripolyphosphate nanoparticles. *Colloids and Surfaces A: Physicochemical and Engineering Aspects, 465,* 137–146.

Anusuya, S., & Banu, K. N. (2016). Silver-chitosan nanoparticles induced biochemical variations of chickpea (*Cicer arietinum* L.). *Biocatalysis and Agricultural Biotechnology, 8,* 39–44.

Behboudi, F., Tahmasebi-Sarvestani, Z., Kassaee, M. Z., Modarres-Sanavy, S. A. M., Sorooshzadeh, A., & Mokhtassi-Bidgoli, A. (2019). Evaluation of chitosan nanoparticles effects with two application methods on wheat under drought stress. *Journal of Plant Nutrition, 42*(13), 1439–1451. doi:10.1080/01904167.2019.1617308

Behboudi, F., Tahmasebi Sarvestani, Z., Kassaee, M. Z., Modares Sanavi, S. A. M., Sorooshzadeh, A., & Ahmadi, S. B. (2018). Evaluation of chitosan nanoparticles effects on yield and yield components of barley (*Hordeum vulgare* L.) under late season drought stress. *Journal of Water and Environmental Nanotechnology, 3*(1), 22–39.

Carella, F., Seck, M., Degli Esposti, L., Diadiou, H., Maienza, A., Baronti, S., ... Adamiano, A. (2021). Thermal conversion of fish bones into fertilizers and biostimulants for plant growth–A low tech valorization process for the development of circular economy in least developed countries. *Journal of Environmental Chemical Engineering, 9*(1), 104815.

Chandra, S., Chakraborty, N., Dasgupta, A., Sarkar, J., Panda, K., & Acharya, K. (2015). Chitosan nanoparticles: A positive modulator of innate immune responses in plants. *Science Report, 5,* 15195.

Chang, T.-H., Lu, Y.-C., Yang, M.-J., Huang, J.-W., Chang, P.-F. L., & Hsueh, H.-Y. (2020). Multibranched flower-like ZnO particles from eco-friendly hydrothermal synthesis as green antimicrobials in agriculture. *Journal of Ccleaner Production, 262,* 121342.

Chen, H., Yuan, X., Xiong, T., Jiang, L., Wang, H., & Wu, Z. (2020). Biochar facilitated hydroxyapatite/calcium silicate hydrate for remediation of heavy metals contaminated soils. *Water, Air, & Soil Pollution, 231*(2). doi:10.1007/s11270-020-4425-1

Choudhary, R. C., Kumaraswamy, R. V., Kumari, S., Sharma, S. S., Pal, A., Raliya, R., … Saharan, V. (2017). Cu-chitosan nanoparticle boost defense responses and plant growth in maize (*Zea mays* L.). *Science Report, 7*(1), 9754. doi:10.1038/s41598-017-08571-0

Cox, A., Venkatachalam, P., Sahi, S., & Sharma, N. (2016). Silver and titanium dioxide nanoparticle toxicity in plants: A review of current research. *Plant Physiology and Biochemistry, 107*, 147–163.

Ehrampoush, M. H., Miria, M., Salmani, M. H., & Mahvi, A. H. (2015). Cadmium removal from aqueous solution by green synthesis iron oxide nanoparticles with tangerine peel extract. *Journal of Environmental Health Science and Engineering, 13*, 84. doi:10.1186/s40201-015-0237-4

Fellet, G., Pilotto, L., Marchiol, L., & Braidot, E. (2021). Tools for nano-enabled agriculture: Fertilizers based on calcium phosphate, silicon, and chitosan nanostructures. *Agronomy, 11*(6), 1239. doi:10.3390/agronomy 11061239

Golbashy, M., Sabahi, H., Allahdadi, I., Nazokdast, H., & Hosseini, M. (2016). Synthesis of highly intercalated urea-clay nanocomposite via domestic montmorillonite as eco-friendly slow-release fertilizer. *Archives of Agronomy and Soil Science, 63*(1), 84–95. doi:10.1080/03650340.2016.1177175

Grillo, R., Rosa, A. H., & Fraceto, L. F. (2013). Poly(ε-caprolactone) nanocapsules carrying the herbicide atrazine: Effect of chitosan-coating agent on physico-chemical stability and herbicide release profile. *International Journal of Environmental Science and Technology, 11*(6), 1691–1700. doi:10.1007/s13762-013-0358-1

Gu, J. F., Zhou, H., Yang, W. T., Peng, P. Q., Zhang, P., Zeng, M., & Liao, B. H. (2018). Effects of an additive (hydroxyapatite-biochar-zeolite) on the chemical speciation of Cd and As in paddy soils and their accumulation and translocation in rice plants. *Environmental Science and Pollution Research, 25*(9), 8608–8619. doi:10.1007/s11356-017-0921-2

Han, Y., Liu, S., Wang, X., & Li, B. (2008). Design of a metamodel-based telecoms modelling language. *2008 9th International Conference on Computer-Aided Industrial Design and Conceptual Design*, Kunming, 2008, pp. 1235 1238, doi: 10.1109/CAIDCD.2008.4730787.

Herlekar, M., Barve, S., & Kumar, R. (2014). Plant-mediated green synthesis of iron nanoparticles. *Journal of Nanoparticles, 2014*, 1–9. doi:10.1155/2014/140614

Jiang, H., Liu, J.-K., Wang, J.-D., Lu, Y., Zhang, M., Yang, X.-H., & Hong, D.-J. (2014). The biotoxicity of hydroxyapatite nanoparticles to the plant growth. *Journal of Hazardous Materials, 270*, 71–81.

Kagan, C. R. (2016). At the nexus of food security and safety: Opportunities for nanoscience and nanotechnology. *ACS Nano, 10*(3), 2985–2986. doi:10.1021/acsnano.6b01483

Kheiri, A., Jorf, S. M., Malihipour, A., Saremi, H., & Nikkhah, M. (2016). Application of chitosan and chitosan nanoparticles for the control of Fusarium head blight of wheat (*Fusarium graminearum*) in vitro and greenhouse. *International Journal of Biological Macromolecules, 93*, 1261–1272.

Kong, M., Chen, X. G., Xing, K., & Park, H. J. (2010). Antimicrobial properties of chitosan and mode of action: A state of the art review. *International Journal of Food Microbiology, 144*(1), 51–63.

Kottegoda, N., Sandaruwan, C., Priyadarshana, G., Siriwardhana, A., Rathnayake, U. A., Berugoda Arachchige, D. M., … Amaratunga, G. A. (2017). Urea-hydroxyapatite nanohybrids for slow release of nitrogen. *ACS Nano, 11*(2), 1214–1221. doi:10.1021/acsnano.6b07781

Kumaraswamy, R., Kumari, S., Choudhary, R. C., Pal, A., Raliya, R., Biswas, P., & Saharan, V. (2018). Engineered chitosan based nanomaterials: Bioactivities, mechanisms and perspectives in plant protection and growth. *International Journal of Biological Macromolecules, 113*, 494–506.

Kumari, R., & Singh, D. P. (2019). Nano-biofertilizer: An emerging eco-friendly approach for sustainable agriculture. *Proceedings of the National Academy of Sciences, India Section B: Biological Sciences, 90*(4), 733–741. doi:10.1007/s40011-019-01133-6

Le, V. T., Bach, L. G., Pham, T. T., Le, N. T. T., Ngoc, U. T. P., Tran, D.-H. N., & Nguyen, D. H. (2019). Synthesis and antifungal activity of chitosan-silver nanocomposite synergize fungicide against Phytophthora capsici. *Journal of Macromolecular Science, Part A, 56*(6), 522–528. doi:10.1080/10601325.2019.1586439

Li, R., Tao, R., Ling, N., & Chu, G. (2017). Chemical, organic and bio-fertilizer management practices effect on soil physicochemical property and antagonistic bacteria abundance of a cotton field: Implications for soil biological quality. *Soil and Tillage Research, 167*, 30–38.

Li, Z., & Huang, J. (2014). Effects of nanoparticle hydroxyapatite on growth and antioxidant system in Pakchoi (*Brassica chinensis* L.) from cadmium-contaminated soil. *Journal of Nanomaterials, 2014*, 1–7. doi:10.1155/2014/470962

Liang, S. X., Xi, X., Ding, L., Chen, Q., & Liu, W. (2019). Immobilization mechanism of nano-hydroxyapatite on lead in the ryegrass rhizosphere soil under root confinement. *Bulletin of Environmental Contamination and Toxicology, 103*(2), 330–335. doi:10.1007/s00128-019-02665-3

Liu, D.-M., Troczynski, T., & Tseng, W. J. (2001). Water-based sol–gel synthesis of hydroxyapatite: Process development. *Biomaterials, 22*(13), 1721–1730.

Ma, C., White, J. C., Dhankher, O. P., & Xing, B. (2015). Metal-based nanotoxity and detoxification pathways in higher plants. *Environental Science Technology, 49*(12), 7109–7122. doi:10.1021/acs.est.5b00685

Ma, X., Geiser-Lee, J., Deng, Y., & Kolmakov, A. (2010). Interactions between engineered nanoparticles (ENPs) and plants: Phytotoxicity, uptake and accumulation. *Science of The Total Environment, 408*(16), 3053–3061.

Maghsoodi, M. R., Ghodszad, L., & Lajayer, B. A. (2020). Dilemma of hydroxyapatite nanoparticles as phosphorus fertilizer: Potentials, challenges and effects on plants. *Environmental Technology & Innovation, 19*, 100869.

Mahapatra, D. M., Satapathy, K. C., & Panda, B. (2022). Biofertilizers and nanofertilizers for sustainable agriculture: Phycoprospects and challenges. *Science of the Total Environment, 803*, 149990.

Makarov, V. V., Makarova, S. S., Love, A. J., Sinitsyna, O. V., Dudnik, A. O., Yaminsky, I. V., ... Kalinina, N. O. (2014). Biosynthesis of stable iron oxide nanoparticles in aqueous extracts of Hordeum vulgare and Rumex acetosa plants. *Langmuir, 30*(20), 5982–5988. doi:10.1021/la5011924

Marchiol, L., Filippi, A., Adamiano, A., Degli Esposti, L., Iafisco, M., Mattiello, A., ... Braidot, E. (2019). Influence of hydroxyapatite nanoparticles on germination and plant metabolism of tomato (*Solanum lycopersicum* L.): Preliminary evidence. *Agronomy, 9*(4), 161. doi:10.3390/agronomy9040161

Maruyama, C. R., Guilger, M., Pascoli, M., Bileshy-Jose, N., Abhilash, P. C., Fraceto, L. F., & de Lima, R. (2016). Nanoparticles based on Chitosan as carriers for the combined herbicides Imazapic and Imazapyr. *Science Report, 6*, 19768. doi:10.1038/srep19768

Mekonnen, T., Mussone, P., & Bressler, D. (2016). Valorization of rendering industry wastes and co-products for industrial chemicals, materials and energy: Review. *Critical Reviews Biotechnology, 36*(1), 120–131. doi:10.3109/07388551.2014.928812

Montazeri, L., Javadpour, J., Shokrgozar, M. A., Bonakdar, S., & Javadian, S. (2010). Hydrothermal synthesis and characterization of hydroxyapatite and fluorhydroxyapatite nano-size powders. *Biomedical Materials, 5*(4), 045004.

Mustafa, G., & Komatsu, S. (2016). Toxicity of heavy metals and metal-containing nanoparticles on plants. *Biochimica et Biophysica Acta (BBA)-Proteins and Proteomics, 1864*(8), 932–944.

Phan, K. S., Nguyen, H. T., Le, T. T. H., Vu, T. T. T., Do, H. D., Vuong, T. K. O., ... Ha, P. T. (2019). Fabrication and activity evaluation on Asparagus officinalis of hydroxyapatite based multimicronutrient nano systems. *Advances in Natural Sciences: Nanoscience and Nanotechnology, 10*(2) , 025011. doi:10.1088/2043-6254/ab21cc

Phumying, S., Labuayai, S., Thomas, C., Amornkitbamrung, V., Swatsitang, E., & Maensiri, S. (2012). Aloe vera plant-extracted solution hydrothermal synthesis and magnetic properties of magnetite (Fe3O4) nanoparticles. *Applied Physics A, 111*(4), 1187–1193. doi:10.1007/s00339-012-7340-5

Poh Yan, L., Gopinath, S. C. B., Subramaniam, S., Chen, Y., Velusamy, P., Chinni, S. V., ... Lebaka, V. R. (2022). Greener synthesis of nanostructured iron oxide for medical and sustainable agro-environmental benefits. *Frontiers in Chemistry, 10*, 984218. doi:10.3389/fchem.2022.984218

Poinern, G. E., Brundavanam, R. K., Mondinos, N., & Jiang, Z.-T. (2009). Synthesis and characterisation of nanohydroxyapatite using an ultrasound assisted method. *Ultrasonics Sonochemistry, 16*(4), 469–474.

Raguraj, S., Wijayathunga, W. M. S., Gunaratne, G. P., Amali, R. K. A., Priyadarshana, G., Sandaruwan, C., ... Kottegoda, N. (2020). Urea–hydroxyapatite nanohybrid as an efficient nutrient source in *Camellia sinensis* (L.) Kuntze (tea). *Journal of Plant Nutrition, 43*(15), 2383–2394. doi:10.1080/01904167.2020.1771576

Rui, M., Ma, C., Hao, Y., Guo, J., Rui, Y., Tang, X., ... Zhu, S. (2016). Iron oxide nanoparticles as a potential iron fertilizer for Peanut (*Arachis hypogaea*). *Frontiers in Plant Science, 7*, 815. doi:10.3389/fpls.2016.00815

Saharan, V., Kumaraswamy, R. V., Choudhary, R. C., Kumari, S., Pal, A., Raliya, R., & Biswas, P. (2016). Cu-Chitosan nanoparticle mediated sustainable approach to enhance seedling growth in maize by mobilizing reserved food. *Journal of Agricultural and Food Chemistry, 64*(31), 6148–6155. doi:10.1021/acs.jafc.6b02239

Saharan, V., Sharma, G., Yadav, M., Choudhary, M. K., Sharma, S., Pal, A., … Biswas, P. (2015). Synthesis and in vitro antifungal efficacy of Cu–chitosan nanoparticles against pathogenic fungi of tomato. *International Journal of Biological Macromolecules, 75*, 346–353.

Sathiyabama, M., & Parthasarathy, R. (2016). Biological preparation of chitosan nanoparticles and its in vitro antifungal efficacy against some phytopathogenic fungi. *Carbohydrate Polymers, 151*, 321–325.

Shankramma, K., Yallappa, S., Shivanna, M. B., & Manjanna, J. (2015). Fe2O3 magnetic nanoparticles to enhance S. lycopersicum (tomato) plant growth and their biomineralization. *Applied Nanoscience, 6*(7), 983–990. doi:10.1007/s13204-015-0510-y

Sipoli, C. C., Santana, N., Shimojo, A. A. M., Azzoni, A., & de la Torre, L. G. (2015). Scalable production of highly concentrated chitosan/TPP nanoparticles in different pHs and evaluation of the in vitro transfection efficiency. *Biochemical Engineering Journal, 94*, 65–73.

Um e, L., Hussain, A., Nazir, A., Shafiq, M., & Firdaus e, B. (2021). Potential application of biochar composite derived from rice straw and animal bones to improve plant growth. *Sustainability, 13*(19), 11104. doi:10.3390/su131911104

van Dijk, M., Morley, T., Rau, M. L., & Saghai, Y. (2021). A meta-analysis of projected global food demand and population at risk of hunger for the period 2010–2050. *Nature Food, 2*(7), 494–501. doi:10.1038/s43016-021-00322-9

Venkateswarlu, S., Rao, Y. S., Balaji, T., Prathima, B., & Jyothi, N. (2013). Biogenic synthesis of Fe3O4 magnetic nanoparticles using plantain peel extract. *Materials Letters, 100*, 241–244.

Yin, H., Li, S., Zhao, X., Du, Y., & Ma, X. (2006). cDNA microarray analysis of gene expression in Brassica napus treated with oligochitosan elicitor. *Plant Physiology and Biochemistry, 44*(11–12), 910–916.

Yoon, H. Y., Lee, J. G., Esposti, L. D., Iafisco, M., Kim, P. J., Shin, S. G., … Adamiano, A. (2020). Synergistic release of crop nutrients and stimulants from hydroxyapatite nanoparticles functionalized with humic substances: Toward a multifunctional nanofertilizer. *ACS Omega, 5*(12), 6598–6610. doi:10.1021/acsomega.9b04354

Zhang, X., Wollenweber, B., Jiang, D., Liu, F., & Zhao, J. (2008). Water deficits and heat shock effects on photosynthesis of a transgenic Arabidopsis thaliana constitutively expressing ABP9, a bZIP transcription factor. *Journal of Experimental Botany, 59*(4), 839–848. doi:10.1093/jxb/erm364

21 Reuse of Plastic Waste in Various Systems

A Review

Vijaykumar Sekar, Baranidharan Sundaram and Brajesh Kumar Dubey

21.1 INTRODUCTION

Plastics are hailed as a revolutionary breakthrough because they could be used to replace most of the inelastic materials used for storage before the twentieth century, such as solidified clay and glass (Shen et al., 2020). Plastics are fibrous least dense synthetic materials and are made from a variety of organic polymers like PE, PA, PVC, PET, PP and nylon (Vikas Madhav et al., 2020). In 1950, the global consumption of plastics was 0.0015 billion tonnes, but by the end of 2018, it had increased to 0.3 billion tonnes. It is predicted that the global consumption of plastics will increase even further, reaching 34 billion metric tonnes by 2050 (Bui et al., 2020). Each year, 10% of plastic garbage is dumped into the oceans and according to the United Nations Environment Programme (UNEP) (Fu & Wang, 2019; Kova et al., 2017; Schmidt et al., 2020), less than 10% of plastic garbage is collected and recycled (Boucher & Friot, 2017; Chandra et al., 2020). Plastics are widely utilized in a wide range of applications, including packaging, electrical and electronic equipment, construction, beverages and textiles, in part because of their low weight and excellent strength-to-weight ratio among other materials (Abejón et al., 2020). Flexible packaging plastics such as PE, PP and PVC are most commonly used in the food, pharmaceuticals, chemical and beverage industries (Soares et al., 2022). Multiple plastic films coated with aluminum foil (Al-foil) make up the bulk of multilayer flexible packaging waste (MLPW). Instead of relying on biological or chemical methods to separate polymeric fractions from aluminum, the EU's recently introduced circular economy strategy relies upon mechanical and thermochemical methods, all of which impose limitations on recycling rates (66 percent), energy consumption, and CO_2 emissions, sustainability, and the inter mixability of recovered polymer blends. Recycling and reuse are the alternative methods that have been practiced across the world to reduce plastic waste littering and are found as alternative sources of energy and bio-oil recovery (Al Rayaan, 2021). Plastic garbage is incredibly durable and difficult for bacteria to break down when it is dumped in the environment. Toxic metals and other contaminants accumulate on the surfaces of plastic garbage because of their long-life spans and durability (Iyare et al., 2020; Yang et al., 2020). It was reported that single-use PE bags made from recycled or biodegradable materials had a lower climate impact than fossil-based PE bags. (Marfella et al., 2024). By enhancing sustainable design, utilizing higher-quality plastics, and encouraging and facilitating reuse, repair, remanufacturing and recycling, a circular plastics economy seeks to preserve the value and utility of products for as long as possible, ensuring that plastics never become waste (Fogh Mortensen et al., 2021).

A developing country like India widely uses a large amount of plastic in every product. As a result, plastic is spread all over the country without proper waste management. The report, the plastic waste generated in India is 9.5 million tons per annum, in which 43% of the plastic is used as packaging

DOI: 10.1201/9781003327646-21

material of various thicknesses, and 40% of the waste is considered littering waste. Due to the exponential proliferation and mismanagement of plastic debris, recent research on plastic pollution, its existence, fate and behaviour in India has become an emerging focus for numerous academicians (Jagadeesh & Sundaram, 2021; Naidu, 2019). Mismanagement of plastic waste causes pollution in most rivers, lakes, oceans and soil, causing significant environmental disruption that must be addressed urgently (Ferronato & Torretta, 2019). Recently the focus on non-recyclable plastics in the Indian environment has not covered much and most of the studies have reported recyclable plastics such as PET, LDPE, HDPE, PP, etc., across various environmental compartments such as marine, freshwater, sediments, salt, atmospheric dust and aquatic species. Billions of people around the world use single-use plastic without realizing the danger it poses to the planet's inhabitants. According to the Central Pollution Control Board (CPCB), India produces 3.3 million metric tonnes of plastic waste each year. There have been decades of efforts to address plastic pollution, but it remains a big issue and continues to provide issues. According to a Hindustan Times report, a survey has been performed in India as only less than 10% of plastics are recycled and in huge quantities of plastics recycling is too expensive to sort and collect. The finally end up result is disposed of in a landfill or dumpsite.

To safeguard the globe from plastic waste and its environmental issues. Researchers and scientists are forming novel solutions endlessly for the circular economy. The concept of the circular economy comes into the picture when the rate of plastic waste generation has increased drastically and is found to be hazardous to the environment (European Commission, 2018). Most circular economy concepts are proposing a motive to utilize the waste materials within a circular loop, where all the energy and raw materials lie within a loop that can be used any number of times (Shamsuyeva & Endres, 2021; European Commission, 2018). Due to the least recycling rate of plastic waste, most of it was dumped into landfill or sent for incineration, causing enormous environmental pollution (Chen et al., 2021). Huge littering of plastic waste results in secondary microplastic pollution in the terrestrial and ecological environment. Even multilayer plastic waste (MLPW) contributes to a combination of chemicals and fossil fuels, which releases greenhouse gases at each stage of their production (Šleiniūtė et al., 2024). LCA of various types of bags such as PP, HDPE, LDPE, jute, and paper was analyzed, and it found that the recycling rate of LDPE and HDPE was less than 10 and 19.8%, whereas the recycling rate of paper and jute bags were more than 50% (Civancik-Uslu et al., 2019). The use of plastics in concrete as a substitution material for natural aggregates has been initiated in earlier stages of the twenty-first century (also plastics used in road construction). Waste plastics are collected, washed, dried and crushed into pieces to make use of them in concrete manufacturing by partially replacing natural aggregates and cement (Ongpeng et al., 2020). Some studies have evidenced using plastic waste as a stabilizer will improve weak soil and expansive soil (Rai et al., 2020). Various design mixes (replacement of 1% to 30%) were prepared by replacing ordinary aggregates and cement with recycled plastics to study the mechanical properties of the concrete cube by varying the percentage of recycled plastics with the original mix (Aghayan & Khafajeh, 2019).

21.2   GLOBAL CONCERN ON RECYCLING OF PLASTIC WASTE

Nowadays, the focus on recycling plastic waste has become a major concern on account of the increase in the usage of huge amounts of plastics for various purposes. In earlier days, the recycling of plastic waste was less and unaccountable, due to the upcoming circular economy concept, the research on recycling of plastic waste has started and increased gradually across the world (Yousef, Eimontas, Zakarauskas, et al., 2021). With the use of online databases such as Science Direct, a comprehensive literature evolution was undertaken. The suitable keywords have been used for exploring the literature 'plastics, waste and recycling' and found in around 223 research articles. It was reported that the landfill and incineration of plastic waste results in huge environmental impacts where CO_2 emissions from landfill disposal of plastic are 253 g/kg plastic. In comparison, incineration outcomes vary between 673 g/kg and 4605 g/kg depending on specific selections (Eriksson & Finnveden, 2009). Reusing or recycling plastic waste will not only reduce environmental pollution and also increase the productivity of materials such as durability will be increased and costs will be reduced.

21.3 SPECIFIC STUDY OF PLASTIC WASTE IN A DIFFERENT SECTOR

Many global studies have developed in recent years for experimenting with plastic waste. The experiment results from the survey indicated some improved properties in materials and stabilization (Table 21.1). Some of the studies on plastic waste (type, shape and content), procedures, methods and tests are discussed below.

21.3.1 BRICK MANUFACTURING

Leela Bharathi et al., (2020) experimented with the manufactured M sand brick using plastic instead of M sand in two different ratios (Waste plastic: M sand) 2.5:2.5 kg (C1 type) and 1.5:3 kg (C2 type). Some of the standard tests were conducted such as adsorption, soundness, hardness and compression test values compared with standard brick. Noticeable changes were absorbed from the result. In the study, the absorption test indicated the improvement of the C2 type of brick from the standard brick and C1 type. Using plastic waste of 1.5 kg (C2 type) in M sand brick has shown an increase in the maximum strength of 55.91 MPa nearly 88.59% increased value from ordinary brick. From the study observation, C2 type bricks demonstrated better results compared with other types. The study concluded that the use of plastic waste in building materials will reduce the cost and the environment hazardous pollution.

Akinwumi et al. (2019) investigated fabricating earth bricks using shredded marine plastic waste in different sizes and percentages as stabilized soil. The soil was obtained from Nigeria and determined the basic property and classified as clayey sand. The shredded plastic waste is prepared in different sizes below 6.5 mm and above 9.6 mm. The plastic waste chips were taken into different divisions from 0, 1, 3 and 7%. The various quantities of shredded plastic waste (below 6.5 mm and above 9.6 mm) and (0, 1, 3 and 7%) were manufactured in earth brick by a hydraulic compacting machine. Standard tests, as per ASTM brick quality, were experimented such as the compressive strength test and durability erosion rate test. An impressive rate of improvement in earth bricks was studied in compressive strength and increased erosion rate to 244.4% and 389.8% at 1% of SPW. The further addition of SPW has studied the progressively decreased value of earth brick's strength. The study concluded the use of 1% SPW will reduce the cost of earth brick, easily manufacturable, reduction of environmental threats, and also environmentally friendly.

Vigneshwar et al. (2021) investigated producing sustainable eco-friendly and low-cost brick. Plastic bottles are filled with soil, steel was tied above for thermal insulation, and molded with two different mixtures of mortar and fly ash brick composition (fly debris 52%, and 25%, gypsum 8%, concrete 10%, and lime 5%). The study resulted from the compressive strength achieved for both plastic molded bricks and plastic molded fly ash bricks. The acid-resistant test resulted in plastic molded brick by addition of H_2SO_4 and HCl was 1.35% and 0.83% lesser than usual bricks and for plastic molded fly ash brick using H_2SO_4 and HCL was 1.12% and 0.779% lesser than usual bricks. The study also estimated the cost of manufacturing a modified brick than normal brick for a 3 m × 3 m × 3 m room. The estimation indicated impressive results in a profit of 31% when making a plastic molded brick for a room and a profit of 80.5% for plastic molded fly ash brick for a room from normal solid bricks. The Study shows the eco-friendly and cost benefits while using plastic waste bottles in bricks.

TABLE 21.1
Recycling of Plastics in a Different Medium

Author	Location	Polymer Type	Polymer Form	Used for	Remarks
(Almeshal et al., 2020)	Palestine	PET	PET shredded and reheated as granules,	Concrete	The unity weight was reduced to 31.6% by the addition of 50%
(Pratap Singh Rajawat et al., 2022)	India	PET	Powder		3% replacement had given good results than regular concrete
(Akinwumi et al., 2019)	Nigeria	Mixed polymer	Shredded plastic waste (<6.3 and >9.6mm	Earth bricks	Addition of 1% SPW had increased 244.4% compressive strength and 389.8% erosion rate
(Leela Bharathi et al., 2020)	India	Mixed polymer	Melted and batch into a brick		2.5% of plastic waste in brick has shown an increase in strength of 88.59%
(Vigneshwar et al., 2021)	India	PET	Plastic waste bottle strips		80.5% cost will be reduced while using plastic-modified bricks
(Ahmed et al., 2011)	Japan	PS	Strip	Soil stabilization	Based on plastic type and size the capillary rise and stiffness increases or decreases in soil
(Rai et al., 2020)	India (Kerala)	Mixed polymer	Granules		0.25% addition of plastic waste shows good results
	India (Punjab)	PET	Strips (25 × 5 mm, 35 × 10 mm & 50 × 15 mm)		Increased CBR value at 6% dosage plastic bottle strip (25mm*5mm)
	Karnataka (India)	PET	Strip		CBR and UCS value improved at 2% of additive in expansive soil
	Kolkata (India)	Mixed plastic	Strip		UCS and CBR increased in 6% of plastic waste
(Abukhettala & Fall, 2021)	Canada	LDPE, HDPE, PET, and PP	ground, flaky, and pellet		Ground LDPE polymer type has less CBR value compared to flaky and pellet
(Gangwar & Tiwari, 2021)	India	PET, PE, and other polymers	Cut into small pieces and passed into a 4.75 mm sieve		The CBR, UCS, and SS have increased by 0.5% of plastic waste
(Amena, 2021)	Ethiopia	PET	Strips		0.75% plastic waste and 30% brick powder had a better result than regular brick
(Koohmishi & Palassi, 2022)	Iran	PET	Strips of 3 cm and pellets		The polymer strip form has higher strength and flexibility compared to the pellet form
(Ahmad et al., 2022)	Turkey	Mixed polymer	Shredded plastic waste	Alker	The addition of plastic has improved compressive strength, flexural strength, shrinkage, and cracks
(Genet et al., 2021)	Ethiopia	LDPE	The shredded -particle is between 2 mm to 5 mm	Flexible pavement	6.5% addition LDPE has increased the stability by 33.67%
(Russo et al., 2022)	Italy	HDPE, LDPE, PP, and PET	Shredded particles passing through 0.063mm	Flexible pavement	The presence of plastic waste has increased the elasticity

21.3.2 Flexible Pavement

Russo et al. (2022) Studied the property of the flexible pavement for roads using plastic waste and waste from jet grouting. Plastic waste like HDPE, LDPE, PP and PET from recycling plants is shredded into small sizes of 0.063 mm. Plastic waste and jet grout waste were used for an asphalt mixture as a base layer. The result shows the presence of plastic waste has increased in elasticity.

Genet et al. (2021) Investigated the various modified bitumen mixes made for highway asphalt. The mix was prepared with LDPE plastic waste with particle sizes of 2 mm to 5 mm. The LDPE was mixed in a proportion of 4, 6, 8 and 10 percent by weight. The mix was tested at various temperatures (160°C, 170°C and 180°C) at various timing from 1, 1.5 and 2 hours. The Marshall test was conducted for various proportions, temperatures and timing. The consistent mix was obtained at 1.5 hours at 170°C with LDPE waste material. The experiments result studied at 6.5% LDPE waste plastic content has maximum stability of 33.67% compared to the usual asphalt mix.

21.3.3 Concrete

Pratap Singh Rajawat et al. (2022) Investigated the concrete strength by partially replacing PET plastic with sand. PET (polyethylene terephthalate) is a broadly used plastic around the world. This study used local PET waste bottles like water bottles and glucose bottles. The bottles are shredded into small parts and heated to a certain temperature to the required fineness and chilled. For this study, Ordinary Portland cement was used in some basic tests experimented on for cement properties such as specific gravity, setting time, consistency test and fineness test. The used concrete design mix was M25 grade with a proportion of 1:1.38:3.02 and a water-cement ratio of 0.45. The PET powder was used as a fine aggregate in 1, 2, 3, 4 and 5 percent by weight. Some of the concrete tests like slump and compressive strength test was tested. The resulting slump cone test showed a decreasing value while increasing plastic content. The best proportion of plastic was obtained at a 3% replacement level.

Almeshal et al. (2020) Experimented with the concrete behaviour by replacing using different proportions starting with 0, 10, 20, 30, 40 and 50%. Concrete tests like unit weight, slump cone, flexural strength, compressive strength, tensile (split test), fire resistance and pundit pulse velocity test were tested for every proportion. The unit weight of concrete is decreased with the increase of plastic waste content. It reduced nearly 31.6% with the addition of 50% of plastic waste. The compressive strength is decreased by 1.2, 4.2, 31, 60 and 90.6 percentage with an increase in the plastic proportion by 10, 20, 30, 40 and 50%. Plastic in concrete decreased flexural strength and increased tensile strength. The pulse velocity test showed a decrease in the quality of concrete with an increase in plastic content. The study concluded with this proportion of concrete can be used on highway pavement and highway medium blocks where the strength is not an important factor.

21.3.4 Soil Stabilization

Gangwar & Tiwari (2021) investigated soil needs to be modified to obtain the required properties so that different stabilizing materials are found in recent years such as fly ash, lime, cement, bitumen, furnace slag, etc. Shredded plastic bottles are used as a stabilizer in research to modify the weak soil to withstand heavy loads shear strength and CBR. The plastic waste bottle chips are mixed in soil in different ratios starting from 0%, 0.5%, 1%, 1.5%, and 2%. Tests like SPC, UCS and CBR were conducted to determine the proportion of good soil to plastic waste. Test results study revealed that CBR has increased to 5.39 and 4.87 at 2.5 mm and 5 mm depth. There was an increase in SS and UCS at 0.5% mixtures values 18.9 kN/m^2 and 37.8 kN/m^2 from 0% plastic waste soil 15.25 kN/m^2 and 30.5 kN/m^2. The study concluded that using plastic waste in the soil as a stabilizing agent will be an economical and eco-friendly method for disposal.

Koohmishi & Palassi (2022) investigated improving the mechanical properties of the clay soil lime stabilizer (with and without) and reinforcement with PET bottle waste (strips and pellets). Various waste bottles were collected from the dumpsite and converted into the size of 3 mm strips and 5 mm pellets and mixed in the clay soil with stabilization and without stabilization of lime in ratios starting from 0.5%, 1%, 1.5%, 2%, 3%, 5% and 10%. Several tests were conducted, namely drop weight impact load test, point load test and standard proctor compaction test. The effect of reinforced PET on the soil strength has followed the same tread line in both unstabilized and stabilized clay. The soil's flexibility has increased in PET and there is an increase in rigidity with lime stabilized with PET. The study resulted from strip PET enhancing higher strength and more flexibility compared to pellet form PET.

Abukhettala & Fall (2021) carried out an experimental investigation to understand the geotechnical influenced character of pavement subgrade soil using plastic waste of different types and shapes. Subgrade (SG) soil tests were conducted as per ASTM standards like water content, California bearing ratio, modified proctor compaction test, unconfined compressive strength, triaxial shear test, resilient modulus test, and falling head permeability test. Experimented with plastic waste in different types of plastic like LDPE, HDPE, PET and PP. The various plastic types were taken in different shapes, flakes produced by shredding, pellets formed by extrusion, and ground. These different types of plastic waste (LDPE, HDPE, PET and PP) in different shapes (flaky (F), ground (G), and pellet (P)) are experimentally studied in different proportions (1 to 10%). The result studied in CBR except for SG-G-LDPE, every other proportion has increased. CBR has shown a variation from 25 to 41% in SG-P-HDPE, SG-F_HDPE, SG-P-PP and SG-P-PET, and from natural SG soil. There is a significant reduction in maximum dry density when there is an addition of plastic material. The increase in plastic content might tend to overlay, resulting in weaker collaboration. In hydraulic conductivity, the permeability had increased in addition to plastic waste. The pellet type plastic has a higher permeability than flaky plastic. Permeability test has revealed the presence of higher pellets will provide more space for water to transport. From the compressive test, the study result has shown LDPE mixture in subgrade soil has low strength compared to other types of subgrade soil. There is a reduction of compressive strength in modified subgrade soil in addition to plastic type, shape and content. Shear strength studies have revealed that modified subgrade soil with plastic had internal friction varied from 20° to 28° and cohesion varied from 39 to 69 kpa. The significant increase in plastic content has a linear reduction in shear strength. The modulus resilient behavior of pavement SG soil had revealed no effect in a variation of recycled plastic waste. The study concluded that partial replacement of plastic waste in the soil demonstrated improved pavement subgrade application.

Amena (2021) investigated the expansive soil in Ethiopia like montmorillonite has complex properties of volume change when it is in a dry and saturated condition. Studies have revealed that to overcome this expansive soil complexity of the volume change, the soil should be replaced or a stabilizer should be used, which leads to expense. Several studies have proven an increased bearing value using construction and demolishing waste and other hazardous waste. An experiment was conducted in different trials using disposed plastic waste bottle strips in (0.25, 0.5 and 0.75) percentages and powdered brick from demolished waste in (20, 30 and 40) percentages. UCS, CBR, SPC and Atterberg's limit were performed in every trial. The result has studied the noticeable strength as an increase in subgrade strength with an increase in plastic waste strips. The swelling of the soil is decreased with an increase in plastic strips and brick powder. The investigation has resulted in the utilization of plastic waste strips and brick powder in 0.75% and 30% have good results in density, strength, shear strength and swelling characteristics. The study concluded with this combination will play a vital role in minimizing environmental pollution and the cost of construction.

Rai et al. (2020) studied the comparison of using plastic waste (granules) in weak and expansive soil. In Kerala (India), plastic granules' proportion of 0.25, 0.5 and 0.75% are added and experimented with weak soil as per the Indian standard soil classification system. A comparison of some soil tests at 0.25% addition of plastic waste shows results. In Punjab (India) plastic waste (bottle strips) was used in 2, 4 and 6% in low compressibility clay soil. The result by ISSCS showed an increase in CBR at 6% dosage plastic bottle strip (25 mm × 5 mm). Also, a soil sample tested in Karnataka (India) using a plastic waste bottle strip showed the effect of increased CBR and UCS value at 2% of additive in expansive soil. Black cotton soil with low compressibility from Kolkata (India) has been experimented with (ISSCS standards) with plastic waste. The result studied have a top value of UCS and CBR in 6% of plastic waste.

Ahmed et al. (2011) studied the effect of the waste plastic tray strip and bassanite (recycled gypsum waste) in the Sandy soil (poorly graded SP). Some study shows the use of plastic waste has reduced its strength. Additional stabilizer for improving compressive strength for weak soil. The author investigated the likelihood of using gypsum waste to increase compressive strength and plastic waste to improve tensile strength. Various tests were conducted, such as capillary rise test, SPC, compressive and split tensile test. Bassanite, used as a stabilizer result, studied the increase in compressive strength. Future study shows the addition of plastic waste strip in gypsum-stabilized soil has enhanced the secant modulus, tensile strength and compressive strength. The plastic type and size can potentially increase or decrease the capillary rise and stiffness of soil.

Recycling plastic as a stabilizer is an inexpensive technology many researchers have proven with results (Table 21.2). Plastic waste has a stabilizer that has improved the soil property from the weak bearing capacity and expansive. Based on the type, content, and shape of the polymer, the soil property is changed.

21.3.5 ALKER PRODUCTION

Ahmad et al. (2022) studied the effect of shredded plastic waste (SPW) in Alker manufacturing and revealed the need for earthen buildings and sustainable materials in developing countries. The experiment procedure was carried out using gypsum and SPW in different percentages. Starting with 0.5%, 1% and 1.5% to repair the granulometry structure and cohesive property. Various tests were conducted, such as x-ray diffraction analysis, flexural test, compressive test, linear shrinkage test and peak signal-noise ratio (PSNR) for crack propagation. It showed that there was a significant change after the 28 days of moulding. The study concluded that plastic waste as a stabilizer in Alker shows an increase in properties such as compressive strength by 30.1%, flexural strength, shrinkage by 1.08%, cracks propagation by 19.13%, and a significant reduction in linear cracks. The study disclosed that there are 30% of the world's people using still earthen buildings, so stabilizing with plastic waste will be a sustainable material.

21.4 UPCYCLING STUDY ON NON-RECYCLABLE PLASTICS

Even though we can recycle some of the plastic wastes such as LDPE, PET HDPE and PP. Some plastics are non-recyclable like mixed and multilayer plastic. Multilayer plastics (MLPs) are the most commonly consumed packaging bags worldwide, which are produced in huge quantities due to their lightweight and lower cost compared with other woven and paper bags (Marfella et al., 2024). Plastic materials, such as pharmaceutical blisters, straws, and most packaging materials, are categorized under MLPs (Yousef et al., 2018). Among various industries, the food industry has got a huge consumption of MLPs, which uses huge amounts of plastic packaging bags (Abejón et al., 2020). All materials that include at least one layer of plastic are considered MLP. Laminated materials like aluminum, plastic and paper are difficult to separate from one other in the MLP. MLP, which is used in food packaging, accounts for 40% of the world's total plastic trash. In the food business, MLP is a desirable material since it preserves the food product goods and so extends their shelf life. India has a wide range of temperatures, and MLP's ability to protect food in humid

TABLE 21.2

Effect of Plastic Waste as a Stabilizer on the Property of the Soil

Author	Location	Soil Type	Polymer Type	Polymer Form	Test	Changes ↓ increase ↑ decrease	Conclusion
(Gangwar & Tiwari, 2021)	India	Silty clay	PET	Strip	SPC UCS CBR	↑	0.5% addition of PET chips
(Koohmishi & Palassi, 2022)	Iran	Clay	PET	Strips pellets	SPC ILT PLT	↓ ↑ ↓	>2% it was observed the increased flexibility
(Amena, 2021)	Ethiopia	Clay	PET	Strip	UCS CBR SPC AL	↑ ↑ ↓ ↓	0.75% plastic swelling character reduced
(Leela Bharathi et al., 2020)	India	Sand	LDPE, PET HDPE, PP	Flake, Pellets, Ground	MPC UCS TT RMT FHP CBR	↓ ↓ ↑ ↓↑ ↑ ↓↑	1–10% addition of plastic. Ground shape and LDPE had shown poor result
(Rai et al., 2020)	Kerala (India)	Clay	Mixed plastic	Granules	CS CBR	↑	0.25% addition of plastic granules
	Punjab (India)	Clay (CL)	PET	Strip (25mm* 5mm)	CBR	↑	6% of the addition of a plastic strip
	Karnataka (India)	Clay (CI)	PET	Strip	CBR UCS	↑	2% of an additional strip
	Kolkata (India)	Silt (ML)	Mixed plastic	Strip	CBR UCS	↑	6% plastic waste
(Ahmed et al., 2011)	Japan	Sand (SP)	HDPE	Strip	CT STT CRT	↑	Plastic waste tray chips

Abbreviation: SPC, standard proctor compaction test; UCS, unconfined compressive strength; CBR, California bearing ratio; AL, Atterberg's limit; ILT; impact load test; PLT, point load test; MPC, modified proctor compaction test; TT, triaxial test; CT, compressive strength test; STT, split tensile test; CRT, capillary rise test; RMT, resilient modulus test; FHP, falling head permeability; ASTM, American Society for Testing and Materials; SPW, shredded plastic waste; SP, poorly graded sand.

conditions makes it ideal for the country. Aside from that, it is light and can be found everywhere from local tea stands to supermarkets in different sizes of packages ranging from small to enormous. This substance can be found everywhere, yet it is tough to dispose of because it is so easy to toss away. Litter and other waste eventually end up in landfills or waterways and become part of our food supply. The collection and segregation of MLPs present several practical difficulties (due to less resale value), including labour-intensive operations, limited storage space due to their volume, etc. This makes recycling MLP commercially infeasible, and it is most likely to wind up in landfills. However, only a few researchers have attempted to handle this waste in India. It was reported that eucalyptus wood along with waste LDPE was subjected to a co-pyrolysis process, where the yield was an energy density of 1.25 and a heat value of 31 MJ/kg (Samal et al., 2021). Another study found that pyrolysis might be used to recover the metallic fractions from the scrapped printed circuit boards and at a temperature of 400 C, 35% weight of combustible gases and 60 %

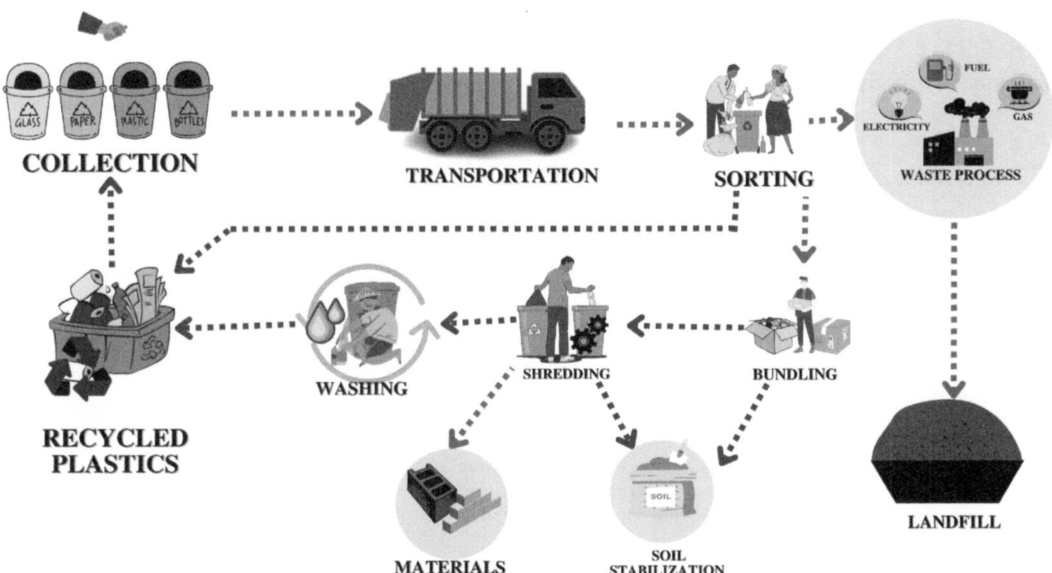

FIGURE 21.1 Systematic plastic material cycle flow.

weight of solid products were produced within a short span (20 minutes) (Jadhao et al., 2020). As an emerging field, a rising emphasis from regulatory authorities makes it a need to establish an eco-friendly approach for MLPW recycling and resource recovery. A fundamental concern with MLPs disposal is that, unlike food and paper wastes, they cannot be eradicated from the environment when left in landfills. Landfilling is the oldest and most used technique of disposing of waste in general; as a result, it accumulates on the planet, thus reducing landfill space and wreaking havoc on the natural ecosystem. Additionally, the existing approaches (mechanical sorting) appear to be far less efficient, thereby increasing the pressure on researchers, and governments to discover innovative and better alternatives. This study aims to make use of the pyrolysis technique to reduce the MLPWs to a potential end product such as char which can be used for carbon dioxide removal from the environment (Figure 21.2). Also, an attempt will be made to quantify the environmental impacts of waste MLPW-derived char. One of the feedstock recycling like pyrolysis is cost-effective for non-recyclable plastic. The critical systematic plastic waste material cycle flow is pictographic (Figure 21.1) created for upcycling on recyclable and non-recyclable plastic waste.

21.4.1 Pyrolysis

It was reported that the major reason to choose pyrolysis for combustion is, it yields the smallest carbon footprint compared to incineration and landfill (Wijesekara et al., 2021). Various types of pyrolysis processes are listed in Table 21.3.

The process of pyrolysis derives precious metals, energy, and bio-oil from MLPW, which occurs at temperatures between 500 and 700°C in the absence of oxygen (Ahamed et al., 2020; Veksha et al., 2020). The released plastic was collected, sorted and samples were made to undergo pyrolysis, which yields various byproducts throughout the process to meet the circular economy and zero waste principle (Yousef, Eimontas, Zakarauskas, et al., 2021). In addition, the byproducts of the pyrolysis process are employed in various civil engineering applications such as adhesives in concrete manufacturing and cement admixtures, and biochar generated from the plastic waste has been employed to produce the least amount of environmental damage (Wijesekara et al., 2021), the char created from the MPFW has been used as a filler in fiberglass epoxy composites, among other things

TABLE 21.3
Various Pyrolysis Types

Type	Thermal Gradient variation	Temp. Range (°C)	Duration	Reference
Hydro Carbonization	Slow	< 350	1–24 hr	(Iñiguez et al., 2019)
Slow Pyrolysis	Slow	300–700	1–2 hr	(Das & Tiwari, 2018)
Flash Pyrolysis	Flash	700–1000	10–60 min	(KLAIMY et al., 2021; Zhou et al., 2020)
Fast Pyrolysis	Faster	>1000	1–5 sec	(Xue et al., 2015)
Gasification	Faster	700–1500	10–30 sec	(Al Rayaan, 2021; Yousef et al., 2022)

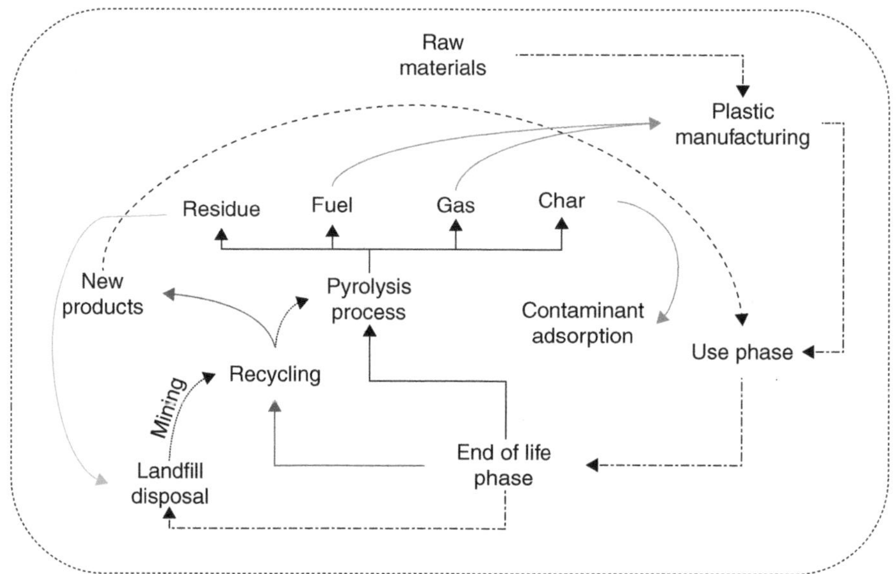

FIGURE 21.2 Schematic flow of plastics in the environment and available knowledge gaps.

(Yousef, Eimontas, Subadra, et al., 2021) and carbon black or char has been used as an adsorbent to remove waste contaminants from wastewater, and the pyrolysis yield of MFPW has been used for biogas production, paraffin wax, precious metals, and carbon black or char has been used as an adsorbent to extract waste contaminants from wastewater (Yousef, Eimontas, Zakarauskas, et al., 2021). It has been discovered that the pyrolysis of flexible packaging plastic waste produces an oil with a calorific value of 45.8 MJ/kg found to be the similar calorific value of diesel and other solid by-products used in the production of multi-walled carbon nanotubes (MWCNTs), which are used in the electrolytic application of oxygen evolution during the oxygen evolution reaction (Veksha et al., 2020). From the studied analysis, the systematic flow chart (Figure 21.2) is prepared for the plastic cycle for pyrolysis.

21.5 CONCLUSION

Recycling plastic waste has become a major concern currently on account of increase in the usage of huge amounts of plastics. This uncontrollable production of plastic waste has been a part of global warming, declining biodiversity, health risk to living organisms, ocean acidification, global sea level rise, toxic airborne, and CO_2 emissions. The proper renewing waste plastic process will reduce the threat to human health and toxicity to the environment and also create a recycling formula for a sustainable plastic environment. To safeguard the environment, there is a need for a sustainable design for the plastic cycle. This study indicated the optimal sorting of plastic waste to improve recycling efficiency and increase the economy from the reuse of different types of plastic in different mediums from the environment, geotechnical and civil engineering perspectives. Some studied result shows the recycling of plastic type LDPE, PET HDPE, PP, etc., in different recycling system and their contribution to pyrolysis, soil stabilization and manufacturing of materials. A change of cycle material flow is needed to stabilize the overpowering production of plastic waste. When the plastic waste material demand is more, there will be a reduction in plastic waste in the environment.

REFERENCES

Abejón, R., Bala, A., Vázquez-Rowe, I., Aldaco, R., & Fullana-i-Palmer, P. (2020). When plastic packaging should be preferred: Life cycle analysis of packages for fruit and vegetable distribution in the Spanish peninsular market. *Resources, Conservation and Recycling, 155* (November 2019), 104666. https://doi.org/10.1016/j.resconrec.2019.104666

Abukhettala, M., & Fall, M. (2021). Geotechnical characterization of plastic waste materials in pavement subgrade applications. *Transportation Geotechnics, 27*, 100472. https://doi.org/10.1016/J.TRGEO.2020.100472

Aghayan, I., & Khafajeh, R. (2019). Recycling of PET in asphalt concrete. In *Use of Recycled Plastics in Eco-Efficient Concrete.* edited by Pacheco-Torgal, F. and Khatib, J. and Colangelo, F. & Tuladhar R. Elsevier Ltd. https://doi.org/10.1016/b978-0-08-102676-2.00012-8

Ahamed, A., Veksha, A., Yin, K., Weerachanchai, P., Giannis, A., & Lisak, G. (2020). Environmental impact assessment of converting flexible packaging plastic waste to pyrolysis oil and multi-walled carbon nanotubes. *Journal of Hazardous Materials, 390*, 121449. https://doi.org/10.1016/j.jhazmat.2019.121449

Ahmad, A., Pekrioglu Balkis, A., & Kurtis Onochie, K. (2022). The use of shredded plastic wastes in Alker production and its effect on compressive strength and shrinkage properties. *Alexandria Engineering Journal, 61*(2), 1563–1570. https://doi.org/10.1016/J.AEJ.2021.06.062

Ahmed, A., Ugai, K., & Kamei, T. (2011). Investigation of recycled gypsum in conjunction with waste plastic trays for ground improvement. *Construction and Building Materials, 25*(1), 208–217. https://doi.org/10.1016/J.CONBUILDMAT.2010.06.036

Akinwumi, I. I., Domo-Spiff, A. H., & Salami, A. (2019). Marine plastic pollution and affordable housing challenge: Shredded waste plastic stabilized soil for producing compressed earth bricks. *Case Studies in Construction Materials, 11*, e00241. https://doi.org/10.1016/J.CSCM.2019.E00241

Al Rayaan, M. B. (2021). Recent advancements of thermochemical conversion of plastic waste to biofuel-A review. *Cleaner Engineering and Technology, 2*, 100062. https://doi.org/10.1016/j.clet.2021.100062

Almeshal, I., Tayeh, B. A., Alyousef, R., Alabduljabbar, H., & Mohamed, A. M. (2020). Eco-friendly concrete containing recycled plastic as partial replacement for sand. *Journal of Materials Research and Technology, 9*(3), 4631–4643. https://doi.org/10.1016/J.JMRT.2020.02.090

Amena, S. (2021). Experimental study on the effect of plastic waste strips and waste brick powder on strength parameters of expansive soils. *Heliyon, 7*(11), e08278. https://doi.org/10.1016/J.HELIYON.2021.E08278

Boucher, J., & Friot, D. (2017). Primary microplastics in the oceans: A global evaluation of sources. In Primary microplastics in the oceans: A global evaluation of sources. https://doi.org/10.2305/iucn.ch.2017.01.en

Bui, X. T., Vo, T. D. H., Nguyen, P. T., Nguyen, V. T., Dao, T. S., & Nguyen, P. D. (2020). Microplastics pollution in wastewater: Characteristics, occurrence and removal technologies. *Environmental Technology and Innovation, 19*, 101013. https://doi.org/10.1016/j.eti.2020.101013

Chandra, P., Enespa, S. & Singh, D. P. (2020). Microplastic degradation by bacteria in aquatic ecosystem. In *Microorganisms for Sustainable Environment and Health.* (pp. 431–467). Elsevier. https://doi.org/10.1016/b978-0-12-819001-2.00022-x

Chen, Y., Awasthi, A. K., Wei, F., Tan, Q., & Li, J. (2021). Single-use plastics: Production, usage, disposal, and adverse impacts. *Science of the Total Environment, 752*, 141772. https://doi.org/10.1016/j.scitotenv.2020.141772

Civancik-Uslu, D., Puig, R., Hauschild, M., & Fullana-i-Palmer, P. (2019). Life cycle assessment of carrier bags and development of a littering indicator. *Science of the Total Environment*, *685*, 621–630. https://doi.org/10.1016/j.scitotenv.2019.05.372

Das, P., & Tiwari, P. (2018). The effect of slow pyrolysis on the conversion of packaging waste plastics (PE and PP) into fuel. *Waste Management*, *79*, 615–624. https://doi.org/10.1016/j.wasman.2018.08.021

Eriksson, O., & Finnveden, G. (2009). Plastic waste as a fuel – CO2-neutral or not? *Energy and Environmental Science*, *2*(9), 907–914. https://doi.org/10.1039/b908135f

European Commission. (2018). A European Strategy for Plastics. *European Comission, July*, 24. https://doi.org/10.1021/acs.est.7b02368

Ferronato, N., & Torretta, V. (2019). Waste mismanagement in developing countries: A review of global issues. *International Journal of Environmental Research and Public Health*, *16*(6). https://doi.org/10.3390/ijerph16061060

Fogh Mortensen, L., Tange, I., Stenmarck, Å., Fråne, A., Nielsen, T., Boberg, N., & Bauer, F. (2021). Plastics, the circular economy and Europe's environment – A priority for action. In *EEA Report* (Issue 18/2020). https://doi.org/10.2800/5847

Fu, Z., & Wang, J. (2019). Current practices and future perspectives of microplastic pollution in freshwater ecosystems in China. *Science of the Total Environment*, *691*, 697–712. https://doi.org/10.1016/j.scitotenv.2019.07.167

Gangwar, P., & Tiwari, S. (2021). Stabilization of soil with waste plastic bottles. *Materials Today: Proceedings*, *47*, 3802–3806. https://doi.org/10.1016/j.matpr.2021.03.010

Genet, M. B., Sendekie, Z. B., & Jembere, A. L. (2021). Investigation and optimization of waste LDPE plastic as a modifier of asphalt mix for highway asphalt: Case of Ethiopian roads. *Case Studies in Chemical and Environmental Engineering*, *4*, 100150. https://doi.org/10.1016/J.CSCEE.2021.100150

Iñiguez, M. E., Conesa, J. A., & Fullana, A. (2019). Hydrothermal carbonization (HTC) of marine plastic debris. *Fuel*, *257*(April), 116033. https://doi.org/10.1016/j.fuel.2019.116033

Iyare, P. U., Ouki, S. K., & Bond, T. (2020). Microplastics removal in wastewater treatment plants: A critical review. *Environmental Science: Water Research and Technology*, *6*(10), 2664–2675. https://doi.org/10.1039/d0ew00397b

Jadhao, P. R., Ahmad, E., Pant, K. K., & Nigam, K. D. P. (2020). Environmentally friendly approach for the recovery of metallic fraction from waste printed circuit boards using pyrolysis and ultrasonication. *Waste Management*, *118*, 150–160. https://doi.org/10.1016/j.wasman.2020.08.028

Jagadeesh, N., & Sundaram, B. (2021). A review of microplastics in wastewater, their persistence, interaction, and fate. *Journal of Environmental Chemical Engineering*, *9*(6), 106846. https://doi.org/10.1016/J.JECE.2021.106846

Klaimy, S., Lamonier, J. F., Casetta, M., Heymans, S., & Duquesne, S. (2021). Recycling of plastic waste using flash pyrolysis – Effect of mixture composition. *Polymer Degradation and Stability*, *187*, 109540. https://doi.org/10.1016/j.polymdegradstab.2021.109540

Koohmishi, M., & Palassi, M. (2022). Mechanical properties of clayey soil reinforced with PET considering the influence of lime-stabilization. *Transportation Geotechnics*, *33*. https://doi.org/10.1016/j.trgeo.2022.100726

Kova, M., Nika, M., Koren, Š., Kr, A., & Peterlin, M. (2017). Microplastics as a vector for the transport of the bacterial fish pathogen species Aeromonas salmonicida. *Mar Pollut Bull*. https://doi.org/10.1016/j.marpolbul.2017.08.024

Leela Bharathi, S. M., Johnpaul, V., Praveen Kumar, R., Surya, R., & Vishnu Kumar, T. (2020). WITHDRAWN: Experimental investigation on compressive behaviour of plastic brick using M Sand as fine aggregate. *Materials Today: Proceedings*. https://doi.org/10.1016/J.MATPR.2020.10.252

Marfella, R., Prattichizzo, F., Sardu, C., Fulgenzi, G., Graciotti, L., Spadoni, T., D'Onofrio, N., Scisciola, L., La Grotta, R., Frigé, C., Pellegrini, V., Municinò, M., Siniscalchi, M., Spinetti, F., Vigliotti, G., Vecchione, C., Carrizzo, A., Accarino, G., Squillante, A., ... Paolisso, G. (2024). Single-use plastic bags and their alternatives. *390*(10), 900–910. http://www.ncbi.nlm.nih.gov/pubmed/38446676

Naidu, S. A. (2019). Preliminary study and first evidence of presence of microplastics and colorants in green mussel, Perna viridis (Linnaeus, 1758), from southeast coast of India. *Marine Pollution Bulletin*, *140*(January), 416–422. https://doi.org/10.1016/j.marpolbul.2019.01.024

Ongpeng, J. M. C., Barra, J., Carampatana, K., Sebastian, C., Yu, J. J., Aviso, K. B., & Tan, R. R. (2020). Strengthening rectangular columns using recycled PET bottle strips. *Engineering Science and Technology, an International Journal*, 1–9. https://doi.org/10.1016/j.jestch.2020.07.006

Pratap Singh Rajawat, S., Singh Rajput, B., & Jain, G. (2022). Concrete strength analysis using waste plastic as a partial replacement for sand. *Materials Today: Proceedings*, *62*(P12), 6824–6831. https://doi.org/10.1016/J.MATPR.2022.04.961

Rai, A. K., Singh, G., & Tiwari, A. K. (2020). Comparative study of soil stabilization with glass powder, plastic and e-waste: A review. *Materials Today: Proceedings*, *32*, 771–776. https://doi.org/10.1016/J.MATPR.2020.03.570

Russo, F., Oreto, C., & Veropalumbo, R. (2022). Promoting resource conservation in road flexible pavement using jet grouting and plastic waste as filler. *Resources, Conservation and Recycling*, *187*, 106633. https://doi.org/10.1016/j.resconrec.2022.106633

Samal, B., Vanapalli, K. R., Dubey, B. K., Bhattacharya, J., Chandra, S., & Medha, I. (2021). Char from the co-pyrolysis of Eucalyptus wood and low-density polyethylene for use as high-quality fuel: Influence of process parameters. *Science of the Total Environment*, *794*, 148723. https://doi.org/10.1016/j.scitotenv.2021.148723

Schmidt, C., Kumar, R., Yang, S., & Büttner, O. (2020). Microplastic particle emission from wastewater treatment plant effluents into river networks in Germany: Loads, spatial patterns of concentrations and potential toxicity. *Science of the Total Environment*, *737*, 139544. https://doi.org/10.1016/j.scitotenv.2020.139544

Shamsuyeva, M., & Endres, H.-J. (2021). Plastics in the context of the circular economy and sustainable plastics recycling: Comprehensive review on research development, standardization and market. *Composites Part C: Open Access*, *6*, 100168. https://doi.org/10.1016/j.jcomc.2021.100168

Shen, M., Song, B., Zhu, Y., Zeng, G., Zhang, Y., Yang, Y., Wen, X., Chen, M., & Yi, H. (2020). Removal of microplastics via drinking water treatment: Current knowledge and future directions. *Chemosphere*, *251*, 126612. https://doi.org/10.1016/j.chemosphere.2020.126612

Šleiniūtė, A., Mumladze, T., Denafas, G., Makarevičius, V., Kriūkienė, R., Antonov, M., & Vasarevičius, S. (2024). Mechanical properties of polymers recovered from multilayer food packaging by nitric acid. *Sustainability*, *16*(5), 2106. https://doi.org/10.3390/su16052106

Soares, C. T. de M., Ek, M., Östmark, E., Gällstedt, M., & Karlsson, S. (2022). Recycling of multi-material multilayer plastic packaging: Current trends and future scenarios. *Resources, Conservation and Recycling*, *176*(April 2021). https://doi.org/10.1016/j.resconrec.2021.105905

Veksha, A., Yin, K., Moo, J. G. S., Oh, W. Da, Ahamed, A., Chen, W. Q., Weerachanchai, P., Giannis, A., & Lisak, G. (2020). Processing of flexible plastic packaging waste into pyrolysis oil and multi-walled carbon nanotubes for electrocatalytic oxygen reduction. *Journal of Hazardous Materials*, *387*, 121256. https://doi.org/10.1016/j.jhazmat.2019.121256

Vigneshwar, P. V., Anuradha, B., Guna, K., & Ashok Kumar, R. (2021). Sustainable eco-friendly fly ash brick using soil filled plastic bottles. *Materials Today: Proceedings*, *81*, 440–442. https://doi.org/10.1016/J.MATPR.2021.03.573

Vikas Madhav, N., Gopinath, K. P., Krishnan, A., Rajendran, N., & Krishnan, A. (2020). A critical review on various trophic transfer routes of microplastics in the context of the Indian coastal ecosystem. *Watershed Ecology and the Environment*, *2*, 25–41. https://doi.org/10.1016/j.wsee.2020.08.001

Wijesekara, D. A., Sargent, P., Ennis, C. J., & Hughes, D. (2021). Prospects of using chars derived from mixed post waste plastic pyrolysis in civil engineering applications. *Journal of Cleaner Production*, *317*(December 2020), 128212. https://doi.org/10.1016/j.jclepro.2021.128212

Xue, Y., Zhou, S., Brown, R. C., Kelkar, A., & Bai, X. (2015). Fast pyrolysis of biomass and waste plastic in a fluidized bed reactor. *Fuel*, *156*, 40–46. https://doi.org/10.1016/j.fuel.2015.04.033

Yang, Y., Liu, W., Zhang, Z., Grossart, H., & Gadd, G. M. (2020). Microplastics provide new microbial niches in aquatic environments. *Applied Microbiology and Biotechnology*, *104*(15), 6501–6511. https://doi.org/10.1007/s00253-020-10704-x.

Yousef, S., Eimontas, J., Striūgas, N., & Abdelnaby, M. A. (2022). Gasification kinetics of char derived from metallised food packaging plastics waste pyrolysis. *Energy*, *239*, 122070. https://doi.org/10.1016/j.energy.2021.122070

Yousef, S., Eimontas, J., Subadra, S. P., & Striūgas, N. (2021). Functionalization of char derived from pyrolysis of metallised food packaging plastics waste and its application as a filler in fiberglass/epoxy composites. *Process Safety and Environmental Protection, 147*, 723–733. https://doi.org/10.1016/j.psep.2021.01.009

Yousef, S., Eimontas, J., Zakarauskas, K., & Striūgas, N. (2021). Microcrystalline paraffin wax, biogas, carbon particles and aluminum recovery from metallised food packaging plastics using pyrolysis, mechanical and chemical treatments. *Journal of Cleaner Production, 290*, 125878. https://doi.org/10.1016/j.jclepro.2021.125878

Yousef, S., Mumladze, T., Tatariants, M., Kriūkienė, R., Makarevicius, V., Bendikiene, R., & Denafas, G. (2018). Cleaner and profitable industrial technology for full recovery of metallic and non-metallic fraction of waste pharmaceutical blisters using switchable hydrophilicity solvents. *Journal of Cleaner Production, 197*, 379–392. https://doi.org/10.1016/j.jclepro.2018.06.154

Zhou, J., Liu, G., Wang, S., Zhang, H., & Xu, F. (2020). TG-FTIR and Py-GC/MS study of the pyrolysis mechanism and composition of volatiles from flash pyrolysis of PVC. *Journal of the Energy Institute, 93*(6), 2362–2370. https://doi.org/10.1016/j.joei.2020.07.009

Index